VOLCANIC ROCK MECHANICS

PAPERS FROM THE 3RD INTERNATIONAL WORKSHOP, PUERTO DE LA CRUZ, TENERIFE (CANARY ISLANDS), SPAIN, 31 MAY–1 JUNE 2010

Volcanic Rock Mechanics

Rock Mechanics and Geo-engineering in Volcanic Environments

Editors

Claudio Olalla
Universidad Politécnica de Madrid, Madrid, Spain

Luis E. Hernández
Área de Laboratorios y Calidad de la Construcción, Tenerife, Spain

J.A. Rodríguez-Losada
University of La Laguna, Tenerife, Spain

Áurea Perucho
Laboratorio de Geotecnia del CEDEX, Madrid, Spain

Javier González-Gallego
Laboratorio de Geotecnia del CEDEX, Madrid, Spain

CRC Press
Taylor & Francis Group
Boca Raton   London   New York   Leiden

CRC Press is an imprint of the
Taylor & Francis Group, an **informa** business

A BALKEMA BOOK

CRC Press/Balkema is an imprint of the Taylor & Francis Group, an informa business

© 2010 Taylor & Francis Group, London, UK

Typeset by Vikatan Publishing Solutions (P) Ltd., Chennai, India
Printed and bound in Great Britain by Antony Rowe (A CPI Group Company), Chippenham, Wiltshire

Published by: CRC Press/Balkema
　　　　　　　P.O. Box 447, 2300 AK Leiden, The Netherlands
　　　　　　　e-mail: Pub.NL@taylorandfrancis.com
　　　　　　　www.crcpress.com – www.taylorandfrancis.co.uk – www.balkema.nl

ISBN: 978-0-415-58478-4 (Pbk)
ISBN: 978-0-203-84238-6 (eBook)

Volcanic Rock Mechanics – Olalla et al. (eds)
© *2010 Taylor & Francis Group, London, ISBN 978-0-415-58478-4*

Table of contents

Preface

The International Society for Rock Mechanics (ISRM) has entrusted the "Sociedad Espanola de Mecanica de Rocas" (the Spanish National Group of the ISRM) with the organization of an ISRM Specialized Conference, the "3rd International Workshop on Rock Mechanics and Geo-engineering in Volcanic Environments" to be held in Puerto de La Cruz, Tenerife, Canary Islands from 31st of May to 3rd of June 2010.

This event is complementary to those technical meetings organized in previous years at Madeira and Azores, in 2002 and 2007, 1st and 2nd Workshop respectively.

This Workshop is included in a multidisciplinary Congress, (Cities on Volcanoes 6—Tenerife 2010), organized by the International Association of Volcanology and Chemistry of Earth Interior. Volcanologists, geologists, engineers, disaster mitigation experts, and other specialized professions are expected to participate.

The third edition of this series of workshops is to highlight the crucial support of the Regional Ministry of Public Works and Transportation of the Government of the Canary Islands. This institution has funded this conference and has spared no effort for its success, aware of the importance to provide a deeper insight in knowledge of the development of infrastructures in volcanic environments.

The Workshop proceedings have been published in a single volume and contain one keynote lecture and 46 papers classified in 3 themes as follows:

– Geomechanical characterization of volcanic materials.
– Instabilities in volcanic islands: Slope stability, large landslides and collapse phenomena.
– Geoengineering and infrastructures in volcanic environments.

The Editors wish to thank contributors for the quality and significant number of papers submitted to this workshop. They also thank all participants for their interest in this event and to all institutions that have contributed to its conclusion.

Volcanic Rock Mechanics – Olalla et al. (eds)
© *2010 Taylor & Francis Group, London, ISBN 978-0-415-58478-4*

Organizing committee

Luis E. Hernández (Chairman)
Claudio Olalla (Co-chairman—SEMR)
Jose A. Rodríguez-Losada (General Secretary)

International Scientific Committee

Alcibíades Serrano (Chairman)
Áurea Perucho (Co-chairman)

Nick Barton (Great Britain)
Juan A. Díez (Spain)
José Estaire (Spain)
Mercedes Ferrer (Spain)
Javier González-Gallego (Spain)
Luis González de Vallejo (Spain)
Nuno Grossman (Portugal-ISRM)
John Hudson (Great Britain-ISRM)
Alejandro Lomoschitz (Spain)
Ana Malhiero (Portugal)
Paul Marinos (Greece)
Ricardo Oliveira (Portugal)
Leoncio Prieto (Spain)
Eda Cuadros (Brasil)
Davor Simic (Spain)

Organization

The 3rd International Workshop on "Rock Mechanics and Geo-engineering in Volcanic Environments" is organised by the Spanish Society for Rock Mechanics (the ISRM National Group), with the collaboration of the Regional Ministry of Works of the Government of the Canary Islands in the frame of the International Congress "Cities on Volcanoes 6—Tenerife 2010", organized by the Cabildo Insular de Tenerife and the Instituto Tecnológico y de Energías Renovables (ITER) with the sponsorship of the International Association of Volcanology and Chemistry of Earth Interior (IAVCE).

Keynote lecture

Volcanic Rock Mechanics – Olalla et al. (eds)
© 2010 Taylor & Francis Group, London, ISBN 978-0-415-58478-4

Low stress and high stress phenomena in basalt flows

N.R. Barton
Nick Barton & Associates, Oslo, Norway

ABSTRACT: Contrasting geophysical, rock mechanics and rock engineering experience in basalts, caused by either exceedingly low or extremely high stress are described, from projects in the USA and Brazil. The first involves a nuclear waste characterization project in Hanford basalts in the USA, and the second describes, in much more detail, stress-fracturing problems in numerous large tunnels at the 1450 MW Ita hydroelectric project in SE Brazil's basalts. Particular phenomena that were noted, include linear stress-strain loading curves when columnar basalt is loaded horizontally, and a k_0 value reaching about 20–25 at Ita HEP.

1 INTRODUCTION

The beauty of columnar basalt, and the huge areal extent of basalt flows across large tracts of many countries, are perhaps the features that characterize basalt most profoundly. The Colombia River basalts in USA, and the Parana Basin basalts of S.E. Brazil, are just two of these major accumulations of 10's of thousands of km^2 of basalt. In this paper, some sophisticated characterization in the first location mentioned, in the hope of finding a nuclear waste disposal candidate, and some major rock engineering problems due to extreme horizontal stress in the second location, will form the core of this paper.

2 STRESS-DEFORMATION CHARACTER

One of the USA's nuclear waste disposal candidates of the mid-eighties was the 900 m deep Cohasset flow of the extensive Colombia River basalts. This was found some distance away at a more convenient *shallow* depth for preliminary but extensive characterization studies, at the so-called Hanford BWIP (basalt waste isolation project).

Some interesting joint deformation effects were caused by the low horizontal stress levels at this (too) shallow location, as revealed in an *in situ* block test, and at larger scale in some cross-hole seismic measurements in a tunnel wall, showing strong EDZ effects. At each scale, behavior was affected in special ways by the anisotropic joint properties and by anisotropic stress levels, particularly the low horizontal stress. The latter could be controlled in the block test, and thermal loading logically caused joint closure: the original state. An unexpected linear stress-deformation behaviour was measured in the block

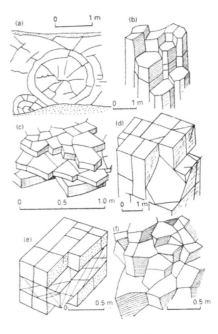

Figure 1. Basalt forms blocks of many shapes and forms.

test, apparently due to the contribution of both shear and normal components of joint deformation.

Some site characterization was performed by the author, along exposures of the candidate Cohasset Flow (Figure 2), which formed impressive cliffs along the distant Colombia River. Both joint properties and rock mass properties were described, in an attempt to evaluate their potential effect on disposal tunnels planned for 900 m depth at the candidate site, and possible tunnel support quantities.

Figure 2. Cohasset flow exposed along the Colombia River.

Figure 3. Cohasset flow sampled at 920 m depth.

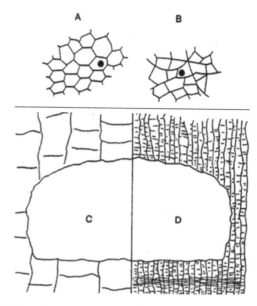

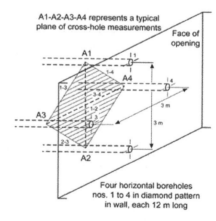

Figure 4. Comparative scales of boreholes in 'massive' rock and tunnels in jointed rock: either columnar or entablature. Stress-fracturing (core-discing or rock-bursting) in cases A, B, and C, but not in case D due to stress re-distribution and greater deformation. Possible $\sigma_{\theta max} \approx 140$ MPa at 900 m depth. with failure initiation at about 0.4 to 0.5 × UCS. Barton (1986).

Figure 5. The borehole layout for cross-hole seismic measurements in the wall of a drill-and-blasted experimental tunnel.

Drilling and stress measurements had indicated strongly anisotropic stresses of approximately 60, 40 and 30 MPa, and some cores, presumably drilled in the midst of columnar basalt, displayed strong core discing (Figure 3). The likely performance of the planned disposal tunnels at the same depth, was therefore of some concern. A conceptual image of excavation 'scale effects' is shown in Figure 4, to illustrate the likely relative effects of massive columnar basalt and the more jointed and irregular entablature, on the performance to be expected in planned disposal tunnels. Barton, (1986), contract report.

King et al., (1986), performed an interesting set of cross-hole seismic measurements, in a part-flow-entablature part-columnar jointed basaltic rock mass, at the BWIP site. The columns were regular but sinuous, 0.15 to 0.36 m in thickness, dipping 70 to 90°, with frequent low angle, discontinuous cross-jointing. The measurements were made between four horizontal boreholes drilled 12 metres into the wall of a drill-and-blasted underground opening, at 46 m depth. The objective was to investigate the effect of blast damage and stress redistribution, i.e., two of the assumed chief components of the EDZ or excavation damage and disturbed zone.

The two diagonal seismic-ray paths #1-4 and #2-4 showed, in contrast, almost identical seismic velocities, with a plateau at about 5-5.5 km/s, and reduction to about 3.6 to 4.4 km/s in the outer 2 to 3 m.

4

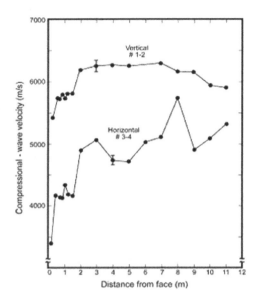

Figure 6. The vertical and horizontal ray-paths showed great contrast due to the low stress across columnar cooling joints.

The large contrasts in Vp values for the vertical path (#1 to #2) and for the horizontal path (#3 to #4) close to the opening (1.5 to 2.0 km/s difference) are the most clear indication of the easily disturbed columnar jointing. There is also some indication of a tangential stress concentration effect: the background (far-field) velocity of about 5.4 to 5.8 km/s appears to be elevated by about 0.5 m/s from about 4 to 8 m depth in the wall, with a lower background velocity.

The authors registered no consistent trend in RQD values with depth, but increased crack density was seen close to the opening. The velocity reductions seem to be a product of blast-damage, stress relief (and redistribution) and possible reduction in moisture content. The authors noted water flow from some of the horizontal holes during the tests, and had originally assumed more or less saturated conditions. However, Figure 7 does show a change in saturation level.

These results are presented in order to emphasise the possibility of drying out of some of the joints, despite water flow from some of the holes. The theoretical analysis of crack density did not appear to be supported by the RQD measurements in general, but is perhaps an expression of joint void ratio changes, with the joints closest to the tunnel wall showing the largest voids and therefore suggesting an apparent (but false) increase in joint density.

A sophisticated heated block test was one of the main components of the *in situ* testing at BWIP. Flat jacks were used to load four sides of the large

jointed block and confinement on a fifth side was available too. Unusually for jointed rock masses, neither concave nor convex load-deformation curves were produced: rather the load-deformation was linear when performed across the part-columnar part entablature jointing.

Figure 8 suggests how this may be due to the combination of joint closure phenomena (concave) and joint shearing tendencies (convex). UDEC-BB

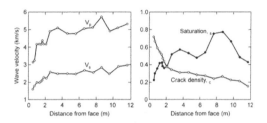

Figure 7. Seismic and other physical measurements in the face of the tunnel in basalt. Zimmermann and King, (1985).

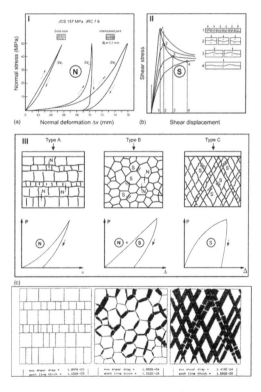

Figure 8. Conceptual explanation of concave, linear and convex load-deformation curves obtained from in situ testing, and UDEC-BB modelling result showing respective joint shearing magnitudes. Barton (1986), and NGI modelling team.

5

Figure 9.   Columnar basalt displaying large-scale sinu-
osity and small-scale undulation due to successive cooling
allowing joint propagation to greater depth. Locations:
Chile and Greenland.

Figure 10.   Mostly small-scale planarity and small-scale
roughness down each column. Suggested Q' classification would be $(90-100)/9 \times (1.5-3)/1$. The six sides of the
hexagons are equivalent to $Jn = 4$.

models of these joint configurations showed such trends.

In Figures 9 and 10, a more simple-minded classification of columnar basalt is suggested, using the first four Q-parameters. Jw and SRF might be extreme.

3 STRESS PROBLEMS AT ITA HYDRO BRAZIL

In the case of the Brazilian 1,450 Mw Ita hydroelectric project, contrasting Q-values in adjacent columnar and entablature flows were the focus of stress-fracturing predictions for various tunnels. It was found that the least jointed flows with high Q-values attracted extremely high stresses in the topographic ridge defining the project location across a meander in the river. Stress fracturing and extensive, many meters deep, 'dog-earing' occurred in the five large 150 m^2 diversion tunnels. In the higher-elevations of five pressure tunnel linings, cracking occurred when contact grouting, specifically in the 3 o'clock and 9 o'clock positions, over total lengths of hundreds of meters. There was also extensive erosion loss of basalt in the first flood-operation of the spillway, which could be attributed to the stress-aligned fracturing. Stress

ratios k_0 as high as 25:1 could be interpreted in the 50 to 100 m deep tunnels.

The first telltale signs of high horizontal stresses and strong stress anisotropy developed gradually as the project itself progressed, during the four years of construction time. Separate phenomena in different locations in the project eventually built a convincing picture of a highly stressed, narrow rock ridge in which the river meander itself presumably had acted like an 'over-coring' agent. (See the satellite photograph reproduced in Figure 12). The assumed regional stress anisotropy was concentrated in the narrow, pillar-like ridge, and with each new excavation, stress concentrations proved to be close to the limit of stress-induced fracturing—and sometimes exceeded the limit, despite the high strength of the basalts.

Popping noises, some thin slab ejection, and larger than expected deformations were recorded during excavation of the *five diversion tunnels*, at depths of only 50 and 100 m beneath the ridge.

Since stress problems and deformations were more notable as the tunnels reached their full height, the previously provided rock bolting in the arch of each top heading proved, in retrospect, to be insufficient, as some areas of excessive scouring in the arch and invert were later experienced following river diversion. Several metres thickness

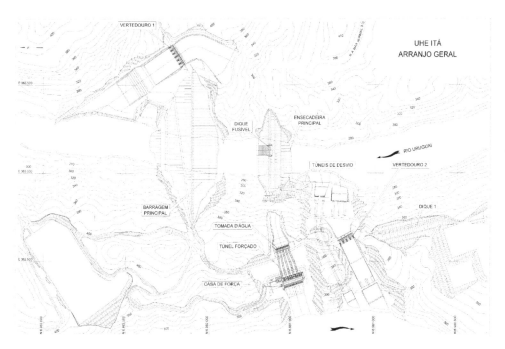

Figure 11a. The 1450 MW Ita hydroelectric project was built across a rock ridge formed by a 12 km meander in the Uruguay River in South East Brazil

Figure 11b. Stress-related problems were noted first in the (Tuneis de Desvio) diversion tunnels, later in the (Tuneis Forcado) pressure tunnels and finally in the spillway.

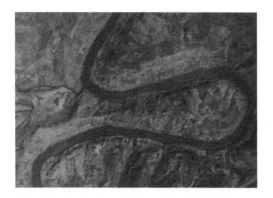

Figure 12. Satellite image of project site. The Uruguai River flows from the bottom to the top of this satellite photograph.

of over-stressed rock were lost in places, in the arch and in the invert.

These lost meters of failed rock will be back-calculated as indicators of stress magnitudes. We will also return soon to the deformations measured in the tunnels, when trying to back-figure the likely levels of stress.

The author's involvement in the project started at the spillway location, with a Q-system based histogram logging of the characteristics of the four basalt flows G, H, I, J that were now well exposed at this downstream location. Large-scale cracking had been noted above the portal of the tunnels, possibly also due to the high, horizontal, ridge-parallel stress.

Most of the diversion tunnels had been excavated in the central, and most massive H and I flows where most of the 'popping' was registered. The Q-logging confirmed the significant difference in the degree of jointing between the basalt flow 'pairs' G and J (above and below) and H and I in the centre. In this case we were dealing with a 'sandwich' with a hard centre, which was perhaps responsible for concentrating horizontal stresses to an even higher level in the N-S oriented ridge (see Figure 12). The relative magnitudes of the Q-parameters in the two pairs of flows were as follows:

Flows G and J: general character:

$$Q = \frac{70-90}{6-9} \times \frac{1.5-2}{1-2} \times \frac{0.66}{1}$$

Flows H and I: general character:

$$Q = \frac{90-100}{3-6} \times \frac{1.5-4}{0.75-1} \times \frac{1}{1}$$

Prior to the assumption of a significant stress differentiation between the two pairs of flows, we can give the following preliminary Q-ranges of 5 to 13, and 30 to 100 respectively. If we assume general high stress for all these flows, and a preliminary SRF ranging from 0.5 to 2, the above ranges are extended to 2.5 to 26, and 15 to 200 respectively.

Correlation of such Q-values with rock mass sparameters such as deformation modulus and seismic velocity are improved, following Barton (1995, 2002), by normalization with the uniaxial strength σ_c.

The normalized value Q_c is estimated as follows:

$$Q_c = \frac{Q \times \sigma_c}{100} \qquad (1)$$

An estimate of P-wave velocity (for verification with site characterization) is given by the following empirical relation for rock of low porosity, and is also shown in Figure 13:

$$VP = \log Q_c + 3.5 \ (km/s) \qquad (2)$$

The basalt at UHE Ita was unusually hard, with a range of uniaxial strengths of 140 to 280 MPa. If we assume a mean of about 200 MPa for convenience, the above Q-value ranges for the two pairs of flows become Q_c estimates of 5 to 52, and 30 to 400 respectively. Ranges of near-surface (nominal 25 m depth) V_P are therefore 4.2 to 5.2, and 5.0 to 6.1 km/s respectively. These ranges proved, quite independently, to show reasonable agreement with the 4.2 to 5.6 km/s range for 'sound rock' measured above the future diversion tunnels many years previously.

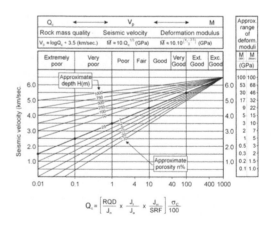

Figure 13. Inter-correlations of Q or Q_c and V_p and E_{mass} which were used for differentiating the basalt flow properties. Barton (2002).

8

The previously referred 'sandwich' of massive flows H and I, in which the diversion tunnels were driven, are likely to have attracted higher levels of horizontal stress than their neighbours, and this can be indirectly assessed by the relative magnitudes of deformation moduli that can be estimated from the following equation, again for near-surface (nominal 25 m depth) and low porosity.

$$E_{mass} = 10\ Q_c^{1/3}\ (GPa) \tag{3}$$

The estimated contrasts in rock mass deformation moduli were perhaps in the range 17 to 37 GPa for flows G and J, and 31 to 74 GPa for flows H and I, in fact roughly a doubling of moduli due to the more massive nature of the central, and eventually very troublesome basalt flows. With greater horizontal stress in the H and I flows, an anisotropic distribution of moduli would probably have been in operation, but this possibility has been ignored in the simple treatment that follows.

4 BACK-CALCULATION OF POSSIBLE STRESS LEVELS

We can first address the magnitude of the deformations actually recorded at up to twenty measurement locations along each of the five diversion tunnels. The convergences were plotted by sadly departed colleague Nelson Infanti, in the approximate $\log_{10} Q$/(span or height) versus $\log_{10}$ (convergence) format of Barton et al., (1994). Even at the top heading stage, the deformations, which ranged from 0.5 to 13 mm, were mostly higher than expected from the central empirical trend of numerous data:

$$\Delta(mm) \approx \frac{SPAN\ or\ HEIGHT}{Q} \tag{4}$$

In the case of TD-5, ten of the twenty instrument locations that were monitored again after benching down to the full 17 m height, showed magnitudes of convergence at the triangular monitoring stations that ranged from 13 to 50 mm, with a median value of 22 mm, and a mean of 25 mm.

Back-calculation according to equation 4 suggested much lower 'stressed' Q-values, 20 mm deformation implying $Q \approx 0.8$, and 50 mm implying $Q \approx 0.3$. So *characterization* prior to tunnel excavation was suggesting Q-values for the massive H and I flows of the order of 15 to 200, while *classification* for tunnel design was, through back-calculation from deformations, suggesting Q-values in the approximate range of 0.3 to 1.5.

We were clearly mostly within the 'stress-slabbing' SRF class (Table 1) of 5–50, Barton and Grimstad, (1994), which implies a σ_c/σ_1 ratio of 5 to 3, or an 'elastic behaviour' tangential stress ratio assumption (σ_θ/σ_c) of 0.5 to 0.66, i.e. a tangential stress high enough to cause failure with rock strength scale effects considered.

In the case considered here, the major principal stress is of course σ_H and the above ratios are suggesting that its value might be in the approximate range 47 to 56 MPa, when using the 140 and 280 MPa uniaxial strengths in the logical way in relation to the above strength/stress ratios of approximately 5 to 3.

Measurements performed at the site with an older LNEC STT (stress tube tensor) method were inconsistent, but maximum stresses of 29, 43 and 54 MPa were recorded, and, significantly, the core removed from the 9 m deep holes above the overcoring sites, showed 'disking', which is a sure sign of strong stress anisotropy and large magnitude. There was also some limited core-disking in deeper parts of two investigation boreholes.

An alternative way of back-calculating the possible horizontal stress level is to use the set of empirical 'depth-of-failure' data assembled in Figure 14. With depths of failure as seen in Figure 15 in the range 2 to 3 m for an average tunnel 'radius' of about 8 m, we see in Figure 14 that ratios of σ max/σ_c of about 0.6 to 0.7 are implied when D_f/a is in the range of $(8 + 2$ or 3 m$)/8 = 1.25$

Table 1. Extract from Q-system SRF, concerning stress-failure of massive rock. Barton and Grimstad (1994).

b) Competent rock, rock stress problems		σ_c/σ_1	σ_θ/σ_c	SRF
H	Low stress, near surface, open joints.	> 200	< 0.01	2.5
J	Medium stress, favourable stress condition.	200-10	0.01-0.3	1
K	High stress, very tight structure. Usually favourable to stability, may be unfavourable for wall stability.	10-5	0.3-0.4	0.5-2
L	Moderate slabbing after > 1 hour in massive rock.	5-3	0.5-0.65	5-50
M	Slabbing and rock burst after a few minutes in massive rock.	3-2	0.65-1	50-200
N	Heavy rock burst (strain-burst) and immediate dynamic deformations in massive rock.	< 2	> 1	200-400

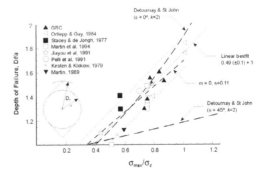

Figure 14. Empirical data for stress-induced depths of failure in relation to stress/strength ratios. Martin et al., (2002).

Figure 15. Consistent stress-induced fracturing of 2 to 3 m depth (also in the invert in places where scoured by flood—flows. These large diversion tunnels measure 15×17 m.

to 1.38. Taking σ_c as an average 200 MPa, the above implies that the maximum tangential stress may have been as high as 120 to 140 MPa. If we further assume relevant vertical stress ranges from about 1.25 to 2.5 MPa from 50 to 100 m overburden depths, and an elastic isotropic theoretical $\sigma_{\varphi(max)} = 3\sigma_H - \sigma_v$, we obtain estimates of σ_H of about 39 to 46 MPa. The implication is therefore that the ratio of principal stresses (σ_H/σ_v) may be as high as approximately 20–25, which of course is exceptional.

4.1 An estimation of negative minimum tangential stresses

As the critical pressure tunnel excavations were at a very preliminary stage, some re-evaluations of potential tangential stress anisotropy was appropriate, for the partly horizontal, partly steeply inclined shafts. For the case of excavation through the massive H and I flows, a similar assumption to the above, of σ_H(max) of 35 to 45 MPa was utilized, together with a vertical stress range assumption of 1 to 2 MPa. This is an unusually extreme stress anisotropy, but phenomena from around the site appear to support it, as we shall see.

Based on the above, and application of simple Kirsch equations, the EW oriented pressure tunnels, like the EW oriented diversion tunnels further

upstream, might have *maximum* tangential stress levels in the range 103 to 135 MPa, and *minimum* tangential stresses as low as (–) 29 to (–) 42 MPa, easily enough to exceed the tensile strength of the basalts.

We know that the former, whatever their real magnitude, had been sufficient to cause stress fracturing in the first tunnels excavated, and loss of 100's up to 1000's of m^3 of stress-fractured rock during river diversion (Figure 14). The latter (tensile stresses) would clearly be large enough to cause tensile fractures on NS sides (or 3 o'clock and 9 o'clock positions) around the pressure tunnels, which were yet to be completed—if these pressure shafts passed through sufficiently massive flows for the above Kirsch elastic isotropic solutions to be relevant (max. tang. stress = 3A-B, min. tang. stress = 3B-A, where A and B are the major and minor principal stresses.).

As it happens it was also discovered during this first involvement with the project that the most massive H and I flows had 'mysteriously' given the highest permeabilities. This mystery is easily explained if the *minimum* horizontal stresses were also of the same order of magnitude as the above vertical stress assumption. Vertical tension cracks along the N and S sides of these parts of the investigation boreholes could readily explain the 'inexplicable' high permeabilities in the most massive rock mass. This is another illustration of the need for separate *characterization* and *classification* for before and after excavation, at whatever scale. Such differentiation when using the Q-system is emphasised in Barton, 2002.

5 CRACKING OF THE PRESSURE TUNNELS

The foregoing 'situation report', which can be summarized effectively by Figure 16, was delivered in 1997, two years before the author's second visit to the site in 1999, following completed excavation and lining of the five pressure shafts/tunnels (mostly a 55 degrees inclined section of 140 m length and a lower horizontal section containing the final steel penstocks). Raised boring of the 'central' core of each inclined shaft had been followed by drill-and-blast excavation of the complex (sometimes double-curved) 9m diameter pressure conduits, which had been temporarily supported with fibre-reinforced shotcrete and rock bolts, followed by about 0.5 m of reinforced concrete—*but with reinforcement only in the lower half of each shaft*, where the water pressure head would vary from 55 to 110 m. This omission of reinforcement proved in the end to be a false economy.

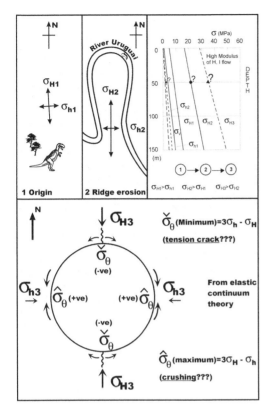

Figure 16. Summary of the probable origin of an elevated horizontal stress and stress anisotropy at UHE Ita, and the reasons for two potential types of rock failure around the tunnel excavations.

Some 'popping' had been recorded during excavation of the shafts, but only the lower part of each shaft was excavated in the massive H and I flows. The number of locations (3, 5, 6, 9 or 11) in the five shafts where popping noises had been recorded during excavation, were given elevated SRF values (2.5, 5 or 25) in the follow-up Q-logging. A concentrated and rather continuous zone of rock noises and 'popping' appears to have occurred in the central section of pressure shaft TF-2, with 8 to 10 close occurrences. Here an SRF of 25 was assumed, giving *classification* Q-values as low as 1.6.

The phenomenon of concern on the occasion of the author's second site visit was intermittent but rather regular cracks along the NS sides (or 3 o'clock and 9 o'clock positions) of the pressure shaft concrete linings. The rather linear, sometimes sporadic, sometimes semi-continuous cracking mostly stretched for some 60 to 90 m down each shaft, and was apparently caused when the contact grouting had been performed, behind the previously sound, cast concrete slip-formed linings.

This grouting had been limited to 0.2 MPa excess pressure. The aperture of these mostly leaking cracks in the concrete was from 0.2 to 2 mm, with many in the range 0.4 to 0.8 mm. The cracks were rough on a small scale (JRC$_0$ about 20 to 25) but remarkably linear on a scale of meters. 'Fortunately' they had occurred before filling the pressure tunnels.

Figure 17 summarizes in graphic form, what is assumed to have happened as a result of the contact grouting in already tension-bearing, and perhaps pre-cracked, but otherwise massive rock. The existing negative total stress (σ_h minimum) at 3 o'clock and 9 o'clock was already probably an even more negative effective stress because of the near-by reservoir filling ($-u$), and this was made even more negative by the grouting pressure ($-\Delta u$). We thus have potentially four stages of crack development, if the final objective of pressure tunnel filling is included, as suggested in Figure 18.

It is clear that this situation was completely unacceptable for pressure tunnel operation, and an extensive repair operation was already underway, using 'epoxy taping' following the method suggested by Andrioli et al., 1998. This repair operation was extended considerably when the unstable nature of the phenomenon was fully appreciated. Of particular concern were of course the consequences of uncontrolled leakage from the pressure

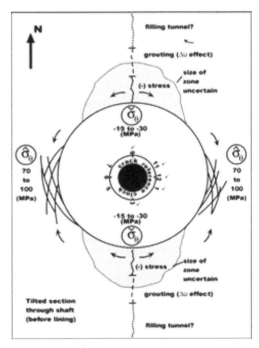

Figure 17. Tensile stress enhancement during contact grouting is assumed to be the reason for the extensive cracking.

11

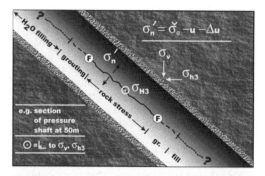

$$\sigma_n' = \breve{\sigma}_\theta - u - \Delta u$$

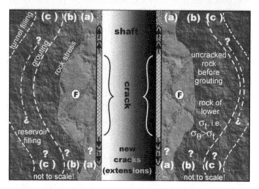

Figure 18. Longitudinal impression of the grouting induced cracking along considerable lengths of each of the inclined pressure shafts, which unfortunately did not have steel reinforcement in their upper 50 to 60 m.

shafts into the rock mass above the deep power-house excavation. An extension of the drainage fans from an extension of the existing drainage gallery was recommended, in order to be able to intersect the above cracks (between each pressure shaft) at more acute angles, to improve drainage efficiency. More fans of drain holes were added. Piezometers were already installed and more were added.

Although some additional drainage capacity was installed, there was preference for the more extensive crack (and potential crack) repair using 'epoxy taping'. Leakage under first filling was well controlled, though was higher than desirable. A scientifically interesting phenomenon was also discovered. There was a very minor rotation of the powerhouse inclinometers, when a pressure tunnel was taken out of operation for inspection, suggesting a coupled (effective stress controlled) deformation of the rock mass surrounding the pressure shafts, in the hillside above the power house. Needless to say, extensive improvement of the grouting in this area was recommended, in order to reduce sensitivity to potential effective stress changes, and more drain holes were drilled from the drainage gallery.

6 AUXILIARY SPILLWAY OPERATION

The fourth, major stress related phenomenon at UHE Ita was experienced in the flat bottomed spillway excavations, in other words right to the surface in excavations intersecting the two massive flows H and I. Partly unlined, or mostly unlined auxiliary spillways have operated with success in Brazilian dam projects built in basalts in the past, and experienced consultants accepted a similar design for UHE Ita. One of the specific reasons that the basaltic rocks strongly resist erosion during the infrequent, but sometimes extreme water flows, is that they have an interlocked, 'jigsaw-like' pattern of vertical and sub-vertical jointing, frequently with curved interlocking roughness (e.g. Figures 9 and 10). Alternatively the joints may have a smaller-scale 'open saw-tooth' roughness, caused by the brittle-ductile cooling front that allows cooling cracks to develop downwards from the surface, but only as an intermittent process, a few centimetres at a time. There are often minor changes of direction with each cooling-joint propagation, which helps to add to the deformation resistance of blocks that are jointed in this way.

The suspicion of a fourth stress-related phenomenon was occasioned by an unprecedented loss of 18,000 m³ of rock from the floor of the 20,000 m³/s auxiliary spillway, but during only 2½ hours (!) of spillway operation with flows of only 800 m³/s for 1 hour, and 1,600 m³/s for 1½ hours in the 2000/2001 rainy season. The author's third visit to the site in the dry season in mid 2001 coincided with the possibility to inspect, in dry conditions, if there was any evidence to suggest stress-enhanced erosion.

Besides a previously hidden 'junta falha' or joint-fault beneath part of the spillway, the most important phenomenon proved to be the existence of numerous, well-oriented tension fractures, which crossed or ran sub-parallel with some existing NE trending joints. The stress-induced fractures had—inevitably—the familiar NS (ridge-parallel) orientation. There was also an equally pervasive development of sub-horizontal and equally fresh (unweathered) tension fractures, which also satisfy a NS maximum principal stress orientation. Thus the basalt had lost its prime property of non-systematic jointing (if we ignore the familiar columnar jointing, which resists erosion with reasonable efficiency).

The now systematic fracturing of the basalt effectively divided existing, irregular-shaped blocks into smaller, more easily eroded units, and as the upper or front parts were removed by traction and/or pore pressure during spillway operation, the next sub-blocks were exposed for a similar treatment. An existing basalt block, if divided by just one vertical and one horizontal fracture, becomes in the process 4 blocks. A finer division of each existing block with two vertical and two horizontal fractures becomes in the process 9 blocks. With respect to erosion resistance this is a catastrophic increase—and was readily observed in the floor of the spillway, sometimes with greater frequency than this, in one of the directions of fracturing.

Several 1 m long cored and instrumented slots had been drilled in the floor of the spillway, following the scouring event, and these showed slow closure when oriented roughly EW (up to 0.5 mm surface-measured closure), while an almost NS aligned slot showed opening, by up to 0.15 mm. Such is broadly consistent with a strong stress anisotropy, but would need to be modelled in three dimensions for interpretation to be meaningful. The measurements could also be influenced by sub-horizontal fracturing.

Interestingly, occasional 'radial' blast-gas induced fracturing seen at the base of some remnant, vertical blast holes, was actually not radial but elliptical—with the long axis inevitably oriented NS, with some particularly extended (gas-and-stress-induced) fractures in this direction. At another location in the spillway floor, a newly stress-fractured slab had lifted (buckled) making a gap of several centimetres, beneath its bridge-like structure. High stress anisotropy was evident in many forms at UHE Ita, and was a valuable learning experience for all parties involved, including the author who was engaged by the consortium of Contractors.

7 CONCLUSIONS

An unexpected linear stress-deformation behaviour was measured in the block test that was performed at the BWIP project in Hanford, USA. This apparently was due to the contribution of both shear and normal components of joint deformation. Joints that are closing normally exhibit concave load-deformation curves, while joints that are loaded in shear exhibit convex load-deformation curves. These contrasting trends appear to 'cancel' and a linear curve results.

Basalt flows encountered near the surface may exhibit low horizontal stress, if there are no topographic or tectonic reasons for stress concentration. This would seem to be related to the tensile nature of joint formation, and is clearest in the case of columnar basalt. At the BWIP site, the thermal and flat-jack applied stress in the heated block test was able to close the vertical columnar, part entablature joints, giving much stiffer behaviour.

Cross-hole seismic performed between four boreholes drilled into the wall of an experimental tunnel, also at shallow depth, exhibited strong contrasts in velocity between horizontal (lowest velocity) and vertical (highest velocity) ray paths.

Figure 19. Views of the site during construction and after reservoir impoundment. The narrowness of the ridge of rock made for a very compact site (compared to 16.6 km of mountain tunnels for a much larger river meander at Jinping II).

Diagonal ray-paths showed intermediate velocities. In all cases there was a strong reduction of velocity of about 2 km/s in the outer 2 to 3 m of the holes, due to an EDZ caused by blast damage, stress reduction, and somewhat reduced moisture content.

At the candidate waste site, the Cohasset flow was encountered at about 900 m depth, and due to strong stress anisotropy and high stresses (60, 40, 30 MPa) core discing was experienced in the more massive columnar-jointed rock. Stress-induced fracturing was predicted for the tunnels in this massiveflow, but joint deformation maybe would have protected tunnels from rock bursting in the more jointed entablature.

At the Ita 1,450 MW HEP in Brazil, an anisotropic horizontal stress distribution of 'normal' magnitude for SE Brazil appears to have been seriously concentrated by maximum-stress-aligned river erosion. This occurred in the narrow 150 m high ridge separating a 12 km river meander, which was chosen for the site of this hydroelectric plant.

Further concentration of horizontal stress in this ridge was caused by general stripping and excavation of large surface structures such as the auxiliary spillway. Within the ridge of basalt are two particularly massive, high Q-value, high modulus flows, which probably concentrated the horizontal stress even more. Tunnelling in these flows produced many surprises, even when tunnel depths were only 50 to 100 m.

Excavation of the five diversion tunnels in an E–W direction beneath this highly stressed ridge apparently increased the already high stresses within the massive flows, to tangential stress levels as high as 120 to 140 MPa, even at 50 m depth. For the large 15 by 17 m temporary diversion tunnels with minimum rock bolting and shotcreting, the resulting 2 to 3 m deep stress-fracturing allowed major erosion, amounting in places to loss of several meters of hard basaltic rock in the invert as a result of river-flood diversion through the tunnels, and of course a similar loss of 2 to 3 m of stress-fractured rock in the arch.

For the five pressure shafts excavated in the same E–W direction through the ridge, the most serious consequence of the extreme stress anisotropy was the highly negative minimum tangential stress, which caused tensile cracking of the rock and later of the concrete lining (in the 3 o'clock and 9 o'clock positions), when even low pressure contact grouting was performed. The location of the cracking followed simple rock mechanics theory, and appears to have been repeated earlier when drilling vertical investigation boreholes, which showed *greatest permeability in the most massive flows*, probably due to N- and S-side tension cracks down the massive-rock parts of the boreholes.

Lessons to be learned include the need for stress measurements in general, when lightly reinforced lined (or unlined) pressure tunnels are contemplated, and topographic reasons suggest insufficient minimum rock stress. The extreme stress anisotropy at Ita HEP would have far exceeded the limits for hydraulic-fracturing based stress measurement— due to drilling-induced tensile cracks that would not have given a break-down pressure nor a shut-in pressure, except at a larger, unknown radius. The maximum principal stress could not then have been estimated in the normal manner.

Extreme, rock stress-induced, systematic tensile fracturing around prospective pressure tunnels, prior to their operation, is rather unusual, and presents a dilemma. High pressure grouting of the rock could probably have helped to eliminate the two regions of negative tangential stress along each pressure shaft, prior to reinforced concrete lining.

Partial 'homogenization' of the tangential stresses and general rock mass improvements through systematic grouting would also have reduced the need for heavy reinforcement of the concrete liner, but would need to be proved by post-treatment permeability and stress measurement, and local cross-hole seismic and more general tunnel wall refraction seismic.

REFERENCES

Andrioli, F.R., Maffei, C.E.M. & Ruiz, M.D. 1998. Improving the lining of the headrace tunnel in Charcani V Power Plant, Rio Chili, Peru. Proc. 5th S. American Conf. on Rock Mech. and 2nd Brazilian Conf on Rock Mech., SAROCKS 98, Santos, Brazil.

Barton, N. 1986. Deformation phenomena in jointed rock. 8th Laurits Bjerrum Memorial Lecture, Oslo. Publ. in Geotechnique, Vol. 36: 2: 147–167.

Barton, N., By, T.L., Chryssanthakis, P., Tunbridge, L., Kristiansen, J., Løset, F., Bhasin, R.K., Westerdahl, H. & Vik, G. 1994. Predicted and Measured Performance of the 62 m span Norwegian Olympic Ice Hockey Cavern at Gjøvik. Int. J. Rock Mech, Min. Sci. & Geomech. Abstr. 31:6: 617–641. Pergamon.

Barton, N. & Grimstad, E. 1994. The Q-system following twenty years of application in NMT support selection. 43rd Geomechanic Colloquy, Salzburg. Felsbau, 6/94. pp. 428–436.

Barton, N. 1995. The Influence of Joint Properties in Modelling Jointed Rock Masses. Keynote Lecture, 8th ISRM Congress, Tokyo, 3; 1023–1032, Balkema, Rotterdam.

Barton, N. 2002. Some new Q-value correlations to assist in site characterization and tunnel design. Int. J. Rock Mech. & Min. Sci. 39/2: 185–216.

Infanti Jr, N., Tassi, P.A., Mazzutti, R., Piller, M. & Mafra, J.M.Q. 1999. Tensões residuais nas obras subterrâneas da UHE Itá. XXXIII Seminário Nacional de Grandes Barragens, Comitê Brasileiro de Barragens. Belo Horizonte, Brazil.

King, M.S., Myer, L.R. & Rezowalli, J.J. 1986. Experimental studies of elastic-wave propagation in a columnar-jointed rock mass. Geophys. Prospect., 34: 1185–1199.

Martin, D.C., Christiansson; R. & Soderhall, J. 2002. Rock stability considerations for siting and constructing a KBS-3 repository, based on experience from Äspö HRL, AECL's HRL, tunnelling and mining. SKB (Swedish Nuclear Fuel Co.) Stockholm, TR-01-38.

Zimmerman, R.W. & King, M.S. 1985. Propagation of acoustic waves through cracked rock. 26th US Symp. On Rock Mechanics, Rapid City SD. Ashworth (ed.). 739–745. Rotterdam: Balkema.

1 *Geomechanical characterization of volcanic materials*

Volcanic Rock Mechanics – Olalla et al. (eds)
© 2010 Taylor & Francis Group, London, ISBN 978-0-415-58478-4

Modeling of the collapse of a macroporous material

Daniel Del Olmo & Alcibíades Serrano
ETSICCP-UPM, Madrid, Spain

ABSTRACT: The macroporous materials are a mix of solid particles, joined together with bridges of materials that may be the same or different of the solid particles. For example, volcanic rocks like volcanic agglomerates. In this way, it is interesting to trying to explain how the collapse of these materials takes place. With the great improvement of the numerical methods and the power of computers it has been possible to carry out a discrete analysis instead of a continuum one, like would had happened with the classical theories of continuum. This article shows the first steps taken in this path of modeling the collapse of macroporous materials in a discrete way.

1 INTRODUCTION

This article is trying to show the first steps in modeling a macroporous material as a discrete media and not as a continuum. That is to say, the macroporous is modeled as a conjunction of particles joined together by a breakable material and not as a homogeneous continuum material.

In this research, the macroporous material had been modeled as a series of spheres particles that are joined with a material of some stiffness with the capacity of breaking. The model created in this way is loaded similarly to a consolidated triaxial test. In the model, it is observed how the contacts get broke and how the collapse of the macroporous material takes place.

The research and the article have been structured in the following way:

- A numerical macroporous material was created.
- The equations of behavior of the material were defined.
- A series of calculus were run, according to the equations and to a criterion of failure of the joining material.
- Finally some conclusions were obtained from the calculus in order to improve the model and take it to the next stage of development.

2 CREATING THE MACROPORUOS MEDIUM

To make the mechanical study of the three dimensional macroporous medium, it is necessary to generate it numerically. To achieve this, the following hypotheses were assumed:

1. The granular material must be contained in a container of prefixed shape and dimensions. The chosen shape was a one meter mat cell. This cell was chosen because from an operative point of view, it is necessary that the boundary conditions are separated enough to obtain results that are not boundary affected. And in addition, a cubic shaped container simplifies the numerical process of filling it with particles.
2. For simplicity in the generation of the material and attending to geometrical criterions only, the particles are spherical shape. In this way, when one sphere is placed over another, the programmed process is simplified, compared with other possible shapes of the particles as ellipsoids, that would have made the process much more complicated for not being a constant radius shape.
3. When the spheres fall into the cell, they roll over the others until a stable position is reached. A position is considered to be stable whenever a sphere is in contact with three other elements or with the bottom of the cell. Any wall or any particle already inside the cell is considered to be one of the three necessary elements to obtain such stable position in addition to the bottom of the cell.
4. The particles or spheres had a continuous grain size distribution. This curve can be seen in Figure 1. In the generation of the medium, with this predetermined distribution, a certain probability of size is established. In this way, the

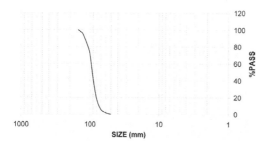

Figure 1. Grain size distribution for the particles that formed the macroporous material.

Figure 3. Example of a numerical macroporous material of 1300 particles.

but a wall-beam model instead. By doing this, it is obtained a joining tablet material in each previously existing contact.

After all this process, a macroporous medium is obtained. In Figures 2 and 3 it is shown some examples of the macroporous created. The particles are in white and the joining material in red. The joining material is not in its real dimensions, it has only been included to observe which particles are in contact and which are not.

Figure 2. Example of a numerical macroporous material of 1300 particles.

probability of size is directly proportional to the percentage retained by the size of each fraction, according to the curve.

5. After this process, a cubic cell filled with particles is obtained. These particles are in contact but a joining material is filling those contacts, keeping the particles joined and cohesioned. In order to introduce this joining material a gap is needed between the particles in contact. To do this, a random uniform distribution between two values was used to reduce the diameter of the particles accordingly to it.

6. An elastic and breakable joining material was inserted in the previously existing particle to particle contacts. The joining material was assimilated to square section prisms with a section side (B) and a length (H) that was determined in the fifth step. The B dimension is randomly created using a random uniform distribution between 2 and 6 times the length. Doing this, a 2 to 6 B/H ratio is obtained. This is necessary because a beam model is not wanted

3 EQUATIONS OF THE MODEL

In this point, the equations that command the behavior of the macroporous material are shown. In general, the macroporous can be under any kind of load in each of its boundaries. But this research is focused in modeling a consolidated triaxial test.

From the point of view of the strength, the response of the medium is conditioned by the joining material and its collapse. To model this material, it has been assumed that a wall-beam, similar to a tablet shaped material, under axial and shear stresses is a good enough approximation. The parameters considered for this material has been the following ones:

$$E = 600 \text{ MPa} \tag{1}$$

$$V = 0.2 \tag{2}$$

$$\sigma_c = 1 \text{ MPa} \tag{3}$$

where: E is the Young's modulus, ν is the Poisson's ratio and σ_c is the unconfined compressive strength.

With this defined properties, a finite element method (FEM) calculus was carried out. In this FEM the joining material was modeled as a tablet loaded with axial and shears forces. The tablet was considered to be broken whenever any fiber of the material reaches the Tresca condition:

$$\sigma_c \geq \sqrt{\sigma_N^2 + 3\tau^2} \qquad (4)$$

where: σ_n is the normal stress and τ is the shear stress. In this way, the failure diagrams (pair of values N; Q, that produces the failure of the tablet) were defined. Each contact will have its own diagram depending on the B/H ratio.

In addition, the normal stiffness (Kn) and the shear stiffness (Kt) were calculated by dividing the applied force against the obtained displacement. The variation of the normal and shear stiffness against the B/H ratio is shown in Figures 4 and 5.

In order to introduce the laws of behaviour in the matrix system, the stiffness formulation for one contact enounced by Kishino (1989) and Tsutsumi and Kaneko (2008) was considered. Being the stiffness expressed as:

$$N = K_N \cdot (\delta_{I,x} - \delta_{J,x}) \qquad (5)$$

$$t = K_t \cdot (\delta_{I,y} - \delta_{J,y}) \qquad (6)$$

$$T = K_T \cdot (\delta_{I,z} - \delta_{J,z}) \qquad (7)$$

where:
{N, t, T} are the forces in local axis, corresponding to the axial force and the two tangential forces respectively, in the union material.
{K_N, K_t, K_T} are non-linear variables that relate the relative displacements between two particles in contact with the mobilized force in each direction.
{$\delta_{I,x}$, $\delta_{I,y}$, $\delta_{I,z}$} is the displacement of the sphere I in the "j" directions of the local axis {x, y, z}.
{$\delta_{J,x}$, $\delta_{J,y}$, $\delta_{J,z}$} is the displacement of the sphere J in the "j" directions of the local axis {x, y, z}.

The global stiffness matrix of the system is the addition of the local stiffness of each contact, projected from the local axis to the global ones. Using the Newtons equations of the static:

$$\Sigma F_X = 0 \qquad (8)$$

$$\Sigma F_Y = 0 \qquad (9)$$

$$\Sigma F_Z = 0 \qquad (10)$$

where: {X, Y, Z} are the global axis. Using these expressions for each particle in the macroporous material it is obtained the following system:

$$[F] \cdot \{\delta\} = \{I\} \qquad (11)$$

where:
[F] is the stiffness matrix of the system.
{δ} is the displacement vector of the particles or spheres.
{I} is the vector that contains the load in the three global axis {X, Y, Z}.

The procedure of calculus is the following:

1. Once the macroporous has been defined geometrically, a normal stiffness (K_N) and a shear stiffness (K_t) are assigned to each existing contact (Fig. 4–5).
2. The local stiffness is projected to the global axis to obtain the matrix [F], using the expressions (5) to (10).
3. The vector {I} is built considering the forces applied on the boundaries of the macroporous material.
4. The matrix system of equations (11) is solved and the displacements of each particle are obtained. Once the displacements are known,

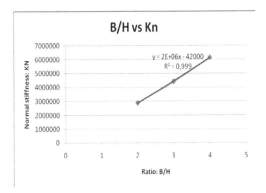

Figure 4. Relation between B/H ratio and normal stiffness Kn.

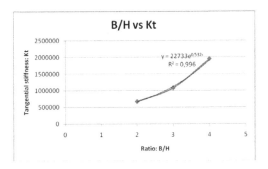

Figure 5. Relation between B/H ratio and the shear stiffness Kt.

they are translated to forces in every joining material of the macroporous.

5. Every contact is checked one by one in order to break those that have exceeded the maximum allowable force by the failure diagrams, (N, Q, pair of values that produces the reach of the Tresca criterion in the tablet joining material) obtained in the FEM calculus. The contacts that had broken are removed from the macroporous by deleting its contribution to the stiffness matrix [F].

By increasing the load progressively further displacements and inner forces are calculated to check how many contacts had broken. In addition, the position (dip and dip direction) of the broken contacts are measured.

This process is repeated until the macroporous cannot reach the equilibrium with the external load. When this situation happens it is said that the collapsed of the macroporous material had taken place.

4 FIRST RESULTS

The model is actually in an early stage of development and needs some improvements. But the first results obtained with this methodology are promising. For example, without having reached the collapse, we have been able to observe the following:

1. In the early steps of loading (point A in Figure 7), the first contacts to break had been the ones with a low dip angle independently

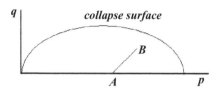

Figure 7. Example of the trajectory of tensions during a collapse calculi.

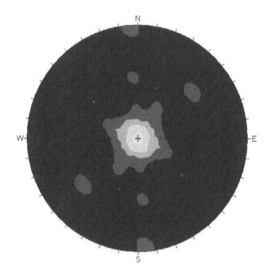

Figure 8. Contour plot of concentration of poles that has broken on the advanced steps of loading.

of the dip direction. That is to say, the more vertical contacts are the firsts to break as can be seen in Figure 6.

2. But when the load is progressively increased (deviatoric), from point A to point B, as can be seen in Figure 7, new contacts broke. These new ones are not horizontal but near to 45° of dip angle instead. This can be seen in Figure 8.

It is thought that in more advanced steps of loading; the concentration of poles among 45° of dip angle will generalize in every dip direction, leading to the collapsed of the macroporous.

5 CONCLUSIONS

In this research, the first steps on how to model the collapse of a macroporous material had been shown. The model is in an early stage of development and this article is only trying to enounce one of the possible ways to model this kind of fragile break.

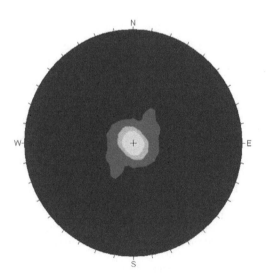

Figure 6. Contour plot of concentration of poles that has broken on the early steps of loading.

The main ideas and conclusions obtained with the proposal formulation are the following:

- A macroporous material had been numerically generated by using random functions. In the contacts a breakable joining material had been inserted.
- It has been presented a general vision on how to set out the system of equations that govern the phenomenon.
- Finally, the first results that have been obtained from the calculus, applying these theories had been shown.

REFERENCES

Del Olmo, D. 2009. Elaboración de un modelo discreto de partículas en 3D para estudio tenso deformacional de materiales gruesos. *Doctoral Thesis. E.T.S.I.C.C.P.—U.P.M.*

Kishino, Y. 1989. Computer Analysis of Dissipation Mechanism in Granular Media, Powders and Grains. A.A. Balkema, Rotterdam (1989) pp. 323–330.

Perucho, A. (2004) "Estudio de deformabilidad de escolleras". *Doctoral Thesis E.T.S.I.C.C.P.—U.P.M.*

Serrano, A.A. 1976. Aglomerados volcánicos en las Islas Canarias. *Mem. Simposio de Rocas Blandas:* Tomo II. pp. 47–53.

Tsutsumi, S. & Kaneko, K. 2008. Constitutive response of idealized granular media under the principal stress axes rotation. *International Journal of Plasticity.*

Uriel, S. & Serrano, A.A. 1973. Geotechnical Properties of two collapsible volcanic Soils of low bulk density at the site of two dams in Canary Island (Spain). *8th Congreso de I.S.S.M.F.E. Moscú,* vol I, pp: 257–264.

Volcanic Rock Mechanics – Olalla et al. (eds)
© 2010 Taylor & Francis Group, London, ISBN 978-0-415-58478-4

Geotechnical description of halloysite clays from La Palma Island (Spain)

J. Estaire, M. Santana & J.A. Díez
Laboratorio de Geotecnia (CEDEX, Mº de Fomento), Madrid, Spain

ABSTRACT: In this paper the geotechnical description of a halloysite clay, coming from boreholes performed in Barlovento Dam, is made. It is important to remark this material had a difficult behaviour during the performance of laboratory tests, clearly different of the materials Soil Mechanics usually treats with. The geotechnical description comprised the following aspects: a) identification tests: the grain size and plasticity tests had to be made with different procedures of the ones described in the technical Spanish standards due to the special characteristic of the material; b) X-Ray diffraction tests; c) chemical tests; d) dry density and natural water content of the material; e) strength tests that comprised dynamic penetration tests (SPT and DPSH test),triaxial and direct shear tests and pressuremeter tests made in the boreholes; f) deformability tests that included oedometer and pressuremeter tests g) permeability and dispersibility tests.

1 INTRODUCTION

In the bottom of Barlovento Dam, situated in the north of La Palma Island, settlements about 2 m were detected. With the aim of determining the causes of such movements (Estaire et al., 2008) a geotechnical investigation campaign was performed. In such study, it was discovered that the ground was mainly formed by halloysite clays.

It is important to remark the laboratory tests made with the halloysite clay material were difficult to perform, due to its clearly different geotechnical behaviour compared with the soils usually studied by Soil Mechanics. (Hurlimann et al., 2001).

2 PROBLEMS DETECTED IN THE DAM

Laguna de Barlovento Dam is situated in the northeast of La Palma Island, in Canary Island Archipelago. The dam was built in 1970 and it has a 3,2 Hm^3 maximum capacity and a 300.000 m^2 area. It was built inside a volcanic cone of elliptical section with 400 and 500 m long axis. In the bottom and in the walls of the caldera, a PVC layer was placed to waterproof the reservoir.

Figure 1 shows a photograph of the reservoir once it was emptied.

In one of the periodical maintenance operations, some cracks in the waterproof layer and great settlements in the bottom of the reservoir were discovered. Figure 2 shows the location of the cracks in the PVC layer and the extension of the zones with same settlements.

The analysis of the previous figure makes it possible to state the following facts:

a. Settlements only affect the bottom of the reservoir and they reflect the elliptical shape of the bottom.

Figure 1. Aerial photograph of the dam, once it was emptied.

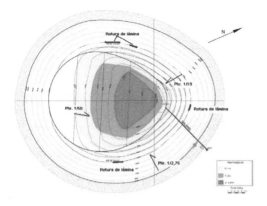

Figure 2. Location of the cracks in the PVC layer and the extension of the zones with same settlements (s): in yellow s < 1 m, in orange 1 < s < 2 m and in pink s > 2 m.

Figure 3. One of the cracks of PVC impermeable layer.

b. Almost 75% of the bottom surface suffered settlements.
c. The maximum settlement was about 2,35 m, situated 40 m distant from the bottom drainage that was the lowest point of the bottom at the end of the construction.
d. The drainage system suffered 65 cm settlements and it was broken due to such movements.
e. In the bottom area far away from the drainage point, settlement gradient is very light as there is 230 m distance between zero and maximum settlement points.

During the inspection, it was checked that the waterproof layer cracks were due to great tilting movements in a concrete piece of the drainage system. Those movements were too localized to be absorbed by the PVC layer without tearing. Figure 3 shows a PVC layer crack.

To discover the origin of the problems, a geotechnical investigation campaign was carried out which consisted in 13 continuous dynamic penetration tests (DPSH) and five boreholes. The borehole length ranged between 40 and 65 m. In one of the boreholes, five pressuremeter tests were performed. With the samples taken in the boreholes, a laboratory test campaign was made.

With all the information obtained in the investigation campaign, some geological-geotechnical profiles were drawn, as the one shown in Figure 4.

The bottom of the volcanic cone was occupied by silty and clayey materials with reddish brown colour which are saturated, very soft and plastic. They can be described as residual soils produced by the alteration of piroclastes and basaltic flows; in some zones, some sand or even gravel size grains were detected, as traces of original piroclastes y

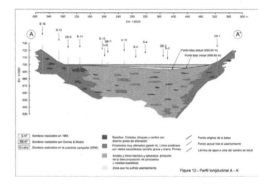

Figure 4. Geological-geotechnical ground profile along the longest axis of the bottom (in blue, basaltic materials; in red, pyroclastic zones and in orange, halloysite clay material).

basalts. Piroclaste zones with different alteration degrees appeared occasionally, where the original rock structure could be seen. The total depth of those clayey materials was unknown, although it is greater to 65 m, depth got by one of the boreholes (S-C).

In the borders of the caldera, pyroclastic zones with little alteration were detected. Those piroclastes are resistant materials and they have basalt blocks inside their structure.

It is important to remark that the material characteristics deduced with the borehole inspection were in accordance with the magnitude of the settlements recorded. For instance, the material recovered in borehole S-C, situated in the zone with the greatest settlements, were the softest ones and the ones with more water content.

3 ORIGIN AND STRUCTURE OF THE HALLOYSITES

The halloysites are mainly formed as the result of weathering of ultramafic rocks, volcanic glass and pumices. The samples analysed in this study came from the alteration of piroclastes of basic characteristic.

They are clay minerals belonging to the group of kaolinites. Its principal characteristic is that the plates are separated by a monolayer of water molecules. This structure is called Halloysite (10Å). As the forces that tie the water molecules with the rest of the structure are weak, the clay can be easily dewatered (by air drying, at vacuum or by a light heating) and be irreversibly transformed in Halloysite (7Å). Due to that reason, Halloysite (7Å) is easily found near the surface while Halloysite (10Å) is detected at great depths. Besides, the manipulation of the samples usually induces that transformation (Joussein et al. 2005). Figure 5 shows the halloysite clay material in its natural state and after being dried in oven. The first sample shows a cluster aspect while the second one has a surface covered with little dewatering cracks. This difference shows the great influence of water content in the material.

Halloysite particles can show a great number of different morphologies. The most usual one is a structure in elongated tubes with an average

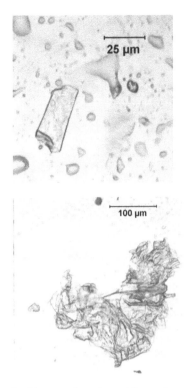

Figure 6. Photographs taken with a microscope of halloysite material: above, in tubular structure and below with the structure resulting from a dehydratation process.

diameter of 0.07 μm. The dehydratation has an influence on the clay morphology: in the spherical structures it provokes diameter reduction of particles, while it increases the diameter of tubular shapes. Figure 6 shows two images, made in the microscope, of an halloysite typical tubular structure and of a structure resulting from a dehydratation process.

4 IDENTIFICATION TESTS

4.1 Grain size distribution tests

The 35 grain distribution tests by sieve that were made showed a percentage of fine particles (particles smaller than 0.08 mm), between 65 and 95%, with a representative average value of 85%. Particles greater than 2 mm were not detected. Figure 7 shows the grain size distribution curves obtained.

In all the samples, grain distribution tests by sedimentation were also performed following the Spanish standard UNE 103102–1995. With the aim of being able to sieve the sample, the material

Figure 5. Photographs of the samples: above, in its natural state and below, after being dried in oven.

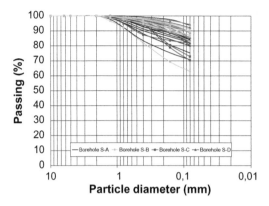

Figure 7. Grain size distribution curves.

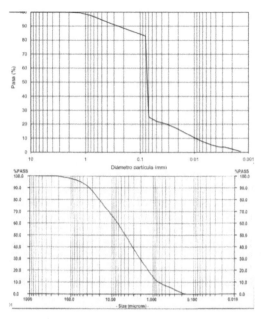

Figure 8. Size distribution curves obtained by a) sieving and sedimentation and b) Laser diffraction method.

was dried in an oven at 50°C so it lost its water content. After that, the material to be tested was placed in a container filled with water and dispersant during 18 hours to disaggregate the existing glomerular particles. Taking into account the tendency of halloysite to form aggregates, the test results were meaningless, as the grain size distribution curve lost its continuity, as it can be seen in Figure 8a).

To avoid this problem some tests were made modifying some of the technical aspects, without getting satisfactory results. The following step was to perform grain size distribution tests with a technique based on the laser diffraction when they go through the sample. In these tests, made with the material passing a 0.40 mm sieve, a continuous grain size distribution curve was obtained as the one showed in Figure 8b), what proved the validity of the test method. The results indicate percentages of clay material (smaller than 2 μm particles) ranging from 20 to 50%, with an average of 35%.

Figure 9 shows the clay content in the different samples analysed. It can be seen that this content is greater in Borehole-C, situated in the area with the greatest settlements.

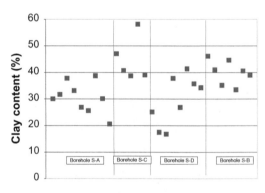

Figure 9. Clay content variation in the different boreholes.

4.2 Atterberg limit tests

The material plasticity was studied by Atterberg Limit tests. According to the relevant Spanish Standard, the material to be used in this kind of tests is the one passing trough a 0.40 mm sieve, so to dry the sample previously is necessary. Once the material is sieved, it is mixed with a certain quantity of water and it is kept during 24 hours to homogenize its water content. Some of the results were anomalous, as the tests indicate the material was no plastic, although the material, in its natural state, was clearly plastic.

This abnormality in the results was caused by the physic-chemist characteristics of halloysite. The cylindrical shape of halloysite particles makes water inflow in the structure to be quite complex and slow, so the standard test method can not be considered as adequate for this kind of material. Due to this particularity it was decided to perform the tests with the material in its natural state, without any posterior manipulation (drying or sieving the material).

Figure 10 shows the results, according to the manipulation method followed with the simples. The values of the liquid limit and the plasticity

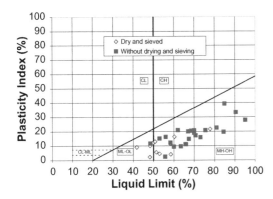

Figure 10. Atterberg limit results.

index obtained with the not manipulated samples are greater and, taking into account the halloysite particularities, are the ones which must be considered the material representative values. The liquid limit values of the 23 tested simples ranged between 48 and 95%, with an average value of 70%; the plasticity index was between 9 and 40%, being 18% the average value. With all these values the material was classified as a High-Plasticity Silt (MH).

4.3 Specific gravity of the soil grains

33 determinations of the specific gravity of the soil grains were performed. The values obtained ranged between 2.52 and 3.07 t/m³, with an average value of 2.90. This value was in accordance with the basaltic origin of the particles.

4.4 Chemical tests

Traces of carbonates and sulphates were not detected in the 32 determinations performed. The percentage of organic material ranged between 0.04 and 3.37%, with a representative average of 1%.

Furthermore five tests to determine the chemical components of the material were performed. Their results indicated that 25% of the material was constituted by aluminium oxides, another 25% by silicates, 20% by ferric oxides, 5% by titanium oxides and the rest (about 25%) was formed by volatile materials lost during the test performance.

5 CURRENT STATE TESTS

The results obtained in the 31 samples tested were:

1. The great majority of water content values ranged between 50 and 70%, with an average value of 60%; no evident correlation with depth was found. Those values were between the liquid and plastic limits.
2. Saturation values were near 100%.
3. Dry density was very low ($\gamma_d = 10.40$ kN/m³) and it showed a certain increase with depth.
4. The great majority of the values of void ratio ranged between 1.4 and 2, with an average value of 1.8; no evident correlation with depth was found.

6 STRENGTH TESTS

6.1 Continuous dynamic penetration tests

In the bottom of the reservoir, 13 continuous dynamic penetration tests (DPSH) were performed up to 20 m of depth. In all the tests and all along the depth analysed, the results obtained were lower than 5 blows/20 cm, as it can be seen in Figure 11; in no single case, refusal was achieved. It can be deduced from those small values that the material is very low resistant and very high deformable.

6.2 SPT tests

23 SPT tests were made at different depths. Their results are shown in Figure 12, except two values of 28 and 100 blows that showed the presence of more resistant intercalated layers. The majority of the values was smaller than 10 blows, being 5 blows/30 cm the representative average value, indicative of a very soft and deformable material. It must be remarked that there were three tests with a result of 0 blows/30 cm, one of them performed at 36 m deep.

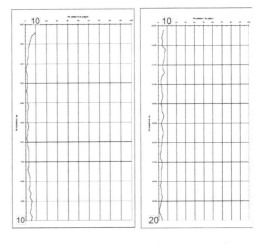

Figure 11. Result of a continuous dynamic penetration test.

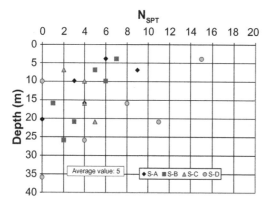

Figure 12. Results of SPT tests performed in the boreholes.

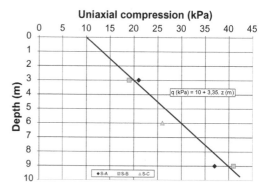

Figure 13. Variation of uniaxial compression test results with depth.

6.3 Uniaxial compression tests

Seven uniaxial compression tests were performed, whose results ranged between 19 and 41 kPa, with an average value of 30 kPa, again indicative of a very low resistant material. The results show a certain increase with depth as it can be seen in Figure 13. In two of the tests, the samples could not be shaped due to the low consistence of the material.

6.4 Direct shear and triaxial tests

Eight direct shear tests were performed, in a 60 mm diameter circular box, with unaltered samples taken at depths between 3 and 9 m. Figure 14 shows the values of normal stress vs. shear stress at failure for all the samples. The best fitting line is also drawn.

The strength parameter values individually deduced from the tests show cohesion between 0 and 59 kPa and friction angle between 18 and 38.5°. The values considered representative of the strength behaviour of this material are 20 kPa of cohesion and 30° of friction angle. Those values are typical of a silty material, which is in accordance with the identification values deduced from grain size distribution and plasticity tests.

One TCU triaxial test was also performed. Cohesion of 30 kPa and friction angle of 29° were deduced, values similar to the ones deduced before.

6.5 Pressuremeter tests

Five pressuremeter tests were performed in one of the boreholes. The main results are collected in Table 1.

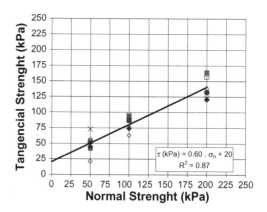

Figure 14. Results of direct shear tests.

Table 1. Results of pressuremeter tests.

Depth	Net yield pressure Pf*	Net limit pressure Pl*	Presuremeter modulus	
(m)	(kPa)	(kPa)	Ep (kPa)	Ep/Pl*
4.50	368	527	4100	7.8
10.20	525	727	3200	4.4
15.20	740	967	5700	5.9
25.20	728	907	5920	6.5
40.20	797	1097	6210	5.7

The yield net and limit net pressure values are very low, clearly indicative of a very low resistant material. It can be remarked a certain increase of the values with depth due to the increase in the confining pressure. The fact that the ratio between limit net pressure and yield net pressure was almost equal in all the tests (values between 1.25 and 1.43) shows that all the tests were performed in the same kind of material.

28

7 DEFORMABILITY TESTS

Six oedometer tests were performed, whose results let highlight the following aspects:

1. Compressibility index (Cc) ranged between 0.18 and 0.48 with an average value of 0.35. The values of swelling index (Cs) were between 0,011 and 0.031, with an average of 0.019. The ratio between both coefficients ranged between 15 and 29, with an average of 20.
2. The values of oedometer modulus, without taking into account the vertical pressure applied, ranged between 1600 y 15600 kPa, which indicates that the material can be classified as soft soil with very high deformability. Those values can be seen in Figure 15.
3. For the interval of vertical stress between 20 and 80 kPa, a clear increase in the value of oedometer modulus due to the vertical stress is noted.
4. The most representative value for global oedometer modulus is 2700 kPa; for the interval between 80 and 300 kPa, the representative value is 5500 kPa while for the upper step between 300 and 1000 kPa, the value increases up to 10500 kPa.
5. Figure 16 shows the variation of those oedometer modulus values with depth taking into account a natural apparent density of 1.65 t/m^3 and the presence of a phreatic level in the ground surface. That variation can be determined by the following expression: E_{oed} (kPa) = 2500 + 75 · z (m).
6. The consolidation coefficients were between 1×10^{-2} and 6.8×10^{-4} cm^2/s, with a representative average of 1.5×10^{-3} cm^2/s, which indicates the material has a deficient drainage, so the consolidation time can be great.

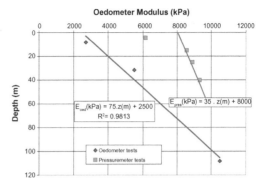

Figure 16. Variation of oedometer modulus, deduced from oedometer and pressuremeter tests, with depth.

The analysis of the deformability values deduced from the five pressuremeter tests makes it possible to state the following:

1. The values of pressuremeter modulus are very low, which indicates tests were performed in a very deformable material.
2. The ratio between pressuremeter modulus and limit pressure (ranging between 4.4 and 7.8) indicates an altered material. Usually, when the values of that ratio are lower than 5, tests can be considered as anomalous and bad performed, due to an excessive ground alteration.
3. The values of oedometer modulus were obtained multiplying the pressuremeter modulus by a constant α with a value of 1.5, in accordance with usual correlations (Briaud, 1992).
4. The values of oedometer modulus, obtained in such a way, are lightly greater than the ones deduced from the oedometer tests, as it can be seen in Figure 16.
5. The variation of oedometer modulus, deduced from pressuremeter tests, can be quantified by the following expression:

$$E_{oed, pres.} (kPa) = 8000 + 35 . z (m)$$

8 OTHER TESTS

One permeability test in triaxial cell and three dispersibility test were performed to complete the characterization of the material. The permeability coefficient obtained was 1.8×10^{-7} cm/s, indicative of a quite impermeable clay material. Permeability coefficient was also calculated, using one dimensional consolidation theory, based on the values of the consolidation coefficient (1.5×10^{-3} cm^2/s) and oedometer modulus (5500 kPa). The result obtained was 2.7×10^{-6} cm/s.

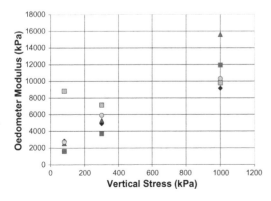

Figure 15. Oedometer modulus deduced form oedometer tests.

Table 2. Representative values of halloysite clays.

Parameters	Representative value
Content of fine particles [Diameter < 80 μ] (%)	85
Content of clays [Diameter < 2 μ] (%)	35
Liquid limit (%)	70
Plasticity limit (%)	52
Plasticity Index (%)	18
Specific gravity of the soil grains	2.90
Carbonate content (%)	0
Sulphate content (%)	Insignificant
Organic material content (%)	1
Water content (%)	60
Dry density (kN/m^3)	10.4
Saturation value (%)	100
Void ratio	1.8
SPT (blows/30 cm)	5
Uniaxial compression (kPa)	30
Cohesion (kPa)	20
Friction angle (°)	30
Compressibility index (Cc)	0.35
Oedometer modulus (kPa)	5500
Consolidation coefficient (cm^2/s)	$1.5 \times (10^{-3}$
Permeability coefficient (cm/s)	$1.8 \times (10^{-7}$

Three dispersibility tests were performed with samples from different boreholes. The results indicate that the material was "not dispersive".

9 FINAL REMARKS

The Barlovento Dam suffered important settlements whose origin was analysed with a geotechnical investigation campaign. The result of in situ testing and laboratory tests showed that the reservoir ground is mainly consisted of clay material with an important proportion of halloysite clays. The representative values of the main geotechnical parameters of those halloysite clays are showed in Table 2.

REFERENCES

Briaud, J.L. 1992. *The pressuremeter.* A.A. Balkema.

Estaire, J., Diez, J.A. & Martínez, J.M. 2008: Análisis de las causas de los asentamientos producidos en la Balsa de Barlovento (Isla de La Palma, Canarias). *2° Cong. Int. sobre Proyecto, Construcción e Impermeabilización de Balsas. Palma de Mallorca (España).*

Hurlimann, M., Ledesma, A. & Martí, J. 2001. Characterisation of a volcanic residual soil and its implications for large landslide phenomena: application to Tenerife, Canary Islands. *Engineering Geology 59. pp. 115–132.*

Joussein, E., Petit, S., Churchman, J., Theng, B., Righi, D. & Delvaux, B. 2005. Halloysite clay minerals—a review. *Clay Minerals*, 40: 383–426.

Volcanic Rock Mechanics – Olalla et al. (eds)
© 2010 Taylor & Francis Group, London, ISBN 978-0-415-58478-4

Geotechnical parameters of basaltic pyroclastics in La Palma Island, based on convergences measured in a tunnel

Miguel Fe Marqués
AEPO, S.A. Ingenieros Consultores., Madrid, Spain

Rodrigo Martínez Zarco
INOCSA Ingeniería, S.L., Madrid, Spain

ABSTRACT: This work presents a back-analysis of a tunnel section based on measured convergences values. The tunnel, located in La Palma Island, was excavated in basaltic pyroclastics. The analysis used FLAC^{2D} and shows the excavation and supporting of the tunnel along with other phases during construction. With this analysis we aimed to obtain realistic values of the geotechnical parameters of pyroclastics. However, its accuracy depends on a great number of variables. Finally, modulus of deformation and parameters c_M and ϕ_M of pyroclastic were estimated. These estimated values are not the only possible solution given the available data, but they represent realistic values in the authors' opinion.

1 INTRODUCTION

The Construction Project of Improvement Works on Highway C-830, stretch Tenagua—Los Sauces (La Palma Island) was made in 1998 by AEPO, S.A. Construction was carried out by FERROVIAL-AGROMÁN under the site direction of Martin Piñar Rodriguez. AEPO—TRAZAS Consortium made the technical assistance to the direction. Construction was completed in 2004.

The Improvement Works includes four tunnel stretches. The so called Tunnel 3 has a length of 250 m and a maximun depth of 116 m over the road level. The internal section has a total width between sidewalls of 9.00 m (2 lanes of 3.5 m,

Figure 2. Far view of the tunnel final portal, showing the pyroclastic cone. Also, an old cavity excavated in the pyroclastics as a warehouse is shown.

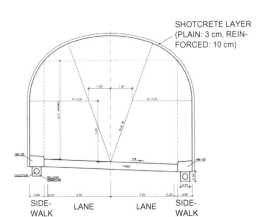

Figure 1. Tunnel 3 typical cross section.

2 hard shoulders of 0.20 m and 2 sidewalks of 0.80 m), straight sidewalls 3.25 m in height and a 3 centre vaulted roof with a maximun inner height of 6.76 m (see figure 1).

More than half of the tunnel length was excavated in basaltic pyroclastics. A geotechnical characterization of this material, based on real data, has been made using the values of measured convergences and the back-analysis of the excavation, installation of support and other construction stages.

2 GENERAL GEOLOGIC DESCRIPTION

Figure 3 shows a longitudinal geologic profile of the tunnel. It shows the existence of a cone of

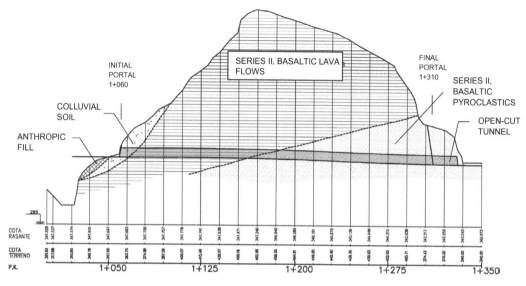

Figure 3. Tunnel 3 longitudinal geological profile.

Figure 4. Detail of the final portal.

Figure 6. Beginning of the tunnel excavation in pyroclastics, in the final portal area.

Figure 5. Pyroclastics detail. The photo was taken in the portal zone, before the beginning of the excavation.

basaltic pyroclastics. These are fossilized by the later basaltic lava flows. These materials belong to the Volcanic Series II, of the four series over the Complejo Basal, defined in La Palma Island.

The pyroclastics are sediments of air projection, like slag, lapilli, bombs and ash. It is a coarse grain size soil, with low density and SPT values between 30 and 60. It's a loose material or slightly welded, but with important overlapping between particles.

The pyroclastic cone is fossilized by an alternation of basaltic and scoria lava flows. They usually are augite-olivine and fluidal types. Terms with augite phenocrysts and olivine-accumulated phenocrysts frequently appear, wich causes a characteristic weathering in the form of coarse-grained sandy material, whose grains are formed from those phenocrysts. Sometime they appear micro-crystalline and vacuolar types lava flows. Basaltic lava flows with enough thickness show disjunction in bowling, or more frequently, columnar disjunction.

The tunnel crosses the basaltic lava flows from the initial portal to chainage 1 + 170 km (approx) where ground gradually turns into piroclastics all the way down to the final portal.

3 CONSTRUCTION STAGES

The tunnel was made by full-face mechanical excavation. Tunnel support has consisted exclusively of shotcrete layers. The different construction stages were the following (see figure 7):

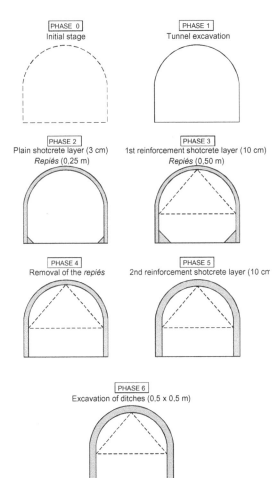

Figure 7.   Constructive phases.

- Full-face excavation and 3 cm thick shotcrete layer (σ_c = 25 MPa). The tunnel stretch in pyroclastics was completed on 01/10/2000 approx. Shortly after, cracks appeared in the shotcrete shell. These were more important in the deeper zone, in which pieces of about 30 cm thick detached from the tunnel crown and the walls (see figure 10).

- On 08/11/2000 a 10 cm thick steel fibre reinforced shotcrete layer was sprayed to the pyroclastics tunnel stretch. Sections measuring convergence were installed. These comprised 3 bolts of convergence (tunnel crown and both walls). The first measurement, or measurement 0, was made on 09/11/2000. In the following months, measurements showed a horizontal deformation rate of 3 mm/month on the deepest sections. However, by February 2001 values showed a trend towards steadiness.

- On 13/02/2001 the *repié* (overwidth of the shotcrete layer at the base of the sidewalls formed by the fall of the rebound material) was removed. The *repiés* have triangular form with a base width of 50 cm approx. and a similar height. Its removal caused an abrupt increase of up to 17 mm in convergence values in a few days (section 1 + 213). Subsequently, a fast stabilization of the convergences took place, with a lower deformation rate than before the removal of *repiés*.

- On 23/04/2001, an additional 10 cm thick steel fibre reinforced shotcrete layer (σ_c = 25 MPa) was sprayed between chainage 1 + 180 and 1 + 220

- On 01/06/2001, two 0.5 × 0.5 m drain ditches were excavated at each sidewall base. This caused a clear rise in the convergences, more gradual than the previous one. Deformation rate peaked at 9 mm/month on section 1 + 213 (see figure 9).

- In the first two weeks of August of 2001 the ditches were filled up, subgrade was formed and pavement laid, which resulted in a new stabilization of the convergences. Convergence value due to the excavation of the ditches was 22 mm on section 1 + 213. Since deformation had not become totally stable as the ditches were filled up, the actual total value was probably somewhat greater.

- After the pavement was laid, convergences got stable. However, a certain deformation remains, the rate of which is steady until now. On section 1 + 213, its value is 0.25 mm /month.

- Figures 8 and 9 show the horizontal convergences evolution with tabs to each stage for the deepest stretch of the tunnel, chainage 1 + 180 and 1 + 213. Overall horizontal convergence value is 31.55 mm on section 1 + 180 and 54.9 mm on section 1 + 213. Figures show how values on both sections are approximately proportional.

33

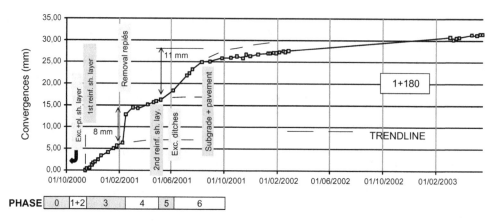

Figure 8. Horizontal convergences in section 1 + 180, with constructive phases.

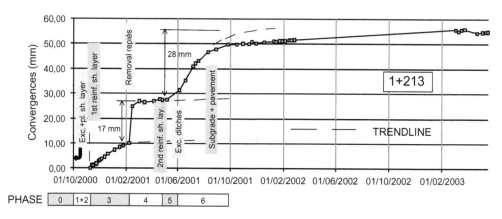

Figure 9. Horizontal convergences in section 1 + 213, with constructive phases.

4 BACK ANALYSIS

4.1 *Analysis method*

A finite differences back-analysis of the tunnel was made using FLAC2D. The model takes into account the different construction stages such as excavation, shotcrete shell installation, removal of *repié* and excavation of ditches.

FLAC2D, version 4.00 (Fast Lagrangian Analysis of Continua), developed by Itasca Consulting Group, is a two-dimensional explicit finite difference program that simulates the behavior of structures built of soil, rock or other materials that may undergo plastic flow when their yield limits are reached. An isotrop elastic-plastic Mohr-Coulomb model has been used.

The analysis aimed to obtain not only horizontal convergence values but also vertical values and observed behaviour of the shotcrete shell.

4.2 *Section of analysis*

The chosen section for analysis belongs to chainage 1 + 213, since it showed the greatest convergences of all sections. Its parameters are:

- Depth: 103.7 m above road level
- Ground: Up to 15 m above road level, soil is formed by pyroclastics. From that level up to the surface, it is formed by basalt and slag. The basalt-pyroclastics interface was modeled as a straight line from the outcrop at the final portal and the section where basalts appeared at tunnel crown.
- Section for convergences data reading: One measurement bolt was placed at the tunnel crown and one at each sidewall, 5 m below crown level.

Horizontal convergences between the two sidewall bolts are shown in figure 9. Fewer data are available for convergences between crown and sidewall. Their values range roughly between 25% and 30% of horizontal convergence.

4.3 Geotechnical characterization of the materials

4.3.1 Basalts

The basalts have been characterized applying the Hoek & Brown failure criterion for rock masses (2002). It is assumed that the shear strength curve approximates a straight line in the basalts stress range. This was done using the program RocLab, version 1.007 (ROCKSCIENCE, 2002). The input data were the following

$m_i = 15$ $\sigma_c = 60$ MPa

$GSI = 50$ Factor $D = 0$

Depth $= 85$ m

The following values of cohesion and friction of the rock mass have been obtained:

$c_M = 0.70$ MPa $\Phi_M = 54,4°$

$E_M = 7746$ MPa

In order to estimate the rock mass modulus of deformation, the well-known expression of Serafim and Pereira (1983), modified by HOEK (1997 and 2002), has been applied:

$$E_M(GPa) = \left(1 - \frac{D}{2}\right)\sqrt{\frac{\sigma_{ci}(MPa)}{100}}\, 10^{\frac{GSI-10}{40}}$$

A density of 2.5 t/m³ and a Poisson ratio, μ, of 0.23 were considered.

4.3.2 Pyroclastics

The geotechnical characterization of pyroclastics is essential in order to obtain realistic results of the back-analysis. Nevertheless, it is a kind of material that is difficult to characterize, due to the impossibility of sampling.

The geotechnical characterization has been made to the article "Propiedades geotécnicas de materiales canarios y problemas de cimentaciones y estabilidad de laderas en obras viarias" (Serrano, A. y Olalla, C, 1999)". According to this article, the shear strength of pyroclastic materials is estimated on the basis of triaxial tests with specimens from Los Campitos, village in the south of La Palma Island. The density of these specimen range from 0.8 to 1.05 t/m³.

The failure criterion is the above mentioned by Hoek & Olalla, applying the change of variable proposed by Serrano and Olalla ($\beta = m\sigma_c/8$, $\zeta = 8s/m^2$).

The shear strength of the pyroclastics is:

$$\frac{\tau_r}{\beta} = \frac{\sqrt{A^2 + 2B\zeta + 2B\dfrac{\sigma_n}{\beta}} - A}{B}$$

where A and B depend on the dilatancy angle, v:

$$A = \frac{1 - \sin v}{\cos v} \quad B = \frac{1}{\cos^2 v}$$

The parameters β_i and ζ_i corresponding to the intact rock need to be lowered for the rock mass according to the following expressions:

$\beta = \beta_i/n_\beta$ $\zeta = \zeta_i/n_\zeta$

$n_\beta = e^{(100-GSI)/28}$ $n_\zeta = e^{(100-GSI)/25.2}$

The shear strength curve has been approximated to a straight line within the considered stress range.

The parameters that govern the shear strength curve are β_i, ζ_i, v and GSI. Adopted values of these parameters are greater to the proposed by Serrano and Olalla because, as explained below, lower values result in parameters that do not suit the observed behavior of the ground and the support. The used parameters are:

$\beta_i = 1500$ KPa $\zeta_i = 0,3$

$v = 10°$ $GSI = 85$

These parameters, within a stress range of 200 to 2000 Kpa, result in the following:

$c_M = 0.30$ MPa $\Phi_M = 28,5°$

A density of 1.25 t/m³ and a Poisson ratio, μ, of 0.25 have been considered for the pyroclastics.

Figure 10. Breakage in the plain shotcrete layer.

As explained below, the modulus of deformation value that results in output data that most accurately match the convergences values is the following:

E = 1700 MPa

4.3.3 Shotcrete
The analysis input parameters are:

$\gamma = 2.3 \ t/m^3$ E = 27,300 MPa

$\mu = 0,20$

The *repié* or overwidth of the shotcrete layer at the base of the sidewalls, has been modeled as a soil material with the concrete geotechnical parameters. This is more realistic than modeling it as beam elements, as it allows for rotational stiffness effect at the wall base.

4.3.4 Other analysis parameters
In addition to ground and support geotechnical and geometric parameters, described above, the following parameters need to be considered in the analysis:

- Lateral earth pressure coefficient K_0: A value of 1.0 has been adopted, for both soil types, basalts and pyroclastics.
- Confinement loss, λ (Panet, 1995): At each stage a fictitious stress applied to the walls has been considered in the calculation. This was defined as $(1-\lambda)\sigma_0$, where σ_0 is the natural stress. The following assumptions have been adopted:
  - Phase 1. Excavation of the tunnel. $\lambda = 0.30$
  - Phase 2. Plain shotcrete layer (3 cm). $\lambda = 0.89$
  - Phase 3: First reinforcement shotcrete layer (10 cm of shotcrete reinforced with steel fibers). $\lambda = 1.0$
  - Following phases: $\lambda = 1.0$

In phase two, the value of λ is lower than one, although installation of the reinforced shotcrete layer was over a month after the excavation and the working face was over 100 m away. This is due to the rheological behavior of the ground, which results in deformations taking place over a significant lapse of time. This is shown in Figures 8 and 9, where it can be seen that, after placing the reinforced layer, deformation continued for several months. The value of λ only reaches 1.0 after a period of four months.

4.4 Calculation results

4.4.1 Horizontal convergences
The table below compares measured horizontal convergences with values obtained from the analysis at each stage. It shows a good correlation between the two values.

The greatest difference was obtained at stage 6, excavation of ditches. When these were filled up, the deformation was not yet stabilized, thus it is higly likely that overall value is greater than measured, matching the analysis value.

4.4.2 Crown-wall convergences
The proportion between convergences crown-wall and wall–wall in the analysis is 28%, which matches the measured values. This percentage is linked to the location of the basalt-pyroclastics interface on the section as well as to the value of K_0. This confirms the adopted assumptions.

4.4.3 Strength of support
Internal forces of the support shell have been checked against its capacity for all shotcrete layers. This derived the following:

- Shotcrete capacity is clearly exceeded at the end of stage 2 (installation of plain shotcrete layer)
- Reinforced layer capacity is close to its limit at the end of stage 4 (removal of *repiés*)
- Capacity of whole support (with all layers) is clearly enough at stage 6 (excavation of ditches)

This fits with the cracking and loosening observed in the plain shotcrete layer and with the subsequent behaviour of the reinforced shotcrete.

In the analysis, a thickness of the plastic zone at the tunnel boundary of about 3 to 4 m was obtained.

4.4.4 Possible variations in the adopted analysis parameters
The validity of the ground parameters assumed in the analysis is largely conditioned by the great number of variables that take part in the calculation, which compels us to adopt a great number of initial hypotheses. The different variables involved are analyzed below:

- The values of c_M and Φ_M of basalts do not have any effect on the calculation. This is due to its high values, that prevent soil failure. Therefore, the accuracy of the calculation value of both parameters does not affect the calculation validity.
- The values of E and μ of the basalt have little effect on the deformations of the tunnel. This is based on the reasonable assumption that its modulus of deformation is clearly greater than the pyroclastics´. This results in an insignificant deformation of the section due to basalts. Therefore, the accuracy of the calculation value of both parameters does not affect, or affects very little, the validity of the calculation.

- The pyroclastics coefficient K_0 affects greatly the calculation results. The validity of the assumed value (1.0) is confirmed by the calculation results of the relation between the crown-wall and wall-wall convergences, similar to the measured one.
- The confinement loss, λ, in the different construction stages has an important effect on the deformation values at each stage, except in the last one (excavation of ditches). In this analysis, the adopted values match the convergences measurements.
- The most determining parameters of the calculation are the parameters of the pyroclastics (c_M, Φ_M, ν, E, μ), especially c_M, Φ_M and E. It is obvious that by changing the adopted values of c_M and Φ_M, a different value of E from indicated in section 4.3.2, concordant with the measured convergences, would have been obtained. The adopted values of c_M, Φ_M and E of the pyroclastics are realistic, in the authors' opinion, as it's explained below, but other combinations of parameters can be equally concordant with the measurements.

4.4.5 Other possible values of c_M, Φ_M and E of the pyroclastics

As explained in section 4.3.2, the adopted values of c_M and Φ_M are high, in the range of the highest expectable values for this type of material. An additional check with lower values of c_M and Φ_M, closer to average, has been made. If adopted values in section 4.3.2. were $\beta_i = 1000$ KPa, $\zeta_i = 0.25$ and GSI = 65, the results would be:

$$c_M = 0.17 \text{ MPa} \qquad \Phi_M = 19.7°$$

With these input data, the modulus of deformation concordant with the measurements of convergences is about 9,000–10,000 MPa. This is a very high value, clearly unreal. It is similar to what could be expected in good quality basaltic lava flows, even though basaltic lava flows and the pyroclastics deformability are clearly different.

Table 1. Horizontal convergences.

SECTION 1+213	Convergences (mm)	
PHASE	Calculation	Measured
1: Excavation	5	–
2: Plain shotcrete layer	18	–
3: 1st reinforcement layer	9	10
4: Removal of the *repiés*	19	17
5: 2nd reinforcement layer	0	0
6: Excavation of ditches	28	22

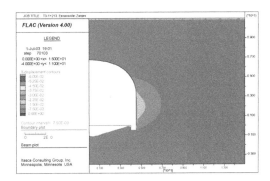

Figure 11. Calculated horizontal displacements after phase 6, excavation of the ditches.

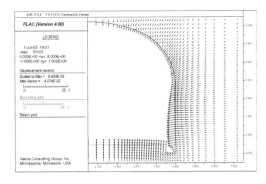

Figure 12. Calculated displacement vectors after phase 6, excavation of the ditches.

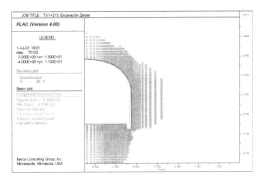

Figure 13. Calculated yielding ground after phase 6, excavation of the ditches.

In addition, this analysis shows a very large plastic zone around the tunnel, with a thickness of 2 tunnel diameters, which does not seem realistic either.

Therefore, it is reasonable to think that, although there are a number of combinations for c_M, Φ_M and E, compatibles with the measurements, the real ground values should not be very different from the ones that have been proposed here ($c_M = 0.30$ MPa, $\Phi_M = 28.5°$ and $E = 1700$ MPa.

5 CONCLUSIONS

Based on the observed behavior of the pyroclastics and the back-analysis carried out, the following conclusions can be drawn:

- The measured convergences are compatible with $K_0 = 1.0$.
- The time evolution of the convergences measurements indicates a rheological behavior of the ground, with a low deformation velocity. The 100% of the deformation after the excavation does not take place after at least 4 months.
- From the behavior of the pyroclastics high strength parameters can be derived. Values proposed on this paper for cohesion and friction ($c_M = 0.30$ MPa, $\Phi_M = 28.5°$) correspond to the application of the failure criterion of Serrano & Olalla (1999) with values $\beta_i = 1500$ KPa, $\zeta_i = 0.30$ and GSI = 85. These are high values for this type of material.
- The elastic modulus of the pyroclastics compatible with the measurements is also high "a priori" for a compact granular material slightly welded or not welded at all. Its value is $E = 1700$ MPa. It must be taken into account that this is the unloading modulus, higher than the loading one, which must be used for foundations.
- There are other possible combinations of values of c_M, Φ_M and E of the pyroclastics, compatible with the observed behavior of the ground. Nevertheless, in the authors′ opinion, the real values of the ground should not be very different from the ones proposed here.

- In all convergences sections in pyroclastics, certain long term flow deformations remains. Though, this flow appears to be constant in time, it actually has a very low speed (between 0.1 and 0.25 mm/month).

ACKNOWLEDGEMENTS

The authors express their gratitude to Martin Piñar Rodriguez, site director of Works, to Emilio Grande de Azpeitia, manager of TRAZAS INGENIERÍA, to †Luis Bascones Alvira (TAGSA), geologist of the Project, to Francisco Pecino (AEPO-TRAZAS Consortium) head of the technical assistance during construction of the tunnel, and to Ivan García Sáez for his contribution to the translation of this paper.

REFERENCES

AFTES (2001). "Recommendations on the convergence—confinement method". Paris, France.
Hoek, E., Carranza-Torres, C & Corkum, B. (2002). "Hoek-Brown failure criterion—2002 edition".
ITASCA Consulting Group, Inc. (2002). FLAC[2D] version 4.0. Minneapolis, USA.
Panet, M. (1995). "Le calcul des tunnels par le méthode des curves convergence-confinement". Presses de l'École Nationale des Ponts et Chaussées. Paris, France.
ROCKSCIENCE Inc. (2002). RockLab versión 1.007. Toronto, Canada.
Serrano, A. & Olalla, C. (1999). "Propiedades geotécnicas de materiales canarios y problemas de cimentaciones y estabilidad de laderas en obras viarias". XXII Semana de la Carretera. Asociación Española de la Carretera. Islas Canarias.

Volcanic Rock Mechanics – Olalla et al. (eds)
© 2010 Taylor & Francis Group, London, ISBN 978-0-415-58478-4

Basic properties of non welded basaltic lapilli and influence on their geotechnical behaviour

A. Lomoschitz & J. Yepes
Departamento de Ingeniería Civil, Universidad de Las Palmas de Gran Canaria, Spain

C. de Santiago
Laboratorio de Geotecnia, CEDEX, Madrid, Spain

ABSTRACT: In the Canary Islands there are hundreds of volcanic cones and extensive blankets of lapilli and civil works are very frequently undertaken in areas containing this material. Furthermore, basic lapilli (of basaltic, basanitic, or tephritic composition) are common in many volcanic regions of the world. They are small pyroclastic fragments (2 to 64 mm in diameter) emitted by Strombolian-type eruptions, very irregular in shape and with many open and closed voids. As a whole it is a light and quite loose granular material. The article covers two related aspects: (1) basic properties, such as texture, unit weight and geochemical composition; and (2) geotechnical parameters and behaviour under different situations: on slopes, under foundations and as granular layer for roads. We conclude that, resulting from their low density, high porosity and angular shape, lapilli particles have a quite different geomechanical response from other granular natural materials.

1 INTRODUCTION

Basaltic lapilli are common in many volcanic regions of the world (e.g., Alaska, the Philippines, Japan, Indonesia, Italy, Greece, and many countries of Central and South America), and they are commonly used in the Canary Islands (Spain) in the construction industry, mainly because of their abundance (more than 800 volcanic cones) and because of the lack of suitable clays from which to make bricks. Three main aspects of lapilli have been studied: (a) basic properties as geological origin, texture, unit weight and geochemical composition; and (b) geotechnical parameters and behaviour under different situations: on slopes, under foundations and as granular layer for roads.

The Canary Islands are located between latitude 28° and 29°N in the Atlantic Ocean (Fig. 1), 110 km off the northwest coast of Africa and about 1,070 km southwest of mainland Spain. There are seven main islands, with a total extension of 7,450 km² and with more than 1.7 million inhabitants. They are mainly volcanic islands located in an intraplate zone.

Lapilli deposits are principally present in Pliocene and Quaternary volcanoes, which are abundant, but only about a quarter of them are exploitable as a result of environmental regulations. However, civil works are undertaken relatively frequently in areas containing this material, and so it is important to

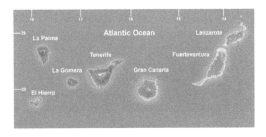

Figure 1. Location of the Canary Islands in the Atlantic Ocean, between latitude 28° and 29° North.

know the geotechnical behaviour of lapilli layers and their properties as aggregate.

Publications on the use of lapilli in construction are scarce in the international literature, and only two specific papers have been found: Makoto et al. (1996) and Lomoschitz et al. (2006). Nevertheless, a number of technical reports (e.g. IGME, 1974a,b, IGME, 1986) related to the uses of lapilli in the Canaries have been written in Spanish.

2 BASIC PROPERTIES

2.1 *Volcanic origin and textures*

Lapilli are small pyroclastic fragments emitted by volcanoes, with a grain size of between 2 and 64 mm and a diverse composition. When they come from a basic magma (of basaltic, basanitic, or tephritic composition) they are usually related to Strombolian-type eruptions and appear either by forming volcanic cones or extensive blankets spread around the vent (Fig. 2).

Close examination reveals that individual particles of lapilli are sharp edged and angular, dark coloured (dark grey to black, or dark red to orange when the iron has oxidized), and have many interconnected vesicles (originally air bubbles in the lava) (Fig. 3). Moreover, fragments have a quite uniform grain size, depending only on the force of the wind and on the distance from the volcano vent.

These lapilli have cryptocrystalline to microporphidic textures and sometimes are hypocrystalline (containing a mixture of glass, microcrystals, and cryptocrystals), but pure glassy texture (obsidian) is uncommon. Moreover, the lapilli used in the cement industry are low weathered and have a low content of fines.

Figure 2. Volcanic basaltic tephra cones are frequent in the Canary Islands and they have abundant basaltic lapilli. Timanfaya National Park in Lanzarote Island.

Figure 3. Basaltic lapilli are sharp edged and angular particles, with a predominant grain size of 2–64 mm.

2.2 Geochemical composition and non reactivity

The lapilli used as aggregate in the Canaries do not contain reactive silica as a result of the basic composition, which is mainly basaltic, basanitic, and nephelinitic (with low content of SiO_2), and because the texture is not glassy. They come from ultrabasic geochemical composition magmas (<45 percent content of silica, SiO_2, and >3 percent of alkali, $NaO+K_2O$), with alkaline trends.

Thus, the lack of reactive silica eliminates the problems of alkali-silica reactivity (ASR) with Portland cement. In contrast, certain acidic volcanic rocks from other regions of the world, such as rhyolite, dacite, or andesite pyroclast or tuff, are able to produce this expansive process. It occurs when the following factors coincide: an aggregate with a high content of silica and glassy or poorly crystalline texture; a wet atmosphere with 80 percent or greater relative humidity, and an alkaline source, normally present in Portland cement.

The ASR process produces a very characteristic external appearance that usually appears between 2 and 5 years; in the Canary Islands there has never been a registered case of a concrete structure with this problem, even though basaltic lapilli have been extensively used for the last 50 years in the production of mass concrete and concrete blocks.

3 POROSITY OF LAPILLI PARTICLES

3.1 Types of porosity

Natural layers of basaltic lapilli show high porosity ranging between 35 and 60 percent, which mainly corresponds to interconnected voids (i.e. open pores). This high porosity is the cause of the low unit weight, but basaltic lapilli particles also include abundant closed voids (i.e. vesicles or closed pores) not interconnected and generated by gas bubbles trapped in the original magma.

These two main typologies of pores (Fig. 4), closed (V_{cp}) and open pores (V_{op}), have to be considered because they strongly determine important properties of this material such as durability, mechanical strength, permeability, absorption properties, etc. (de Santiago & Raya, 2008).

Helium pycnometry and Mercury porosimetry are adequate techniques for such a porosity determination. Santana et al. (2008) have defined four types of pore structures (reticular, vacuolar, mixed and matrix) and among them the most common in ultrabasic pyroclastic rocks, which in the case of basaltic lapilli tuffs are vacuolar and mixed pore structures. Although the consideration of a basaltic lapilli layer as a rock would need some degree of welding or consolidation among lapilli particles, which are not always present, the irregular

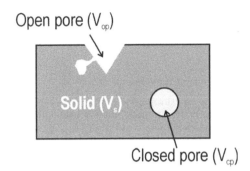

Total Volume $(V_t) = V_s + V_{op} + V_{cp}$

Figure 4. Schematic diagram of a porous material showing open pores and closed pores in a solid matrix.

shape of particles makes a "package" effect on each layer.

3.2 New sampling and lab tests

Lapilli long-term behaviour is highly influenced by the pore size distribution and the ratio of open to closed porosity. It is very difficult to quantify these micropores by water absorption tests because of two main reasons: some pores are too small to be penetrated by water and a certain percentage of the pores are isolated.

The use of helium pycnometry allows to detect and quantify this closed porosity made of non-connected pores, as well as to describe the porous net of this material.

To this end, basanite to basalt lapilli from three Pleistocene and well preserved cinder cones of Gran Canaria Island have been sampled: Jinámar (JN-4), Bandama (BN-10) and Monte Lentiscal (ML-11). The selected samples have in common that they were taken from points which are close (<30 m) to the volcanic crater (or central eruption vent) and sampling was done on recent slope cuts and at a depth of 2–5 m, from non weathered zones.

All the samples were prepared to be analysed: solid rock fragments crushed to different particle sizes: <10 mm, 5 mm (5 UNE sieve), <2 mm (2 UNE sieve), <0.63 mm (0.63 UNE sieve) and <0.080 mm (0.080 UNE sieve) were dried in an oven until constant mass was reached.

Specific gravity values (γ) were obtained following the procedure ASTM num. 5550–94 in an Auto Pycnometer 1330.

A gas pycnometer operates by detecting the pressure change resulting from displacement of gas by a solid object. The accuracy and precision of the

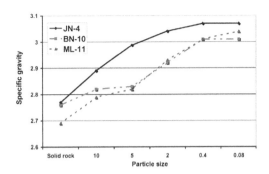

Figure 5. Variation between particle size and specific gravity (G) in the three lapilli samples.

gas pycnometer in the determination of skeletal volume and density can be quite high, but relies on the use of a pure analysis gas being free of moisture. For this reason, the gas is a pure gas or dry air, and the sample is pre-treated in a vacuum oven to remove volatiles. Helium typically is the gas used because it readily diffuses into small pores.

Although helium pycnometry is an analytical method to determine the material volumes or densities by gas displacement, porosity information is a by-product of volume determinations.

First of all, dry bulk density (γ_d) was determined for each sample. Secondly, the specific gravity values were obtained by Helium pycnometry (Fig. 5). The specific gravity of the sample with particle size under 0.080 mm is the highest and it is considered to be the real one (γ_s). The specific gravity of the rest of samples, where enclosed air-filled pores exist and Helium does not penetrate, is smaller than the real one. Those values are called "apparent" ($\gamma_{s,ap}$) or skeletal.

3.3 Calculation procedure

The total porosity would be defined as the sum of open pores and closed pores related to the total volume of each lapilli sample:

$$n_{total} = \frac{V_{op} + V_{cp}}{V_t} = \frac{\gamma_s - \gamma_d}{\gamma_s} \qquad (1)$$

For the particles under 0.080 mm the whole total porosity is equal to the open porosity, while the rest of particles show a certain percentage of isolated pores which make the result at the pycnometer to be lower than the real one. From this "apparent specific gravity" ($\gamma_{s,ap}$), open porosity is calculated.

$$n_{open} = \frac{V_{op}}{V_t} = \frac{\gamma_{s,ap} - \gamma_d}{\gamma_{s,ap}} \qquad (2)$$

Finally the closed porosity results from the difference between total and open porosity

$$n_{closed} = n_{total} - n_{open} \qquad (3)$$

Alternatively, another way of expressing the closed porosity is relating the total closed pores volume not to the total volume but to the skeleton volume, being the skeleton volume the sum of solid volume and the closed pores volume ($V_{skel} = V_s + V_{cp}$).

Table 1. Helium pycnometry results for sample ML-11.

Particle Size (mm)	Rock	<10	<5.0	<2.0	<0.4	<0.08
γ_s	3.04	=	3.04	=	3.04	=
γ_d	1.364	=	1.364	=	1.364	=
γ_{skel}	2.69	2.79	2.82	2.93	3.01	3.04
n_{total} (%)	55.1	55.1	55.1	55.1	55.1	55.1
n_{open} (%)	49.3	51.1	51.6	53.4	54.7	55.1
n_{closed} (%)	5.8	4.0	3.5	1.7	0.4	0.0
n_{skel} (%)	11.5	8.2	7.2	3.6	1.0	0.0

= constant value.

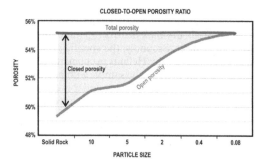

Figure 6. Relationship between particle size and closed-to-open porosity ratio.

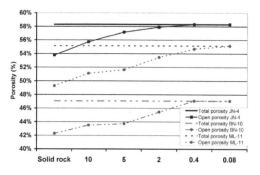

Figure 7. Relationship between particle size and close-to-open porosities in the three lapilli samples.

$$n_{skel} = \frac{V_{cp}}{V_{skel}} = \frac{V_{cp}}{V_s + V_{cp}} = \frac{\gamma_s - \gamma_{s,ap}}{\gamma_s} \qquad (4)$$

3.4 Results and interpretation

Taking as an example the sample ML-11, Helium pycnometry results and porosity calculations are summarized in Table 1 and shown in Figure 6.

There is a close connection between the lapilli particle size and the open and closed porosity, and this trend is present the tree representative samples (Fig. 7) and can be attributable to the fact that closed porosity decreases when the particle size also decreases.

4 GEOTECHNICAL PARAMETERS AND BEHAVIOUR

Basaltic lapilli are the most common type of lapilli. Table 2 shows their weight–volume representative values and Table 3 their geotechnical properties.

The principal conclusions are the following:

1. lapilli, as a whole, are essentially a granular material with high angles of friction (34–37°)
2. The values of unconfined compressive strength (qu) are relatively high, but usual bearing capacity (q) values used for superficial foundation design are low, about 100 to 200 kPa.
3. High slope angles (even a 1H/5V relation) are stable in the short term because of the high irregularity of the particles, together with a certain degree of welding or cementation; however, in the long term and as a result of superficial erosion, the slope angles decrease considerably and even reach the angle of repose. In this

Table 2. Unit weight values of basaltic lapilli.

Parameter	Sample place	Values, KN/m³
Specific gravity, G	General[1]	22.17–29.3
	Canary Is.[2]	25.1–27.2
Dry unit weight, γ_d	Canary Is.[3]	11.87–12.16
	Gran Canaria[4]	7.94–9.51
	Canary Is.[5]	7.8–10.4
Moist unit weight, γ	Lanzarote[6]	10.86–15.77
	Gran Canaria[2]	12.26–16.38
Saturated unit weight, γ_{sat}	Canary Is.[3]	20.79–24.13

[1] Blyth and de Freitas (1984), [2] IGME (1974a), [3] IGME (1974b), [4] ENADIMSA, unpubl., [5] Serrano et al. (2002), [6] Ministerio de Agricultura, unpubl.

Table 3. Basic geotechnical parameters of lapilli.

Parameter	Symbol (Units)	Values
Friction angle	$\emptyset^\circ$	34–37[1]
Cohesion	c (kN/m²)	137.3–176.6[1]
Compressive strength	q_u (MN/m²)	0.39–0.54[1]
Bearing capacity	q (kPa)	100–200[2]
Short term slope angle	α°	70–85[2]
Long term slope angle	β°	37–40°[2]

[1]Modified from IGME (1974a) and [2]Lomoschitz (1996).

work only loose or weakly welded lapilli have been considered, although from a geotechnical perspective, the degree of compaction and welding among the particles noticeably affect the quality of these materials, as Serrano et al. (2002) have indicated.

5 LAPILLI AS AGGREGATE USED IN CONSTRUCTION

5.1 *Lapilli as aggregate in mass concrete*

Guigou Fernández (unpubl.) carried out a number of tests with mass concrete using different ratios among three groups of aggregate: fine sand, coarse sand, and four types of gravel from volcanic crushed rock (basalt, phonolite, basaltic pyroclast, and phonolitic pumice tuff). These gravels correspond to three size intervals: 5/10, 10/20, and 20/40 mm. Using Cylinder Test methods with concrete that contains lapilli as coarse aggregate, the following results have been obtained:

1. As a result of the low values of the moist unit weight of lapilli, a very high aggregate/cement ratio is necessary (0.7 to 0.8 compared with the habitual rates of 0.55 to 0.65), and only a water content of between 230 and 260 l/m³ (23–26 percent in volume) provides a suitable consistency of concrete;
2. the average values of the dry unit weight, γ_d = 17.08 kN/m³, and of the saturated unit weight, γ_{sat} = 19.12 kN/m³ of the concretes with lapilli are lower than the standard average value, γ = 22.56 kN/m³, used in Spain for mass concrete;
3. the best aggregate ratio, known as type 4D, includes Portland cement (14 percent); fine natural sand (11 percent); coarse sand from crushed rocks (33 percent); and gravels 5/10 (17 percent) and 10/20 (25 percent) (d/D in mm) made of lapilli;
4. the mean compressive strength (f_{cm}) of concrete with lapilli reaches 25.97 MPa at 28 days

and, as a result, produces satisfactory 25 MPa concrete.

5.2 *Lapilli as granular material for highways*

Aggregates that are used in the Canaries for road construction are obtained from three types of deposits (ENADIMSA, unpubl.): (a) natural sand and gravel from alluvial deposits of rivers and ravines (Canary "barranco" deposits); (b) lapilli from volcanic cones with abundant pyroclasts, tephra in a general sense (Canary "picón" deposits); and (c) basalt, phonolite, basanite, and tephrite lava flows. These are massive formations that yield aggregates of excellent quality. Their unit weights are usually high, 27.5–28.5 kN/m³, and their use is very common.

Lapilli have been traditionally used in the Canary Islands for minor road works (rural paths, forest tracks, etc.). However, current standards in Spain e.g. FOM/3460/2003 (Ministerio de Fomento, 2003) require a series of properties for the aggregates used in highways, and to these requirements there have to be added those of the European regulations on aggregate, included in The Construction Products Directive (European Economic Council, 1989).

Lapilli and, in general, pyroclastic deposits (i.e. tephra) are not suitable materials for the construction of embankments because of their low density, their high water absorption capacity, and, consequently, their difficult compaction. However, when a highway has to cross a zone with thick strata of lapilli and the embankment of the highway has to be constructed, then common soil stabilization methods are frequently used, such as mixing the soil with cement. Lapilli have been also used as a subbase granular material in flexible pavements below the bituminous layer (Fig. 8), with satisfactory results.

Figure 8. Basaltic lapilli used as granular material for a highway in La Palma Island.

6 CONCLUSIONS

The main conclusions that have been obtained in this study are related to (a) basic geological and geotechnical properties; (b) properties as aggregate of concrete; and (c) properties as granular material for highways. We conclude that, due to their low density, high porosity and angular shape, lapilli particles have a quite different geomechanical response than other granular natural materials. The uses we have made of lapilli in the Canary Islands could serve as a useful guide for its use in other volcanic areas of the world. It is especially interesting for regions with recent basaltic lapilli deposits and a climate that is dry or slightly humid, where weathering processes are not very intense.

REFERENCES

Blyth, F.G.H. & de Freitas, M.H. 1984. *A Geology for Engineers*. London: Edward Arnold.

European Economic Council, 1989. *The Construction Products Directive*. Brussels: Council Directive 89/106/EEC of 21 December 1988 on the approximation of laws, regulations and administrative provisions of the Member States relating to construction products.

ENADIMSA unpubl. *Inventario y catalogación de estructuras de lapilli en las Islas Canarias (La Palma, Fuerteventura y El Hierro)*. Unpublished report, 1987. Madrid: Empresa Nacional de Investigaciones Mineras, S.A.

Guigou Fernández, C. unpubl. *Influencia de las características petrográficas de los áridos en el hormigón*: Unpublished PhD, 1990. Universidad de Las Palmas de Gran Canaria: Departamento de Construcción Arquitectónica. E.T.S. de Arquitectura.

IGME 1974a. *Mapa de Rocas Industriales, escala 1:200.000, Las Palmas de Gran Canaria*. Madrid: Instituto Geológico y Minero de España.

IGME, 1974b. *Mapa Geotécnico General, escala 1:200.000, Las Palmas de Gran Canaria*. Madrid: Instituto Geológico y Minero de España.

IGME 1986. *Bases para la ordenación minera y ambiental de la extracción de picón en las Canarias (Tenerife, Lanzarote y Gran Canaria)*. Madrid: Instituto Geológico y Minero de España.

Lomoschitz, A. 1996. *Caracterización geotécnica del terreno, con ejemplos de Gran Canaria y Tenerife*. Las Palmas de Gran Canaria: Departamento de Construcción Arquitectónica. E.T.S. de Arquitectura—ULPGC.

Lomoschitz, A. Jiménez, JR. Yepes, J. Pérez-Luzardo, JM. Macías-Machín, A. Socorro, M. Hernández, L. Rodríguez, JA. & Olalla C. 2006. Basaltic Lapilli used for Construction Purposes in the Canary Islands, Spain. *Environmental and Engineering Geoscience* (12): 327–336.

Makoto, K. Hiroshi, T. & Wataru, I.1996. Properties of concrete using lapilli as coarse aggregate. *Journal of the Society of Materials Science*, Japan. 45(9):1008–1013.

Ministerio de Agricultura unpubl. *Los Lapillis de la Isla de Lanzarote*. Unpublished report, 1983. Instituto Nacional para la Conservación de la Naturaleza (I.C.O.N.A.) Servicio Provincial de Las Palmas, Las Palmas de Gran Canaria.

Ministerio de Fomento 2003. Spanish Standard 6.1-IC: *Secciones de firme de la Instrucción de Carreteras*. Orden FOM/3460/2003, de 28 de noviembre: 44274–44292.

Santana, M. de Santiago, C. Perucho, A. & Serrano. A. 2008. Relación entre características químico-mineralógicas y propiedades geotécnicas de piroclastos canarios. Geo-Temas, Sociedad Geológica de España 10:947–950.

de Santiago, C. & Raya, M. 2008. Study of the porosity features of lightweight expanded clay aggregates by means of Helium pycnometry and Mercury porosimetry. In AEGAIN (ed.), *Cities and their underground environment. Proc. of II European Conference of the International Association for Engineering Geology. EUROENGEO 2008*. Madrid, 15–19 September 2008.

Serrano, A. Olalla C. & Perucho A. 2002, Evaluation of non-linear strength laws for volcanic agglomerates. In Dinis da Gama and L. Ribeiro e Sousa (eds.). *Workshop on Volcanic Rocks. EUROCK 2002*. Funchal, 27 November 2002.

Volcanic Rock Mechanics – Olalla et al. (eds)
© *2010 Taylor & Francis Group, London, ISBN 978-0-415-58478-4*

Geotechnical characterization of volcanic rocks and soils of Madeira Island

J.C. Lourenço, J.M. Brito, J. Santos, S.P.P. Rosa, V.C. Rodrigues & R. Oliva
Cenorgeo, Geotechnical Engineering, Lisbon, Portugal

ABSTRACT: This paper focuses on the description and the characterisation of volcanic rocks and soils of Madeira Island, based on data from expertise judgment, field survey and laboratory tests. The objective has been the compilation of data from geotechnical designs for Madeira Island, in the last 20 years, in order to describe the geological conditions and to evaluate the geotechnical parameters of the main volcanic formations such as: basalts, breccias and tuffs.

1 STRUCTURE AND GEOLOGY OF MADEIRA ISLAND

1.1 Introduction

The archipelago of Madeira is composed by a group of four islands located on the Atlantic Ocean: Madeira, Porto Santo, Desertas and Selvagens (Fig. 1).

The Madeira Island's shape is elongated with a maximum length of about 57 km (W–E), between Ponta do Pargo and São Lourenço, and a width of about 23 km (N–S), between Ponta de São Jorge and Ponta da Cruz (Fig. 2). It has an area of about 730 km² which corresponds to 90% of the total area of the archipelago. It has an average altitude of 700 m being the maximum equal to 1819 m.

The island was originated from submarine volcanic eruptions, mostly explosive, during the Vinbondonian period (Zbyszewsky quoted by Rosa, 1995). This initial phase was followed by others more effusive with prolific emission of basaltic lavas.

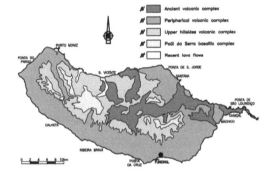

Figure 2. Simplified plan of the volcanic complexes (proposed by Zbyszewsky *et al.*, quoted by Rosa, 1995).

1.2 Volcanic complexes

1.2.1 Introduction

Madeira Island comprises five volcanic complexes as shown in Figure 2. The division between complexes is questionable because it is mainly based on morphological and geometrical criteria being presented in this paper the division proposed by Zbyszewsky *et al* (quoted by Rosa, 1995). The main complexes are β^1 and β^2.

1.2.2 Ancient volcanic complex (β^1)

This complex, dated from the Mio-Pliocenic period, occurs mainly in the central part of the island namely in Curral das Freiras, Serra de Água, Vale de S. Vicente, Vale da Boaventura, Ruivo and Areeiro mountains, São Roque, Porto da Cruz, Machico and Ponta de São Lourenço. It is believed that this complex is older than 5.2 Ma due to the existence of some calcareous fossils dated from this period.

It consists mainly of diversified pyroclastic material with some intercalations of basaltic lava flows mostly very weathered.

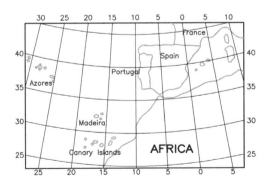

Figure 1. Geographical location of Madeira archipelago.

The pyroclastic formations are coarse, with big angular blocks, volcanic bombs, lapilli and volcanic ash. The lava flows are composed of alkali, olivine, basanite and hawaiite basalts (Aires Barros quoted by Rosa, 1995).

1.2.3 *Peripherical volcanic complex* (β^2)

This volcanic complex dated from the Post-Pliocenic period extends to the majority of Madeira.

Island, namely to Ribeira da Janela, Porto Moniz, Ponta do Pargo, Ribeira Brava, Câmara de Lobos, Funchal, Camacha, Santa Cruz and Santana. It is the volcanic complex occurring in Madeira's capital city (Funchal), where several construction's works were made, therefore being the most well geotechnically characterized.

This complex β^2 is very heterogeneous and consists mainly of alternated and irregular layers of disaggregated basalts and compact or disaggregated autoclastic disaggregated breccias with variable thickness, interbedded with less significant compact tuffs.

1.2.4 *Upper hillsides volcanic complex* (β^3)

This Post-Miocenic period complex occur in the high altitude areas that separate the north region from the south region of the island, namely in Paúl da Serra, Achada da Pinta, Lombada das Vacas, Lombada do Meio, Cabeços de Fajã dos Vinháticos and Terreiro da Luta. These formations are composed by an alternation of little thick lava flow and of pyroclastic materials usually yellow-reddish. The volcanic flows are predominant suggesting that they were originated from a more effusive than explosive phase.

1.2.5 *Paúl da Serra basaltic complex* (β^4)

This superior period basaltic complex occurs in Paúl da Serra, Chão dos Balcões, Poiso and Achada do Teixeira. It is constituted mainly by compact basaltic lava flows with some extensive levels of pyroclastic fine material. It is the result of a mainly effusive phase from fissure eruptions and secondary volcanic structures having the lava flows covered the flattened surfaces.

1.2.6 *Recent lava flows* (β^5)

This modern period basaltic complex occurs in Porto Moniz and in the valleys of São Vicente and Seixal. It corresponds to the last volcanic phase that occurred on the island and it is aged between 0.2 and 0.5 Ma. These flows occupy valleys previously excavated by erosion, near the coastline.

1.3 *Other volcanic formations*

The island is crossed by a dense network of dykes. Although this network presents some inflections in its main orientation it converges to the major volcanic structures located at the centre of the island, where its structure is well defined. Its most important orientations are WNW–ESE and NW–SE.

2 GEOLOGICAL FORMATIONS

Most of the geological formations of the island are constituted by extrusive volcanic rocks. These are divided into two main groups: the lava rocks and pyroclastic rocks. The lava rocks are mainly basaltic alkalyne type. Its origin is effusive and it may occur more compact or more vacuolar in layers with variable thickness. The lava rocks are abundant throughout the island.

The pyroclastic rocks are mainly of basaltic nature originated by explosive volcanic activity. They are more abundant in the central area of the island, where the main volcanic structures are located, usually in a random spatial distribution. The formations are little compact and present a breccious aspect.

The base volcanic complex β^1, the older one, is represented by different types of pyroclastic materials, with intercalations of basaltic lava, always weathered to very weathered. Pyroclastic formations are coarse, with big angular blocks, volcanic bombs, lapilli and volcanic ash. Sometimes the fine pyroclastic materials (tuffs with bombs) are dominant (Fig. 3).

The Post-Miocenic volcanic complex β^2 consists mainly of alternated and irregular layers of compact basalts and compact and disaggregated autoclastic breccias interbedded with significant thickness. In general, the thick basalts correspond to the most representative geological formation and concern basaltic rocks, of high resistance, with gradual variation to vacuolar basalts, from slightly to very vacuolar and more or less breccious. They occur generally slightly weathered or fresh, with moderate to wide fractures (Fig. 4).

Figure 3. β^1 tuffs with bombs.

Figure 4. β^2 alternated and irregular layers of compact basalts.

Figure 5. β^2 compact breccias.

Figure 6. β^2 moderately weathered tuffs.

Disaggregated and weathered basalts are, in general, from vacuolar or breccious formations characterised by some degree of disaggregation presenting weathering signals along the discontinuities. Compact breccias occurr generally from moderately to very weathered, with moderate to close fractures. Disaggregated breccias occur from completely to highly weathered, with very close fractures (Fig. 5).

Tuffs are less representative and occur from moderately to highly weathered. Sometimes the compact and disaggregated breccias and the tuffs present significant thickness (Fig. 6).

3 GEOLOGICAL CHARACTERISATION

The characterisation of the mechanical properties of the rock masses is a very important stage in geotechnical design. This characterisation is made by considering both geological criteria as well as mechanical properties (quantified by in-situ and laboratorial tests).

Volcanic environments, due to their origin, have a great heterogeneity of rocks and soils, characterised by different mechanical properties which usually present random distributions. However, due to economical limitations, it is not possible to establish extensive ground investigations for a better characterisation of the rock mass.

Considering these two issues, the main initial aim of the design is to conduct the geotechnical investigation towards grouping the rock mass in geotechnical zones, based on their mechanical properties, in order to reduce the number of tests and the number of design parameters, thus simplifying the problem.

The mass is first divided into soils and rocks. Rocks include as main units: compact basalts, basalts sometimes vacuolar to moderately weathered, very weathered to fractured basalts, compact breccias and slightly weathered breccias and very weathered breccias. Soils include tuffs (fine pyroclast) breccious, granular tuffs, clayey tuffs and coarse pyroclasts.

Usually in volcanic environments the zoning might not be very accurate due to the significant spatial heterogeneity genesis and mechanical characteristics. However this zoning is imperative in design stage and so it must be done carefully and with the knowledge of how each formation is going to control the global response of the rock mass. Usually, observations and measurements of the behavior of the structure and its surrounding ground should be made to identify any need for remedial measures or adjustments to the construction procedures defined in the project.

4 GEOTECHNICAL CHARACTERISATION

After completing the geological characterisation it is necessary to define the design geotechnical parameters.

In regard to the rock masses, the mechanical behavior is assessed by approaches that are based on the mechanical characterisation of the intact rock samples combined with geotechnical classification systems. These approaches aim to determine the rock mass properties, before and after the excavation.

The strength of the rock matrix can be determined from triaxial compression tests (only for highly important works due to its cost), uniaxial compression tests (the most commonly made), point load tests or tensile tests. The stiffness can be determined in laboratory by uniaxial compression tests or seismic tests, and in situ by plate load tests, dilatometer tests, flat jack tests or seismic tests.

Classification methods, such as the Hoek-Brown criterion (in development since 1980), have made clear that the strength of the jointed rock mass depends not only on the properties of the rock matrix, but also on the freedom of the intact rock pieces to rotate and translate among them under certain stress conditions (Hoek et al., 2002). This movement is controlled by the geometrical and mechanical characteristics of the joints. The Geological Strength Index classification was developed to describe the global behavior of the rock mass (Marinos & Hoek, 2000).

From the mechanical properties of the matrix rock and GSI (and since recently, the massif's perturbation induced by the construction process) it is possible to estimate the design parameters of the rock mass by the Hoek-Brown criterion. This criterion is only applicable to the rock masses in which the joint pattern is reduced (and therefore the mass behavior is controlled mainly by the rock matrix) or fractured rock masses with a random distribution of joints so the rock mass can be considered homogenous and isotropic.

In addition to this criterion, other geomechanical classifications can be found in the literature. In accordance with the assessment of a limited number of geotechnical parameters which, in general, can be determined from simple laboratory tests, from the field observations or from the visual inspection of the samples obtained from the borehole, the rock mass is classified into different classes. Among these, one of the best known is the Rock Mass Rating (RMR), proposed by Bieniawsky in 1973 and revised in 1989 (Bieniawsky, 1989), which allows the evaluation of the rock mass quality and the estimation the rock mass strength and stiffness through correlations. Other common classification is the quality classification system for rock masses, also known as Q system, proposed by Barton, Lien and Lunde (1974). It is based on the experience obtained in more than 200 underground work cases, and also aims to provide the classification of massives for later use in tunnels's design.

The mechanical behavior of soils is normally assessed by in situ tests, such as the standard penetration tests, cone penetrometer tests, pressuremeter tests, and by laboratory tests, such as identification tests, direct shear tests and triaxial tests.

5 GEOMECHANICAL PROPERTIES OF VOLCANIC ROCKS OF MADEIRA

5.1 Collected data

A considerable amount of data was collected from 48 tunnels' designs made in the last 20 years in the volcanic complexes β^1 and β^2. The greater amount of laboratory tests was performed in β^2 complex mainly because of the great difficulty in obtaining good quality samples in the β^1 complex, due to its high level of alteration.

After reviewing all the design's data, it was set the goal to systematize the information by dividing the basalts into three categories: i) Fresh to slightly weathered basalts; ii) Moderately weathered, partially vacuolar basalts; iii) Highly weathered vacuolar basalts. Rocky pyroclastic materials were divided into: i) Fresh to slightly weathered breccias; ii) Moderatelly weathered partially vacuolar breccias; iii) Highly weathered vacuolar disaggregated breccias.

Some typical properties are proposed afterwards for each category in Table 1.

5.2 Uniaxial compression tests results

A set of 14 uniaxial compression tests were performed in samples from complex β^1 and 136 tests from complex β^2. The test results are presented in the first chart of Figure 7. A set of 65 uniaxial compression tests were made in breccias from complex β^2 (second chart of Fig. 7).

Analysing all the available information it was established criteria to define the basalts and breccias categories based on their uniaxial compressive strength (σ_{ci}). Basalts were divided into the following categories: i) fresh to slightly weathered basalts with $\sigma_{ci} > 100$ MPa; ii) moderately weathered, partially vacuolar basalts with $10 \leq \sigma_{ci} < 100$ MPa; iii) highly weathered, vacuolar basalts $\sigma_{ci} < 10$ MPa. Each category is equivalent to 43%, 49% and 8%, respectively of the tested rock cores (Fig. 8). For the rocky pyroclastic materials it were established the following categories: i) fresh to slightly weathered breccias with $10 \leq \sigma_{ci} < 100$ MPa; ii) moderately weathered partially vacuolar breccias with $\sigma_{ci} < 10$ MPa. Each category is equivalent, respectively, to 63% and 37% of the tested rock cores. It is noted that no tests were made in the highly weathered, vacuolar disaggregated breccias due to the impossibility of sampling intact cores for testing.

It is also noted that in 64% of the test results analysed the value of poisson's ratio of the rock matrix lies between 0.20 and 0.25. In breccias, for 62% of the tests the value of poisson's ratio lies between 0.15 and 0.25.

Table 1. Expected parameters values for the different volcanic materials.

Description	Weathering grade	Spacing dimensions	RQD	GSI	Unit weigth [kN/m³]	$\sigma_{c,intact}$ [MPa]	E_{intact} [GPa]
Fresh to slightly weathered basalts	W_{1-2}	F_{2-3}	80–100	70–90	28–31	100–1000	50–100
Moderately weathered, partially vacuolar basalts	W_{2-3}	F_{3-4}	40–80	55–75	25–28	10–100	10–75
Highly weathered, vacuolar basalts	W_{3-4}	F_{4-5}	10–50	40–60	23–25	1–10	1–6
Fresh to slightly weathered breccias	W_{2-3}	F_{3-4}	30–90	45–75	–	8–60	5–65
Moderatelly weathered partially vacuolar breccias	W_{4-3}	F_{4-5}	20–70	35–55	–	1–8	0.1–8
Highly weathered, vacuolar disaggregated breccias	W_5–N/A	N/A	0–20	N/A	–	–	–
Breccious tuffs	W_{4-5}	–	30–90	N/A	–	0.5–7	0.1–2
Granular tuffs	N/A	N/A		N/A	–	1.5–10	0.2–1
Silty-clayey or clayey-silty tuffs	N/A	N/A		N/A	–	0.5–10	0.3–4

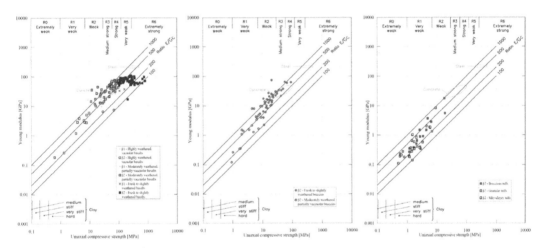

Figure 7. Uniaxial compression tests results.

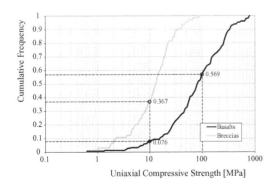

Figure 8. Cumulative frequencies of the uniaxial compressive tests on basalts and breccias.

5.3 Geological strength index

Based on the experience gained from the geotechnical investigations and the design projects involving these formations, it was possible to define ranges for GSI typical values. The proposed values for the different types of basalts and breccias are presented in Figure 9. Although most of tuffs are transition materials between soft rocks and compact residual soils and therefore the Hoek-Brown criterion should not be applied, this figure presents some possible values of GSI for tuffs.

5.4 Dilatometer results

Only 8 dilatometer test results are available in complex β², 4 from basalts and other 4 from breccias.

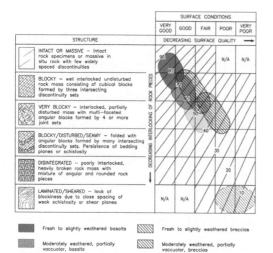

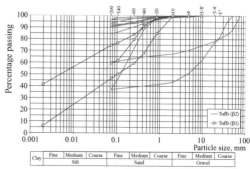

Figure 10. Grading curves of tuffs.

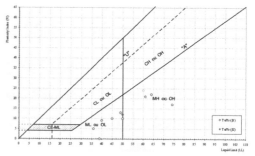

Figure 9. Common values of GSI of Madeira's volcanic rocks.

The dilatometer modulus present values between 1.6 and 5.6 GPa in basalts and between 0.7 and 5.2 GPa in breccias. Due to the limited number of results collected these values cannot be taken as representative though they do not contradict the values presented afterwards in the abacus of Figure 14.

6 GEOMECHANICAL PROPERTIES OF VOLCANIC SOIL DEPOSITS OF MADEIRA

6.1 Index properties

For the soil deposits the results from 13 tests for particle size distribution test results and 12 tests for determining the Atterberg limits are presented in Figure 10 and in Figure 11.

As shown in Figure 10 the tuffs can present a coarse or silty matrix and are classified based on particle size as breccious tuffs, granular tuffs or silty clayey tuffs. It should also be noted that in terms of plasticity, the fines are classified as silt, with medium to high plasticity.

6.2 Shear strength tests

A set of 9 triaxial compression test results and 2 direct shear test results from β^2 tuffs were selected. The cohesion values are plotted against angle of shearing resistance in Figure 12.

6.3 Unconfined compression tests

Figure 7 presents the compression test results in tuffs: 58 results from samples collected in β^2 and

Figure 11. Casagrande's plasticity chart.

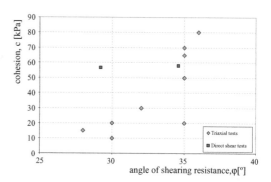

Figure 12. Shear strength results on β^2 tuffs.

only 2 from β^1. It can be concluded that the three categories of tuffs present similar values although the silty clayey tuffs are slightly more resistant than the others. Analysing the total data of the tuffs it can be concluded that most of the compression strength values varies between 0.3 and 6.9 MPa and the Young modulus between 0.1 and 3.9 GPa.

6.4 Standard penetration tests

The standard penetration test (SPT) is widely used in geotechnical designs. The main advantage of the

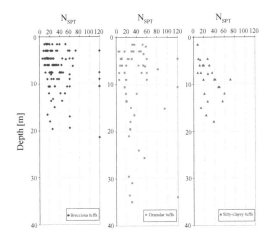

Figure 13. Standard penetration test results on β¹ tuffs.

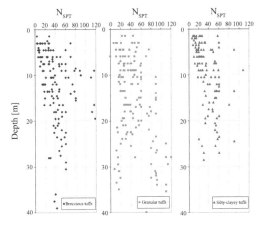

Figure 14. Standard penetration test results on β² tuffs.

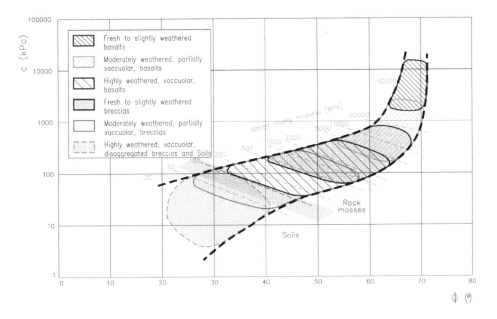

Figure 15. Madeira's volcanic formations strength parameters for preliminary designing purposes.

test is it world wide use which allowed the creation of a large database of correlations, providing a general idea about the strength and the relative density of the soil.

The SPT results on β¹ and β² tuffs are presented in Figures 13 and 14. In all cases the values are widely scattered but a slightly tendency of increase in the value of N_{SPT} with depth can be noted. This aspect is, in most cases, related to the geostatic stress distribution which, due to these soils' genesis, can be quite irregular.

7 RESULTS COMPILATION AND FINAL CONSIDERATIONS

Table 1 presents a compilation of the expected parameters values for the different materials studied. It is based on the available data information and its interpretation presented above and also in other studies from literature, and that can be used for preliminary design purposes.

In addition to this table, the abacus of Figure 15 was prepared to show the overall strength and

Table 2. Range of Hoek-Brown criterion values used for the parametric study.

Description	$\sigma_{c,intact}$ [MPa]	GSI	mi	D	E_{intact} [GPa]
Fresh to slightly	105	70	20	**0**	60
weathered basalts	**241**	**80**	**25**	0.5	**73**
	651	90	30	0.8	100
Moderately weathered,	16	55	20	**0**	10
partially vacuolar	**50**	**65**	**25**	0.5	**34**
basalts	95	75	30	0.8	59
Highly weathered,	1	40	20	**0**	0.2
vacuolar basalts	**7**	**50**	**25**	0.5	**3**
	9	60	30	0.8	6
Fresh to slightly	11	45	14	**0**	8
weathered breccias	**18**	**60**	**19**	0.5	**16**
	55	75	24	0.8	62
Moderatelly weathered	1	25	14	**0**	0.3
partially vacuolar	**5**	**40**	**19**	0.5	**3**
breccias	10	55	24	0.8	8

stiffness parameters for the different formations that can be used as a guideline for preliminary design purposes. The parameters of the rock masses were obtained from the Hoek-Brown criterion, based on a parametric study considering the possible range of values for each parameter, as presented in Table 2. The soil parameters are mainly based on the Figure 12 and on values available from the literature.

REFERENCES

Bieniawski, Z.T. 1989. *Engineering rock mass classifications*. New York: John Wiley & Sons.

Hoek, E., Carranza-Torres, C. and Corkum, B. 2002. Hoek-Brown criterion—2002 edition. *Proc. NARMS-TAC Conference,* Toronto, 2002, 1, 267–273.

Hoek, E. and Diederichs, M.S. 2006. Empirical estimation of rock mass modulus. *International Journal of Rock Mechanics and Mining Sciences*, 43, 203–215.

Marinos, P. and Hoek, E. 2000 GSI—A geologically friendly tool for rock mass strength estimation. *Proc. GeoEng2000 Conference*, Melbourne. 1422–1442.

Rosa, S. 1995. *Geological and geotechnical characterization of Madeira's volcanic formations—Geotechnical plan of the North area of Funchal*. M.Sc. Thesis in Geological Engineering (in Portuguese). Universidade Nova de Lisboa.

Volcanic Rock Mechanics – Olalla et al. (eds)
© 2010 Taylor & Francis Group, London, ISBN 978-0-415-58478-4

Contribution to geotechnical characterization of Basaltic pyroclasts

A.M. Malheiro
Civil Engineering Regional Laboratory, Azores, Portugal

J.F.V. Sousa
Civil Engineering Regional Laboratory, Madeira, Portugal

F.M. Marques
Civil Engineering Regional Laboratory, Azores, Portugal

D.M. Sousa
Civil Engineering Regional Laboratory, Madeira, Portugal

ABSTRACT: Both the Azores and Madeira islands, located in North Atlantic Ocean, are of volcanic nature. The present work focuses on the geotechnical characterisation of basaltic pyroclasts from the Azores and Madeira, in order to get some comparison among them. In order to characterize, evaluate the geomechanical properties and get some more geotechnical data about basaltic pyroclasts, some samples were collected in both archipelagos, to do some laboratory tests. In situ tests were also made with these materials. Results include data on SPT tests, plate load tests, Los Angeles tests, in situ dry density and specific weight tests, determination of particle size distribution, compaction and CBR tests and consolidated drained (CD) direct shear tests. Some correlations between several properties are presented, namely between the strength and the deformability of volcanic materials. Finally, some considerations are made about the potentially utilizations and problems related to engineering applications.

1 INTRODUCTION

Being the Madeira and the Azores archipelagos of volcanic origin, volcanic materials have always been, not only the main construction material for the local buildings, but also the only foundation terrain of all the constructions in these islands.

The increasing development in civil engineering works has shown the need to know better the geotechnical characteristics of volcanic materials, as they have some particularities and differences relatively to the non volcanic materials and besides, studies on that subject are very scarce. One of the existing volcanic materials that assume particular importance due to its applicability in the scope of Civil Engineering is basaltic pyroclasts. These materials are often used in the Azores, but its exploration is conditioned in Madeira archipelago, due to its reduced expression.

However, as it is of great interest to get a deeper knowledge of its geotechnical characteristics, it was considered important to make a survey of some tests executed with this type of material in both archipelagos, in order to be able to make some eventual comparisons.

In this context, this paper aims to contribute for a better knowledge of some geotechnical aspects of basaltic pyroclasts, given its specificities, aiming an eventual wider application, as well as to improve the current uses.

2 SOME GEOLOGICAL CARACTERISTICS

Fragments of rock thrown out by volcanic explosions are called *ejecta*, and accumulations of such fragments are known as *pyroclastic rocks* (MacDonald 1972). This type of material has also other designations, such *tephra, volcanic scoria*, "*bagacina*", the common name given at the Azores islands or even "*areão*" as it is known at Madeira.

Pyroclastic deposits form directly from the fragmentation of magma and rock by explosive volcanic activity. There are several ways to classify pyroclastic rocks. The most important classification of tephra is the one based on the size of the fragments (MacDonald 1972). MacDonald (1972) defined Bombs (round to subangular) and Blocks (angular) as ejecta lager than 64 mm, Lapilli if the average diameter is between 2–64 mm and Ashes if the diameter is <2 mm.

They also can be grouped into three genetic types according to their mode of transportation and deposition: Falls, Flows and Surges.

Volcanic scoria are integrated in the first group.

Explosive activity builds scoria, cinder or spatter cones, or both, at the vent, with scoria-fall deposits of limited aerial extent and volume being deposited around and downwind of the vent. Those cones well defined, are usually symmetric, with a height that rarely exceeds a few hundreds meters. The inclination of the slopes of these cones is normally close to 33° (natural slope angle of loose scoria), value that changes with the time, due to erosion (Fraga 1988).

Scoria-fall deposits are composed largely of vesiculated basalt to near-basaltic magma. These are the deposits characteristic of strombolian explosive activity. Such eruptions eject scoria and relatively fluid lava spatter, and are often accompanied by the simultaneous effusion of lava.

Deposits of scoria cones often consist of rather poorly bedded, very coarse-grained with metre-sized ballistic bombs and blocks.

Scoria is typically dark gray to black in color, mostly due to its high iron content. Sometimes, oxidation (of iron) by hot gases streaming preferentially through the central part of the volcanic edifice, leads to a deep reddish-brown color (Scmincke 2006).

On what concerns mineralogical composition, some comparisons were made between X ray diffractometry analyses made with both Madeira (LREC Madeira) and Azorean samples (Fraga 1988). In both cases, mineralogy was similar: the most abundant mineral was a plagioclase (anorthite), in second place, pyroxens (augite) and then olivines (Forsterite). In the samples of the red basaltic pyroclasts, there was still the presence of hematite (resultant from the iron oxidation).

3 GEOTECNICAL CHARACTERISTICS OF PYROCLASTS MATERIALS

Due to its volcanic nature, basaltic pyroclasts are a frequently heterogeneous material, non plastic, porous and with low density, with high levels of water absorption, being more or less consolidated.

The loose pyroclastic deposits are, therefore, a group of specific materials (on what regards its geotechnical characteristics) as they don't have the typical behaviour of a rock or of a soil; so they should be analysed as an independent geotechnical group. Being a rock, from a geological point of view, its geotechnical behaviour approaches, however, to that of a soil, with the particularity of being almost indifferent to the water content. Especially due to this characteristic and because

they are formed by grains, volcanic scoria should be treated as an aggregate (non crushed natural aggregate) (Fraga 2009).

Its properties depends on the size of the grain, the shape, the porosity and petrological composition as well as the degree of packing among the particles, compaction state of the deposit and resistance of the referred particles (Vallejo et al., 2006).

Some geotechnical characteristics of the tested materials are presented below. Several Madeira samples of basaltic pyroclasts were tested as well as some samples from the Azores.

The following laboratory tests were performed with the Madeira samples: tests for grain size distribution, specific weight, compaction and CBR tests (Californian Bearing Ratio) and consolidated drained (CD) direct shear tests.

Consolidated drained direct shear tests were made on 7 samples, being tested separately the fractions with sizes inferior to 2.00 mm (sieve ASTM n° 10) and with the fractions of sizes inferior to 4.75 mm (sieve ASTM n° 4), in a shear box with the following dimensions: $100 \times 100 \times 25$ mm³.

For two of the samples, consolidated drained direct shear tests were performed on the entire samples, in shear box with the dimensions $300 \times 600 \times 200$ m³, existing in the Laboratory of Construction Materials of the Faculty of Engineering of the University of Porto.

Some in situ tests were performed to determine the dry density, and plate load tests to determine the characteristics of natural deposits of basaltic scoria.

In Azorean samples the following laboratory tests were performed: tests for grain size distribution, Los Angeles, CBR and specific weight tests.

The following in situ tests were also performed: SPT, plate load tests and *in situ* dry density.

One should remark that the in situ tests in Madeira were performed on natural deposits, while the ones performed in the Azores were executed on landfill material, with exception of the SPT tests which were performed on natural terrain.

3.1 *Results of the tests performed*

3.1.1 *Madeira samples*
On what concerns the pyroclastic materials of Madeira Island, the values obtained for the main geotechnical properties are: specific weight = 24.7–29.1 kN/m³, dry density = 12–14 N/m³, cohesion = 0–0.1 MPa, friction angle = 29–50°.

One should emphasize that though these materials are loose, in consolidated drained direct shear tests a "cohesion" value was obtained which is associated to the form and imbrication of the grains and to the need of an increase of the volume of the specimens to provide the shear failure.

The values of friction angle obtained in the consolidated drained direct shear tests, with the fractions passed through the 2.00 mm sieve and through the 4.75 mm sieve are similar, while the obtained values for the cohesion show a wide variety (Figure 1).

The consolidated drained direct shear tests results on the two entire samples led to higher values, both in friction angles and cohesion, when compared to previous results, which are obviously related to the bigger dimensions of the particles and the correspondent increase volume during the shear failure phase.

Figure 2 shows an example of a test made with the entire grading on the big shear box.

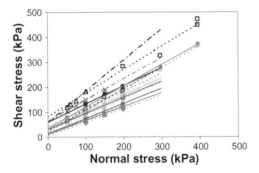

Figure 1. Results obtained with the consolidated drained direct shear tests on Madeira samples.

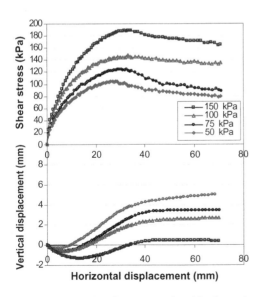

Figure 2. Example of a test made with the entire grading in the big shear box.

Compaction and CBR tests were also performed, and values for the maximum dry density were obtained between 13.5 and 17.1 kN/m³, for the optimum water content values between 4.5 and 17.7%. For the CBR the values varied between 40 and 96%.

Seven in situ plate load tests were performed with a 0.6 m diameter plate. In these tests a wide dispersion in the values of the deformability secant modulus was obtained, which varied between 10 and 250 MPa, for a tension of 300 kPa.

The values of the deformability modulus obtained in situ are closer to the values obtained in the laboratory tests, presented by Serrano et al. (2007), and are more different from the values presented by Uriel & Serrano (1976, *in* Serrano et al., 2007), which were obtained with in situ plate load tests, using a 1 × 1 m² plate.

3.1.2 *Azores samples*

Tests for grain size distribution, specific weight determination, CBR and Los Angeles were performed in laboratory on these materials.

Particle size distribution results are presented in Figure 5, being possible to see that they are mainly well graduated gravel. On what concerns the particle specific weight, the values are between 19.6 and 21.6 kN/m³, and the water absorption range between 3.3 and 14.8%

Results from Los Angeles on the 18 samples tested are presented in Figure 3, showing a high variation of values, between 25% to 70%.

According to Fraga (1988) the coefficient of Los Angeles in these materials varies with the fraction of material that is tested. That researcher found that the same material had a value of Los Angeles coefficient that was lower when tested in the finer fractions and higher in the coarser fractions. He also found that this variation was linearly with the fineness modulus. The behaviour of this material

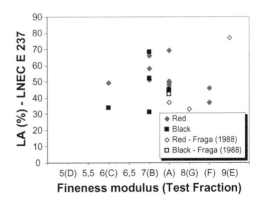

Figure 3. Los Angeles tests done in Azorean samples.

thus differs from that observed in basalts, in which the results are much the same regardless to particle size fraction of material that is tested.

The compaction tests made with the Azorean samples exhibit values for the maximum dry density between 12.1 and 14.1 kN/m³, for the optimum water content values between 15 and 35%.

In the laboratory the mechanic resistance of the basaltic pyroclasts was evaluated through CBR tests, having obtained values that varied between 42–73%. It was possible to verify that this mechanic resistance is much better if the tested granulometric fraction is finer.

On what regards the in situ tests, as it was already mentioned, the presented results were obtained in different tests executed in different landfills, such as in situ dry density, determined in experimental landfills, and plate load tests (PLT).

The values obtained for the in situ dry density in landfills varied between 12.2 and 18.7 kN/m³, and it was possible to verify that their value increases with the decrease of the particle dimensions, showing the opposite on what regards of water absorption (Fraga 2009).

Nineteen plate load tests were performed with 0.3 and 0.6 m diameter plates, obtaining values that varied between 70 Pa and 140 Pa, for the deformability secant modulus for a tension of 300 kPa, and between 120 and 480 Pa in the recharge (Figure 4).

Besides the tests performed in landfills, some SPT tests were executed in natural deposits, and N_{SPT} values between 4 and 20 were obtained, with a medium value of 12. These low values in a rocky material show the loose and uncompact state of these formations in nature due to their genesis.

3.2 *Comparison of test results*

Figure 5 shows the results of the tests for particle size distribution of samples. The grading of Madeira basaltic pyroclasts are presented with a continuous line while the ones of the Azorean samples are presented with a hatched line.

According to the ASTM D2487 standard, the majority of Madeira samples are classified as sand and those of the Azores as gravel.

For a better comparison of the results obtained in the tests for the determination of specific weight, dry density, CBR, friction angle, and cohesion, a table was elaborated (Table 1) where these values are presented for the Madeira, Azores and Canarias archipelagos.

On what regards the specific weights values, it is possible to verify that the Azorean samples are the ones that presents lower values followed by those of Canarias, being the ones from Madeira, those with higher specific weights.

On what concerns the results obtained for the in situ dry density, it is possible to verify that there is a range of values common to Madeira and Canarias archipelagos, though Canarias presents a wider range. The values of maximum dry density found in the Azorean pyroclasts are lower than to those obtained in the laboratory for the Madeira pyroclasts as observed for the specific weight.

Data related to CBR tests, friction angle and cohesion are similar in the existing samples. However, it's possible to verify the existence of higher CBR values for Madeira samples.

Regarding the results obtained in the plate load tests, it is possible to see that with the Azorean pyroclasts the dispersion of value is smaller than the Madeira ones, which is expectable, once the values refer to compacted landfills and in Madeira samples, the values refer to natural deposits with different degrees of compaction.

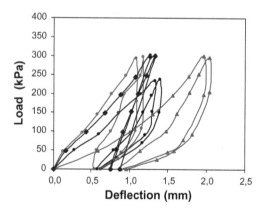

Figure 4. Typical plate load test curves obtained on landfills made with basaltic pyroclastic rocks.

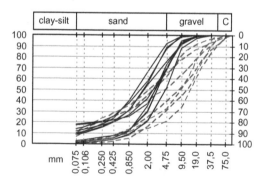

Figure 5. Comparison of the particle size distribution curves obtained on Madeira and Azores pyroclastic rocks.

Table 1. Results from tests made on Madeira, Azores and Canary Islands samples.

Tests	Units	Madeira	Azores	Canarias*
Specific Weight	kN/m³	25–29	20–22	23–25
In situ Dry Density	kN/m³	12–14	–	8–15
Maximum Dry Density	kN/m³	14–17	12–14	–
CBR	%	40–96	42–73	–
Friction Angle	°	29–50	–	30–45
Cohesion	MPa	0–0.1	–	0–0.1

* Gonzalez Vallejo et al., 2007.

4 SOME APPLICATIONS OF BASALTIC PYROCLASTIC ROCKS ON CIVIL ENGINEERING WORKS

As volcanic scoria are plentiful in the Azores, specially in S. Miguel Island, and due to its geotechnical characteristics, they assume particular importance because of its use in different kinds of civil engineering works.

One of the uses of this material is in the execution of landfill (improvement of the soil foundations, landfills on roads, etc), and they can be used in any part of these.

Because scoria is very hard and porous, it makes a good base for roads.

They have been used traditionally as a material for pavement layers of roads, as well as used in pavement beds, and for the improvement of excavation areas. In most cases it is used only as sub-base layer, but sometimes as base and wear coarse layer in roads with little traffic (Fraga 1988).

In both these applications, once its grain size distribution suffers changes by particles breakage and crush, particularly when compacted, the control of this material in civil engineering works should only be executed considering the grain size distribution that the materials shows after compaction.

The industry of making masonry bricks also uses volcanic scoria as row material.

Crushed and screened to specific sizes, the open structure and excellent drainage properties of volcanic scoria creates a truly versatile product for both landscaping purposes and as a bedding material for under soil drainage applications, and behind retaining walls.

5 CONCLUSIVE REMARKS

As it was possible to see by the results presented in this chapter, there are several similarities between the Madeira and the Azores basaltic pyroclasts.

In the first case, they are finer, with slightly bigger specific weights as well as maximum dry density. CBR values presents a wider range than the Azorean ones.

Though it's possible to take some conclusions with the existing data, it's necessary to proceed the investigation with the execution of more tests to understand better the geotechnical behavior of these materials in order to improve their applications in civil engineering works.

AKNOWLEDGEMENTS

Most of the tests presented in this paper were made in LREC Madeira and LREC Azores to whom we are grateful. We also would like to thank Prof. Maria de Lurdes Lopes, Prof. Castorina Vieira and Master Rui Silvano, from Faculty of Engineering of the University of Porto, for the support given in the execution of tests in big dimension shear box.

REFERENCES

ASTM D2487-06e1 Standard Practice for Classification of Soils for Engineering Purposes (Unified Soil Classification System), in *Book of Standards Volume: 04.08, Soil and Rock*, ASTM International, W. Conshohochen, PA. ASTM International.

Cas, R. & Wright, J. 1987. *Volcanic Successions*. London: Ed. Allen & Unwin.

Fraga, C. 1988. *Caracterização Geotécnica de Escórias Vulcânicas*. Master Thesis. Univ. Nova de Lisboa, 146 p.

Fraga, C. 2009. Especificações Técnicas para Aplicação de Bagacinas em Sub-bases de Pavimentos Rodoviários. NT 88/2009, 17 p, LREC, P. Delgada.

LNEC E 237 1970. Agregados. Ensaio de Desgaste pela Máquina de Los Angeles. Lisboa: LNEC.

MacDonald, G. 1972. *Volcanoes*. New Jersey: Prentice-Hall.

Scmincke, H. 2006. Volcanism, Berlin: Springer.

Serrano, A., Olalla, C., Perucho, A. & Hernandez-Gutierrez, L. 2007. *Strength and deformability of low density pyroclasts*. In A. Malheiro & J.Nunes (eds), *Volcanic Rocks*: 35–43. London, Taylor & Francis/Balkema.

Vallejo, L., Hijazo, T., Ferrer, M. & Seisdedos, J. 2006. *Caracterización geomecánica de los materiales volcánicos de Tenerife*—Madrid. Ed. Vallejo e Ferrer, Instituto Geológico y Minero de España.

Vallejo, L., Hijazo, T., Ferrer, M. & Seisdedos, J. 2007. *Geomechical characterization of volcanic materials in Tenerife*. In A. Malheiro & J. Nunes (eds), *Volcanic Rocks*: 21–28. London, Taylor & Francis/Balkema.

Volcanic Rock Mechanics – Olalla et al. (eds)

Deformational behaviour of pyroclastic rocks beneath the upper reservoir of the hydro-wind plant at El Hierro

M.R. Martín-Gómez, F. Fernández-Baniela & J.J. Arribas-Pérez de Obanos
IDOM Internacional Geotechnical Department, S.A., Madrid, Spain

A. Soriano
Universidad Politécnica, Madrid, Spain

ABSTRACT: El Hierro Hydro-wind Plant is a singular project, as its aim is to make the island energy self-sufficient. It involves constructing two reservoirs—the higher of which began being constructed in September 2009. This article attempts to reflect the peculiarity of the behaviour of the materials at the bottom of this reservoir, which will have to support a significant hydrostatic load, and the difficulty of characterising these in laboratories and design offices in order to assign geotechnical parameters to them which will reflect their geotechnical performance in situ. The studies undertaken to achieve this characterisation are described and the expected settlement of the materials is analysed. Finally, the article describes in detail the ground treatment proposed in order to minimize settlements, whose estimated values are compared with those obtained at the works site.

1 DESCRIPTION OF THE "EL HIERRO HYDRO-WIND PLANT" PROJECT

Before very long, the Canary Island known as El Hierro will become the first place in the world capable of being self-sufficient in electricity from renewable sources of energy.

In order to achieve this, the Regional Government of El Hierro (Cabildo de El Hierro), Unelco S.A. and ITC (the Technical Institute of the Canary Islands), which together form the company "GORONA DEL VIENTO", planned the project "Aprovechamiento Hidroeólico de El Hierro", which was presented on the 28th of February 2007, and whose detailed engineering was commissioned to Ingeniería Idom Internacional, S.A.

The project covers the design, development, construction and start-up of a hydro-wind system formed by:

- Two water reservoirs constructed at different heights: the upper one will be located in the area known as "La Caldera", with the maximum water level at el. 713.90 m and an active storage estimated at 556,000 m³; the lower reservoir will be constructed by means of a closing dyke which will make use of a stream bed close to the location of the existing diesel plant at Llanos Blancos, and will have a maximum water level at el. 56.50 m and an active storage of 200,000 m³ of water.
- A wind power park consisting of five wind generators with the latest technologies, located on

the "Pico de los Espárragos", which will produce a total power output of 11.5 MW,
- The pumping and turbine stations, located next to the lower reservoir,
- The pressure pipes which will connect the upper reservoir with the hydro-plants,
- And an electrical sub-station interconnecting the hydraulic plants and the wind generators.

The process of this reversible system (Fig. 1) consists in using wind energy (1) to pump water from the lower reservoir (2) to the upper reservoir (3), making use of the turbines (4) to generate

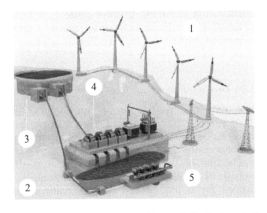

Figure 1. Sketch of the workings of the hydro-wind plant.

energy as the water flows from one reservoir down to the other, thus providing immediate and continuous electrical energy (5).

The existing diesel engine station at Llanos Blancos will supplement power generation in exceptional circumstances, when wind and water are insufficient to meet energy demands.

Besides being self-sufficient with clean energy, the island will also have a guaranteed supply of fresh water, one of the most prized resources, as the system will work with water that has previously been desalinated.

This project will turn the island of El Hierro into a worldwide reference in terms of alternatives to fossil fuels.

2 GEOLOGICAL SETTING

2.1 Geological setting

The age of the volcanic island of El Hierro is estimated at around 1.2 million years, making it the youngest island in the Canary Archipelago. It was formed by the successive overlapping of two significant volcanic structures (Tiñor, 1.1–0.9 Myrs old, and El Golfo, 0.5–0.1 Myrs old), and a latter dorsal ridge (Carracedo, 2001), with three axes that give the island its characteristic "Y" shape.

The upper reservoir is being built at "La Caldera", a depression made up of volcanic materials associated with the Tiñor structure, more specifically with its middle unit (the "tabular" unit, formed by lava extrusions that dip 20–30° towards the E–SE) and upper layers (the "Ventejís" volcano group, made up of pyroclastic rocks with a low lava proportion).

Petrologically, the abovementioned extrusions correspond to basaltic rocks. The geo-chemical analysis of different samples of rock taken from the perimeter of "La Caldera" have enabled their classification according to their content of SiO_2, as basic or ultra-basic rocks; with regard to their content of MgO, they are classified as basalts or trachybasalts.

2.2 Geological model of "La Caldera"

The study and investigation of "La Caldera" for the purposes of the construction of the upper reservoir has given rise to at least two geological theories concerning its origins.

The first of these attributes its origins to the collapse of a volcanic tube, with the materials that formed the ceiling of this tube currently being found in a fragmented condition as fill at the bottom of the caldera.

The second theory explains the conical shape of "La Caldera" and the characteristics of the deposits located in the eastern half of its perimeter, as being the result of a phreatomagmatic eruption that came about when the dyke that runs from North to South came into contact with phreatic water that had accumulated in the area.

3 SITE INVESTIGATION

3.1 Site investigation works

Two investigation campaigns were carried out for the Geological-Geotechnical Study, the majority of the site investigations being dedicated to studying the locations for the upper and lower reservoirs. (Table 1).

3.2 Boreholes and in-situ testing carried out at the upper reservoir

Eleven boreholes were drilled for the study of the upper reservoir, using rotation for continuous core recovery, whilst also carrying out pressuremeter tests using an Elastemeter-2, model 4181 equipment (from Oyo Corporation, Japan) (Table 2) and permeability tests. Seven of the boreholes were distributed across the bottom of "La Caldera" (Fig. 2) to study the fill materials and estimate the position of the bedrock; and the four remaining boreholes were drilled in the outer part of the eastern rim in order to analyse the stability of the rim where previous geological cartographies had revealed unfavourable geological characteristics, such as the dipping of the layers towards the slope and the existence of at least two layers of almagra interspersed amongst these lava layers.

Table 1. Prospections carried out at the chosen locations for the reservoirs and buildings, and the route of the pressure pipe.

Location	Type of investigations*					
	S.	C.	DPSH	P.S.R	T.E.	S.E.V
Upper reservoir	11	11	9	9	2	6
Lower reservoir	9	18	7	16	4	5
Pumping station	6	6	5	–	–	–
Turbine plant	4	2	–	–	–	–
Pressure pipe	–	–	–	–	–	3

*S, borehole; C, trial pit; DPSH, dynamic penetration; P.S.R., seismic refraction survey; T.E., electrical tomography profile; S.E.V., vertical electrical sounding.

Table 2. Pressuremeter tests carried out in the boreholes at the upper reservoir.

Borehole		Pressuremeter tests			
Name	Depth (m)	Depth (m)	Lithology*	Ep** (MPa)	Ep*(3) (MPa)
SC-1	20.6	–	–	–	–
SC-2	36.2	4.20	Qc	4.9	–
		28.70	B	679.8	–
		35.90	E	29.3	–
SC-3	32.6	2.00	Qc	2.1	–
		5.70	Mh	4.2	–
		10.20	Mh	15.6	–
SC-5	20.6	–	–	–	–
SC-6	25.6	3.20	Qc	2.8	–
		7.90	Mh	8.3	–
		14.70	Mh	11	–
		19.70	Mh	8.1	–
SC-7	50.0	–	–	–	–
SC-8	25.5	26.20	E	64.9	–
SC-9	45.0	9.70	Ec	122.6	–
		21.20	A	23.1	–
		29.70	B	25	–
		39.20	B	744.4	–
SC-10	32.0	–	–	–	–
SC-12	20.6	–	–	–	–
SC-13	25.0	–	–	–	–
S-1	30.0	10.55	Mh	15.7	31.1
		15.00	Mh	14.4	14.4
		22.00	Mh	2.3	2.9
		30.00	Mh	27.2	34.4
S-2	30.0	9.50	Qc	10.5	10.5
S-3	30.0	10.50	Mh	14.7	31.5
		15.00	Mh	3.6	6.1
		22.50	Mh	14.7	17.0
		30.00	Mh	36.7	70.7
S-4	30.0	11.00	Mh	15.3	16.7
		16.50	Mh	19.0	28.5
		22.50	Mh	16.7	17.4
		30.00	Mh	72.9	92.6
S-5	30.0	8.50	Qc	5.9	33.3
		15.50	Mh	15.3	15.3
		20.50	Mh	3.9	18.6
		30.00	Mh	35.8	42.7

* Qc, colluvial deposit; Mh, bottom fill material at "La Caldera"; B, basaltic lava layers; E, scoria; Ec, consolidated scoria; A, almagra.
** Ep, pressuremeter modulus obtained from first load.
*(3) Ep, pressuremeter modulus obtained from reload.

4 GENERAL DESCRIPTION OF THE PROBLEM AND SOLUTION ADOPTED

The location of the upper reservoir is in a spot known as "La Caldera". The materials supporting the basin are made up of fill material and colluvials. On the basis of the models that have been obtained with geophysical techniques (electrical tomography and seismic refraction) the position of the bedrock has been estimated at a depth varying between 25 and 40 m, with the deepest areas being in southern part of "La Caldera" at around 35–40 m. This substratum is made up of alternate lava flow layers and pyroclastic levels, predominantly scoria.

The process of obtaining undisturbed samples from these materials was complicated, as drilling caused their crumbling; these materials therefore had to be studied with in-situ tests: pressuremeter (Table 2) and penetration resistance (Fig. 4), with the conclusion that the foundation of the basin had low load bearing capacity and high deformability.

On the basis of the investigation, the following stratigraphy was adopted in order to estimate the settlements:

– From 0 to 5–6 m deep: colluvial deposits made up of material that had fallen from the slopes (sand and gravel size, with pebbles and even metric blocks of basalt), and altered in situ (Qc).
– From 5–6 to 15 m deep: material making up the fill of the bottom of "La Caldera", with grading and geotechnical properties closer to the group of lapilli (Mhs).
– From 15 to 25 m in depth: fill material with granulometry and geotechnical properties closer to the scoria group (Mhi).
– At a depth varying between 25 and 40 m continuous competent levels were detected, which have been interpreted as basaltic flow layers. These are considered to be non-deformable (B+E).

The deformation moduli were obtained from the in situ tests.

4.1 Estimation of settlement caused by the filling of the basin

The settlement due to the filling of the basin was calculated using the elastic method, taking into account the sole treatment of the removal of the loosest colluvial material, which would be subsequently re-placed and compacted, which was assigned a resulting modulus of 20 MPa.

The settlement associated with the filling in of the reservoir, using elastic moduli deduced from the compacity of the material and derived from the pressuremeter tests was 35 and 25 cm, respectively.

Although this settlement could possibly take place without causing the failure of the basin impermeable membrane, it is worth limiting the differential settlement. In order to do this it was considered necessary to treat the ground to limit the total settlement to 10–12 cm.

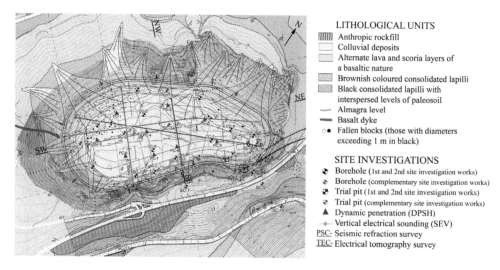

LITHOLOGICAL UNITS

⫿⫿⫿ Anthropic rockfill
☐ Colluvial deposits
▨ Alternate lava and scoria layers of
a basaltic nature
▨ Brownish coloured consolidated lapilli
▨ Black consolidated lapilli with
interspersed levels of paleosoil
⟋ Almagra level
▬ Basalt dyke
○● Fallen blocks (those with diameters
exceeding 1 m in black)

SITE INVESTIGATIONS

♠ Borehole (1st and 2nd site investigation works)
♠ Borehole (complementary site investigation works)
▣ Trial pit (1st and 2nd site investigation works)
▣ Trial pit (complementary site investigation works)
▲ Dynamic penetration (DPSH)
—●— Vertical electrical sounding (SEV)
PSC- Seismic refraction survey
TEC- Electrical tomography survey

Figure 2.   Geological plant of the upper reservoir showing the location of the site investigations carried out.

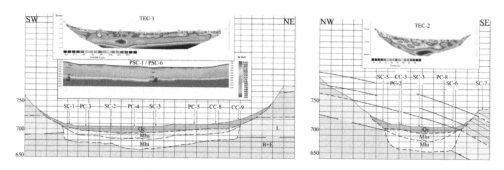

Figure 3.   Geological profiles of "La Caldera" (B+E, basaltic lava and scoria layers; L, consolidated lapilli; Mh, fill material; Qc, colluvial deposits).

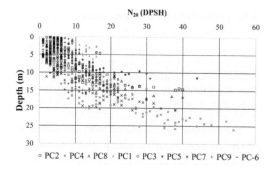

Figure 4.   Values of the N_{20} index with depth.

4.2   *Treatments analysed and solution proposed*

Different treatments were evaluated to correct the settlement:

– The execution of stone columns (of gravel) supported on the basaltic layer; this treatment was

rejected because of the impossibility of avoiding potential basaltic blocks or scoria nodules embedded in the loose material.

– Carrying out dynamic compaction, an ideal treatment for this type of granular ground, with fragile soil particles and a large volume of voids. However, in order to reach down to the 30 m of material to be treated, impact energies in the order of 1600 t·m would be necessary. It is for this reason that, although the method could be very efficient and appropriate from a technical point of view, the unavailability of a large enough crane on the island of El Hierro meant that this solution was rejected.

– Carrying out pre-loading; this treatment is a good solution, as it reduces and homogenises the settlement, provided that the material does not exhibit a markedly elastic behaviour.

In view of the above considerations, the chosen solution was pre-loading. The surface to be preloaded has an area of 40,000 m², which requires a very high volume of earthwork. Given the environmental uniqueness of the island, which was declared a Biosphere Reserve in 2000, it is important that the balance of earthwork be adjusted as much as possible, so it was decided to carry out a partial pre-loading using the volume of excavated material. The planned pre-loading consisted in moving some 100,000 m³ of earth inside "La Caldera" in different stages, to eventually cover the whole of the surface of the reservoir. This solution tends to take very long to execute, as it requires waiting for the settlement to take place before the preloading embankment is moved.

By placing the embankment prior to the filling of the reservoir, the filling will correspond to a reloading process. The reduction of settlement obtained with this pre-loading compared to the estimated settlement if no treatment was undertaken corresponds to the quotient between the reloading deformation modulus with respect to that of the virgin compression, but affected by a coefficient α, which corresponds to the relationship between the increase in stress at depth and the load that is spread at the surface. This coefficient enables a comparison to be made between the effect of the complete preload ($\alpha = 1$) and the partial preload. The average value of α in the 30 m of compressible material worked out to 0.98. Therefore, this partial preload would achieve a similar effect to that of the complete preload.

Before analysing the trial embankment there was no information regarding the possible moduli for the unloading or reloading of the ground. In any case, it was supposed that the relationship between the deformation modulus for the virgin compression line and that corresponding to recompression would be in the order of 4–5.

It was thus considered that the settlements would roughly be a quarter to a fifth of those that would result if there were no treatment, and that the total expected settlement would therefore be about 6–8 cm, which would be admissible.

5 EVOLUTION OF THE TRIAL EMBANKMENT

In order to complement the studies relating to the deformability of the upper layers of the ground, the project included the construction of a trial embankment in the southern sector of "La Caldera", where the greatest thickness of loose materials had been estimated (location and dimensions in Figure 5). The embankment was equipped with nine settlement plates to measure the response of the foundation to load variations and variations in the time taken for the foundations to consolidate.

The trial embankment is the best means for obtaining the ratio between the deformation modulus for virgin compression and that corresponding to recompression, with the latter tending to equate that of the unloading stage. At the same time, it provides the velocity of the response from the foundation material and enables the waiting times during the preloading to be adjusted.

A settlement of 56 cm was estimated for the trial embankment, which was located in the most unfavourable area (overlying a deformable ground 40 m thick), considering that the upper five meters were not treated (Table 3).

5.1 Results of the preloading test

The graphic in Figure 6 shows the average settlements measured on the settlement plates during the loading stage, the stabilisation of the settlement (waiting period) and during the unloading stage.

The average settlement during the tests was between 40% and 60% greater than that estimated at the design stage, with a deferred settlement averaging 22 cm.

The ratio between the settlement of the foundation and the height of the embankment has been close to 0.1, a very high value, with a settlement of 99 cm being recorded in the centre plate (n° 5) and an average value of 87.5 cm.

These results could be due to two factors:

– A deformability of the foundation material larger than expected,
– Or the material considered to be a "rigid layer" being actually deformable.

There was scarcely any rebound (between 4 and 8 cm) when the unloading was carried out, which indicates a permanent deformation of the ground, thus confirming that the type of treatment chosen for the foundation is appropriate.

Figure 5.   Location and definition of the trial embankment.

Table 3.   Comparison between estimated and measured settlement.

Layer	Thickness of layer (m)	Project estimate		Adjustment after trial preload	
		E* (MPa)	Settlement (cm)	E** (MPa)	Settlement (cm)
Layer 1	5	3.00	27	3.00	27
Layer 2	5	5.00	16	6.00	13
Layer 3	10	20.00	8	14.00	11
Layer 4	10	30.00	5	14.50	11
Layer 5	10	30.00	5	47.50	3
Instantaneous settlement			56		66
Deferred settlement/creep			5.6 *[3]		22 *[4]
Total settlement			62		88

* E, moduli deduced from standard penetration tests (SPT) and dynamic penetration tests (DPSH).
** E, pressuremeter moduli obtained from first loading.
*[3] Post-construction settlements due to creep (10% of instantaneous settlement).
*[4] Deferred settlement due to creep (33% of instantaneous settlement).

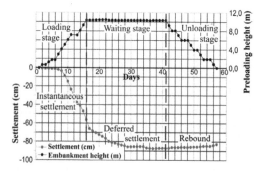

Figure 6.   Results obtained from the trial embankment.

5.2   Analysis of the results

As can be seen in Figure 7, the stress-deformation relationship is not lineal, but varies with the load value, as indicated by the break in the slope produced at a vertical pressure of 70 kPa.

During stoppages due to staff rest periods in weekends there was a settlement of the foundation without any increase in the load. Following this rest period there was an increase in the slope of the stress-deformation curve, which could be likened to the effect of a pre-consolidation pressure, which means that the stress-deformation relationship also depends on how long the load is applied.

Under vertical pressures below 70 kPa, where the stress-deformation relationship is equivalent to that of the unloading phase, the ground behaves in an elastic way. At a stress of 70 kPa there is a reduction in slope of the stress-deformation curve, which might possibly be due to the breakage of the structure of the particles (reduction of intragranular or vacuolar porosity).

The materials that fill the bottom of "La Caldera" are characterised by their low density, which is due to their high intergranular and intragranular porosities. Values of isotropic collapse pressures have been set out for lapilli-sized pyroclasts (Serrano et al. 2007) that vary between 0.8 and 11 MPa. Once the material resistance has been exceeded, its collapse takes place and its mechanical properties degrade to values similar to those of soils with equivalent density. This change brings about a significant reduction in volume.

When these materials are altered, their skeleton loses its structure under small loads. This hypothesis appears to be corroborated by analysing the ground granulometry before and after the load is placed. Grading curves for materials from the foundation of the basin show 12% of fines content, specifically in the tests carried out on the material from the foundation of the load embankment (CC10+SC12) which turned out not to be plastic with a percentage of fines not exceeding 12%. Once the material had been subjected to the loading process, three tests were carried out (C-5, C-6 and C-7) showing an increase in the fines content (60% with a maximum value of 92% and a minimum of 21%). They also showed an average plasticity index of 14.

The yield pressure or pressure of initial collapse (σ_c) deduced for the materials of the embankment foundation in this test is just under 70 kPa. For higher pressures, settlement owing to the collapse of particles is significant, and results in substantial reductions of volume.

In addition to instantaneous settlement, a deferred settlement occurs due to the rheology of the particles of the foundation materials. An analysis—in the

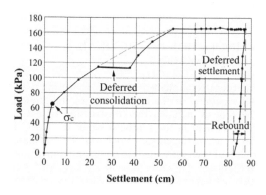

Figure 7.   Stress-deformation relationship of the supporting ground.

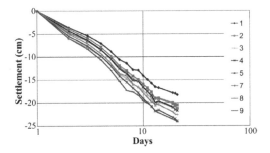

Figure 8.   Average creep during the waiting stage.

different plates and in average values—has been made of settlement corresponding to secondary consolidation due to material creep, which showed that the deferred settlement is the equivalent of 33% of the instantaneous settlement. This value far exceeds the 10% settlement that had been assumed in a first calculation and which constitutes an upper limit for granular materials.

Deferred settlement tends to be expressed as a law such as the following:

$$\frac{s}{H} = \alpha \log\left(\frac{t}{t_o}\right) \tag{1}$$

The phenomenon of creep has been studied during the "waiting stage", establishing (t_0) as day 1, being the day following the completion of the embankment.

The value s corresponds to the increase in settlement that occurs when time t is increased tenfold, and an average value of 18 cm has been obtained. As the settlement value (s) varies according to the compressible thickness (H), to make it possible to compare different locations, a post-constructive index (α) is employed, which is the ratio between s and H (Soriano, 1999).

$$\alpha = \frac{s}{H \cdot \log\left(\frac{t}{t_o}\right)} = \frac{180 \text{ mm}}{40 \text{ m}} = 4.5\%_o \tag{2}$$

The index of post-construction settlement obtained for the embankment support (as can be seen in Equation 2) is much larger than that which was supposed at the design stage, 1.25‰.

Finally, the creep values measured during the test have been used to compare the settlement estimated in the design calculations with those obtained after construction of the embankment (Table 3). The moduli used to adjust the calculation of the instantaneous settlement during the construction of the trial embankment were the pressiometric moduli

obtained during the first loading (or initial modulus), with a rheological factor of 1, a factor value that is commonly used for peat, another material with a high porosity.

6   CONCLUSIONS

The elastic behaviour of materials resulting from the alteration of pyroclasts can be established as follows:

– When subjected to stresses that are less than the yield stress, the soil skeleton has an elastic behaviour,
– When said stress is exceeded, the particles break up and the soil undergoes excessive creep deformation due to the high intragranular porosity of these materials.

The initial settlement was instrumental in carrying out an acceptable calibration of instantaneous settlement, which has been shown to be 20% greater than expected. However, the creep value could not be foreseen under these load levels.

The post-construction settlement value that had been supposed in the initial analysis (10% of the instantaneous settlement) establishes an upper bound for granular materials and was therefore a conservative value. However, the deferred settlement due to creep has turned out to be in the order of 33% of the instantaneous settlement, far greater than that initially supposed.

In addition, it has been seen that the instantaneous settlements adjust fairly well to those calculated through the elastic method, using the pressuremeter moduli obtained from the first loading. A rheological factor of 1 has been used in order to obtain the deformation modulus from the pressuremeter modulus. This is the value that is recommended for peat, another material that is highly compressible and which also presents a structure with a high porosity.

Ground heave has scarcely been observed in the unloading stage: rebound oscillated between 4 and 8 cm, which indicate a permanent deformation of the ground. The carrying out of preloading is therefore considered to be an appropriate treatment for this type of ground.

ACKNOWLEDGEMENTS

This work constitutes a part of the geotechnical conclusions extracted from the investigation, analysis and calculations carried out for the "Estudio Geologico-Geotécnico del Depósito Superior", presented by Idom Internacional in

November 2008. This study, in turn, belongs to the engineering project "Aprovechamiento Hidroeólico de El Hierro", developed by Gorona del Viento and 65% financed by IDAE (Spanish Institute for Energy Diversification and Saving).

The authors acknowledge the confidence shown by Gorona del Viento in IDOM and the support that they have been given in developing the project engineering. They also thank Daniel Arias Prieto (Geology Department at the University of Oviedo) for his contributions and assessment during the project. Finally, they thank Inés Galindo (IGME) for her explanation about the genesis of "La Caldera".

REFERENCES

Carracedo, J., Badiola, E., Guillou, H., Nuez, J. & Pérez, F. 2001. Geology and volcanology of La Palma and El Hierro, Western Canaries. *Estudios Geológicos* 57: 175–273.

Serrano, A., Perucho, A., Olalla, C. & Estaire, J., 2007. Foundations in volcanic areas. *XIV European Conference on Soil Mechanics and Geotechnical Engineering. Geotechnical Engineering in Urban Environments Madrid,* 24–27 September, 2007.

Soriano, A. & Sanchez, F.J. 1999. Settlements of railroad high embankments. *Proceedings Geotechnical Engineering for Transportation Insfrastructures,* Amsterdam 1999. Balkema.

Volcanic Rock Mechanics – Olalla et al. (eds)
© 2010 Taylor & Francis Group, London, ISBN 978-0-415-58478-4

Rock mass classification schemes in volcanic rocks

M. Muñiz Menéndez
Master Mecánica de Suelos e Ingeniería de Cimentaciones (CEDEX), Madrid, Spain

J. González-Gallego
Laboratorio de Geotecnia (CEDEX), Madrid, Spain

ABSTRACT: Volcanic massifs, due to their origin, present certain characteristics that make their study through a classic geomechanical classification not always appropriate. The presence of discontinuities, the peculiar block shape and the presence of voids condition the behavior of lavatic massifs. Behavior of pyroclastic massifs depends mainly on their matrix rock, making the use of existing geomechanical classifications inadequate. This work represents a first step in the development of a geomechanical classification specific to volcanic massifs, accounting for all the properties that actually condition their behavior.

1 INTRODUCTION

From the second half of the 20th century onwards, many geomechanical classifications have appeared, with the original aim of easing decision-making when considering support in subterranean excavations.

Nowadays, geomechanical classifications are widespread in every undertaking related to a rock massif, and are used in the design stages of a project as well as in the construction stage. This makes it necessary to know the limitations and difficulties entailed by such classifications.

Most of these classifications are based on quantifying several of the massif's parameters and variously computing them to obtain a quality index for the massif, thus obtaining an approximation to the geomechanical behavior of the massif through direct observations and simple field tests.

Numerical and empirical methods employing values obtained from one of these classifications as an entry parameter are more abundant every day, forcing classifications to be very finely tuned, and every one of their parameters carefully considered.

Volcanic rocks, due to their origin, show very distinctive geomechanical characteristics. Volcanic rock massifs are very heterogeneous in their structural and lithological characteristics, making the application of geomechanical classifications on them especially difficult and, sometimes, impossible.

2 VOLCANIC ROCKS

The geomechanical properties and geotechnical behavior of these materials are completely different to those of non-volcanic materials (González de Vallejo, L.I. *et al.*, 2006). This calls for very attentive and cautious study of these massifs, following a methodology different to the usually employed.

The main distinguishing features of volcanic massifs are their high structural and lithological heterogeneities, with alternating materials of very different properties, the presence of discontinuities of varied origin, the existence of cavities of different sizes and a great variability in layer thickness.

From a geotechnical perspective, volcanic rocks are classified in two main groups: lavatic rocks, formed by the cooling of lava flows, and pyroclastic rocks.

2.1 Lavatic rocks

The contact of a lava flow with the air can cause its surface to break due to rapid cooling, creating a drossy surface, while the inside of the flow cools more slowly and presents a more orderly structure.

Different volcanic episodes usually create a series of lava flows separated by scoria or even soils. This alternating of materials increases the heterogeneity of the volcanic massif.

The main distinguishing features in these massifs are their discontinuities, the presence of cavities and their resistant properties.

2.1.1 Discontinuities

Volcanic massifs are affected by discontinuities created by the different geological processes that affect them.

These discontinuities can be classified according to the geological process that originated them into the following categories (González de Vallejo, L.I. *et al.*, 2006):

– Discontinuities of thermal origin.
– Discontinuities of tectonic origin.
– Discontinuities created by intrusive structures.
– Discontinuities due to the contact of different formations.
– Voids/cavities (see below).

Discontinuities of thermal origin are, undoubtedly, one of the main characteristics of lava flows. They are created by the retraction suffered by volcanic materials when cooled. These discontinuities can present a great opening; their walls are usually smooth and show little filling.

The most characteristic of these discontinuities are the columnar ones (see Figure 1), produced by the cooling of massive lavatic materials. These columns usually present polygonal structures of three to twelve sides, although usually limited to between five and seven sides. They usually show perpendicular fractures with an even spacing. Length of the columns can vary enormously from one site to another, from 3 cm to 130 m (Lyell 1871). Diameter can also vary, from more than 3 meters to 3 centimeters or less.

Columns grow perpendicular to the lava cooling surface, thus appearing usually vertical though they can also be horizontal, leaning or even curved. The orientation of columns can change from one part of the lava flow to another.

Columns can sometimes show a radial distribution that further hinders their study.

Spheroid discontinuities appear when water penetrates the inside of a lava flow. If the flow is of great thickness, horizontal retraction planes may appear, usually located at third of the thickness closest to the base (González de Vallejo, L.I. et al. 2006).

Discontinuities created by intrusions can be of importance in slope instability processes, mainly due to their great continuity.

Discontinuities due to the contact of different formations appear usually in the contact between a lava flow and a pyroclastic deposit, being the contact of erosive or depositional origin.

2.1.2 *Voids*

One of the main features of volcanic massifs is the presence of voids generated during the eruptive process, with sizes ranging from microns to several kilometers.

These cavities can be of several different origins, the main three being discussed below.

The succession of volcanic episodes in time can produce an alternating of materials of different compositions. Drossy or breccioid materials can appear in between more or less massive lavatic materials. In many occasions these materials can give way to voids at the base and top of the lavatic materials (see Figure 2).

Magma contains great quantities of dissolved gases. On cooling, these gases tend to separate from the magma (degasification). Given enough lava flow viscosity, the gases can be trapped, forming vacuoles of different sizes. Examples of these structures can be found in vacuolar basalts.

In the cooling process of a very fluid lava flow, areas in contact with open air are cooled rapidly, creating a crust under which lava keeps on flowing, due to the low thermal conductivity of basaltic rocks (González de Vallejo, L.I. *et al.*, 2006). Sometimes, these crusts remain after the flow has left and become hollow structures that can run for several kilometers, called lava tubes. Examples of these structures are: the Cueva del Viento in Tenerife, which runs for more than 17 kilometers; the Túnel de la Atlántida in Lanzarote, with more

Figure 1. Columnar disjunction in Castellfollit de la Roca (Spain). Image courtesy of Irene López.

Figure 2. Scoriaceus contact between pyroclastic deposits and a basaltic flow with decimeter-sized cavities. Garrotxa Natural Park (Spain). Image courtesy of Irene López.

than 7 km and which includes the Jameos del agua and the Cueva de los verdes.

After studying all three cave formation processes, two types of cave can be classified according to their position relative to lava flows:

– Type I: caves at the top or the base of a lava flow.
– Type II: caves inside a lava flow.

Millimetric or centimetric cavities are treated as part of the rock matrix when considering its characteristics and therefore do not need a specific study. The size of lava tubes will usually make it necessary to study them individually and apart from the rest of the lavatic massif. Decimetric and metric cavities cannot be studied as part of the matrix and their size does not allow an individualized study; these cavities must then be included in a geotechnical classification of volcanic massifs.

A way of assessing the importance of these middle-sized cavities is through the cave index (Estaire, J. *et al.*, 2008), defined as the quotient between the sum of the volumes of the massif's connected cavities and the total volume of the massif. This index allows the quantitative assessment of cavity occurrence and, with it, cavity importance when evaluating the geotechnical properties of the massif as a whole.

2.1.3 *Lava flow strength*

Compression strength values of volcanic rocks are high, between strong rock and extremely strong rock.

The high strength of this kind of rocks means that the stability of a volcanic massif will depend mainly on the shear strength of the discontinuities.

Due to the behavior of liquid lava, flow structures may appear in these rocks, mainly because of mineral orientation. These structures generate anisotropies that affect the geotechnical properties of the rock.

The possibility of the existence of flow structures must, then, be accounted for when measuring its strength.

Drossy layers, from a geotechnical point of view, behave like a granular soil, and their compactibility will depend, among other factors, on any consolidation due to overlaying lava flows and to cementation by circulating fluids (Peiró 1997).

2.2 *Pyroclastic rocks*

Pyroclastic rocks show a complex behavior, sometimes closer to that of a soil than to that of a rock. Their geotechnical properties have little in common with the rest of rock types. The behavior of these massifs will depend mainly on the strength of the matrix rock and, as opposite to most rock massifs, very little on the fracture system.

2.2.1 *Pyroclastic rock strength*

Pyroclasts can be bound by two different processes: welding of the clasts at high temperatures and cementation by interstitial fluids.

The mechanic behavior of this kind of rocks will depend on five factors (Serrano, A. *et al.* 2002):

– Compaction.
– Degree of particle welding.
– Imbrication of particles.
– Particle intrinsic strength.
– Alteration.

The first three parameters are usually related: the higher the imbrication and welding are, the higher the compaction will be. This is not always true, however, and sometimes low density deposits with a very high degree of welding can be found, or very compacted deposits with a low level of particle joining.

The degree of compaction is the most influential factor on the strength of a pyroclastic massif, and close attention must be paid to this parameter when studying massif properties. The best way would be to measure the density of the material on site.

Welding degree and imbrication must be assessed together. Imbrication degree is related to the number of contact points among particles, while welding degree is related to the percentage of total particle area that is in welded contact with another particle. These parameters can only be measured by direct observation on the field.

Compression strength values in pyroclastic rocks show a very wide range. Obtaining laboratory data for this type of rocks is difficult in many cases, since they are composed of a conjoining of many irregular fragments, making it difficult to sculpt testing specimens. Data dispersion in this tests is high, and many specimens are needed to obtain reliable values (Serrano, A. *et al.*, 2008).

Low density volcanic agglomerates have a very specific mechanical behavior. At low tensile strengths they behave like rocks with very high deformation modulus, but at high tensile strengths they crumble to dust and behave like soils, with an important increase in their deformability. This is known as a mechanical collapse (Serrano, A. 1976).

Mechanical collapse can be explained by the high porosity usually present in these materials and the strong unions among grains due to their high-temperature welding.

Unions among particles are highly rigid for low loads but at higher loads these unions progressively break until the material becomes a low density soil, thus explaining its high secondary deformability.

The tensile strength level at which this difference in behavior (from rock to soil) occurs is directly related to the starting density of the material (Serrano, A. *et al.*, 2008).

The presence of water greatly affects the resistant properties of volcanic tuffs. Vásárhelyi, B. (2002) showed, from the study of several Hungarian tuffs, that the compressive strength of a saturated tuff is approximately a 70% of that same tuff's compressive strength when dry. Studies in Filipino tuffs by Catane, S. and colleagues (2006) give percentages closer to 90%.

2.2.2 *Discontinuities in tuff massifs*

Most pyroclastic rocks (tuffs) form very homogeneous and continuous massifs, very little affected by discontinuities.

The strength of these massifs, therefore, is not determined by the strength of the discontinuities, but that of the matrix rock.

This notwithstanding, tuff massifs, like other massifs, can be affected by tectonic processes, and tectonic discontinuities can thus appear in them.

3 ROCK MASS CLASSIFICATION SCHEMES IN VOLCANIC ROCKS

As previously seen geomechanical properties and geotechnical behavior of these materials are entirely different from non-volcanic materials. Therefore you should consider whether the instruments used to study other types of rock masses are applicable in the volcanic massifs. If used, should be done with certain precautions and paying special attention to the distinguishing features of this massifs.

The existence of intercalated layers, as lava, scoria, autoclastic breccias, etc., requires a sectorization of the rock mass due to the setting of the different categories of it (González-Gallego, J. 2008).

3.1 *Lavatic rocks*

The behavior of this type of rock is similar to other compounds by resistant intact rock. Geotechnical properties depend mainly on the shear strength of its discontinuities. However, there are certain differences, the rest of the massifs, which must be taken into account. These differences are mainly: the great heterogeneity of alternating materials of different behaviors, characteristics of the discontinuities, the shape of the blocks, the presence of voids and the presence of flow structures.

The main difficulties of studying lavatic massifs with actual classifications are as follows:

3.1.1 *RMR*

This classification mainly ponders the state of fracturing of the massif by measurement of RQD, spacing, characteristics of the discontinuities and their orientation. In volcanic rocks the estimation of these parameters can be difficult:

RQD is a parameter highly dependent on the direction of measurement. The discontinuities of thermal origin typical of such rocks show a markedly one-dimensional distribution, with typical columnar blocks. With this blocometric configuration RQD measurement can be very complicated, requiring numerous steps in different directions. In these cases the block size as directly through the block size index (Vb) is highly recommended and easily.

The spacing of discontinuities is a parameter difficult to assess in massifs affected by thermal discontinuities. In addressing the measurement of the spacing in an area with columnar or spheroidal disjunction is complex decide the direction in which the measures are to be made.

The characteristic of the discontinuities, especially its persistence, is complex to assess because of its morphology and possible differences if this is measured in either direction.

The orientation of the discontinuities on the excavation is very complex to evaluate in this type of solid, because sometimes it is very variable.

Another important parameter that ignores RMR is the presence of voids, which can have an important role in the behavior of the massif.

In this classification exists a correlation between the point load test and the compressive strength of the rock, however, in volcanic rocks has shown that the relationship between both parameters show a wide dispersion (Romana, M. 1996) and therefore should avoid using this test, being recommendable in any case, the use of laboratory tests to assess the strength of the rock matrix.

3.1.2 *Q system*

As in the RMR, the estimation of the parameters that make up Q can be tricky:

The RQD has the same problems as in the previous paragraph.

The rate of jointing in the Q system is based on measuring the number of families that presented the massive discontinuities. The discontinuities of thermal origin are not easily assimilated families. In a columnar or spheroidal disjunction is difficult to decide the number of families present and appreciation of different people can differ too.

As in the RMR the presence of voids is not taken into account.

Of the 212 historical cases studied by Barton for the development of this classification, only one of them corresponds to a solid lava, basalt rock specifically.

3.1.3 *GSI*

This classification only evaluates the degree of jointing of the massif and the state of the discontinuities.

There are more parameters that influence the behavior of massifs and that this classification does not account, in particular the presence of holes and the shape of the blocks.

It also ignores the simple compressive strength of rock.

The heterogeneity of this type of solid GSI makes the classification difficult and often inadequate. However, the amendments made by Marinos, P. *et al.* (2001) for type flysch massifs and Hoek, E. *et al.* (2005) for solid molasses type can be much more appropriate.

3.2 *Pyroclastic rocks*

The resistance of the pyroclastic massifs does not depend on the discontinuities but on the resistance of the rock matrix (Gonzalez-Gallego, J. 2008).

Geomechanical classifications, most commonly used today (RMR, Q and GSI) are designed for the study of rock masses composed of high strength rock controlled therefore by the properties of the fracturing system. These classifications do not consider adequately the strength of the rock matrix which is the most important parameter in pyroclastic beds.

3.3 *Recommendations for the design of new classifications*

In view of all previous data are discussed below the basic features to follow for the proper design of geomechanical classifications to study volcanic massifs.

3.3.1 *Lavatic rocks*

The parameters to be evaluated:

The compressive strength of the rock matrix, which should be evaluated by laboratory tests mainly, discourage the use of point load test given his difficult

Table 1. Parameters to be considered in the classification of lavatic massifs.

Parameter	Importance
Intact rock compressive strength	+
Block size and shape	++
Cave index	++
Discontinuities characteristics	++++
Water presence	+

relationship with compressive strength in this type of rock. Be taken into account the possible anisotropy of the rock by the presence of flow structures.

The size and shape of block, using the block volume (Vb), less influenced by the direction as the RQD.

The presence of voids, the cave index can be used in caves.

The characteristics of the discontinuities, being primarily responsible behavior of the massif, with regard to the aperture, roughness, presence of fill and alteration of the walls.

The presence of water must also be evaluated.

Below is a summary table that shows the importance of each parameter.

The importance of each parameter, assigned a priori, must be evaluated carefully. The historical case studies should lead to the quantification of the relative importance to obtain reliable and useful classification in the geotechnical work.

3.3.2 *Pyroclastic rocks*

Previously we have seen that the behavior of these massifs depends primarily on:

Compaction degree. It is the most influential factor in the resistance of the massif. The best method is to determine the in-situ density of the material.

The degree of welding and the imbrication of particles. This parameter can be evaluated using the classification proposed by Serrano, A. *et al.* (2002).

The degree of alteration.

The presence of water may decrease in up to 30% the strength of the rock (Vásárhelyi 2002).

The intrinsic strength of the particles presents difficulties in its extent, then it may be advisable to disregard this parameter.

Discontinuities. Although not normally present in these materials, their possible presence must be taken into account.

Below is a summary table that shows the importance of each parameter.

Table 2. Degree of imbrication and welding (IW) (Serrano *et al.,* 2002).

	Welding			
Inbrication	Loose <5%	Not very welded 5–15%	Welded 15–45%	Extremly Welded >45%
Low <8 contacts	IW 1-1	IW 1-2	IW 1-3	IW 1-4
Medium 8–12 contacts	IW 2-1	IW 2-2	IW 2-3	IW 2-4
High >12 contacts	IW 3-1	IW 3-2	IW 3-3	IW 3-4

Table 3. Parameters to be considered in the classification of pyroclastic massifs.

Parameter	Importance
In-situ density	+++
Welding and overlapping	+++
Alteration	++
Presence of water	+
Discontinuities	+

As in the previous section, the weight of each parameter has to be carefully evaluated by in-situ studies, supported by laboratory tests and by studying historical cases.

4 SUMMARY AND CONCLUSIONS

Volcanic rocks have different geotechnical behavior of other rock types.

The use of geomechanical classifications must be done carefully, paying particular attention to their distinctive characteristics.

The behavior of the lavatic rocks depends mainly on its discontinuities, both in its system of fracturing as the presence of holes. It should also be taken into account the great heterogeneity present in these materials. The special morphology of this fracture system makes specific indices must be used to consider adequately.

The pyroclastic massifs are mainly dependent behavior of the resistant characteristics of the rock matrix and hence the use of existing geotechnical classifications may not be suitable.

REFERENCES

Barton, N., Lien, R. & Lunde, J. 1974. Engineering classification of rock masses for the design of tunnel support. In Rock Mechanics, 6 (4), pp. 189–236.

Catane, S., Orense, R. & Tsuda, N. 2006. Effect of water saturation on the compressive strength and failure modes of the Diliman tuff. Unpublished.

Estaire Gepp, J., Serrano, A. & Perucho, A. 2008. Cimentaciones superficiales en rocas con cavernas. II Jornadas Canarias de Geotecnia.

González de Vallejo, L.I, Guhazi, T. & Ferrer, M. 2008. Engineering geological properties of the volcanic Rocks and soils of the Canary Islands In Soils and Rock. Sao Paulo, 31 (1), pp. 3–13.

González de Vallejo, L.I., Hijazo, T., Ferrer, M. & Seisdedos, J. 2006. Caracterización geomecánica de los materiales volcánicos de Tenerife, p. 40.

González-Gallego, J. 2008. Clasificaciones Geomecánicas (Aplicación a Rocas Volcánicas). II Jornadas Canarias de Geotecnia.

Hoek, E., Marinos, P. & Marinos, V. 2005. Characterization and engineering properties of tectonically undisturbed but lithologically varied sedimentary rock masses. International journal of rock mechanics & mining sciences, 42, pp. 277–285.

Lyell, C. 1839. Elements of Geology. London.

Marinos, P. & Hoek, E. 2001. Estimating the geotechnical properties of heterogeneous rock masses such as Flysch. In Bull. Engg. Geol. Env., 60, pp. 85–92.

Muñiz Menéndez, M. 2009. Clasificaciones geomecánicas en roca volcánicas. Master en Mecánica de Suelos e Ingeniería Geotécnica. CEDEX, Madrid.

Peiró, R. 1997. Caracterización geotécnica de los materiales volcánicos del archipiélago canario. In Tierra y tecnología, pp. 45–49.

Romana Ruiz, M. 1996. El ensayo de compresión puntual de Franklin. Ingeniería civil/CEDEX. N. 102 (abr.-jun) pp. 116–120.

Serrano, A. 1976. Aglomerados volcánicos en las islas Canarias. In Mem. simp. nac. de rocas blandas. Tomo 2. A-10, Madrid.

Serrano, A., Olalla, C. & Perucho, A. 2002. Evaluation of non-linear strength laws for volcanic agglomerates. In *ISRM International Symposium on Rock Engineering for Mountainous Regions and Workshop on Volcanic Rocks*, Eurock 2002. Funchal, 27 Nov.

Serrano, A., Olalla, C., Perucho, A. & Hernández, L. 2008. Resistencia y deformabilidad de piroclastos de baja densidad. In II Jornadas canarias de geotécnia, Tacoronte, Tenerife.

Vásárhelyi, B. 2002. Influence of the water saturation on the strength of volcanic tuffs. In ISRM International Symposium on Rock Engineering for Mountainous Regions and Workshop on Volcanic Rocks, Eurock 2002. Funchal, 27 Nov.

Volcanic Rock Mechanics – Olalla et al. (eds)
© 2010 Taylor & Francis Group, London, ISBN 978-0-415-58478-4

Relationships between porosity and physical mechanical properties in weathered volcanic rocks

A. Pola, G.B. Crosta, R. Castellanza, F. Agliardi, N. Fusi, V. Barberini, G. Norini & A. Villa
Dipartimento di Scienze Geologiche e Geotecnologie, Università di Milano-Bicocca, Milano, Italy

ABSTRACT: Volcanic rocks are frequently found under weathered/altered conditions. Degradation and transformation can occur both at the surface and at large depth causing a progressive change in the physical mechanical properties. Degradation can cause an increase in porosity and this can control the rock behavior. In this paper we discuss the relationships between porosity characteristics, micro-structure and texture, and the mechanical behavior of lava at different degrees of weathering (lavas from the Campi Flegrei, Italy). The performed laboratory tests include: uniaxial compression, indirect tension, and uniaxial compression with ultrasonic wave measurements. A description of the mechanical behavior is obtained and a detailed description is performed through a series of pre and post failure non destructive analyses.

Porosity values have been related to stress and strain relationship, in addition pore size characterization is presented in a companion abstract/manuscript. Results are interpreted in the key of degree of weathering and its related characteristics. An empirical linking between the change in strength with the degree of alteration is presented and discussed.

1 INTRODUCTION

The analysis of the stability of volcanic edifices or rock masses in volcanic rocks is often problematic because of the variability of the materials (e.g. individual lava flows, pyroclastic deposits, and interbedded units), their heterogeneity, the presence of abundant voids and variable degree of cementation. These characteristics make sometimes extremely difficult to reach a correct and representative characterization of the physical mechanical behavior.

Volcanic rocks are frequently composed of both matrix material and pores, and they are often found in altered/weathered conditions because of the highly active volcanic environment and the presence of hydrothermal conditions. Generally, both the strength, the deformability and stiffness of these rocks shows a dependence on the porosity. Porosity can be formed by voids, between grains or minerals, of different size and shape, with a particular frequency distribution of size and it can be interconnected or disconnected. Various researchers investigated the physical mechanical behaviour of rocks as a function of their porosity.

Dolostones, sandstones, limestones, dolerite and granites are mainly characterised by microporosity. On the other hand, volcanic rocks (e.g. basalts, scorirae, lithophysae-rich tuffs, tuffs and pyroclastic deposits in general, breccias) often present a brecciated, porous or vesicular texture characterised by abundant vesicles and pores, with different sizes, that sometimes are filled with secondary minerals (Al-Harthi et al., 1999, Tillerson and Nimick, 1984, Hudyma et al., 2004). Previous studies on similar rock lithologies suggest that the compressive strength is controlled by total porosity, the abundance of macropores and the pore structure and size distribution as well as the type of forming particles (Luping, 1986; Nimick, 1988; Al-Harthi et al., 1999; Price et al., 1994; Aversa and Evangelista, 1998; Avar et al., 2003; Avar and Hudyma, 2007; Hudyma et al., 2004).

In this paper we present the initial results of a study concerning the effect of alteration and porosity characteristics on the physical mechanical behaviour of a volcanic rock. This paper is directly related to a companion paper presented at this conference (Pola et al.) on the characterization of porosity by means of different methodologies.

2 STUDY AREA AND MATERIALS

The tested materials have all been collected from the Solfatara volcano in the Campi Flegrei area (Campanian volcanic province, Italy). Large lava samples, characterized by diverse degree of alteration, have been collected and resampled in the laboratory to perform physical mechanical tests.

A mineralogical and petrographical description of the materials have been completed through

Table 1. Summary of the main petrographical and mineralogical observations.

	Fresh	Slightly	Moderately	Highly	Totally
Plagioclase	+++	++	+	–	?
Potassic feldspar	+++	+++	++	+	–
Pyroxene	++	+	+	–	?
Biotite	+	+	±	–	–
Mineral oxidation	±	+	++	±	–
Matrix oxidation	±	+	++	±	–
Matrix argilization	–	±	++	+++	+++

+++: abundant; ++: few; +: rare; –: absent; ?: difficult or uncertain identification.

observations at the outcrop and sample scale, through optical microscopy and by X-ray diffraction analyses.

On this bases we subdivided the samples in five main grades of alteration increasing with the ordering number (SL1, 2, 3, 4 and 5).

Summarizing the main characteristics the major constituents (see Table 1) are sodic plagioclase and potassic feldspar with minor amounts of pyroxene and biotite. The fresh—less weathered material has a porphyritic to trachytic texture with euhedral plagioclase phenocrysts and a matrix with pilotaxic texture. Alteration is present prevalently as: oxidation along the boundaries of crystals and as stains, and as argilization both of the matrix and crystals.

Abundant alunite is evidenced by XRD analyses X-ray power diffraction for grade 4 materials argilization and silicification are also observed. Complete substitution of the matrix by argilization and presence of small silica-amorphous minerals is observed for grade 5 materials and the porosity characteristics can be observed for example on thin sections (see Figure 1 and the companion paper for more detailed description).

3 METHODS

Three different physical-mechanical parameters were studied: compression and splitting tensile strength, ultrasonic wave velocity both before and during testing. Uniaxial compression strength (according to ASTM D 2938) and splitting tensile strength (ASTM D 3967) were determined.

23 cylindrical samples (18 mm and 54 mm in diameter) were prepared, from the five differently altered rock materials, to perform compression test. Small samples have been tested only to compute some of the values of uniaxial compressive strength.

27 samples, 54 mm in diameter, were prepared for the brazilian tests. In general all the samples have a size much larger than both the larger and average particles/crystal and pore at their interior.

Uniaxial compression and Brazilian tests were carried out in a servo-controlled hydraulic testing frame (GDS). A constant 5 mm/hr displacement rate was imposed during the test. Axial and radial deformations were measured by means of strain gauges (ASTM D 3148). The splitting tensile strength of the specimens was calculated according the correspondent ASTM standard.

Two P-waves ultrasonic transducers were installed in contact with the upper and lower load bearings so to perform continuous measurements during the test under loading conditions. These measurements should be useful to verify the progressive changes sustained by the samples during the tests. Furthermore, ultrasonic velocities have been measured for all the samples under no applied load both in dry and wet conditions. For the case of no load conditions both the compression and shear waves have been measured, whereas only P waves have been measured during the loading tests.

Concerning the porosity, total and interconnected, this has been evaluated for all the samples by different techniques (Saturation-drying, Hg-porosimetry, pycnometer, X-ray Micro and Medical CT, thin sections). The resulting values and their trend with the alteration grade are reported in the companion paper (this conference, Pola et al.).

4 RESULTS

All test results demonstrate an evident relationship with the alteration grade (SL1 to SL5, see also Figure 1) as attributed to each sample on the basis of the mineralogical and petrographycal description.

Failure mechanism in fresh samples (SL1) follow the sub-parallel arrangement of phenocrysts (prevalently plagioclase). Dominant failure mechanism for SL2 samples is the reactivation of micro-fractures where the oxidation process is concentrated. On the other hand, failure mechanism

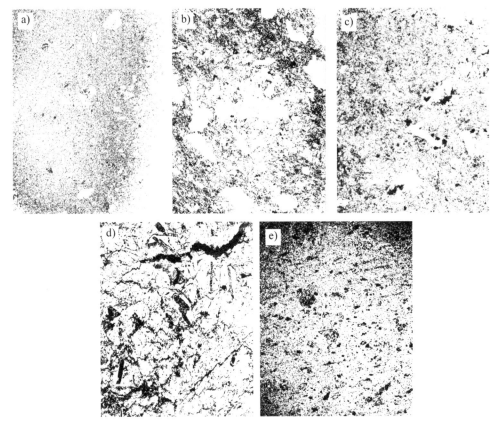

Figure 1. Representation of porosity in black color from thin sections (scale: 3 cm in width) for different alteration grades. Alteration grade increases from a) to e). The matrix material is shown in white.

changes considerably in samples SL3 and SL4. This could be controlled by the frequency and distribution of interconnected pores. In the most altered rocks (SL5) failure mechanism is typical of an homogeneous material. This could be the result of the almost complete replacement of the matrix crystals by amorphous minerals and through argilization and silicization processes.

The geometry of the stress vs strain curves (Figure 2) shows some characteristic features:

− low to very low gradient in the initial part,
− progressive decrease of slope with increasing alteration grade,
− multiple peaked rise limb for the more altered samples (SL4 and SL5),
− sharp peak to smoothed and/or irregular peak passing from low to high alteration grades,
− fragile to less fragile peak/post peak behavior from fresh to altered lavas,
− observable post peak behavior for grade 4 samples.

To summarize the results from the various tests the unconfined compressive strength, the tensile strength and the Et_{50} are plotted against the alteration grade.

The dispersion of the results in terms both of uniaxial compressive strength and tensile strength is relatively small (see Figure 3a, b). In particular dispersion is small (± 5–10 MPa) when the uniaxial compressive strength values are considered for all the alteration grades and sample sizes (see Figure 3a).

The values of the measured P-wave velocity under different conditions follow the same type of trend both for loaded (in uniaxial compression tests) and unloaded samples. Figure 3c shows also the increase in the P-wave velocity observed during uniaxial compression tests for samples with alteration grade from 3 to 5.

Figure 4 presents directly the results in terms of the dependence on the total porosity values, as obtained from the micro CT analyses. An exponential decay of the properties is observed when average

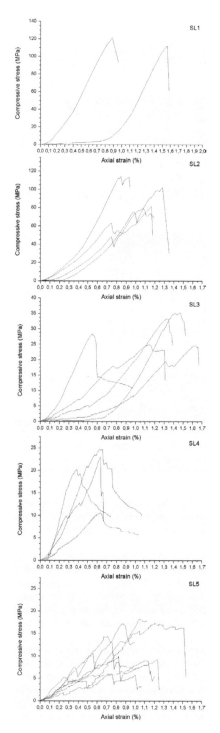

Figure 2. Stress vs displacement curves for the tested samples presented according to their alteration grade (1 to 5).

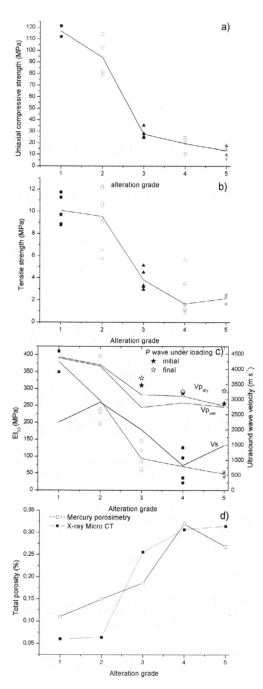

Figure 3. Test results plotted with respect to the alteration grade a) uniaxial compressive b) indirect tensile tests, c) Elastic modulus, Et_{50} (various symbols) and ultrasonic P-wave (dry and wet) and S-wave velocity (lines) under unloaded and loaded conditions. The filled and empty stars represent the initial and final P wave velocity values during compression tests, respectively; d) total porosity.

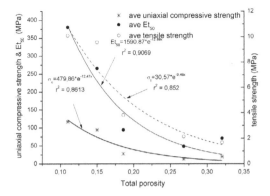

Figure 4. Dependency of elastic modulus Et_{50}, uniaxial compressive and tensile strength on total porosity as from Micro CT image analysis. Plotted data are average (ave) values. Data shows higher variation of modulus at third grade of alteration (SL3 sample). Best fitting exponential relationships are shown together with their coefficient of determination (R^2).

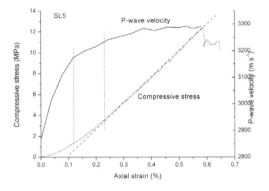

Figure 5. Example of the relationship between recorded P-wave velocity values at different strain levels and the stress vs strain curve for one sample characterized by alteration grade 5.

values for the uniaxial compressive strength, tensile strength and Et_{50} are compared with the average porosity value. The computed best fitting exponential curves are characterized by values of the coefficient of determination, R^2, that are relatively high and ranging between 0.85 and 0.90.

These relationships follow a general trend presented in the technical literature for other lithologies at varying degree of weathering/degradation. The most rapid rate of decrease is observed for the Et_{50} and the compressive strength values between values of the total porosity ranging from 10 to 20%.

Finally, we show in Figure 5 the relationship between the recorded P wave velocity and strain. As we can see there is an initial step increase in P wave velocity that ends for a strain level around 0.1. This behavior corresponds to the more concave upward part of the stress vs strain curve. After this, the P wave velocity increases at a much lower rate till a plateau or slightly increasing trend part in the P wave vs strain curve. The point where this plateau begins is approximately coincident with the onset of the rectilinear part of the stress vs strain curve. In some cases a slight decreasing trend is observed just before of the peak stress level.

5 CONCLUSIONS

In this paper, we have presented the initial results regarding the physical mechanical behavior of a lava affected by different degrees of alteration. Alteration induces strong changes in physical properties and behavior of any rock material, and this is particularly true for volcanic rocks subjected to intense hydrothermal alteration.

All the lava specimens were characterized by their total porosity (see companion paper). The total porosity for the tuff specimens ranges between approximately 6% and 32% according to the Micro CT study.

The uniaxial compressive strength decreased non-linearly with increasing total porosity with a minimal to moderate data dispersion that is quite reduced for high alteration grade samples.

Uniaxial compressive strength decreases by almost an order of magnitude passing from fresh to totally weathered conditions. Tensile strength decreases of about 5 times to 1 order of magnitude.

In general, a strong decrease in properties is observed passing from grade 2 to grade 3 (15% to 20% increase in total porosity) whereas a very small increase in strength is sometimes observed when moving to grade 5 (total alteration). In any case this does not seem to be a systematic behavior.

If we compare both the tensile and compressive strength with respect to the fractal dimensions computed in the companion manuscript we observe an increase with fractal dimension. Again, this trend is not visible in the most altered sample (SL5).

Furthermore, results suggest a notable decrease in stiffness with the level of degradation. This could be associated to a most diffuse argilization of the matrix and the increasing frequency of medium to large pores. The increase in large pores frequency seems associated with the increasing alteration grade till grade 5. For this last grade we observe a decrease in large pores and a relative increase of very small pores. As mentioned in the companion paper, this trend is associated with the increasing content in amorphous minerals resulting from hydrothermal deposition.

Dispersion of the strength values, both compressive and tensile, is minimal for the case of maximum alteration grade. This seems reasonable if we consider that the porosity increases but generally by minor pore sizes and so excluding large imperfections and weaknesses that instead are more frequent at lower alteration grades (see companion paper).

The progressive increase in the P wave velocity that has been observed for some of the samples could be associated to the progressive compaction of the sample with a slight decrease in pore volumes.

Finally, Figure 6 shows the general trend followed by the various physical mechanical properties as a function of the pore size fractal dimension determined by Micro CT image analyses (see companion paper). In general, an increase in the values of compressive and tensile strength is observed for increasing fractal dimensions.

This trend is inverted for high fractal dimension values that in our case are relative to the more altered grade (5). This is probably the result of an increase in frequency of small pores with respect to large ones (see increase in total porosity in figure 6, and the decrease in pore size of figure 1) and a decrease in strength of the material due to argilization and silicization, and precipitation of amorphous minerals within large preexisting pores.

This study confirms the exceptional role played by alteration in degrading mechanical properties of volcanic rocks. As a consequence, this suggests that a correct and complete physical-mechanical modeling is required when stability problems involving volcanic and highly altered/weathered rocks are performed.

Presently, a more complete physical-mechanical characterization is under way for the described lithology so to develop a more complete geomechanical constitutive model to be introduced in numerical models. In particular, triaxial tests, isotropic compression and soft oedometer tests are currently being performed.

REFERENCES

Al-Harthi, A.A., Al-Amri, R.M. & Shehata, W.M. 1999. The porosity and engineering properties of vesicular basalt in Saudi Arabia. *Engineering Geology* 54, 313–320.

ASTM. Standard D 3148-02. *Standard test method for elastic moduli of intact rock core specimens in uniaxial compression*. In: Annual book of ASTM standards. West Conshohocken, PA: American Society of Testing Materials, 1993.

ASTM. Standard D 3967-95a. *Standard test method for splitting tensile strength of intact rock core specimens*. In: Annual book of ASTM standards. West Conshohocken, PA: American Society of Testing Materials; 2001.

Avar, B. & Hudyma, N. 2007. Observations on the influence of lithophysae on elastic (Young's) modulus and uniaxial compressive strength of Topopah Spring Tuff at Yucca Mountain, Nevada, USA. *International Journal of Rock Mechanics and Mining Sciences* 44, 2, 266–270.

Avar, B.B., Hudyma, N. & Karakouzian, M. 2003. Porosity dependence of the elastic modulus of lithophysae-rich tuff: numerical and experimental investigations. *International Journal of Rock Mechanics and Mining Sciences* 40, 919–928.

Aversa, S. & Evangelista, A. 1998. The Mechanical Behaviour of a Pyroclastic Rock: Yield Strength and "Destructuration" Effects. *Rock Mech. Rock Engng.*, 31 (1), 25–42.

Hudyma, N., Burcin, B. & Avar, Karakouzian, M. 2004. Compressive strength and failure modes of lithophysae-rich Topopah Spring Tuff specimens and analog models containing cavities. *Engineering Geology* 73 (2004), 179–190.

Luping, T.N. 1986. A study of the quantitative relationship between strength and pore size distribution of porous materials. *Cement and Concrete Research* 16, 87–96.

Nimick, F.B. 1988. Empirical relationships between porosity and the mechanical properties of tuff. In: Cudall, Sterling, Starfield (Eds.), *Questions in Rock Mechanics*. Balkema, Rotterdam, pp. 741–742.

Pola, A., Crosta, G.B., Agliardi, F., Fusi, N., Barberini, V., Galimberti, L. & De Ponti, E. 2010. Characterization and comparison of pore distribution in weathered volcanic rocks by different techniques. *3rd International Workshop in Volcanic Rock Mechanics*, Tenerife. Balkema, Rotterdam, pp. 79–86.

Price, R.H., Boyd, P.J., Noel, J.S. & Martin, R.J. 1994. Relationship between static and dynamic properties in welded and non welded tuff. In: Nelson, P.P., Laubach, S.E. (Eds.), *Rock Mechanics: Models and Measurements Challenges from Industry*, Proceedings of the First North American Rock Mechanics Symposium. A.A. Balkema, pp. 505–512.

Tillerson, J.R. & Nimick, F.B. 1984. Geoengineering properties of potential repository units at Yucca Mountain, southern Nevada. *Sandia National Labs Report*, SAND84-0221.

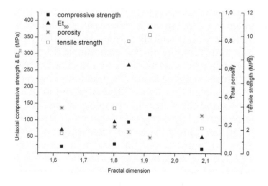

Figure 6. General trend observed for the change of the physical mechanical properties with the average fractal dimension as evaluated by porosity values determined by Micro CT analyses (see companion paper).

Volcanic Rock Mechanics – Olalla et al. (eds)
© 2010 Taylor & Francis Group, London, ISBN 978-0-415-58478-4

Characterization and comparison of pore distribution in weathered volcanic rocks by different techniques

A. Pola, G.B. Crosta, F. Agliardi, N. Fusi, V. Barberini & L. Galimberti
Dipartimento di Scienze Geologiche e Geotecnologie, Università di Milano–Bicocca, Milano, Italy

E. De Ponti
Struttura Complessa di Fisica Sanitaria, Azienda Ospedaliera San Gerardo, Monza, Italy

ABSTRACT: Volcanic rocks are widespread in different geological contexts and exhibit complex behaviors, from hard to extremely soft rocks, depending on mineralogy, porosity, and weathering. In particular, their mechanical properties are influenced by the size and shape of pores. We report the results of porosity characterization by different techniques, performed on lavas with different degrees of alteration, sampled in the Campi Flegrei area (Italy) and characterized through petrographic analyses. Bulk-specific weight measurements, water immersion and Hg-porosimetry gave total and interconnected porosity values. Analysis of thin sections provided 2D pore size and shape estimates and insight in pore relationships with rock matrix and weathering. X-ray Computer Tomography allowed complete 3D reconstruction of rock pores. Different image processing methods for data extraction and analysis have been used to develop a standard analysis procedure. Results in terms of measured porosity and pore size and shape distributions obtained by different techniques are discussed.

1 INTRODUCTION

Volcanic rocks are widespread in different geological contexts and exhibit complex behaviors ranging from hard to extremely soft rocks, depending on mineralogy, cementation, porosity, and alteration/weathering. Nevertheless, despite their influence on the geotechnical behavior of engineering structures and natural slopes, the relationships among physical and mechanical characteristics of these rocks are complex and still poorly understood. Mechanical properties of weathered/altered volcanic rocks are greatly influenced by the size and shape of pores. Recently, X-ray tomography and two-dimensional images have been widely used to characterize the structure and sizes of different particles (e.g. clasts, and minerals). Some works describe automated methods to extract grain characteristics (Butler et al., 2000), some others applied image methodologies to quantify heterogeneity in clastic rocks (Geiger et al., 2009). In this study we describe results of porosity characterization by different techniques: bulk weight and specific weight measurements, water immersion, mercury porosimetry, image analyses of thin sections, and X-ray tomography.

Weighting, water immersion and mercury porosimetry give information about total and interconnected porosity. Analysis of thin sections provides 2D estimates of porosity with data concerning pore size and distribution and their relationship with rock matrix and weathering.

X-ray tomography allows a complete 3D reconstruction of porosity distribution both before and after the performance of geomechanical tests. 3D reconstruction of pores from X-ray tomography has been performed for this study at different resolution (from 5 to 650 μm). Pore geometry, interconnection and distribution can be analyzed and introduced in numerical models or used to interpret rock behavior observed in situ and during laboratory tests.

Different methods of image analyses have been used in this study and they could be easily transported to other cases. The entire procedure includes: image processing (noise reduction and filtering), data extraction (grey-scale thresholding and particle separation), and data analysis (measurements and selection). Results are in terms of measured porosity, geometrical distribution, differences among the various adopted methods. Advantages and disadvantages are presented and discussed.

Quantification of porosity and its geometry (size, shape, distribution and frequency) is of primary importance to evaluate and interpret rock behavior observed in situ and during laboratory tests. In fact, physical-mechanical properties of a rock such as compressive strength and modulus of elasticity are greatly influenced by pore structure and are presented in a companion abstract/paper.

In the following, we discuss the adopted procedures to characterize total and effective porosity, pore types and commet about their applicability and accuracy.

2 STUDY AREA AND MATERIALS

Lava samples characterized by different degrees of alteration were collected from the Campi Flegrei (Italy). The Campi Flegrei provide an excellent setting to study the evolution of geometry and topology of pore network and grain size and shape in weathered volcanic rocks. Campi Flegrei are located near the city of Naples and the volcanic area belongs to the Campanian province, which represents the southernmost sector of the plioquaternary volcanic belt along the Italian peninsula (Washington, 1987). Campi Flegrei comprises the volcano of Solfatara where the sampled outcrops are situated. The Solfatara is located in Pozzuoli and is part of the last period of intense volcanic activity of Campi Flegrei occurred between 4.8 and 3.8 Kyr B.P. (Di Vito et al., 1999).

During field work a total of 10 block samples were collected. Sampling was focused on covering the entire range of alteration and each sample consists of a block weighting between 30 to 50 kg and with a minimum thickness of 15 cm to allow coring of samples at least 5.4 cm in diameter and 13 cm high. The operational procedures and initial description of alteration followed the BS standard methods (BS 5930, 1999). Samples were marked according to the origin and alteration grade. Mineralogical and petrographical changes in samples, which represent different and progressively weathering grades were examined by optical microscopy, moreover, all descriptions were completed by X-ray diffraction data.

2.1 Solfatara Lava (SL)

The lava rock mass is heavily fractured, joints are often infilled of loose material and are strongly altered suggesting that widespread fumarolic and thermal springs activity could be following preferential pathways. Discontinuities are often of very small size and sometimes not visible with a naked eye. Outcrop observations show that altered lava varies significantly on a short distance especially when approaching to fumarolic activity. We easily identified the effects produced by hydrothermal alteration: the rock fabric and texture in some places are completely lost making the in situ identification of the original rock very difficult. In the following we describe the petrographical characterization of the altered samples.

2.1.1 Fresh lava SL1

This represents the less altered sample of the studied sequence. Its major constituents are sodic plagioclase (oligoclase and andesine) and potassic feldspar with minor amounts of pyroxene and a small amount of biotite. The sample has a porphyritic texture in which predominantly euhedral phenocrysts are plagioclase with an average size of 3 mm. The matrix presents a sub-parallel arrangement of micro-pyroxenes and micro-plagioclases (pilotaxic texture). Two main types of alteration are observed: oxidation within the boundary of most minerals, affecting all biotite crystals, and as stains into the macles of plagioclases; argilization is presented around minerals as blurred stains.

2.1.2 Slightly weathered lava SL2

In the second grade of alteration the major constituents are sanidine with minor amount of plagioclase and a small amount of pyroxene and biotite. The sample has a trachytic texture in which predominantly euhedral phenocrysts are sanidine. The matrix is composed principally by micro-sanidines. In this sample we can observe argilization and oxidation along microfractures and within the boundary of minerals. Biotite is almost totally replaced.

2.1.3 Moderately weathered lava SL3

This sample represents the third grade of alteration. Its major constituents are potassic feldspar (sanidine) with minor amount of sodic plagioclase and a small amount of pyroxene; biotite is almost missing. Relatively large crystals of sanidine (2 mm of ave. size) completely bounded by its characteristic faces are surrounded by a major amount of micro-sanidines and micro-plagioclses (porphyritic texture). In this sample oxidation increases affecting all crystals boundaries whereas biotite and pyroxene are almost completely replaced (Figure 1). Argilization is presented into the matrix and all around minerals as blurred stains.

2.1.4 Highly weathered lava SL4

This sample represents the fourth grade of alteration. Although all minerals are altered, sanidines prevail over pyroxenes and plagioclases. X-ray power diffraction shows that alunite (derived from acid alteration of potassic feldspar) is very abundant. Matrix is totally replaced by argilization and a new process of silicification can be observed into the potassic feldspar and within the matrix. Small silica-amorphous minerals appear between the grains.

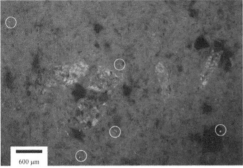

Figure 1. Mineralogy and fabric characteristics of weathered lava at different degrees. Upper photo: sample SL3. Pyroxene (image centre) is almost totally replaced by oxides. Argilization is presented as blurred stain all around matrix. Lower photo: sample SL5. Complete alteration of all minerals is observed and only some pyroxenes can be recognized from their partially-distorted geometry (right hand side). Matrix is totally replaced by argilization and small silica-amorphous minerals can be observed (within circles).

2.1.5 *Totally weathered lava SL5*

Complete alteration of all minerals occurs at this alteration grade, but some pyroxenes could be recognized from its partially-distorted geometry. Matrix is totally replaced by argilization and small silica-amorphous minerals can be observed (Figure 1).

3 METHODS

Aim of this study is to tests different methods to quantify porosity of volcanic materials. Effective porosity (n_o) was initially obtained following the standard test procedure by ISRM (1972). Procedure consists in calculating dry density (γ_{dry}) and saturated density (γ_w) of cylindrical core samples through measure of their volume and weight. Effective porosity can be obtained by the following equation: $n = e/1 + e$, where e is the void ratio, $e = (G\gamma_w/\gamma_{dry}) - 1$, and G the apparent specific gravity of the material. In order to measure the degree of saturation and water absorption, six cylindrical specimens (53 mm × 130 mm) from ten different volcanic rocks were submerged in distilled water under a constant air vacuum pressure. Measurements average values of the experimental data are summarized in Table 1.

Porosity (n) was also obtained indirectly by pycnometer tests, following the standard test procedures described in Germaine & Germaine (2009). The method defines the specific gravity of the material as the ratio of the mass of a given volume of soil particles to the mass of an equal volume of distilled water ($G_s = \gamma_s/\gamma_w$) where γ_s is mass density of solids. Then porosity value is determined through the phase relationships, which include water content, density, void ratio (or porosity), and saturation (see Bardet 1997).

Mercury intrusion porosimetry is a more advanced and frequently adopted technique. It consists in applying a set of increasing pressure steps to a dry specimen and measuring the corresponding mercury intrusion volume. The pore size intruded at each step is determined by the pressure applied to force mercury into a pore against the opposing force of the liquid's surface tension. Pore size frequency and connected porosity, and mean pore size were obtained by a Pascal 140/240 Thermo-Fisher mercury porosimeter.

Two-dimensional digital image analysis is another way to estimate and quantify porosity (n). In this study, digital images were obtained by scanning 13 thin sections using a photogrammetric scanner at 600 dpi resolution (Figure 2). In order to distinguish pore spaces and improve the contrast with transparent minerals, the images were acquired with a color background. Pores

Table 1. Summary of all porosity values from different techniques for different alteration grades.

Sample and alteration grade	Total porosity (n)			Effective porosity (n_o)	
	Pycnom	X-ray	Thin_ sect	Hg-por.	ISRM
Fresh (SL1)	0.146	0.060	0.128	0.110	0.060
Slightly (SL2)	0.151	0.064	0.169	0.150	0.054
Moderately (SL3)	0.285	0.256	0.185	0.186	0.023
Highly (SL4)	0.331	0.307	0.150	0.320	0.210
Totally (SL5)	0.303	0.315	0.095	0.268	0.193

identification, definition and differentiation from minerals or other clasts (image segmentation) was carried out by using Adobe Photoshop software. Pores have been isolated by using a color-scheme selection tool. This step for pore isolation was supported and confirmed by thin section analysis and petrographical description (mineral constituents, rock texture, crystal size range, and porosity types). Final processing included a more accurate image segmentation and image calibration performed by the ImageJ code (NIH) (Figure 2). Once each pore has been identified, porosity shape parameters (location, perimeter, surface area, circularity and aspect ratio) have been automatically extracted.

Image analysis on thin sections results in a 2D description of pore distribution and it is important to derive a 3D porosity description. Different techniques to characterize spatial structure and sizes of randomly distributed particles/voids from two-dimensional sections are available and can result

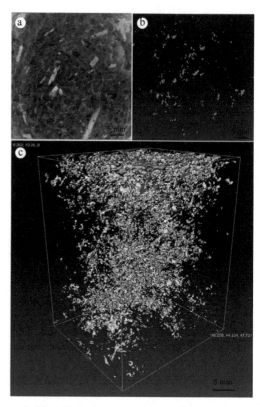

Figure 3. (a) Example of a filtered image reconstructed from a tomographic slice through sample SL4. (b) pre distribution obtained from thresholding of image (a). (c) 3D pore-system reconstruction (Avizo 6).

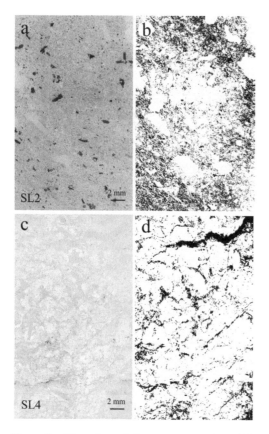

Figure 2. Illustration of the two steps performed in image processing of thin sections by ImageJ (NIH). (a, c) original scanned images of thin sections of SL1 and SL4 samples, respectively. (b, d) images as obtained after the segmentation process.

in different values. Stereological methods could allow to determine the number of particles for each particular size and shape enclosed by a given volume from the observed number of particle cuts, of a particular size and shape, on a randomly oriented cross-section through the volume (Sahagian and Proussevitch, 1998). Stereological conversion techniques have been used for geosciences fields (Higgins 2000, and Morgan and Jerram 2006).

Higgins (2000) have applied stereology to quantify textural aspects of igneous rocks (e.g. crystals sizes and distribution) by processing a sequence of thin-section images. We followed a simpler approach proposed by Farmer et al. (1991) to compute the 3D porosity by 2D shapes by applying the two relationships: $V = Ab$ and $V = [A(a + b)/2]$ where A is pore area, a and b are the major and minor axes, respectively. Because of the subspherical geometry a and b are similar and the results of the two results are comparable.

A more complete three dimensional reconstruction of rock sample structure and porosity characteristics can be obtained from a set of contiguous two dimensional X-ray computerized tomography images (CT) at different resolutions (5 to 650 µm). X-ray CT reconstructs internal images based on the distribution of the X-ray linear absorption coefficient deduced from the projection of X-rays through a sample (Ohtani et al., 2000). X-ray CT images have been obtained by means of a GE D-600 medical CT hybrid scanner and a BIR Actis 130/150 Micro CT/DR system. The advantages of medical scanner include high image acquisition velocity and configuration versatility. On the other side, a Micro CT/DR system allows the acquisition of images at higher resolution (40–60 µm) for small samples (<2 cm). Image processing, data extraction and data analysis were elaborated by a 3D visualization software (Avizo 6) (Figure 3). The results of image processing are: a geometrical, morphological and topological description of the features inside the investigated volume, the identification and isolation of elements of interest (e.g. porosity, minerals or particles and voids distribution). In any case, the quality of final results depends on a series of noise reduction, filtering, thresholding, and particle separation steps (Gualda and Rivers, 2006; Ketcham, 2005).

4 RESULTS

The adopted classification scheme, based on mineralogical and structural characteristics, allowed a preliminary classification of the collected samples in base of their grade of alteration from fresh to slightly, moderately and highly altered lava (SL1, SL2, SL3, SL4, SL5, respectively).

Effective porosity was obtained from bulk-specific weight measurements, water immersion and Hg-porosimetry (Table 1 and Figure 4). According to porosimetry results, a direct relationship between the effective porosity and alteration/weathering degree of samples exists. On the other hand, effective porosity from bulk-specific weight measurements seems to have no clear relationship with weathering grade. The reason could be the percentage and sizes of interconnected pores and fractures contained in each samples. It means that porous system connectivity do not increase progressively with weathering degrees. On the contrary, a strong connectivity increase is found between second and third weathering grade. The decrease in the computed porosity by imbibition could also be the result of a decrease in size of the pores with the increased alteration and then with a consequent difficulty in their saturation under low air vacuum conditions.

Total porosity was obtained from thin sections, X-ray tomographic image analysis and pycnometer test method (Figure 5). Image analysis represents a rapid and precise method to obtain pore area, volume, shape, frequency, and spatial distribution.

Results from X-ray tomography and pycnometer tests reveal that total porosity increase progressively with weathering grade with some minor changes for the fifth alteration grade.

Differences in the results from Micro-CT and Medical CT can result from the different resolution and from the consequent averaging. In particular, results reported in Figure 5 are obtained starting from an evaluated density values from which total porosity is computed by knowing the specific weight of the solid phase.

This could also depend on the required manual thresholding process and image sharpening, which

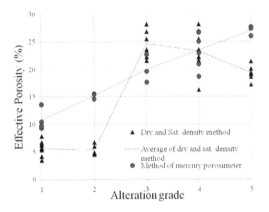

Figure 4. Effective porosity obtained from bulk specific weight measurements, water immersion and mercury porosimeter.

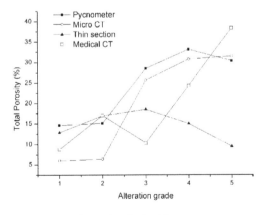

Figure 5. Total porosity obtained from pycnometer test, thin section analysis and X-ray tomography by Medical scanner and Micro CT. Porosity from Medical scanner images has been obtained from bulk density estimates and specific weight (assumed constant value: 2.78).

in turn could depend on the modes of acquisition and the heterogeneous nature of the rock. Pores size distributions shows a clear linear trend on a log-log plot (Figure 6) allowing to determine the exponent of the best fitting power law relationship. Pore shape description, which includes circularity and aspect ratio, reveals that pores are slightly elongated with aspect ratio values ranging mainly from 1.5 to 3.5.

To assess the influence of alteration grade on pore size distribution we applied the fractal dimen-

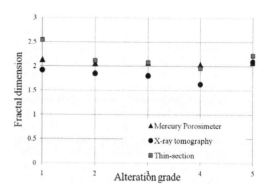

Figure 7. Fractal dimension calculated from pores size distributions obtained by different techniques (see legend).

sion method (Turcotte, 1992; Crosta et al., 2007) to pore distribution values obtained by mercury porosimeter, X-ray tomography, and thin section analysis.

The values of the fractal dimension range between 1.62 and 2.54, and they decrease progressively from intact rocks (SL1) to the fourth grade (SL4) of alteration (Figure 7) suggesting an increase in large pore frequency. The fractal dimension corresponding to the most altered rock (SL5) deviate from this trend.

An increment in the fractal dimension is observed and suggests an increment of the relative frequency of smaller pores (see also Figure 6). This could be due to the fact that new small pores have been generated by hydrothermal processes, as well as, big pores have been filled by new minerals (amorphous silica and clay minerals) (Figure 1). This hypothesis is supported by thin section observations and XRD analyses showing a strong increase in amorphous and semi-amorphous minerals.

5 CONCLUSIONS

Five different procedures to quantify porosity of altered/weathered volcanic materials have been implemented and compared. These procedures allow the definition and quantification of total and effective porosity, spatial pore structure and pore size distribution. Results demonstrate the relationship between alteration grade and porosity distribution in the same rock lithology.

The most relevant conclusions are as follow:

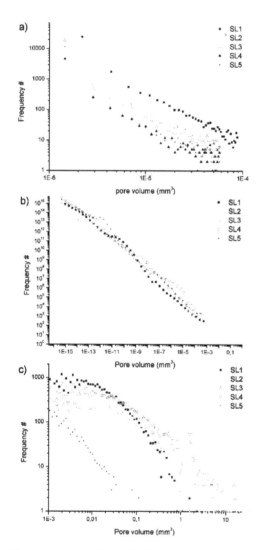

Figure 6. Pore volume distributions obtained from (a) thin section analysis, (b) mercury porosimetry, (c) X-ray tomography.

- a significant relationship exists between porosity and alteration/weathering grade for all the samples. Total porosity increases with grade.

- Pore frequency distribution suggests a fractal behavior of porosity with values between 1.62 and 2.54. This implies that an increase in alteration, from fresh to highly altered lava, is linked to a relative increase in frequency of large pores.
- Totally altered rock is characterized by an increase in small pore frequency (see Figure 6) with respect to large ones, resulting in increasing fractal dimension (see Figure 7). This could be the result of hydrothermal deposition of amorphous minerals within large pores and more alteration.
- Thin section image analysis allow to describe pore shape characteristics on a plane (e.g. area, perimeter, circularity, and roughness). Thin section preparation, orientation and size could influence the final estimates.
- Pycnometer results are generally slightly higher than X-ray tomography data. This could result by forced intrusion and damaging or opening of small fractures.
- Connectivity and effective porosity estimates can be obtained from bulk specific weight and mercury porosimeter measurements.
- 3D values of porosity can be computed starting from 2D data obtained by means of analysis of images of thin sections (e.g. see Farmer et al., 1991). The final value is controlled by the adopted transformation relationship.
- X-ray tomography is the fastest and more precise techniques to obtain 3D textural information. This method allows to measure pore size and pore distribution.
- Porosity values computed from X-ray medical data density values for different slices are in this case influenced by the coarser resolution with respect to micro CT and the adopted specific weight value of the solid phase used in the calculations. At the same time medical CT allows to evaluate the total porosity on large rock core samples so it is able to provide a porosity value at a larger scale.

REFERENCES

BS5930 1999. *Code of practice for site investigations.* British Standards Institution, London.

Bardet, J.P. 1997. *Experimental Soil Mechanics.* 1st ed. Prentice Hall. 583 p.

Butler, J., Lane, B. Stuart, N. & Chandler, J.H. 2001. Automated extraction of grain-size data from gravel surfaces using digital image processing. *Journal of Hydraulic Research*, vol. 39, no. 4.

Crosta, G.B., Frattini, P. & Fusi, N. 2007. Fragmentation in Val Pola rock avalanche, Iatalian Alps. *Journal of Geophysical research.* 112: F01006, doi:10.1029/2005 JF000455.

Di Vito, M.A., Isaia, R., Orsi, G., Southon, J., deVita, S., D'Antonio, M., Pappalardo, L. & Piochi, M. 1999. Volcanic and deformational history of the Campi Flegrei caldera in the past 12ka. *Journal of Volcanology and Geothermal Research* 91: 221–246.

Farmer, I.W., Kemeny, J.M. & McDoniel. 1991. Analysis of rock fragmentation in bench blasting using digital image processing. In: *Proc. Int Cong. Rock Mech.* 2: 1037–1042.

Geiger, J., Hunyadfalvi, Z. & Bogner, P. 2009. Analysis of small-scale heterogeneity in clastic rocks by using computerized x-ray tomography (CT). *Engineering Geology* 103: 112–118.

Germaine, J.T. & Germaine, A.V. 2009. *Geotechnical Laboratory Measurements for Engineers.* John Wiley & Sons, Inc., Hoboken, New Jersey. 351 p.

Gillot, P.Y., Chiesa, S., Pasquale, G. & Vezzoli, L. 1982. 33000 yr K/Ar dating of the volcano-tectonic horst of the Isle of Ischia, Gulf of Naples. *Nature* 299: 242–245.

Gualda, G. & Rivers, M. 2006. Quantitative 3D petrography using x-ray tomography: Application to Bishop Tuff pumice clasts. *Journal of Volcanology and Geothermal Research* 154: 48–62.

Higgins, M.D. 2000. *Measurement of crystal size distributions.* American Mineralogist. vol. 85, 1105–1116.

ISRM Committee on Laboratory Tests 1972. Suggested methods for determining water content, porosity, density, absorption and related properties and swelling and slake-durability index properties. Document no. 2: 1–36.

Ketcham, R.A. 2005. Computational methods for quantitative analysis of three-dimensional features in geological specimens. *Geosphere* 1: 32–41.

Metz, F. & Fnöfel, D. 1992. Systematic mercury porosimetry investigations on sandstones. *Materials and Structures.* 25: 127–136.

Morgan, D.J., Jerram, D.A. 2006. *On estimating crystal shape for crystal size distribution analysis. Journal of Hydraulic.* Journal of Volcanology and Geothermal Research. 154: 1–7.

Ohtani, T., Nakashima, Y. & Muraoka, H. 2000. Three-dimensional miarolitic cavity distribution in the Kakkonda granite from borehole WD-1ausing X-ray computerized tomography. *Engineering Geology* 56: 1–9.

Orsi, G., De Vita, S. & di Vito, M. 1996. The restless, resurgent Campi Flegrei nested caldera (Italy): constraints on its evolution and configuration. *Journal of Volcanology and Geothermal Research* 74: 179–214.

Sahagian, D.L. & Proussevitch, A. 1998. *3D particle size distributions from 2D observations: stereology for natural applications.* Journal of Volcanology and Geothermal Research. 84: 173–196.

Turcotte, D.L. 1992. *Fractal and Chaos in Geology and Geophysics.* Cambridge University Press, New York. 412 p.

Washington, H.S. 1906. The Roman comagmatic region. *Carnegie Institution of Washington*, publication no. 57: 199 p.

Volcanic Rock Mechanics – Olalla et al. (eds)
© *2010 Taylor & Francis Group, London, ISBN 978-0-415-58478-4*

Lightweight aggregate and lightweight concrete and its application in the improvement of the thermal properties of volcanic lightweight aggregate concrete blocks from Canary Islands

E. Rodríguez Cadenas & A. García Santos
Innovative and Sustainable Building Techniques Research Group,
Department of Architectural Building and Construction, Superior Technical School of Architecture (UPM),
Madrid, Spain

ABSTRACT: Concrete blocks with volcanic aggregates currently produced in the Canary Islands (BHIC) have a high coefficient of thermal conductivity. The application of Spanish Building and Construction Regulations, known as CTE, leads to the need for multilayer construction systems in external building walls. These systems are less efficient from an economic and environmental point of view. This paper focus on the improvement of thermal properties of the BHIC, so that the external building walls can be executed in the islands using single-leaf masonry without having to add thermal insulation.

1 INTRODUCTION

Lightweight concrete has its own technology; the standard regulations for composition, mixture and compaction of normal concrete do not apply to lightweight concrete. Therefore, there is not an ideal granular composition for lightweight concrete, as in normal density concrete. Accordingly, an experimental study has been undertaken in order to better understand the influence of the granular fraction of different volcanic aggregates on the thermal properties of the resulting lightweight concrete.

Concrete blocks are frequently used in the Canary Islands owing to clay shortage in this volcanic region and they are manufactured with volcanic lightweight aggregate concrete. The reference mixture is composed of volcanic materials: lapilli aggregate and volcanic slag, cement, sand and water. This lightweight concrete has a dry absolute density of 1500–1900 kg/m³: the first value is for the island of Tenerife and the second one, for the island of Gran Canaria. All the available information on lightweight concrete indicates that the thermal properties of these mixtures can be improved. This study examines, on the one hand, the influence of volcanic material additions such as the expanded perlite aggregate (APE 0–6 mm) and the natural pozzolanic aggregate (APN 0–5 mm) and, on the other hand, the influence of different granular fractions of volcanic lapilli aggregate (AEV 0–12 mm and 0–8 mm) and a thermal mortar

called Thermocal™ in the physical properties of lightweight concrete.

Mouli & Khelafi (2008) have demonstrated the convenience of using pre-soaked aggregates, in the case of aggregates with a high water absorption coefficient like APE and APN, in order to guarantee that the water in the mixture will not be absorbed by the aggregates. Due to the absorption of these aggregates, the water absorbed during mixing helps to hydrate the cement particles in the interfacial zone, which also increases the bond between the aggregate and the mortar phase. Guigou Fernández (1997) studied lightweight concrete with canarian volcanic aggregates. His study showed that there are two methods to decrease the density of lightweight concrete: to eliminate the fine aggregate, or to use only one fraction of aggregate (monogranular). According to Hummel (1966) the lightweight concrete has its own technology. This author emphasizes the relationship between porosity, density and the thermal insulation properties of the materials. Among the procedures he proposes in order to achieve low apparent densities, he appoints the BHIC lightweight concrete, referred to as "Lightweight concrete with clustered pores and pores in the granules", which consists of the employment of porous aggregates bound with a small amount of blending material so that interstitial pores are formed between large-sized granules. The blending of the volcanic aggregates is produced through the welding of the cement paste in the meeting points between the aggregates. Hence the importance of the aggregate-cement interfaces in these kinds of concretes and the water/cement ratio in the strength properties.

Topçu (2008) studied the use of APE both as an aggregate and as a cement addition in order to improve density and strength in lightweight concrete blocks. The perlite powder has a pozzolanic effect on concrete mixtures despite the potential alkali-aggregate reaction.

2 EXPERIMENTAL STUDY

2.1 Materials

2.1.1 Portland cement with pozzolan

For this study portland cement with pozzolanic additions type CEM II/ A-P 32,5R was used. This is the most commonly used cement in masonry production in the Canary Islands. This cement has 80–94% of clinker and 6–20% of natural pozzolan in accordance with UNE EN 197-1:2000 regulations. This type of cement reduces hydraulic retraction and heat hydration in lightweight concrete mixtures, and therefore improves the concrete block surface. The use of cement with less compressive strength entails an increase in the amount of cement needed to get a certain mechanical strength from the concrete block, which improves the bathability of the aggregates. This is important in lightweight concretes with volcanic slag aggregates (VSA) that are produced using dry mixtures. Soria (1972) demonstrated that pozzolan contribution is greater in poor blended and in high cement-water ratio concrete mixtures, always with a moist cure. This is why, in the manufacture of BHIC, cement with pozzolanic additions is the most appropriate for mixtures having a small quantity of cement.

2.1.2 Thermocal™

Thermocal™ (THER) is an ecologic mortar which acts as a thermal and acoustic corrector and is used as a covering for external building walls. Its main property is its low density and resistance to water permeability. Tables 1 and 2 show the physical and mechanical properties of the Thermocal™ used in this study.

2.1.3 Mixture water

The mixture water employed in this study is in accordance with the Spanish Structural Concrete Code (EHE-08). The water is the same as the one used in the precast concrete industry where the tests were carried out.

2.1.4 Natural yellow sand

In geological terms, sand is a granular material with a variable size between 0.063 and 4 mm. It is mainly composed of silica, generally in quartz

Table 1. Physics characteristics of the Thermocal™ mortar (GP-CSIII-W2-T1).

Physical properties	Units	Value
λ Thermic conductivity	W/m · K	0.068
Steam permeability	ml/cm^2 48 h	<8.46
Water absorption	kg/mm^2 · min · 0,5	0.2
Specific heat (MDSC)	kJ/ kg · K	0.823
Powder density	kg/m^3	350
Mixer water	%	54
Consistence	mm	135
Fire resistance	class	A1

Table 2. Mechanical characteristics of the Thermocal™ mortar (GP-CSIII-W2-T1).

Mechanical properties	Units	Value
Compressive strength (14 days)	N/mm^2	2.4
Compressive strength (28 days)	N/mm^2	4.3
Flexural strength (14 days)	N/mm^2	0.1
Flexural strength (28 days)	N/mm^2	1
Adhesion	N/mm^2	>0.1

Table 3. Physical test on 0–4 mm natural yellow sand.

Physical properties	Units	Value
Fines percentage	%	18.3
Friability coefficient	F.A	30
Sand equivalent	Index	29
Absorption	%	0.8
Dry real density	g/cm^3	2.7
Water content	%	3.4

form. However, its composition varies depending on the local rock conditions. In general, the fine aggregate or sand will not contain clay, lime, alkali, organic material and other harmful substances, in accordance with regulations. The sand or fine aggregate used in the study is a polygenic and partially calcareous sand. A granular test was made in accordance with UNE-EN 933-1 and UNE EN 933-2:1996 regulations. See Table 3.

2.1.5 Volcanic lapilli aggregate (AEV 0–12 mm and AEV 0–8 mm)

Volcanic slag aggregate (AEV), also called "lapilli", are small, dark-coloured pyroclastic fragments, with 8.5% to 33% of pores. This high porosity is the cause of their low density. They come from basic magma and they are usually related to strombolian-type eruptions (Lomoschitz et al. 2006). The lack of reactive silica eliminates the problems of alkali-silica reactivity with portland cement. They are sharp edged, angular and typically reddish to black in colour, mostly due to their high iron content (Demirdag & Gündüz. 2008). 85% of these

Table 4. Physical test for the AEV 0–12 mm and AEV 0–8 mm fraction.

Physical test	Units	0–12 mm	0–8 mm
Specific gravity	g/cm^3	2.6	2.6
Bulk density	kg/dm^3	963	894

Table 5. Chemical composition of the APN 0–5 mm.

Chemical composition (%)	Value
SiO$_2$	56.12
Al$_2$O$_3$	16.54
Fe$_2$O$_3$	4.99
CaO	2.61
MgO	1.11
SO$_3$	0.2
P.F.	7.2
R.I.	50.57
Reactive SiO$_2$	32.8

particles fit within the interval of 2–64 mm, and the rest are minor particles. For this experimental study VSA in the 0–12 mm and 0–8 mm size fractions were used. Granular tests were made in accordance with UNE-EN 933-1 and UNE EN 933-2:1996 regulations. Table 4 shows the results of the density tests for both kinds of fractions.

2.1.6 *Natural pozzolanic aggregate (APN 0–5 mm)*

Natural pozzolanic aggregate (APN) is a type of aluminosilicate mineral containing large quantities of reactive SiO$_2$ and Al$_2$O$_3$. It is widely used in the cement industry in the Canary Islands as an active cement admixture. The properties of pozzolan materials depend on their chemical composition and internal structure. Normally these materials do not have blending properties by themselves, but when mixed in finer sizes with lime or cement and in presence of water, the active oxides will gradually have a secondary reaction with Ca(OH) and will form basic calcium hydrosilicates in the blended paste, as well as the so-called pozzolanic reaction, forming low basic calcium hydrosilicates. As a result, there will be an increase not only in the quality of the hydrates but in the quantity as well, and the strength of the cement paste and other properties can be greatly improved (Xincheng Pu. 1999).

In this study APN 0–5 mm size fraction from Arguineguin (Gran Canaria) was used. The physical properties of APN can be summarized in two parameters: the relative density of 2.35 gr/cm^3 and the high moisture content of 13–16%. It is a rather weak rock, with a low compressive strength of about 40 kp/cm^2. Tables 5 and 6 show the chemical

Table 6. Mechanical characteristics of the APN 0–5 mm.

Mechanical properties	Units	Value
Compressive strength (7 days)	N/mm^2	3.3
Compressive strength (28 days)	N/mm^2	7.6
Compressive strength (90 days)	N/mm^2	10.1
Compressive strength (180 days)	N/mm^2	10.9

Table 7. Physics characteristics of the APE 0–6 mm.

Physical properties	Units	Value
λ Thermic conductivity	W/m · K	0.05
Ignition less	%	<2
Balance of moisture	mm	1.5
Specific gravity	g/cm^3	2.2
Bulk density	kg/dm^3	90 ± 20%Vol
PH		7
Fire reaction	class	A1

Table 8. Chemical composition of the APE 0–6 mm.

Chemical composition (%)	Value
SiO$_2$	74,0 ± 2,0
Al$_2$O$_3$	13,0 ± 1,5
Fe$_2$O$_3$	1,0 ± 0,3
CaO	1,4 ± 0,4
MgO	0,6 ± 0,3
K$_2$O	3,5 ± 1,5
Na$_2$O	4,5 ± 1,0
H$_2$O combined	<3

and mechanical properties of APN used in this study. The granular tests were made in accordance with UNE-EN 933-1 and UNE EN 933-2:1996 regulations.

2.1.7 *Expanded perlite aggregate (APE 0–6 mm)*

Perlite is a siliceous volcanic rock. Upon rapid heating, above 871°C, this aggregate transforms into a low-density cellular material increasing up to twenty times its original volume. This expansion is due to the 2–6% of water in the natural perlite aggregate. This material is referred to as *expanded perlite*. Expanded perlite aggregate (APE) is defined as a light granular thermal insulation material, with a white or grayish colour The physical and mechanical properties of 0–6 mm APE used in this experimental study are shown in Table 7 and in Table 8. The Spanish Concrete Structural Regulations (EHE-08) recommend that the density for each granular fraction and its corresponding volume should be determined, and they also exempt APE from fulfilling water absorption limits.

2.2 Experimental method

2.2.1 Mixture design

Several mixture batches were cast and tested using the selected materials. The addition of natural pozzolanic aggregate (APN) to the 0–5 mm size fraction as a substitute for 10% cement by volume was analyzed in order to improve the mechanical properties of the blends. The APE, AEV and the APN have a high absorption coefficient. Steiger & Hurd (1978) observed that when the concrete density increased 1% due to water absorption, the thermal conductivity of concrete decreased up to 5%. The consequences of adding Thermocal™ as a substitute for Portlant cement in 10%, 20% and 30% by volume were also analyzed in order to improve the absorption coefficient.

The Test Plan was divided into two stages, and the results were then compared to the manufacture or reference mixture (Table 9), which was taken from a precast concrete manufacturing industry located in Gran Canaria. In Stage 1 the initial mixture was adjusted, and the influence of the compaction method and the water-cement ratio in eight mixtures was analyzed.

In order to correlate the data obtained in manufactured in the block machine, a vibrocompacting device was designed and built. The machine was used to mould and vibrocompact the 30 × 30 cm square plaque shaped specimens of different widths, as well as to compact the 40 × 40 × 160 mm prismatic specimens. This device was also used during the Stage 2 of the test plan. The main conclusions of this stage 1 are that the vibrocompacting method is the most appropriate for reproducing in the laboratory a lightweight concrete with similar characteristics as the reference lightweight concrete, and that a 1.2 W/C ratio is the most appropriate to produce in the laboratory lightweight concrete mixtures with similar density characteristics as the reference lightweight concrete. Fifteen batches of lightweight concrete were analyzed during Stage 2 after having incorporated the different selected materials by weight and by volume. The water: cement ratio was 1.2 and the compaction was made using the vibrocompacting machine.

Table 9. Lightweight concrete manufacture (or reference) batch from the manufacture of BHIC.

Materials	kg/m³
Cement CEM II/A-P 32.5	240
Water	96
Natural yellow sand 0–4 mm	400
Volcanic Lapilli aggregate 0–12 mm	1700
W/C ratio	0.4

2.2.2 Specimens preparation

Twenty three blends (23) and 435 specimens were cast altogether, making a total of 1753 tests. Between 12 and 18 specimens (40 × 40 × 160 mm) were cast for the density and capillary absorption tests, and three (3) 300 × 300 mm plaque shaped specimens with widths varying from 50 mm to 70 mm were used in the density, water absorption and thermal conductivity tests. For each mixture the specimens were cast and cured in laboratory conditions until the moment of testing (7 and 28 days). After the curing, the specimens were tested in laboratory conditions in accordance with UNE EN 772-11:2001, UNE EN 12350-6:2006, UNE EN 12390-7:2001 and UNE EN 12664:2001 regulations. The specimens were produced in a vibromoulding machine designed to reproduce the same vibration conditions of a block machine.

3 RESULTS AND DISCUSSION

3.1 Density

Figures 1 and 2 show that as we add new materials to the mixture by weight, the density values decrease about 7% in the 4 × 4 × 16 cm specimens as compared to those obtained for the same batches in the 30 × 30 × 5 cm specimens.

The +10%APE 0–6 mm batch showed the greatest percent difference in density in 28 days: a reduction of 20.83% in the 4 × 4 × 16 cm specimens and of 12.80% in the 30 × 30 × 5 cm specimens. To replace 100% by weight of the AEV 0–12 mm fraction with the 0–8 mm fraction did not entail a significant reduction in density. Only in the 4 × 4 × 16 cm specimens a slight decrease of about 5% was observed.

Figures 3 and 4 show that the influence of the specimens' size on these batches by volume results in 6% decrease in the density values of the 4 × 4 × 16 cm specimens, as compared to those obtained for the same batches in the 30 × 30 × 5 cm specimens. In the 4 × 4 × 16 cm specimens, the +30%APE 0–6 mm batch obtained a greater percent difference in density in 28 days, with a reduction of 21.41%.

However, in the 30 × 30 × 5 cm specimens it was the 90%CEM 10%THER batch that obtained a greater percent difference in density in 28 days with a decrease of 19.13%.

3.2 Capillary water absorption

With regard to the Capillary Water Absorption Coefficient in weight batches, in Figures 5 and 6 we realize that in the standard batch, such as the 100%AEV 0–12 mm (W/C = 1.2), the absorp-

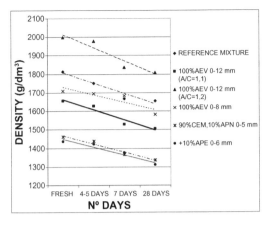

Figure 1. Density tests results in the batches produced by weight in $4 \times 4 \times 16$ cm specimens.

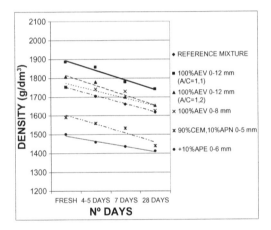

Figure 2. Density tests results in the batches produced by weight in $30 \times 30 \times 5$ cm specimens.

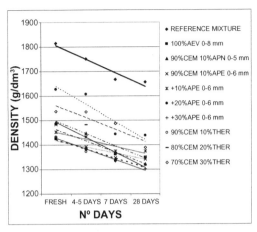

Figure 3. Density tests results in the batches produced by volume in $4 \times 4 \times 16$ cm specimens.

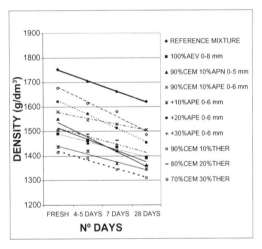

Figure 4. Density tests results in the batches produced by volume in $30 \times 30 \times 5$ cm specimens.

tion coefficient decreases as the density increases. On the whole, for the $4 \times 4 \times 16$ cm specimens greater values were registered as compared to the $30 \times 30 \times 5$ cm specimens.

In the batches by volume (Fig. 7), the Standard Batch 100%AEV 0–12 mm has a slightly lower absorption coefficient than the 100%AEV 0–8 mm. Therefore, it can be concluded that the AEV 0–8 mm does not improve the capillary water absorption coefficient.

In the $30 \times 30 \times 5$ cm specimens (Fig. 8) the addition of APE 0–6 mm results in a reduction of the absorption coefficient of up to 25% in the +30%APE 0–6 mm. On the other hand, the gradual replacement of the cement with Thermocal™ results in a reduction of the absorption coefficient of 87% for the 70%CEM 30%THER batch.

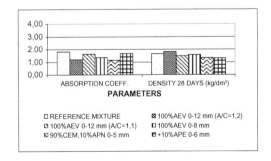

Figure 5. Absorption coefficient ratio—Density ratio in 28 days in batches by weight ($4 \times 4 \times 16$ cm specimens).

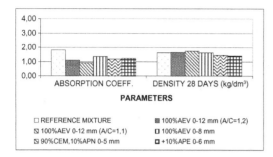

Figure 6. Absorption coefficient ratio—Density ratio in 28 days in batches by weight (30 × 30 × 5 cm specimens).

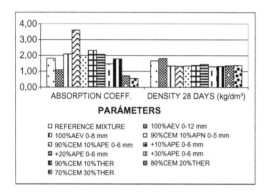

Figure 7. Absorption coefficient ratio—Density ratio in 28 days in batches by volume (4 × 4 × 16 cm specimens).

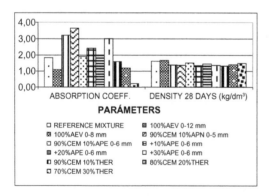

Figure 8. Absorption coefficient ratio—Density ratio in 28 days in batches by volume (30 × 30 × 5 cm specimens).

3.3 Coefficient of thermal conductivity

With regard to the coefficient of thermal conductivity tests, the +10%APE 0–6 mm obtained the lowest λ coefficient of thermal conductivity in the batches by weight, with a percent reduction

Table 10. Coefficient of thermal conductivity results in the batches by weight (30 × 30 × 5 cm specimens).

Batch by weight	λ Thermal conductivity (W/m · K)
REFERENCE MIXTURE	0.460
100%AEV 0–12 mm (A/C = 1.2)	0.425
100%AEV 0–12 mm (A/C = 1.1)	0.420
100%AEV 0–8 mm	0.410
90%CEM 10%APN 0–5 mm	0.420
+10 APE 0–6 mm	0.365

Table 11. Coefficient of thermal conductivity results in the batches by volume (30 × 30 × 5 cm specimens).

Batch by weight	λ Thermal conductivity (W/m ·K)
REFERENCE MIXTURE	0.460
100%AEV 0–8 mm	0.305
90%CEM 10%APN 0–5 mm	0.260
90%CEM 10%APE 0–6 mm	0.325
+10 APE 0–6 mm	0.255
+20 APE 0–6 mm	0.265
+30 APE 0–6 mm	0.275
90%CEM 10%THER	0.295
80%CEM 20%THER	0.260
70%CEM 30%THER	0.310

of 16% as compared to the Reference Batch. If we compare the results of the 100%AEV 0–12 mm (W/C = 1.2) batches with the 100%AEV 0–8 mm batch, we observe that the replacement of 100% of the thick aggregate by volume with a finer fraction does not produce a significant improvement in the coefficient of thermal conductivity λ (Table 10).

The batches by volume (Table 11) that obtained the highest percent reduction in the coefficient of thermal conductivity λ were the +10%APE 0–6 mm, with a reduction of 43.33%, the 80%CEM 20%THER and the 90%CEM 10%APN 0–5 mm, with a reduction of 42.22%. On the other hand, although the addition of these materials results in an improvement in the coefficient of conductivity in all batches, the APE 0–6 mm is the material that presents a greater influence in this parameter, as the optimum percentage is 10% by volume.

4 CONCLUSIONS

Tables 12 to 14 show a summary chart with the conclusions of the physical tests obtained in stage 2, which have been explained in the results and discussion section. The chart displays the analyzed batches both by weight and by volume. The decrease (negative) or increase (positive) percentages for the different parameters are shown as compared to the reference batch, except the coefficient of capillary water absorption value in the batches by weight, which was calculated as compared to the value obtained for the AEV 0–12 mm standard batch with a W/C = 1.2 ratio.

In view of the conclusions obtained in this study and the information that has been presented, we can establish that:

– The batches by volume seem to be more appropriate for the production of concretes with volcanic slag light aggregates.
– To replace the AEV 0–12 mm with 0–8 mm fraction by volume improves the thermal characteristics of the lightweight concrete in the BHIC.
– In view of the improving effect that the addition of the Thermocal™ mortar, which replaces cement in different percentages, has on the coefficient of absorption, we can conclude that the +10%APE 0–6 mm and the 90%CEM 10%APE 0–6 mm batches can present improvements on this parameter by adding Thermocal™ mortar to these mixtures.
– The addition of 20%APE 0–6 mm produces an important and significant improvement in the lightweight aggregate used in the manufacture of BHIC. This percentage corresponds with the conclusions of several researchers, such as Demirboğa (Demirboğa & Gül, 2003), Hossain (Hossain & Lachemi, 2007) (Hossain, 2003), who in their numerous studies stated that the optimum percentage of replacement of cement with pozzolanic additions in lightweight concrete is 20%.

Table 12. Summary chart for the dry density (28 days) results.

Batch	Dosage type	Δ Dry density (±%)
REFERENCE MIXTURE	by weight	Ref. Value
100%AEV 0–12 mm (A/C=1.2)	by weight	±0
100%AEV 0–8 mm	by weight	−5.00
100%AEV 0–8 mm	by volume	−14.00
90%CEM 10%APN 0–5 mm	by weight	−11.00
90%CEM 10%APN 0–5 mm	by volume	−16.00
90%CEM 10%APE 0–6 mm	by volume	−8.00
+10 APE 0–6 mm	by weight	−12.80
+10 APE 0–6 mm	by volume	−16.96
+20 APE 0–6 mm	by volume	−7.00
+30 APE 0–6 mm	by volume	−21.41
90%CEM 10%THER	by volume	−20.00
80%CEM 20%THER	by volume	−15.00
70%CEM 30%THER	by volume	−16.00

Table 13. Summary chart for the capillary water absorption coefficient.

Batch	Dosage type	Δ Water ab-sor. (±%)
REFERENCE MIXTURE	by weight	Ref. Value
100%AEV 0–12 mm (A/C=1.2)	by weight	Ref. Value
100%AEV 0–8 mm	by weight	±0.00
100%AEV 0–8 mm	by volume	±0.00
90%CEM 10%APN 0–5 mm	by weight	+8.00
90%CEM 10%APN 0–5 mm	by volume	+200.00
90%CEM 10%APE 0–6 mm	by volume	+5.00
+10 APE 0–6 mm	by weight	+12.0
+10 APE 0–6 mm	by volume	+30.00
+20 APE 0–6 mm	by volume	+14.00
+30 APE 0–6 mm	by volume	+65.00
90%CEM 10%THER	by volume	−12.00
80%CEM 20%THER	by volume	−33.00
70%CEM 30%THER	by volume	−87.00

Table 14. Summary chart for the thermal conductivity coefficient.

Batch	Dosage type	Δ λ (±%)
REFERENCE MIXTURE	by weight	Ref. Value
100%AEV 0–12 mm (A/C=1.2)	by weight	−5.56
100%AEV 0–8 mm	by weight	−8.89
100%AEV 0–8 mm	by volume	−32.22
90%CEM 10%APN 0–5 mm	by weight	−6.67
90%CEM 10%APN 0–5 mm	by volume	+42.22
90%CEM 10%APE 0–6 mm	by volume	+27.78
+10 APE 0–6 mm	by weight	+18.89
+10 APE 0–6 mm	by volume	−43.33
+20 APE 0–6 mm	by volume	−41.11
+30 APE 0–6 mm	by volume	−42.22
90%CEM 10%THER	by volume	−34.44
80%CEM 20%THER	by volume	−42.22
70%CEM 30%THER	by volume	−31.11

– The 70%CEM 30%THER batch is the most suitable mixture to be used in exposed BHIC non structural walls.

REFERENCES

Demirboğa & Gül. 2003. The effect of expanded perlite aggregate, silica fume and fly ash on the thermal conductivity of lightweight concrete. *Cement and Concrete Research* 33: 723–727.
Demirdag, S. & Gündüz, l. 2008. Strength properties of volcanic slag aggregate lightweight concrete for high performance masonry units. *Construction and Building Materials* 22: 135–142.
Guigou Fernández, C. 1997. *Los hormigones cavernosos con áridos canarios.* Las Palmas de Gran Canaria: Universidad de Las Palmas de Gran Canaria.

Hossain. 2003. Blended cement using volcanic ash and pumice. *Cement and Concrete Research* 33: 1601–1605.

Hossain & Lachemi. 2007. Strength durability and microstructural aspects of high performance volcanic ash concrete. *Cement and Concrete Research* 37: 759–766.

Hummel, Alfred. 1966. *Prontuario del hormigón. Hormigones normales. Hormigones ligeros.* 2° Edición española. Barcelona: Editores Técnicos Asociados, S.A.

Lomoschitz, A. et al. 2006. Basaltic Lapilli Used for Construction Purposes in the Canary Islands, Spain. *Environment & Engineering Geoscience*, Vol. XII, n° 4: 327–336.

Mouli, M. & Khelafi, H. 2008. Performance characteristics of lightweight aggregate concrete containing natural pozzolan. *Building and Environment* 43: 31–36.

Soria, F. 1972. *Conglomerantes hidráulicos. Estudio de materiales IV.* Madrid: Instituto Técnico de la Construcción y del Cemento, Instituto Eduardo Torroja de la Construcción y del Cemento.

Steiger, R.W. & Hurd, M.K. 1978. Lightweight insulating concrete for floors and roof decks. *Concrete Construction* 23 (7): 411–422.

Topçu, I.B. & Işikdağ, B. 2008. Effect of expanded perlite aggregate on the properties of lightweight concrete. *Journal of materials processing technology* 204: 34–38.

Xincheng Pu. 1999. Investigation on pozzolanic effect of mineral additives in cement and concrete by specific strength index. *Cement and Concrete Research* 29: 951–955.

Volcanic Rock Mechanics – Olalla et al. (eds)
© 2010 Taylor & Francis Group, London, ISBN 978-0-415-58478-4

Volcanic dikes engineering properties for storing and regulation of the underground water resources in volcanic islands

J.C. Santamarta Cerezal
Escuela Técnica Superior de Ingeniería Civil e Industrial, Universidad de La Laguna (ULL)
Avenida Astrofísico Francisco Sánchez, La Laguna, Tenerife, Spain

L.E. Hernández
Consejería de Obras Públicas y Transportes, Gobierno de Canarias, Spain

J.A. Rodríguez-Losada
Departamento Edafología y Geología, Universidad de La Laguna (ULL)
Avenida Astrofísico Francisco Sánchez, La Laguna, Tenerife, Spain

ABSTRACT: The main feature of aquifers in volcanic islands with high rainfall rates and steep topography is that they are on raised, mainly due to the presence of volcanic dikes. Dikes are igneous bodies with very high aspect ratio, which means that their thickness is usually much smaller than the other two dimensions and tend to be vertical or of high dip angle. It can be considered as impermeable and interconnected walls where aquifers, which are recharged by rain, raise the water table between the dikes. This is especially important in volcanic rift zones. The method to obtain water in these volcanic areas involves digging horizontal galleries with explosives. The galleries cross the dikes and drain the water from the water saturated area. The main problem of this type of perforation is to obtain the water continually, without the possibility of water regulation. This question has been solved by means of the reconstruction of several technically viable dikes, to enable the store of water resources through channels in order to regulate the water wealth of the built galleries. Methods, results and viability on the use of the volcanic materials as relevant works of civil engineering highlights on the following paper.

1 INTRODUCTION

1.1 *Volcanic areas*

The geology around the Canary Islands is dominated almost entirely by a succession of volcanic materials and structures. Sequences of lava emissions and pyroclastic deposits of highly variable composition, that present extreme contrasts from the standpoint of lithology, environment, landscape and weather.

1.2 *Volcanic dikes and hydrology*

Dikes are as nearly vertical walls or thin layers formed from a dense and compact rock, with little width in most cases (1 to 6 m), which play the role of impervious or semi-permeable screens. The numerous dikes intact, more or less parallel, hampering movement and the presence of transverse open cracks, that favor the longitudinal flow axis, give particular hydrogeological behavior. In addition, dikes raised the island aquifer, which subsequently determines the utilization of water resources through horizontal galleries.

In the oceanic islands, dikes usually appear grouped in the form of swarms of subparallel dikes intruding along structural axes known as rift zones (Walker 1992). In the Canary Archipelago rift zones were defined in different islands as structures playing a key role in processes such as mass wasting and destruction of ancient oceanic volcanoes or in the structural control on the growth of large oceanic-island volcanoes (Carracedo 1994, Rodriguez-Losada *et al.*, 2000).

1.3 *Groundwater resources in volcanic areas*

In volcanic island systems, we can establish two types of aquifers, the first watershed aquifer which is set approximately at elevation 400 to near the higher elevations of the islands and ridges, this aquifer is tapped through mainly horizontal galleries, although there are some cases where the aquifer is tapped by drilling wells or high altitude.

Figure 1. Dike across lava flows.

Figure 2. Sectional view a water gallery on the Hierro island, Canary Islands, Spain.

The other type of aquifer, most exploited, mainly because the population centers and land uses are established near the sea, is called the coastal aquifer, this presents an important difference that, it is affected by the tides, so we have a fresh water body, which "floats" by its density above the body of salt water, denser, the border between the two phases, is a fragile mix zone (interface). The latter aquifer is more exploited than the first and here is where mainly the effects of seawater intrusion. There aren't dikes in this part of the aquifer.

1.4 *Water mining*

The horizontal galleries for obtaining underground water is a method widely millennium. From ancient times was known as far away as China, Persia, Spain and Latin America.

The Canary archipelago gets much of its water resources primarily from underground water, mainly in the Western Isles our case study, in the eastern islands, by the water scarcity, little rain, erosion and evapotranspiration, eastern islands are mainly supplied by desalination of water sea, in fact the first desalination plant was installed in Lanzarote in the 60s, although as has been found there in the last century in the archipelago of Malta and more recently in the twentieth century in Gibraltar and Turkey were the first desalinations plants in Europa.

In the late nineteenth century began piercing galleries in those parts where it was clearer evidence of the existence of groundwater, ie, in areas already existing natural springs.

The horizontal gallery type, most common, has a small section of 1.80 by 1.80 m, mainly were built by private initiative, although built by the Administration have come to section 4 and 2.50 m, with a length of up to 6 km, where water are obtained by gravity. These horizontal galleries are oriented towards dorsals. Generally have produced water

flows to 200 l/s, representing about 6 hm 3 per year.

The function of a gallery is twofold since, besides acting as water collector, also serves as a transport medium for this. The galleries are perforated usually by explosives because of the hardness of the basalt.

As discussed, a fundamental concept for understanding the aquifer exploitation by horizontal galleries located at considerable heights above sea, is that, the aquifer is raised by dikes. Dikes are cells where the water is stored; these dikes create a staggering basaltic aquifer with dynamic hydraulic gradients. The highest concentrations of dykes are set at the ridges of the islands. These ridges are higher rainfall areas of the islands, with vertical rainfall and convective rain, they are the areas where the aquifer is higher.

The galleries have a singular constructive system, are basically oriented toward the ridges of the islands, based on what was said previously, Is not possible to apply continental areas drilling techniques, such as tunnel boring machines drilling machines, because mainly to the following issues: the first is a gallery of water has no outlet and the other reason is that volcanic material is very heterogeneous, that creates many problems in tunnel boring machines drilling, in case a problem of calibration of force may even rupture, since in the front can be anything from a layer of slag or basalt.

The choice of construction method to be used in any underground works will be conditioned, first, by the geotechnical characteristics of the land to drilling and secondly, by the geometry and size of the excavation.

It is possible to estimate the properties of the terrain and geological materials by studing technical literature works and projects in similar perforations.

Figure 3. Concrete dike and hatch access to the next section of the gallery of water.

With regard to the geotechnical characteristics of the terrain to be excavated, usually is a massive material, like basalt, no stability problems. Scoriaceous material In part, the pyroclastic often result in more rounded section, before it can be used to support systems can be a good option use shotcrete cement, with the task of arming the field.

The gallery is usually constructed with an straight line, although at times, and because the different materials are encountered with different geotechnic properties, there may be changes in direction, at certain times of the excavation.

The first kilometers usually run dry, when we enter the into the aquifer, the infrastructure is introduced into the saturated zone, water flow is not adjustable, leaves in continuous due to the slope built into the exit of the gallery.

Along the trace of the water gallery covers several dikes which are listed with Roman numerals to the left of the trace, is usually choose later, for geomechanical closures, depending on the quality of the dike, width, minimum of 6 m, finally the water that dikes can store and its area of influence, these closures represent a significant innovation in this type of drilling, because the geomechanical closures can store water in the aquifer and regulate their use.

2 UNDERGROUND WATER USING DIKES

2.1 Stating the problem

As has been said the main problem of the galleries, is the lack of water resources regulation, which means that once it has penetrated into the aquifer, the water flows continuously therefore cannot regulate the water flows.

This poses a problem, especially in winter, when there is a good rainy season, at that time of year there is surplus water in the galleries, part of this flow cannot be stored in irrigation pools. The following problems are generated.

The high cost of pipelines from catchment to the raft. Flow losses in the transport of water (approximately 25–30%). Evaporation losses once stored in the raft.

The underground drilling once it enters the saturated zone, is draining the water, once enough has been drilled to obtain a target flow, planned or economically and socially profitable, is preparing to channel the flow through the pipe or channel to areas of consumption. The average length of drilling in the Western Isles is between 1,500 and 2,500 meters, each island has its gallery average length as a function of their geological characteristics, while currently on the island of El Hierro are under the mean lengths in the island of Tenerife has higher average lengths some galleries with over 4,000 water meters.

In the case of a borehole or a well no more complex the regulation of water flows, is sufficient to disconnect the pump.

2.2 Construction method

Once selected the dikes, following the methodology of the engineer Soler Liceras in 1996, dug three feet wide to include the reinforcement of concrete and steel bolts that will make this artificial structure in solidarity with the basalt dike.

The assembly of the reinforcement of concrete needs to replace cast iron pipes that are responsible for draining the water from one area to another and remove water from the gallery, we must record that can be done several dikes in a single operation in according to the capacity of the aquifer and geologic structures that are available.

We must also give way to vent pipe for further drilling.

Once it is finished metal structure, is followed by the concrete pump. With the concrete we have to consider the environment in which to build the structure, leaving space for the door be left large enough for the operation of the gallery according to the initial dimensions.

We have to bear in mind that, if cracks form our gallery exploitation is destroyed by high pressure water, the most contentious in this regard is the gate, so we have to include some wetsuits and test the installation.

The gate must withstand the high pressures exerted by water at extrados, which in the case studies have reached 72 m of water column, as is the case in the gallery of Los Padrones on the island of El Hierro (Spain). This gate is made of steel and made to measure, although in the last gallery, where this infrastructure has been included in

Figure 4. Final state and pipeline construction with open gate.

La Gomera (Spain), the door has been made by IPE profiles and metal plates, because this size is large.

Another advantage of this innovation, is that water is stored in the aquifer, so it is beyond any kind of pollution and disposal of applications needed.

There are experiences of this kind of innovation in continental soils, such as a gallery of Cuzco in Peru, but these have the following features, in speaking of a sedimentary.

Drainage galleries in Peru are very similar to the Canary Islands in Spain, although those in sedimentary soils in Peru, geological features, have focussed on exploitation of groundwater into sedimentary rock (sandstone) fractured, although there are work experiences highly fractured volcanic rocks. Water galleries have been built with a section of 2 × 2 m and the methodology has been the conventional drilling using a pneumatic drill powered by compressed air supplied by an air compressor, unlike drilling for explosives in the Canary case.

The sealing of the dike system in Peru is not as large as the dimensions of the case in the Canaries.

3 CONCLUSIONS

The water storage system within the aquifer, through the reconstruction of volcanic dikes has solved the problem of regulation of the water galleries. Since 1996, in the Canary Islands in Spain, are being used on the island of El Hierro, and are currently concluding the work of reconstruction of dikes on the island of Gomera.

Geo-mechanical systems of regulation groundwater in volcanic soils encountered in this study, continue to operate effectively and ensure good management of water resources.

The systems described in this article can be used in any volcanic and island system as the archipelago of Hawaii, Macaronesian islands or the Galapagos Islands in South America and many other.

REFERENCES

Carracedo, J.C. 1994. *The Canary islands: an example of structural control on the growth of large oceanic-island volcanoes.* J. Volcanol. Geotherm. Res. 60: 225–241.

Rodriguez-Losada, J.A., Martinez-Frias, J., Bustillo, M.A., Delgado, A., Hernandez-Pacheco, A. & de la Fuente Krauss, J.V. 2000. *The hydrothermally altered ankaramite basalts of Punta Poyata (Tenerife, Canary Islands).* J. Volcanol. Geotherm. Res. 103: 367–376.

Santamarta Cerezal, J.C. & Rodriguez, J. 2008. *Singularidades de las obras hidráulicas para abastecimiento de agua potable en medios volcánicos. El caso del archipiélago Canario.* Spain. *Congreso Internacional sobre gestión y tratamiento del agua.* Córdoba, Argentina.

Santamarta Cerezal, J.C. & Hernandez, C. 2008. *La isla de El Hierro (Archipiélago Canario, España); solución a la intrusión marina futuro de la gestión hidráulica de la isla.* Publicaciones del Instituto Geológico Minero de España (ed.), *Los acuíferos costeros. Retos y soluciones.* Volumen I: 1085–1095. Madrid. Spain.

Santamarta Cerezal, J.C. 2009. *La minería del agua en el archipiélago canario. De Re Metallica. Sociedad Española para la Defensa del Patrimonio Geológico y Minero.* De Re Metallica n° 12: 1–8.

Walker, G.P.L. 1992. *Coherent intrusion complexes in large basaltic volcanos—a new structural model.* J. Volcanol. Geotherm. Res. 50(1–2): 41–54.

Volcanic Rock Mechanics – Olalla et al. (eds)
© 2010 Taylor & Francis Group, London, ISBN 978-0-415-58478-4

Geotechnical properties of volcanic materials of the Mount Erciyes

S. Yüksek & A. Demirci

Engineering Faculty, Mining Engineering Department, Cumhuriyet University, Sivas, Turkey

ABSTRACT: The Mount Erciyes (3917 m) is the largest volcanic mountain of Central Anatolia (Turkey) and situated 15 km south of Kayseri. Mt. Erciyes its volcanic groups have produced calc-alkaline and pyroclastic rocks such as basalt, andesite, tuff, ignimbrite, dacite, rhyodacite and pumice. Thickness of these rocks varies from one to several hundred meters and these rocks constitute of whole The Cappadocia Region comprising Nevşehir, Kayseri and Niğde provinces of Central Anatolia. These rocks are mined in more than a hundred quarries and used construction purposes. In this study several geotechnical parameters related to these rocks are determined in the laboratory. There have been found good relationship between these products such as P-wave velocities versus UCS, P-wave velocities versus thermal conductivity coefficient, UCS versus thermal conductivity coefficient. These parameters show these rocks quite suitable for construction purposes.

1 INTRODUCTION

1.1 *A General view for Volcanic Mountains of Turkey*

Anadolu (Turkey) is one of the young volcanic areas within the Alpine orogenic belt. Volcanic activities began in the Tertiary era and continued up to the Early Holocene time. During these period numerous volcanic cones were formed; such as Big Ararat Mt.(5137 m), Small Ararat Mt. (3896 m), Tendürek Mt.(3533 m), Suphan M.(4058 m), Nemrut Mt.(3050 m), Kisir Mt. (3197 m) and Akbaba Mt.(3040 m) in the Eastern Anatolia; Karaca Mt.(1954 m) in the South-Eastern Anatolia; Erciyes Mt. (3917 m), Hasandag Mt. (3268 m), Melendiz Mt. (2963 m), Karaca Mt. (2288 m) and Kara Mt (2288 m) in the Central Anatolia.

Mount Erciyes is the huge stratovolcano mountain (3300 km²) of Central Anatolian Volcanic Province (CAVP) in Turkey, with a basal diameter of 55–60 km. (Şen et al., 2003). Around the central cone there are sixty-eight monogenetic vents with a diameter of 600 m–3000 m being situated on its flanks (Ketin, 1985). Mt. Erciyes is situated about 15 km south of Kayseri and its materials are distributed in large areas (~30,000 km²). First volcanic activity of Mt. Erciyes had been started 4.39 Ma ago (Ayranci, 1969). Numerous studies (Innocenti et al., 1975, Pasquare et al., 1988, Notsu et al., 1995, Kurkcuoglu et al., 1998, Toprak, 1998, Sen et al., 2003) have been carried out for the evaluation of Mt. Erciyes and its vicinity.

Mt. Erciyes and its volcanic groups have produced calc-alkaline and pyroclastic rocks such as basalt, andesite, tuff, lavas, ignimbrite, dacite, rhyodacite and pumice. Thickness of these rocks varies from one to several hundred meters and these rocks constitute of whole The Cappadocia Region comprising Nevşehir, Kayseri, Aksaray and Niğde provinces of Central Anatolia. Actually resources of the materials of Cappadocian Region are not only from Mt. Erciyes volcano but also Mt. Hasandag and Mt. Melendiz volcano as well. However majority of the materials of the region originate from the Mt. Erciyes.

1.2 *The Cappadocia Region*

The Cappadocia Region is one of the historical sites in the World Heritage List and a famous touristic site of Turkey due to its spactacular and unique landforms, fairy chimneys and historical heritages as seen Figure 1. Various geomorphological landforms like columns, towers, pillars with chimney caps (named fairy chimneys) and obelisks are developed owing to differential weathering and erosion activities. Cappadocia has been continuously inhabited from the Neolithic era up to the present day. It was known as Hattie in the late Bronze Age, and was the homeland of the Hittite power centered at Hattusa. After the fall of The Hittite Empire, Cappadocia was fallen into the power of the Persian Empire. In this reason monastery, hermitages, shrines and even dwellings were carved in the tuffs (Fig. 2).

Figure 1. Pasabaglari chimneys (Demir 2008).

Figure 2. A group of churches carved into rock at Goreme (Umar, 1998).

The carved grottoes reflect Byzantine architecture, style and painting. Most of the dwellings and chapels date from the 10th–13th centuries. (Sari & Özsoy 2009). Easy carving and thermal isolation properties of the soft tuffs have been the main reasons for the extensive multi–purpose underground settlement in the Cappadocia region from past to present (Aydan, 2003).

2 A BRIEF GEOLOGY OF MOUNT ERCIYES

Mount Erciyes is the largest stratovolcano in the eastern part of the Central Anatolian Volcanic Province. A great many studies have been carried out on the Central Anatolian volcanism (Keller, 1974, Batum, 1978, Besang, et al., 1977, Pasquaré et al., 1988, Bigazzi et al. 1993, Druitt et al., 1995, Le Pennec et al., 1994, Froger et al., 1998, Kürkçüoğlu, 1988, Toprak 1998, Şen et al., 2003). The Central Anatolian volcanism is influenced by the intercontinental convergence associated with collision of the African plate with the Eurasian plate and produced many polygenetic volcanoes, such as Hasandag, Erciyes, Melendiz, Keciboyduran; monogenetic volcanoes, such as Golludag as domes, Acigol, Cora as maars and several hundreds of scoria cones and extensive ignimbrite sheets (Gencealioglu & Geneli, 2008). This volcanic belt extends about 300 km along a NE-SW direction. Mount Erciyes is situated in the eastern part of this belt. The related geological map of Mount Erciyes and its calderas, domes and cones are shown in Figure 3. The volcanic evolution of Mount Erciyes is divided into two stages named as Kocdag and New Erciyes (Kürkçüoğlu, 1998 and Şen, 2003). The stratigraphical column is given in Figure 3. Kocdag is mainly composed of lava flows of alkaline basalt, andesite and basaltic andesites. They constitute the eastern flank of the Mount Erciyes volcanic complex. The first explosive activity is the ignimbrite eruption, 2.8 Ma ago (Innocenti et al., 1975) and this eruption was followed by a caldera collapse. The new Erciyes stage represents different basaltic, andesitic, dacitic, rhyodacitic lava generations and associated pyroclastics. The volcanic products are related to central and adventive monogenetic vents (domes and cones). The last eruption dated 0.083 Ma (Notsu et al., 1995) corresponds to the fourth dacitic lava generation in the volcano-stratigraphical column (Figure. 4). The emplacement of the pyroclastics prior to rhyodacitic dome and debris avalanche deposits is the best known and most recent products of the volcano (Kürkçüoğlu, 1998).

3 CONSTRUCTION MATERIALS RELATED TO THE MOUNT ERCIYES

The rocks of Mt. Erciyes, andesite, ignimbrite and basalt were widely used as construction materials in low storeyed buildings and especially in historical monuments in the past. However, nowadays andesite, basalt, ignimbrite and different colored tuffs are used for decorative purposes and paving. Additionally basalts are used as concrete and road construction aggregate. The other rocks of Mt. Erciyes are pumice and perlite which are used making light weight bricks and concrete. In the Cappadocia Volcanic Province there are two hundred and five quarries related to volcanic rocks

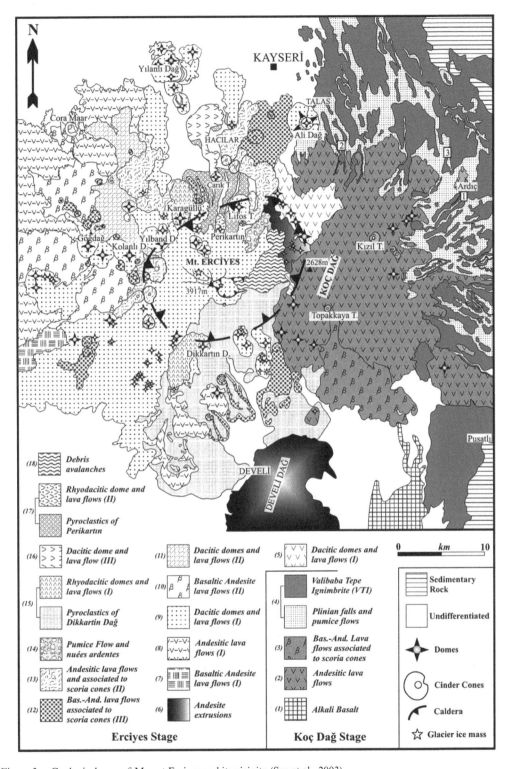

Figure 3. Geological map of Mount Erciyes and its vicinity (Şen et al., 2003).

101

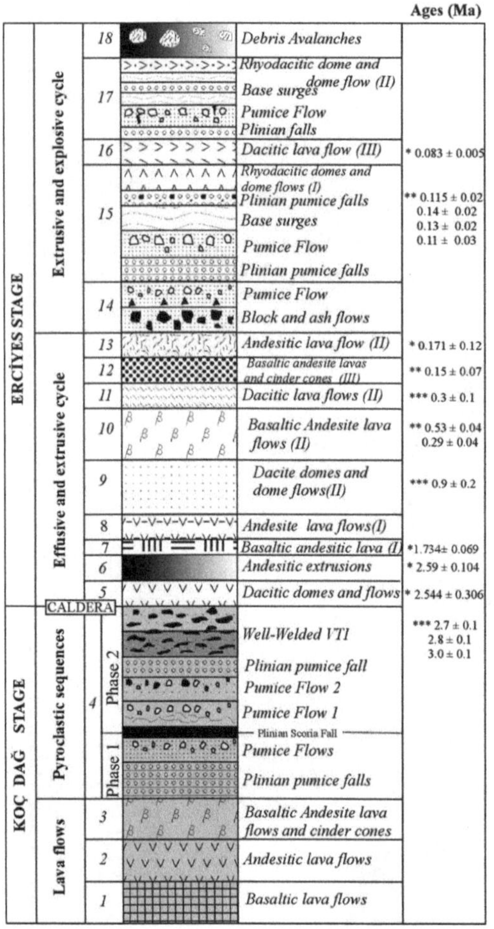

Figure 4. Stratigraphic column of Mount Erciyes stratovolcano. (Şen et all. 2003) Ages from: *** Innocenti et al., 1975; Ercan et al., 1994; Notsu et al., 1995. Italik numbers refer to the stratigraphic position of the products.

(Migem 2010). One hundred and nine quarries of them lies in Kayseri province and the rocks of their quarries originated Mt. Erciyes resources. In this study the information and their lacations in the Cappadocia region compiled. Following this activity geotechnical properties of the rocks are summarized.

3.1 Geotechnical properties of Mt. Erciyes rocks

Apart from geological investigation of Central Anadolian volcanic province and studies on the evaluation of Mt. Erciyes, several engineering geological and conservation studies on Cappadocian tuffs and Mt. Erciyes materials have been carried out (Erguvanlı, 1977, Topal et al., 1997, Aydan & Ulusay 2003, Ulusay et al., 2006). These studies mainly deal with the deterioration and durability of the tuffs. Other few studies are performed on material geology and investigation of usability of these rocks as building stone (Erdogan, 1986 and 1989, Korkanç, 2007, Temur et al., 2007).

3.2 Some laboratory experiments and their results on the building materials of Kayseri province quarries within Mount Erciyes

According to the MIGEM (Turkish Republic Ministry of Energy and Natural Resources–General Diractorate of Mining Affairs) records there are one hundred and nine quarries in Kayseri province at present. Forty one of them are pumice stone quarries. Total rock materials mined from the current quarries amounts to 772,957 tones in the year 2009. 279,523 tones of these products composes of andesite, 28,115 tones of basalt, 13,488 tones ignimbrite, 389116 tones pumice and 62,715 tones tuff respectively.

Volcanic rocks to be tested in our laboratory are taken from six different quarries in Kayseri province. Cylindrical NX size specimens are tested for some physical and mechanical properties of them. Tests were carried out in accordance with procedures laid out in the ISRM (1985) Suggested Methods and TS 699. The results of experiments performed are summarized in Table 1. In this table the rock categories are as follows: Rock type I: pumice; Rock type II: andesite; Rock type III: beige tuff; Rock type IV: basalt; Rock type V: ignimbrite; Rock type VI: red-tuff. The geotechnical properties summarized consist of dry unit weight (g_d) in kN/m^3, saturated unit weight (g_{sat}) in kN/m^3, porosity (n) in%, P-wave velocity-dry and saturated in m/s, coefficient of thermal conductivity (k) in W/mK, and uniaxial compressive strength (UCS) in MPa.

The parameters given in Table 1 are used to describe the relationship as given in Figures 8, 9, 10, 11, 12, 13. The Figure 5 depicts the dry and saturated unit weight of different rock types. As seen in this figure highest value of unit weight is given by basalts (IV) and lowest value by red-tuffs (VI). The Figure 6 gives the P-wave velocities of the sample rocks. As seen in this figure basalts (IV) despicts the highest P-wave value, whereby the pumice (I) and red tuffs (VI) the lowest values. The Figure 7 shows the thermal conductivity coefficient values as W/mK. These values are optained in our laboratory steady state test apparatus designed and produced in Cumhuriyet University. The tests are carried out in 100°C (373 K) temperature range of this apparatus. As seen in this figure the thermal conductivity coefficient of these rocks vary between 0.630 and 0.924 W/mK. The Figures 8, 9, 10, 11, 12 and 13 represent the relationships between P-wave

102

Table 1. Summarized some physical and mechanical properties of Mt. Erciyes products.

Properties*	Rock type					
	I	II	III	IV	V	VI
g_d (kN/m³)	18.78	18.10	18.80	26.86	19.79	15.42
$g_{sat.}$ (kN/m³)	20.38	19.96	20.40	27.14	21.35	18.60
Porosity (%)	16.37	19.11	16.43	2.83	15.96	26.72
P-wave Velocity-dry (m/s)	1921	3369	2312	4760	2765	1954
P-wave Velocity-saturated (m/s)	1619	3399	1657	5339	2727	1742
Thermal conductivity coefficient (W/mK)	0.767	0.924	0.838	0.850	0.63	0.784
UCS (MPa)	36.71	52.31	42.1	175.19	73.22	24.95

* Mean Values (I: Pumice; II: Andesite; III: Beige tuff; IV: Basalt; V: Ignimbrite; IV: Red tuff).

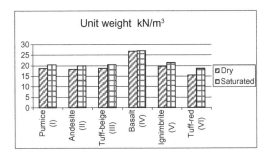

Figure 5. Dry and saturated unit weights of the rocks.

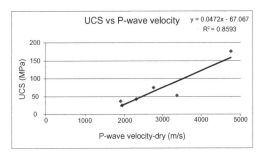

Figure 8. Relationship between P-wave velocity and UCS.

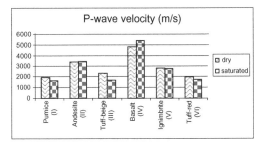

Figure 6. Dry and saturated P-wave velocities of the rocks.

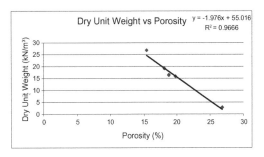

Figure 9. Relationship between porosity and dry unit weight.

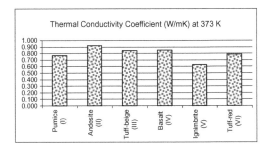

Figure 7. Thermal conductivity coefficient of the rocks.

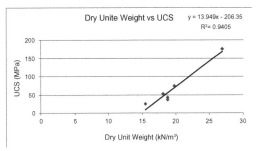

Figure 10. Relationship between Thermal conductivity and UCS.

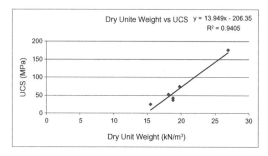

Figure 11. Relationship between dry unit weight and UCS.

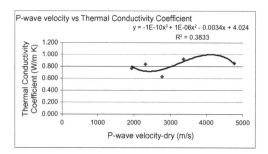

Figure 12. Relationship between P-wave porosity and thermal conductivity coefficient.

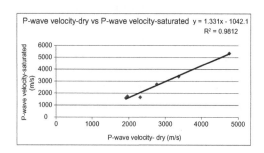

Figure 13. Relationship between P-wave velocity-dry and P-wave velocity-saturated.

velocities versus UCS; porosity versus unit weight; UCS versus thermal conductivity coefficient; unit weight-dry versus UCS; P-wave velocities-dry versus P-wave velocities-saturated. The correlation coefficients R^2 given in the figures show good relationship between given geotechnical parameters. The Figure 8 shows a positive linear relation between P-wave velocities and UCS. In Figure 9 is seen a negative linear relationship between porosity and unit weight-dry. Figure 10 gives a polynomial relationship between UCS and thermal conductivity coefficient. In Figure 11 is seen a linear relationship between unit weight-dry and UCS.

The relationship between P-wave velocities and thermal conductivity coefficient is of a concave polynomial character as given in Figure 12. Figure 13 depicts the linear relationship between P-wave velocity-dry and P-wave velocity-saturated rocks.

4 CONCLUSION

The Mount Erciyes region comprises an area in the range of about 30,000 km². There are several rock types such as basalt, andesite, pumice, ignimbrite and tuff in this region. These rocks are mined in more than a hundred quarries and used construction purposes. In this study several geotechnical parameters related to these rocks are determined in the laboratory of our university. There have been found good relationship between these products such as P-wave velocities versus UCS, P-wave velocities versus thermal conductivity coefficient, density versus porosity, UCS versus thermal conductivity coefficient, density versus UCS and P-wave velocities-dry versus P-wave velocities-saturated. These parameters show these rocks quite suitable for construction purposes especially the thermal conductivity coefficient hints to the fact that most of these rocks have a property for energy saving in case being used in house constructions. Further investigations are necessary to be carried out for more reliable conclusion.

REFERENCES

Aydan, Ö. & Ulusay, R. 2003. Geotechnical and geoenvironmental characteristics of man-made underground structures in Cappadocia, Turkey. *Eng. Geol.* 69:245–272.

Ayrancı, B. 1969. Zur Petrologie und Geologie des Erciyes Vulkangebietes bei Kayseri—Zentral Anatolien/ Turkei. *Inaugural-Dr. Diss.*, Universitat Wurzburg.

Batum, I. 1978. Geology and petrography of Acigol and Golludag volcanics at southwest of Nevsehir (Central Anatolia, (Turkey). *Yerbilimleri* (Publ. Inst. Earth Sci. HacettepeUniv.) 4, 50–69.

Besang, C., Eckhardt, F.J., Harre, W., Kreuzer, H. & Mueller, P. 1977. Radiometrische Altersbestimmungen an neogenen Eruptivgesteinen der Turkei. *Geol. Jahrb. B* 25, 3–36.

Bigazzi, G., Yegingil, Z., Ercan, T., Oddone, M. & Ozdogan, M. 1993. New data for the chronology of Central and Northern Anatolia by fission track dating of obsidians. *Bull. Volcanol.* 55, 588–595.

Demir, Ö. 2008. *Cappadocia, Cradle of History*, 12th. Revised Edition, Pelin Ofset, Ankara. (in Turkish).

Druitt, T.H., Brenchley, P.J., Gokten, Y.E. & Francaviglia, V. 1995. Late Quaternary rhyolitic eruptions from the AcigolComplex, central Turkey. *J. Geol. Soc.* (London) 152, 655–667.

Erdogan, M. 1986. *Nevsehir-Ürgüp yöresi tüflerinin malzeme jeolojisi açısından araştırılması* (in Turkish). Doctoral thesis, I.T.U.

Erguvanlı, A.K. & Yüzer, A.E. 1977. Past and present use of underground openings excavated in volcanic tuffs at Cappadocia area. *In. Proceedings of the International Symposium on Storage in Excavated Rock Caverns*, Stockholm, Sweden, Bergman M. (ed). Pergamon Press, Oxford, 15–17.

Froger, J.L. Lenat, J.F. Chorowicz, J. Le Pennec, J.L. Bourdier, J.L., Kose, O., Zimitoglu, O., Gundogdu, N. & Gourgaud A. 1998. Hidden calderas evidenced by multisource geophysical data; example of Cappadocian Calderas, Central Anatolia. *J. Volcanol. Geotherm. Res.* 85, 99–128.

Innocenti, F., Mazzuoli, R., Pasquaré, G., Radicati, F. & Villari, L. 1975. Neogene calc-alcaline volcanism of Central Anatolia: geochronological data on Kayseri-Nigde area. *Geol. Mag.* 112, 349–360.

ISRM 1981. *Rock characterization, testing and monitoring, International Society for Rock Mechanics suggested methods*. Pergamon Press, Oxford.

Keller, J. 1974. Quaternary maar volcanism near Karapinar in Central Anatolia. *Bull. Volcanol.* 38, 378–396.

Ketin, İ. 1983. *Türkiye Jeolojisine Genel Bir Bakış Istanbul ITU publ. (in Turkish)*.

Korkanç, M. 2007. The effect of geomechanical properties of ignimbrites on their usage as building stone: Nevsehir stone, *Geological Engineeing* (in Turkish) 31(1): 49–60.

Kurkcuoglu, B., Şen, E., Aydar, E., Gourgaud, A. & Gundogdu, N. 1998. Geochemical approach to magmatic evolution of Mt. Erciyes stratovolcano, Central Anatolia, Turkey. *J. Volcanol. Geotherm. Res.* 85, 473–494.

Kuscu, G.G. & Geneli, F. 2008. Review of post-collisional volcanism in the Central anatolian Volcanic Province (turkey), with spezial reference to the Tepekoy Volcanic Complex, *Int. J. Earth Sci,* DOI 10.1007/s00531-008-0402-4.

Le Pennec, J.-L., Bourdier, J.-L., Froger, J.-L., Temel, A., Camus, G. & Gourgaud, A. 1994. Neogene ignimbrite of the Nevsehir Plateau (Central Turkey): stratigraphy, distribution and source constraints. *J. Volcanol. Geotherm. Res.* 63, 59–87.

Migem 2010. Turkish Republic Ministry of Energy and Natural Resources–General Diractorate of Mining Affairs.

Notsu, K., Fujitani, T., Ui, T., Matsuda, J. & Ercan, T. 1995. Geochemical features of collision-related volcanic rocks in Central and Eastern Anatolia, Turkey. *J. Volcanol. Geotherm. Res.* 64, 171–192.

Pasquare, G., Poli, S., Vezzoli, L. & Zanchi, A. 1988. Continental arc volcanism and tectonic setting in Central Anatolia, Turkey. *Tectonophysics* 146, 217–230.

Sari, F.Ö. & Özsoy, M. 2010. Cappadocia (Kapadokya), *Natral Heritage from East to West*, N. Evelpidou et al. (eds.), Springer-Verlag Berlin Heidelberg.

Sen, E., Kurkcuoglu, B., Aydar, E., Gourgaud, A. & Vincent P.M. 2003. *Journal of Volcanology and Geothermal Research* 125, 225–246.

Temur, S. Temur, Y. & Kansun, G. 2007. Geological petrographical and technological investigation of Erkilet basalt, Kayseri, Central Anatolia, *Geological Engineeing* (in Turkish) 31(1): 1–13.

Topal, T. & Doyuran, V. 1995. Effect of discontinuities on the development of fairy chimneys in the Cappadocia region (Central Anatolia-Turkey). *Turk J Earth Sci* 4: 49–54.

Toprak, V. 1998. Vent distribution and its relation to regional tectonics, Cappadocian volcanics, Turkey. *J. Volcanol. Geotherm. Res.* 85, 55–67.

TS699 1978. Dogal yapı taslarının muayene ve deney metodları (in Turkish). Türk Standartları Enstitüsü. Ankara.

Umar, B. 1998. *Kappadokia, Bir Tarihsel Coğrafya Araştılması ve Gezi Rehberi*, Tükelmat A.Ş. İzmir. (in Turkish).

Volcanic Rock Mechanics – Olalla et al. (eds)
© 2010 Taylor & Francis Group, London, ISBN 978-0-415-58478-4

A sensitive analysis on Mohr-Coulomb and Hoek-Brown parameters effective in ground response curve

A.R. Kargara & R. Rahmanejad
Department of Mining Engineering, Shahid Bahonar University of Kerman, Kerman, Iran

ABSTRACT: Convergence-confinement is one of the most popular methods that is applied for analyzing the interaction of a circular opening in rock masses. It is assumed that circular tunnel excavated in a continuous, homogeneous, isotropic, initially elastic rock mass subjected to a hydrostatic stress p_o.

Selecting appropriate failure criteria is very important in the analysis since it affects on plastic zone and on the resulted displacement and stress field around the opening. Some closed-form solutions have suggested for the ground reaction curve, although they are driven based on elastic-perfectly plastic or elastic-brittle-plastic models of rock mass behavior.

Brown et al. (1983) proposed a stepwise procedure based on Hoek-Brown criterion to solve stress and displacement around the circular opening for elastic-strain softening model of rock mass behavior. A similar stepwise procedure was extracted in this study for Mohr-Coulomb criterion. Finally a sensitive analysis was implemented for Mohr-Coulomb and Hoek-Brown criteria in respect to their parameters.

By comparison of the relative displacement caused by changing strength parameters in Hoek-Brown and Mohr-Coulomb criteria, it can be concluded that Mohr-Coulomb criterion is more sensitive in respect of variation of strength parameters of rock mass than Hoek-Brown.

1 INTRODUCTION

Analysis of stresses and displacements around circular opening that excavated in isotropic rock masses has been one of the fundamental problems in geotechnical engineering. Provided that the initial stress field is hydrostatic, the problem may be regarded as axisymmetric and an analytical solution can be found. This solution is useful in various situations that include the validation of constitutive models, the stability assessments of circular openings such as borehole and TBM excavated tunnel, the verification of numerical codes, the construction of ground-support reaction curves, etc. In order to obtain ground response curves for circular tunnels, a number of analytical solutions have been presented by considering the elastic-perfectly plastic and elastic–brittle–plastic models of material behavior with the linear Mohr-Coulomb (M-C) and nonlinear Hoek-Brown (H-B) criteria (Brown et al., 1983; Detournay, 1986; Wang, 1996; Carranza-Torres and Fairhurst, 1999; Sharan, 2003, 2005; Carranza-Torres, 2004; Park and Kim, 2006).

For an elastic-strain softening model, Brown et al. (1983) presented a numerical stepwise procedure for the stresses and displacements in the H-B media by assuming the constant value of elastic strain in the plastic region such as that at the elastic–plastic interface and the constant dilatancy angle in the strain softening zone. Alonso et al. (2003) proposed the self-similarity solution by solving the system of ordinary differential equations of equilibrium, persistence, and radial displacement velocity and flow rule. In routine engineering application, the dilatancy of the rock is assumed to be constant.

The aim of the present study is to apply a sensitive analysis to strength parameters of Hoek-Brown and Mohr-Coulomb criteria and investigating their effects on the ground reaction curve.

2 DEFINITION OF THE PROBLEM

Figure 1 shows a circular tunnel is excavated in a continuous, homogeneous, isotropic, initially elastic rock mass subjected to a hydrostatic stress p_o. The tunnel surface is subjected to an internal pressure p_i. As p_i is gradually reduced, the radial displacement occurs and a plastic region develops around the tunnel when p_i is less than the initial yield stress.

The material behaviour of elastic-strain softening model used in this study is shown in Figure 2. There are three different zones around the tunnel: the elastic zone, the softening zone, and the residual zone. After initial yielding, the strength of rock drops gradually with increasing strain and follows the

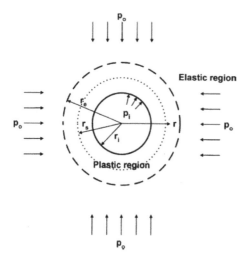

Figure 1.   A circular tunnel in an infinite medium.

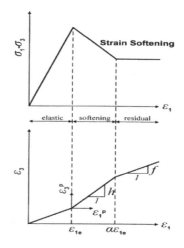

Figure 2.   Elastic-strain softening model (Brown et al., 1983).

post-yield softening behaviour. It is required to solve for the stresses and displacements in the plastic region to obtain the ground reaction curve.

Hoek-Brown yield criterion is implemented in this study as nonlinear yield function:

$$\sigma_\theta = \sigma_r + (m\sigma_c\sigma_r + s)^{0.5} \qquad (1)$$

in which σ_c = the uniaxial compressive strength of the intact rock material, m and s = material constants which depend on the properties of the rock and on the extent to which it has been broken before being subject to the stresses.

Mohr-Coulomb yield criterion is used here as linear yield function:

$$\sigma_\theta = \sigma_c + \sigma_r K(\varphi) \qquad (2)$$

where $K(\varphi)$ and σ_c is defined as following:

$$K(\varphi) = \frac{1+\sin\varphi}{1-\sin\varphi} \qquad (3)$$

φ = the friction angle of rock mass and c is the rock mass cohesive strength. Strength parameters change gradually from peak to residual values for both criteria as following:

$$\bar{m} = m_p - \frac{(m_p - m_r)\gamma^p}{\gamma^{p*}} \qquad (5)$$

$$\bar{s} = s_p - \frac{(s_p - s_r)\gamma^p}{\gamma^{p*}} \qquad (6)$$

$$\bar{\varphi} = \varphi_p - \frac{(\varphi_p - \varphi_r)\gamma^p}{\gamma^{p*}} \qquad (7)$$

$$\bar{C} = C_p - \frac{(C_p - C_r)\gamma^p}{\gamma^{p*}} \qquad (8)$$

γ^p = the softening parameter, and γ^{p*} = the value of the softening parameter controlling the transition between the softening and residual stages. In this study the softening parameter is defined as the difference between the major and minor principal plastic strains, which reflects the plastic shear strain:

$$\gamma^p = \varepsilon_\theta^p - \varepsilon_r^p = (\varepsilon_\theta - \varepsilon_\theta^e) - (\varepsilon_r - \varepsilon_r^e) \qquad (9)$$

there are different way of apply variable dilatancy for example the constant dilatancy (Brown et al., 1983), the linear decrease of dilatancy (Alonso et al., 2003) and an exponential decreasing of dilatancy with plasticity (Detournay, 1986; Alejano and Alonso, 2005), can be considered. In this study dilatancy angle changes linearly from peak to residual values as following:

$$\bar{\psi} = \psi_p - \frac{(\psi_p - \psi_r)\gamma^p}{\gamma^{p*}} \qquad (10)$$

2.1   Stepwise procedure for elastic-strain softening rock

Assuming a state of plane strain and axial symmetry around the tunnel opening, the equilibrium equation and strain–displacement relation in polar coordinate system are given by

$$\frac{d\sigma_r}{dr} + \frac{\sigma_r - \sigma_\theta}{r} = 0 \qquad (11)$$

108

$$\varepsilon_r = -\frac{du}{dr} ...., ...\varepsilon_\theta = -\frac{u}{r} \qquad (12)$$

By dividing the plastic region with a number of thin annular rings, one can obtain the radial stress at r_j for the H-B yield criterion (Brown et al., 1983).

$$\sigma_{r(j)} = b - \sqrt{b^2 - a}. \qquad (13)$$

where

$$a = \sigma_{r(j-1)} - 4k_1\left[\frac{1}{2}\bar{m}\sigma_c\sigma_{r(j-1)} + \bar{s}\sigma_c^2\right] \qquad (14)$$

$$b = \sigma_{r(j-1)} + k_1 m\sigma_c \qquad (15)$$

$$k_1 = \left(\frac{r_{j-1} - r_j}{r_{j-1} - r_j}\right)^2 \qquad (16)$$

A similar stepwise procedure was extracted in this study for Mohr-Coulomb criterion. The radial stress at r_j for the M-C yield criterion is obtained as following:

$$\sigma_{r(j)} = \frac{1}{1 + k_2(1 - K(\varphi))} \qquad (17)$$
$$[k_2(\sigma_c + \sigma_{\theta(j-1)} - \sigma_{r(j-1)}) + \sigma_{r(j-1)}]$$

$$k_2 = \left(\frac{r_{j-1} - r_j}{r_{j-1} + r_j}\right) \qquad (18)$$

$$k(\bar{\varphi}) = \frac{1 + \sin\bar{\varphi}}{1 - \sin\bar{\varphi}} \qquad (19)$$

$$\bar{\sigma}_c = \frac{2\bar{c}\cos\bar{\varphi}}{1 - \sin\bar{\varphi}} \qquad (20)$$

using the radii given by

$$\frac{r_j}{r_{j-1}} = \frac{2\varepsilon_{\theta(j-1)} - \varepsilon_{r(j-1)} - \varepsilon_{r(j)}}{2\varepsilon_{\theta(j)} - \varepsilon_{r(j-1)} - \varepsilon_{r(j)}} \qquad (21)$$

Let the circumferential strain increment be:

$$d\varepsilon_\theta = \varepsilon_{\theta(j)} - \varepsilon_{\theta(j-1)} \qquad (22)$$

Then, the radial strain increment at rj can be obtained as

$$d\varepsilon_r^p = -\beta d\varepsilon_\theta^p \qquad (23)$$

$$d\varepsilon_{r(j)} = d\varepsilon_{r(j-1)}^\ell - \beta(d\varepsilon_{\theta(j)}^\ell - d\varepsilon_{\theta(j-1)}^\ell) \qquad (24)$$

Where $d\varepsilon_{r(j-1)}^\ell, d\varepsilon_{\theta(j-1)}^\ell$ = the radial and circumferential elastic strain increments at r_{j-1}, respectively, and $\beta = \frac{1+\sin\bar{\psi}}{1-\sin\bar{\psi}}$.

By selecting an arbitrary small value for $d\varepsilon_\theta$, the values of $\varepsilon_{\theta(j)}$, $\varepsilon_{r(j)}$, $\varepsilon_{\theta(j)}, r_{j-1}$ and u_j can be calculated using Eqs. (22), (23), (21) and (12) in that order. The process is then repeated several times to determine completely the strains and displacements in the plastic region. The elastic strains at rj can be obtained by using the elastic stress–strain relationship,

$$\varepsilon_{r(j)}^\ell = \frac{1}{2G}[(1-v)(\sigma_{r(j)} - p_o) - v(\sigma_{\theta(j)} - p_o)] \qquad (25)$$

$$\varepsilon_{\theta(j)}^\ell = \frac{1}{2G}[(1-v)(\sigma_{\theta(j)} - p_o) - v(\sigma_{r(j)} - p_o)] \qquad (26)$$

where G = the shear modulus and v = the Poisson's ratio of the rock.

The radial stress at the elastic–plastic interface, in which $r = r_e$, is given for H-B criterion by

$$\sigma_{re} = p_o - M\sigma_c \qquad (27)$$

$$M = \frac{1}{2}\left[\left(\frac{m_p}{4}\right)^2 + \frac{m_p p_o}{\sigma_c} + s_p\right] - \frac{m_p}{8} \qquad (28)$$

It is obtained for M-C criterion as following:

$$\sigma_{re} = \frac{2p_o - \sigma_c}{1 + K(\varphi)} \qquad (29)$$

Using Eq. 27 and 29 as the starting point, successive values of $\sigma_{r(j)}$ can be calculated from Eq. 13 and 17 for the radii determined from Eq. 21.

Critical deviatoric plastic strain at rj can be approximately estimated as

$$\gamma^p = \varepsilon_{\theta(j-1)} - \varepsilon_{r(j-1)} + \frac{1}{2G}\left\{\delta_{r(j-1)} - \delta_{\theta(j-1)}\right\} \qquad (30)$$

if γ^p exceeds γ^{p*} strength of rock mass reaches residual condition.

3 APPLICATION

Data sets 1–1, 1–2 (related to rock masses with elastic-strain-softening behavior) and 2 (related to a rock mass with elastic-brittle-plastic behaviour) are applied in this study.

In the first stage the sensitive analysis is implemented for Hoek-Brown criterion by changing m and s parameters.

Table 1. Data sets (H-B and M-C criterion).

	Data set 1–1	Data set 1–2	Data set 2
Young's modulus, E (Mpa)	1380	1380	40000
Poisson's ratio, v	0.25	0.25	0.2
Initial stress, po (MPa)	3.31	10*	108
Radius of tunnel, r_i (m)	5.35	5.35	4
δ_c (Mpa)	27.6	30	300
m_p	0.5	4.5	7.5
s_p	0.001	0.02	0.1
m_r	0.1	0.45*	1
s_r	0	0.002*	0.01
φ_p	33*	43*	46*
φ_r	21*	24*	29*
c_p (Mpa)	0.38*	1.61*	23.17*
c_r (Mpa)	0.19*	0.73*	11.34*
ψ_p	19.47	27*	35*
ψ_r	5.22	0	0
γ^*	0.004742	0.004742*	0.004*

* Assumed value.

Table 2. Relative deformation by variation of H-B and M-C parameters.

	Data set 1–1	Data set 1–2	Data set 2
$s_r = \dfrac{s_p}{3} \sim \dfrac{2}{3}s_p$	10.8	12.8	8.6
$s_r = \dfrac{2}{3}s_p \sim s_p$	5.6	6.8	4.3
$n_r = \dfrac{m_p}{3} \sim \dfrac{2}{3}m$	25.5	12.6	8.9
$m_r = \dfrac{2}{3}m_p \sim m_p$	8.5	5.9	6.6
$c_r = \dfrac{c_p}{3} \sim \dfrac{2}{3}c_p$	116.5	59.4	42.5
$c_r = \dfrac{2}{3}c_p \sim c_p$	26.7	18.2	13.8
$\varphi_r = \dfrac{\varphi_p}{3} \sim \dfrac{2}{3}\varphi_p$	282.3	76.8	39.5
$\varphi_r = \dfrac{2}{3}\varphi_p \sim \varphi_p$	54.5	33.6	24.6

Suppose that s_r parameter changes gradually into $\frac{s_p}{3}, \frac{2}{3}s_p$ and s_p. As it is observed from figure 3a, increase of s_r parameter reduces displacement at the opening surface for data set 1–1, this behavior is the same for data set 1–2 and 2. Suppose that the relative displacement is obtained by division of the displacement at special $s_r(\frac{s_p}{3}, \frac{2}{3}s_p$ or $s_p)$ to the displacement in which s_r becomes equal to its real value of rock mass (the relative deformation for other rock mass parameters are calculated in the same way). The relative deformation for a rock mass with elastic-brittle plastic behavior (data set 2) is smaller than that of strain softening rock mass.

In the next stage parameter m_r of rock mass changes into $\frac{m_p}{3}, \frac{2}{3m_p}$ and m_p. Figure 3b shows that increase of m_r parameter makes a reduction with displacement for data set 1–1. This behavior is repeated for data set 1–2 and 2. It is inferred based on table 2 that relative displacement for data set 1–1 is larger than that of data set 1–2 and for data set 1–2 is also larger than that of data set 2. It shows that variation of m_r parameter causes smaller relative displacement with increase of rock mass quality.

It is observed from table 2 that keeping m and s constant is more important for medium quality rock masses. Specially reduction of m_r and s_r parameters into values less than $\frac{2}{3m_p}$ and $\frac{2}{3}s_p$

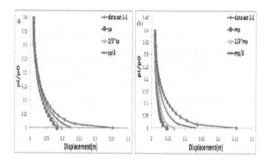

Figure 3. Effect of variation of s_r (a) and m_r (b) parameters on GRC for data set 1–1.

causes smaller relative deformation in rock masses in elastic-strain softening behavior. Therefore, it should be attempted to keep m and s parameters on values larger than $\frac{2}{3m_p}$ and $\frac{2}{3}s_p$ by means of mechanical excavation or controlled blasting.

Sensitive analysis of Mohr-Coulomb parameters is investigated in this stage. First c_r parameter changes into $\frac{c_p}{3}, \frac{2}{3c_p}$, and c_p. Figure 4a shows that increment of c_r parameter causes a reduction in rock mass deformation for data set 1–1, this behavior is the same for data set 1–2 and 2. It is observed from table 2 that relative displacement for data set 1–1 is larger than that of data set 1–2 and for data set 1–2 is also larger than that of data set 2.

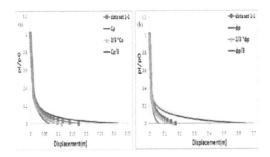

Figure 4. Effect of variation of c_r (a) and φ_r (b) parameters on GRC for data set 1–1.

It shows that sensitivity of rock mass deformation decreases in terms of variation of the c_r parameter with increasing rock mass strength.

In the next stage it was managed to change parameter φ_r into $\frac{\varphi_p}{3}, \frac{2}{3}\varphi_p$ and φ_p. It is observed from figure 4b that increasing φ_r parameter causes a reduction in rock mass deformation for data set 1–1. Table 2 shows the same procedure for data set 1–2 and 2. It can be inferred by comparing these three set of data that relative displacement of rock mass decreases vs. variation of parameter φ_r from data set 1–1 to 2 based on table 2. In other word, relative rock mass displacement reduces in term of variation of φ_r parameter with increasing strength of rock mass. It is also observed that increment of φ_r and c_r parameters cause a reduction in rock mass deformation. However these variations are more intense for low values of φ_r and c_r.

It can be concluded from table 2 that rock mass deformation is more sensitive to the variation of Mohr-Coulomb parameters than Hoek-Brown ones.

4 CONCLUSION

- By comparison of the relative displacement caused by changing strength parameters in Hoek-Brown and Mohr-Coulomb criteria, it can be concluded that the sensitivity of rock mass deformation is higher for Mohr-Coulomb criterion than Hoek-Brown in respect of variation of strength parameters.
- It is inferred from analyses that with improving rock mass quality, the sensitivity of rock mass deformation reduces in respect of strength parameter variation for Mohr-Coulomb criterion.
- It is observed from table 2 that variation of rock mass strength parameters causes more relative deformation for elastic-strain softening rock mass than elastic-brittle-plastic. Therefore m and

s parameters are more important for rock mass of medium quality than good quality. Reduction of m_r and s_r parameters less than $\frac{2}{3m_p}$ and $\frac{2}{3}s_p$ causes more deformation in the rock mass, therefore it should be managed to retain m_r and s_r parameters more than $\frac{2}{3}m_p$ and $\frac{2}{3}s_p$ by means of controlled blasting or mechanical excavation.

- According to above mentioned results, it is concluded that the method of excavation should be designed so that it causes the least reduction in strength parameters of rock mass first for medium quality (often with elastic-strain softening behavior) and in the second order for good quality of rock masses (often with elastic-brittle-plastic behavior).

REFERENCES

Alonso, E., Alejano, L.R., Varas, F., Fdez-Manin, G., Carranza-Torres, C., 2003. Ground response curves for rock masses exhibiting strainsoftening behavior. *Int. J. Numer. Anal. Meth. Geomech.* 27, 1153–1185.

Brady, B.H.G., Brown, E.T., 1993. *Rock Mechanics for Underground Mining.* Chapman & Hall.

Brown, E.T., Bray, J.W., Ladanyi, B., Hoek, E., 1983. Ground response curves for rock tunnels. *J. Geotech. Eng. ASCE* 109, 15–39.

Carranza-Torres, C., Fairhurst, C., 1999. The elasto-plastic response of underground excavations in rock masses that satisfy the hoek-brown failure criterion. *Int. J. Rock Mech. Min. sci.* 36, 777–809.

Detournay, E., 1986. Elastoplastic model of a deep tunnel for a rock with variable dilatancy. *Rock Mech. Rock Eng.* 19, 99–108.

Duncan Fama, M.E., Trueman, R., Craig, M.S., 1995. Two and three dimensional elasto-plastic analysis for coal pillar design and its application to highwall mining. *Int. J. Rock Mech. Min. sci. & Geomech. Abstract.* 32, 215–225.

Florence, A.L., Schwer, L.E., 1978. Axisymmetric compression of a Mohr- Coulomb medium around a circular hole. *Int. J. Number. Anal. Meth. Geomech.* 2, 367–379.

Guan, Z., Jiang, Y., Tanabasi, Y., 2007. Ground reaction analyses in conventional tunneling excavation. *Tunnel. Under. space Technol.* 22, 230–237.

Hoek, E., Brown, E.T., 1980. Underground Excavation in rock. *Institution of Mining and Metallurgy*, London.

Kargar, A.R., Rahmanejd, R., 2009. The effect of various constitutive models on ground response curve. *M.Sc. Thesis*, Bahonar University of Kerman.

Lee, Y.-K., Pietruszczak, S., 2008. A new numerical procedure for elasto-plastic analysis of a circular opening excavated in a strain-softening rock mass. *Tunnelling and Underground Space Technology* 23 (2008) 588–599.

Park, K.-H., Tontavanich, B., Lee, J.-G., 2007. A simple procedure for ground response curve of circular tunnel in elastic-strain softening rock masses. *Tunnelling and Underground Space Technology* 23 (2008) 151–159.

Volcanic Rock Mechanics – Olalla et al. (eds)
© 2010 Taylor & Francis Group, London, ISBN 978-0-415-58478-4

Isotropic collapse load as a function of the macroporosity of volcanic pyroclasts

A. Serrano
E.T.S.I.C.C.P., Universidad Politécnica de Madrid, Madrid, Spain

A. Perucho & M. Conde
Laboratorio de Geotecnia, CEDEX, Madrid, Spain

ABSTRACT: Two main types of "macroporosity" in pyroclastic volcanic rocks can be distinguished: "reticular" and "vacuolar". The first type is produced when the large pores are located between grain particles. The second type is produced when large pores are located inside a vitreous rock mass. However, a mixed type can be defined in most cases, when both kinds of void are present. At other times, the pyroclasts do not exhibit any kind of "macroporosity". An extensive study of how the type of porosity may affect the strength of the material is being carried out at CEDEX geotechnical laboratory. Samples with different types of "macroporosity" have been tested under isotropic loads. As a result of a theoretical study, an expression of the isotropic collapse load has been obtained, for any type of "macroporosity", and compared with test results.

1 PORE MACROSTRUCTURE

Two basic pore macrostructures can mainly be distinguished (CEDEX, 2007; Serrano et al, 2007a & b; Santana et al, 2008):

a. reticular structure (Fig. 1a)
b. vacuolar structure (Fig. 1b)

The reticular type corresponds to the pore structure of the rock made up of an aggregate of particles joined together either by the heat action produced in their formation, or by some type of cement. The vacuolar type corresponds to the pores of a rock which are all vacuolar in type, formed due to the gas formed inside the viscous lava as it is expelled from the volcano.

The vacuolar pores generally have no connection between them while the reticular ones are generally connected between them.

In general rocks, and particularly macroporous volcanic rocks contain pores of both types. In this case they are said to have a mixed pore structure (Fig. 1c), and the total porosity is:

$$n = n_R + n_V$$

where n_R = reticular porosity and n_V = vacuolar porosity.

More complex cases and structures of pores may exist where the grains are in turn formed by micrograin agglomerates which in turn have microvacuolas.

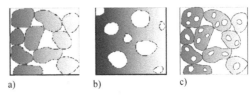

a) b) c)

Figure 1. Pore structures: a) Reticular; b) Vacuolar; c) Mixed.

2 ISOTROPIC COLLAPSE LOAD

2.1 *Rocks with reticular pore structure*

Take a ball of macroporous rock with a reticular pore structure and subject it to an external isotropic pressure, p (Fig. 2a).

The amount of work supplied to the ball by the external pressure p is:

$$Te = p\Delta V$$

ΔV is the total change of volume experimented by the sample which can be decomposed into two addends, corresponding to the rock (ΔV_i) and the reticulated pores (ΔV_R):

$$\Delta V = \Delta V_i + \Delta V_R$$

The work T_i carried out by the internal stresses of the intact rock is:

$$T_i = \sigma_i \Delta V_i$$

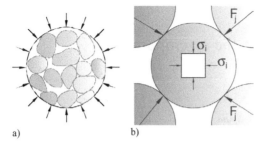

a)   b)

Figure 2. Isotropic compression test. (a) Reticular structure. (b) Detail of particle with the contact forces.

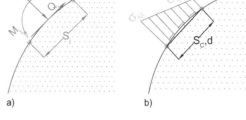

a)   b)

Figure 3. Layout of the forces on a contact (a); Stress in the contact.

where σ_i is an average isotropic stress to which the intact rock is subjected as the result of an effect of the exterior isotropic pressure, p.

Both works have to be equal, $T_i = T_e$, so:

$$\sigma_i \Delta V_i = p \Delta V$$

then:

$$\sigma_i = p \frac{\Delta V}{\Delta V_i}$$

Admitting that $\Delta V_i / V_i = \Delta V / V$ (volumetric deformation of the intact rock equals the total), gives:

$$\sigma_i = p \frac{V}{V_i} = p \frac{1}{1 - n_R}$$

The average isotropic stress, σ_i, to which the intact rock particles were submitted are transferred through forces F_j which operate on the contacts between the other particles connected to them (Fig. 2b). Each F_j can be broken down into a normal force, P_j, a tangential force, Q_j, and a flexural moment, M_j (Fig. 3).

Coefficients λ and μ are defined as:

$$Q_j = \lambda_j P_j$$

$$M_j = \mu_j D P_j$$

where D is the size of the particle.

P is displaced a distance μD from the centre of the contact.

If N is the coordination number (number of contacts of the particle), then:

$$NP = S\sigma_i$$

where P is the average value of the normal components of the contact forces. Then:

$$P = \frac{S}{N}\sigma_i = \frac{S}{N}\left(\frac{p}{1 - n_R}\right) \quad (1)$$

If λ and μ are average coefficients, then:

$$\lambda P = Q$$

$$\mu D P = M$$

where Q and P are the transverse and normal force respectively and M is the average flexural moment in the contact.

In the contact between particles there is an average stress, σ_c, defined by:

$$\sigma_c = \frac{P}{S_c} = \frac{S}{S_c N}\left(\frac{p}{1 - n_R}\right) = \frac{p}{w(1 - n_R)} \quad (2)$$

where S_c = contact area (Fig. 3b); w = welding coefficient defined as:

$$w = \frac{N S_c}{S}$$

Maximum and minimum normal stresses (σ_M and σ_m, respectively) are:

$$\sigma_M = \frac{P}{S_c}\left(1 + \frac{\mu D}{d}\right) = \sigma_c\left(1 + \mu \frac{D}{d}\right) \quad (3)$$

$$\sigma_m = \frac{P}{S_c}\left(1 - \frac{\mu D}{d}\right) = \sigma_c\left(1 - \mu \frac{D}{d}\right) \quad (4)$$

where d = diameter of the contact.

The maximum shear in the contact is:

$$\tau_M = \frac{3}{2}\lambda \sigma_c \quad (5)$$

According to Equations (2), (3) (4) and (5):

$$\sigma_M = \left(1 + \mu \frac{D}{d}\right)\frac{p}{w(1 - n_R)}$$

$$\sigma_m = \left(1 - \mu\frac{D}{d}\right)\frac{p}{w(1-n_R)}$$

$$\tau_M = \frac{3}{2}\lambda\frac{p}{w(1-n_R)}$$

Contact can be broken as a result of compression, traction, shear or a combination of these three types of stress. By simplification the contact can be said to break when:

$$\sigma_{cri}(K_{cri})\frac{p}{w(1-n_R)}$$

where σ_{cri} the stress of what ever type will produce failure.

K_{cri} is a function of λ and μ:

$$K_{cri} = K_{cri}(\lambda, \mu)$$

The λ and μ coefficients depend on the overlap of the particles. A high overlap means many contacts and the shear stresses and moments in these contacts are very small in relation to the normal components. On the contrary however, in very open structures with few contacts the moments and shear stresses increase relatively.

All of these reasons suggest a ratio of the type:

$$K_{cri} = K(n)$$

or better still:

$$K_{cir} = K\left(\frac{n}{1-n}\right)$$

The simplest function that can be adopted is:

$$K_{cri} = \frac{1}{a_R}\left(\frac{n}{1-n}\right)^{a_R}$$

Finally, when external pressure p is so big that it reaches σ_{cri}, the collapse of the macroporous structure is produced. Thus:

$$p_{ci} = \sigma_{cru}a_R w(1-n_R)\left(\frac{1-n_R}{n_R}\right)^{\alpha_R} \qquad (6)$$

If the specific weight of the intact rock is G and the specific weight of the rock with pores is γ, then:

$$\gamma = (1 - n_R)G$$

Subsequently Equation (6) can be expressed the following way:

$$p_{ci} = \left(\sigma_{cri}\frac{a_R w}{G}\right)\gamma\left(\frac{\gamma}{G-\gamma}\right)^{\alpha_R}$$

$$= L_R\gamma\left(\frac{\gamma}{G-\gamma}\right)^{\alpha_R} \qquad (7)$$

where L_R is a parameter with longitudinal dimension:

$$L_R = \sigma_{cri}\frac{a_R w}{G}$$

L_R and α_R depend on:

- the critical strength σ_{cri};
- the welding coefficient w;
- the overlap a_R.

2.2 Rocks with vacuolar pore structures

As previously seen, when a macroporous rock is subjected to an isotropic external pressure, p (Fig. 4), it undergoes isotropic stress, σ_i:

$$\sigma_i = p\left(\frac{1}{1-n_v}\right)$$

where n_v is the vacuolar porosity in this case.

The σ_i stress is an average stress.

There are concentrations of circumferential stresses on the edges of the vacuolas (Fig. 4), which will be larger the closer the vacuolas are and the larger they are in relation to the intervacuolar space. Peak stresses, σ_{ic}, on the edge are:

$$\sigma_{ic} = k\sigma_i$$

thus

$$\sigma_{ic} = k\left(\frac{p}{1-n_V}\right)$$

where k is a concentration factor.

The rock will collapse when σ_{ic} reaches a certain critical value σ_{cri}, which is thought to be

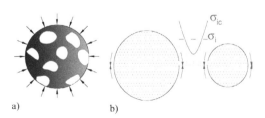

a)   b)

Figure 4. Isotropic compression test (a) Vacuolar structure. (b) Detail of the intervacuolar stress.

115

related to the unconfined strength of the intact rock.

Function k needs to check that for very low porosities the collapse pressure tends to be infinite and for porosities close to the unit this pressure is negligible.

The simplest function meeting the limit conditions is of the type:

$$k = \frac{1}{a_V}\left(\frac{n_V}{1-n_V}\right)^{\alpha_V}$$

The result is that the isotropic collapse load of rocks with a vacuolar porous structure takes on the form:

$$p_{ci} = a_V\left(1-n_V\right)\left(\frac{1-n_V}{n_V}\right)^{\alpha_V}\sigma_{cir}$$

in other words,

$$p_{ci} = a_V\frac{\sigma_{cri}}{G}\gamma\left(\frac{\gamma}{G-\gamma}\right)^{\alpha_V} = a_V\gamma\left(\frac{\gamma}{G-\gamma}\right)^{\alpha_V} \quad (8)$$

where L_R is a parameter with longitudinal dimension:

$$L_V = \sigma_{cri}\frac{a_V}{G}$$

L_V and α_V depend on:

– another intrinsic strength of the intact rock, σ_{criV} (linked to unconfined strength);
– the size and proximity of the pores;
– the shape and distribution of the pores.

3 PRACTICAL CONCLUSION

Equations (7) and (8) have the same structure, differing only in the values for parameters L and α.

It is extremely difficult, not to say impossible to determine separately parameters σ_{cri}, w, a_R and a_V. On the other hand it is relatively easy to determine parameters L_R and L_V, as also exponents α_R and α_V thus using isotropic collapse tests adapting Equations (7) and (8) to tests on the same rock and different densities.

Generally speaking, there will not be a clear reticular or vacuolar structure but rather a mixed one as a result of a single equation is proposed for the isotropic collapse load of macroporous rocks:

$$p_{ci} = L\gamma\left(\frac{\gamma}{G-\gamma}\right)^{\alpha} \quad (9)$$

G, L and α are to be determined with tests.

Figures 5 and 6 show correlation between Isotropic collapse pressure and dry density, and Statistical analysis for different values of α, respectively.

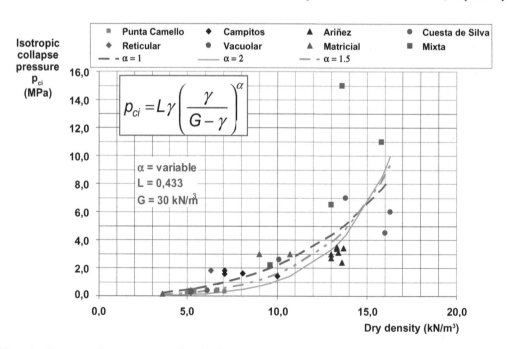

Figure 5.   Isotropic collapse pressure vs. Dry density.

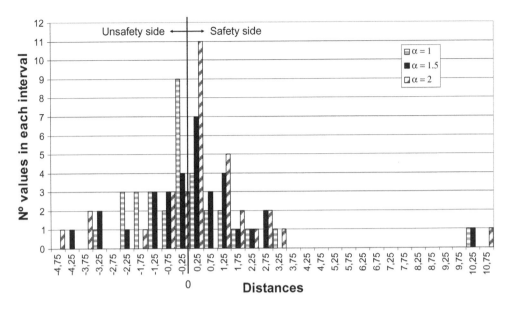

Figure 6. Statistical analysis for different values of α. The distance is the difference between real and calculated pressure values.

In principle the suggestion is to adopt:

$G = 30$ kN/m³, $\alpha = 1$ (for best fit) or $\alpha = 2$ (safest value for applications) and $L = 0.433$, as from the fit carried out on the pyroclastic samples from the Canary Islands.

At the CEDEX Geotechnics Laboratory research is being carried out on Canary Islands pyroclastic samples and on artificial samples with a view to obtaining greater reliability in these parameters.

REFERENCES

CEDEX, 2007. Caracterización geotécnica de los piroclastos canarios débilmente cementados. *Final report* April, 2007.

Serrano, A. Olalla, C., Perucho, A. & Hernández, L. 2007. Strength and deformability of low density pyroclasts. *ISRM International Workshop on Volcanic Rocks*. Ponta Delgada, Azores, 14 July, 2007.

Serrano, A., Perucho, A., Olalla, C. & Estaire, J. 2007. Foundations in Volcanic Areas. *XIV European Conference on Soil Mechanics and Geotechnical Engineering. Geotechnical Engineering in Urban Environments.* Madrid, 24–27 September, 2007.

Santana, M de Santiago, C., Perucho, A. & Serrano, A. 2008. Relación entre características químico-mineralógicas y propiedades geotécnicas de piroclastos canarios. *VII Congreso Geológico de España*. Geo-Temas 10. Las Palmas de Gran Canaria, July, 2008.

Volcanic Rock Mechanics – Olalla et al. (eds)

General method for estimating the active and passive earth pressures on retaining walls assuming different strength criteria

A. Serrano
E.T.S.I.C.C.P., Universidad Politécnica de Madrid, Madrid, Spain

A. Perucho & M. Conde
Laboratorio de Geotecnia, CEDEX, Madrid, Spain

ABSTRACT: A new method for estimating the earth pressures on retaining walls has been developed. It is an extension of Coulomb's earth pressure theory for non cohesive materials that can follow a non-linear strength criterion. This was previously done by the authors (Serrano et al, 2007) for some basic assumptions that have now been extended. The method is valid for materials that may have either a linear or non-linear strength criterion (parabolic or Hoek-Brown), a non-horizontal surface and an earth-wall friction angle. The method considers the material dilatancy. Moreover, the failure surface does not need to be plane, as in previously developed methods, but its shape is obtained as a result of the calculus, by applying Euler's variational method that obtains the extremal force.

1 INTRODUCTION

This study is an extension of coulomb's classic method of calculating wall pressures on incoherent materials with a non linear strength criterion

At the present time the calculation of pressures on these materials requires a previous linearization of the failure criterion. This linearization is always problematical so its success depends on the election of a carefully chosen range of stresses.

The fundamental hypothesis of the Coulomb method was abandoned -the adoption of a plane for the failure surface- and now this surface is obtained directly by Euler's variational method making the force extremal. Therefore it is necessary to dispose of a calculation method for non-linear media able of incorporating these new failure surface.

The results of this study allow determination of the pressures on walls due to materials such as rockfills, highly fractured rock masses, pyroclasts, etc., whose mechanical behaviour is clearly non linear.

2 BASIC HYPOTHESES

2.1 *Geometrical (cf Fig. 1a)*

1. The wall is assumed to be indefinite so that it is a two-dimensional problem in plane deformation condition.

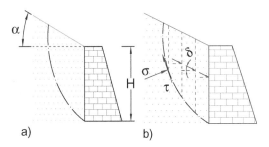

Figure 1. Wall geometry (a) and geomechanical hypotheses (b).

2. The earth surface is plane, forming an α angle with the horizontal.
3. The wall has a vertical backfilling.

2.2 *Geomechanical (cf Fig. 1b)*

1. A wedge of earth limited by a surface passing through the foot of the wall leaning against the wall.
2. The earth is dry, i.e., pore pressures are not taken into account. The calculation is done on effective pressures.
3. The earth has a specific weight of γ.
4. The normal (σ) and tangential (τ) stresses acting on the surface of the wedge can verify a failure criterion of the type (Fig. 2):

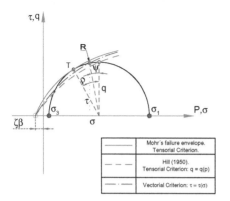

Figure 2. Tensorial and vectorial types failure criteria.

- vectorial, Coulomb type: $\tau = \tau(\sigma)$
- tensorial, Mohr type: $q = q(p)$. In this case it is necessary to know the flow rule in plane deformation to determine the failure plane and then the stresses acting on it.

$$q = \frac{\sigma_1 - \sigma_3}{2}; \quad p = \frac{\sigma_1 + \sigma_3}{2}$$

Flow rule is:

$$\sin \psi = N(\sin \rho)$$

where ρ = angle of instant friction; ψ = angle of dilatancy; $N(\sin \rho)$ the dilatancy function.
   Both failure criteria are related:

$$\sin \rho = \frac{dq}{dp}$$

$$\tau = q \cos \psi; \quad \sigma = p - q \cos \psi$$

where $q = q(\rho)$ and $\sin \psi = N \sin \rho$.
5. The total pressure on any vertical plane of the wedge forms a constant δ angle with the horizontal. This angle is called the earth-wall-friction angle and is defined by its tangent:

$$\mu = \tan \delta$$

6. The failure surface produces the extremal force on the wall (maximum and minimum in the active and passive case respectively).
7. For the maximum force on the wall to be produced the wall has to be moved away from the earth while for the minimum the wall has to be moved towards it. In these situations the wedge attempts to lower or rise itself and thus there are established the directions of the reactions on the failure surface of the wedge.

3 EQUILIBRIUM EQUATIONS

3.1 *Differential system*

The horizontal and vertical balance of forces and the balance of momentum let to get the following differential system (Fig. 3):

$$\frac{dn}{dx} = \sigma - \tau y' \tag{1}$$

$$\frac{dt}{dx} = \gamma h y' - (\sigma y' + \tau) \tag{2}$$

$$\frac{dm}{dx} = n - t y' \tag{3}$$

where:

$$t = \mu n; \quad m = nB; \quad h = x - y \tan \alpha$$

being μ a constant.
   Basic convention: from now onwards all the developments will be done adimensionally. To achieve this the dimensional measuring units in Table 1 are adopted:
   Having solved any problem the real results would have to be obtained by multiplying them by the respective physical units.
   This will give the basic differential system:

$$\frac{dn}{dx} = \sigma - \tau y' \tag{4}$$

$$\mu \frac{dn}{dx} + \left(\frac{nd\mu}{dx}\right) = (x - ytg\alpha) y' - (\sigma y' + \tau) \tag{5}$$

$$\frac{dm}{dx} = (1 - \mu y') n \tag{6}$$

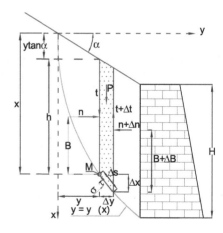

Figure 3. Forces in a slice.

Table 1. Dimensional measuring units.

Parameter	Unit
Densities: γ	kN/m³
Pressures: β	kN/m²
Lengths: $L = \beta/\gamma$	m
Force: $L\,\beta = \beta^2/\gamma$	kN/m
Moments: $L^2\,\beta = \beta^3/\gamma^2$	kN

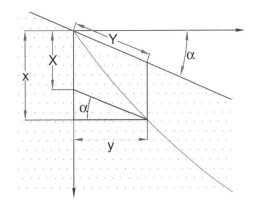

Figure 4. Oblique axes (X, Y).

Combining (4) and (5) and expressing the equations in oblique axes (Fig. 4) gives the system:

$$\frac{dn}{dX} = \sigma + (\sigma \sin\alpha - \tau \cos\alpha)Y' = A_0 + A_1 \frac{dY}{dX} \qquad (7)$$

$$XY' = \frac{\mu\sigma + \tau}{\cos\alpha} + [\sigma(\mu \tan\alpha + 1) \\ + \tau(\tan\alpha - \mu)]|Y' = B_0 + B_1 \frac{dY}{dX} \qquad (8)$$

$$\frac{dm}{dX} = [1 + (\sin\alpha - \mu\cos\alpha)Y']n = \left(1 + C_1 \frac{dY}{dX}\right)n \qquad (9)$$

where:

$$A_0 = \sigma, \; B_0 = \frac{\mu\sigma + \tau}{\cos\alpha}$$

$$A_1 = \sigma \sin\alpha - \tau \cos\alpha; \; C_1 = \sin\alpha - \mu\cos\alpha \qquad (10)$$

$$B_1 = \sigma(1 + \mu \tan\alpha) + \tau(\tan\alpha - \mu)$$

3.2 Basic results

The final aim is to determine n (horizontal component of the force on the wall) and m (tilting moment acting on the foot on the wall) for X = H, where H is the height of the wall.

$$n = \int_0^H (A_0 + A_1 Y')dX \qquad (11)$$

The Y(X) function must be found, which defines the wedge surface making the integral maximum with the condition of the link $XY' = B_0 + B_1 Y'$. Having obtained Y', Y(X) and n determines m.

4 FAILURE LINE

If the failure criterion is expressed parametrically it gives:

$$\tau = \tau(u); \; \sigma = \sigma(u) \qquad (12)$$

where u is a parameter that can be the stress, σ, or any other.

Equation (11), taking into account equations (10) and (12), adopts the form:

$$n = \int_0^H F\left(u, \frac{dY}{dX}\right)dX \qquad (13)$$

Thus, Equation (8) leads to:

$$u = U\left(X, \frac{dY}{dX}\right) \qquad (14)$$

Taking parameter u from Equation (14) to Equation (13) it is finally obtained:

$$n = \int_0^H F\left(X, \frac{dY}{dX}\right)dX \qquad (15)$$

In other words, n is a functional of the failure line Y = Y(X).

In this integral the function Y(X) must be sought, which will make it extremal, in other words the maximum or minimum of the functional must be found depending on whether active or passive pressure is sought. This can be done by applying Euler's variational method.

In this case the Euler condition is simply:

$$F_{y'} = \frac{\partial F}{\partial Y'} = k$$

being k a constant.

Developing the equations it is finally obtained a second degree equation in Y' (Equation 16).

121

$$Y'^2 - 2SY' + P = 0 \qquad (16)$$

where:

$$S = -\frac{1}{2} \frac{A_1 B_{0u} + A_{1u} B_0}{A_1 B_{1u}}$$

$$P = \frac{A_{0u} B_0}{A_1 B_{1u}}$$

$$A_{0u} = \frac{dA_0}{du} = \sigma_u;$$

$$A_{1u} = \frac{dA_1}{du} = \sigma_u \sin\alpha - \tau_u \cos\alpha$$

$$B_{0u} = \frac{dB_0}{du} = \frac{\tau_u + \mu\sigma_u}{\cos\alpha};$$

$$B_{1u} = \frac{dB_1}{du} = \sigma_u(1 + \mu\tan\alpha) + \tau_u(\tan\alpha - \mu)$$

being: $\sigma_u = \dfrac{d\sigma}{du} \quad \tau_u = \dfrac{d\tau}{du}$

The two solutions of which define the failure line in the case of active (+) and passive (–) pressure, respectively:

$$Y_2' = S \pm \sqrt{S^2 - P}$$

5 CALCULATION PROCEDURE

5.1 Starting data for the calculation and adimensionalisation

The intrinsic physical units referred to in Table 2 are taken to be able to work adimensionally. These units depend on the geotechnical characteristics of the earth.

The intrinsic longitude is defined, $L = \beta/\gamma$.

To treat as a group the three criteria, it can be put forward:

Table 2. Basic data (for Mohr-Coulomb linear or Hoek-Brown non-linear criteria)

Data		Value	
ε	Constant	± 1	sign + active/ – pasive
α	Slope of the earth surface on the wall backfilling (°)		
μ	Earth-wall friction (°)	$\mu = \tan\delta$	
ψ	Dilatancy (°)		
ϕ	Internal friction angle of the material		Mohr-Coulomb linear criterion
β	Strength modulus (kN/m²)		
H^*	Real wall height (m)		H: adimens. height. H^*/L
GSI	Geological Strength Index		
γ^*	Specific earth mass (kN/m³)		γ: adimens. weight. = $\gamma^*/\gamma = 1$
m_0	Hoek & Brown parameter	<25	
a	Hoek & Brown constant	14 or 28	
σ_c	Unconfined strength of the material (kN/m²)		

$$\sigma = k_C \left(\frac{u}{\tan\phi} \right) + k_P (u^{\frac{1}{b}})$$
$$+ k_H \left(\left[1 + (1-n) \left(\frac{u}{\cos\psi} \right) \right]^k \frac{u}{\cos\psi} - u\tan\psi \right)$$

being:

$$k_C = \frac{(i-2)\cdot(i-3)}{2}$$

$$k_p = (3-i)(i-2)$$

$$k_H = \frac{(i-1)\cdot(i-2)}{2}$$

where $i = 1,2,3$ to Coulomb, Parabolic, or Hoek & Brown's Criterion, respectively.

The obtention of every magnitude necessary for the design of a wall can be programmed in a spreadsheet, in columns, in the following manner:

1. u	2. τ (u)	3. σ(u)	4. τ'(u)	5. σ'(u)	6. A_0	7. A_1	8. B_0	9. B_1
10. C_1	11. A_{0u}	12. A_{1u}	13. B_{0u}	14. B_{1u}	15. S	16. P	17. Y'	18. X
19. ΔX	20. ΔY	21. Y	22. Δn	23. n	24. λ	25. Δm	26. m	27. B = m/n
28. x	29. y	30. t	31. E					

The columns of interest are:

- Col. 18: variable X, which is the height of the wall, H.
- Col. 23: variable n, which is the horizontal component of the force, E_h.
- Col. 24: variable λ, which is the force coefficient (horizontal component):

$$n = E_h = \frac{1}{2} \lambda H^2$$

- Col. 26: variable m, moment with respect to foot of wall.
- Col. 27: variable B, height of force application point.
- Cols. 28 and 29: variables x and y, which define the failure line in orthogonal axes.

The final objective is to obtain the laws λ = λ(H), B = B(H) and y = y(x), as shown in Figures 5, 6,7 and 8.

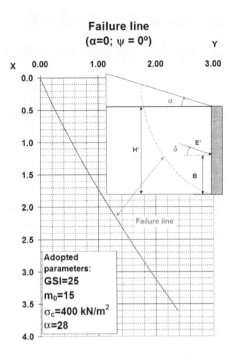

Figure 6.   Shape of failure line.

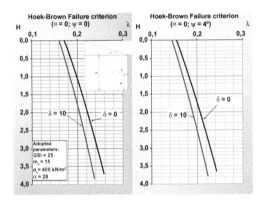

Figure 5a.   Values for force coefficient, λ, as a function of the adimensional wall, H, for α = 0°; ψ = 0° y 4°.

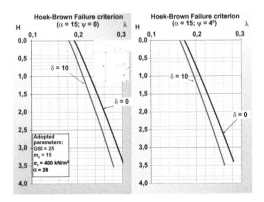

Figure 5b. Values for force coefficient, λ, as a function of the adimensional wall, H, for α = 15°; ψ = 0° y 4°.

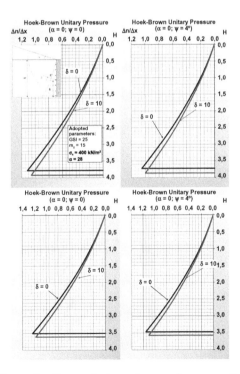

Figure 7.   Shape of the unitary earth-pressure for different hypotheses.

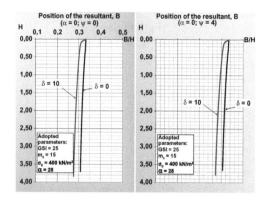

Figure 8. Position of the resultant of the force. H – (B/H) law, with the indicated parameters and for different hypotheses.

5.2 Example

Taking the case of a wall with: $\alpha = 0$, $\delta = 0$ and $H^* = 10$ m. Adopting the H&B criterion with $\psi = 0$, $\zeta = 0$, $\beta = 200$ kN/m², $\gamma = 10$ kN/m³, then: $L = \beta/\gamma = 20$ m

$H = H^*/L = 0,5$. Figures 5a and 8 give $\lambda = 0.185$ and $B \approx 0,164$.

The horizontal component of the force (E_h) and the momentum at the foot of the wall (M) are:

$$E_h = \left(\frac{1}{2}\lambda H^2\right) * Lb = 92.5 \text{ kN/m}$$

$$M = \left(E_h B\right) * L = 303.4 \text{ kNm/m}$$

REFERENCES

Serrano, A., Olalla, C. & Perucho A. 2007. Active and passive earth pressures on retaining wall assuming a non-linear strength criterion and constant dilatancy. *ISRM 11th International Congress on the Second Half Century of Rock Mechanics*. Lisbon, 9–13.
Hill, R. (1950). *The mathematical theory of plasticity*. Clarendon Press. Oxford.

Volcanic Rock Mechanics – Olalla et al. (eds)
© *2010 Taylor & Francis Group, London, ISBN 978-0-415-58478-4*

Natural stone from the Azores archipelago: Relationship between lithology and physical-mechanical behaviour

João B.P. Silva
Research Unit Geobiotec, FCT, Universidade de Aveiro, Aveiro, Portugal

Cristina Carvalho
LNEG (National Laboratory for Energy and Geology), U.C.T.M. – Lab, Infesta de Mamede, Portugal

Sérgio Diogo Caetano
ARENA (Regional Agency for Energy and Environment of the Azores Autonomous Region),
São Miguel, Portugal

Celso Gomes
Research Unit Geobiotec, FCT, Universidade de Aveiro, Aveiro, Portugal

ABSTRACT: Since the early days of the settlement in the Azores archipelago, in the 15th century, natural stone has been used in the construction of residences, religious monuments and public buildings, because of its local abundance. The purpose of this study was the characterization of the main commercial types of natural stone of the archipelago of Azores, in terms of their petrographic, mineralogical, chemical and physico-mechanical properties. This characterization allowed the appraisal of their suitability for the different types of applications. In a preliminary stage, the study comprised seven varieties of natural stones exploited in severals quarries and processing plants of São Miguel Island and Santa Maria Island—two of the nine islands that form the Azores archipelago. Petrographic studies and chemical analysis were performed in all the selected varieties as well as the following physical-mechanical tests (carried out according to European Standards): uniaxial compressive strength, flexural strength under concentrated load, apparent density, water absorption at atmospheric pressure, open porosity (or apparent porosity), linear thermal expansion coefficient, abrasion resistance (with Capon machine) and rupture energy (commonly known as impact resistance). Finally, relationships were established between the main physical-mechanical properties of the natural stones under study, and their main lithological and textural characteristics.

1 INTRODUCTION

On a national range, natural stone is a resource whose current importance, in economical terms, has been accompanied by growing scientific and technological knowledge (Moura, 1991, Costa *et al.*, 1995).

On a regional scale, regarding Azores archipelago, extracting, manufacturing and trading natural stone industries on São Miguel Island, and to a lesser extent on Santa Maria Island, have great social, cultural and economical interest for their communities, whose demand for local raw materials has grown in recent years.

The study of the main physical, chemical and technological properties of natural stone is designed to optimize the use of this resource in the construction industry, putting in perspective the establishment of quality criteria, both in the resource exploitation and in the resource use.

Since the beginning of the Azores archipelago settlement in the 15th century, lavic and pyroclastic volcanic stones, both common in this archipelago, are important mineral resources used for the construction of houses, monuments and public works, among other applications.

On this work seven rock types of natural stone were studied. The selection criteria were based upon its representativeness, and also on the fact of being widely used, both in quality and in quantity on the islands of São Miguel (5) and Santa Maria (2). In Azores arquipelago there is a trade and popular nomenclature used to characterize natural stone which comprises hardness, cohesion, mechanical strength, carving and cutting mode: dimensional stone, farming or "lavoura" stone and sawed or lumber stone.

2 GEOLOGICAL SETTING AND LOCATION OF NATURAL STONE QUARRIES

The Azores archipelago is composed of nine islands and several islets that are located in the North Atlantic Ocean between latitudes 37° and 40°N and longitudes 25° and 31° W, spread over 610 km long in a direction WNW-ESE. Due to its geographic distribution, the archipelago is organized into three groups: the Western Group which includes the islands of Flores and Corvo, the Central Group is composed of the islands of Terceira, Graciosa, São Jorge, Pico and Faial, and the Eastern Group which includes the islands of São Miguel and Santa Maria, and the islets of Formigas.

Azores is located in an area where three major tectonic plates converge—the Eurasian, African and North American—and where other minor tectonic structures occur. These are arranged along a set of tectonic alignments of general orientation WNW-ESSE (França et al., 2003). This general trend is usually also found, in the shape of most of the islands.

The dynamics of its tectonic Framework (Figure 1) is the element responsible for the petrological and geochemical characteristics of lavas ejected by regional volcanic apparatus (França et al., 2003).

Regional volcanism determines that stones from Azores are primarily basaltic, although there are also acidic stones, with less representation (França, et al., 2003). While Santa Maria, São Jorge and Pico islands are mainly formed by basalts and alkali basalts, the other islands have a variety of lithologies, from alkali basalt to trachyte,

through the hawaiite and mugearite (França, et al., 2003).

Most of the geological formations that occur in the Azores archipelago from which extrusive rocks are used as natural stone, are distributed by two very different sets:

a. The first set comprises compact and/or porous and vacuolar lavic rocks represented by basaltic flows (s.l.) and trachytic flows (s.l.).
b. The second set comprises pyroclastic rocks resulting from explosive activity. In fact, there is a wide variety of materials, from huge blocks to very fine ash, through intermediate terms as "bagacina", "gravel", "lapilli" and "areão" (coarse sand) exhibiting vesicular and spongy structure, among which bombs frequently can be found. However, these pyroclasts exhibit some consolidation (welded materials), and in some cases they look like heterometric breccia and in other cases they look like tuff (Gomes & Silva, 1997).

The quarries of natural stone in the island of São Miguel are more numerous than those in the island of Santa Maria.

In the island of São Miguel, given the demand for natural stone of diverse lithology, natural stone from distinctive volcanic complexes, such as: Complexo Vulcânico dos Picos (1), Complexo Vulcânico do Fogo (2), Complexo Vulcânico das Furnas (1) and Complexo Vulcânico da Povoação (1), has been studied.

With the exception of the Complexo Vulcânico dos Picos of fissural nature, all the others being studied comprise central volcanic structures (Figure 2).

In the case of the island of Santa Maria, the two studied natural stones are included in the Complexo Vulcânico dos Anjos which occupies most of the west sector of the island and consists of basaltic lava flows of metric thickness, sometimes alternating with thin layers of pyroclasts. Veins are frequently seen cutting the basaltic flows.

Table 1 contains the references, the sources, the technical-commercial names, and the exploiting

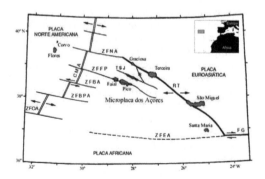

Figure 1. Geodynamic context of the Azores (modified from França, et al., 2003).

Figure 2. Location in the island of São Miguel of the quarries where natural stone has been sampled.

126

Figure 3. Location, in the island of Santa Maria, of the quarries where natural stone has been sampled.

Table 1. References, sources, trade names and exploiting enterprises.

S. Miguel Island

R1-Pedra do Porto Formoso	Herdeiros Agostinho F. Medeiros, Lda
R2-Pedra da Povoação	Herdeiros Agostinho F. Medeiros, Lda
R3-Basalto da Chã das Gatas	Herdeiros Agostinho F. Medeiros, Lda
R4-Basalto do Bacharel	Herdeiros Agostinho F. Medeiros, Lda
R5-Pedra do Dâmaso	José Dâmaso e Filhas, Lda

Sta. Maria Island

R6-Pedra da Vila do Porto	Mário Batista
R7-Pedra do Anal	Mário Batista

enterprises involved in the extraction of the seven types of natural stone being studied.

3 GENERAL AND MACROSCOPIC CHARACTERIZATION OF THE NATURAL STONE FROM THE AZORES

The ornamental value of an igneous rock depends on its colour and textural homogeneity. These characteristics determine to a large extent the rock ornamental and commercial value.

The colour exhibited by a natural stone, to a large extent, is associated with the colour of the dominant minerals, with the granularity of the same minerals, and with minerals alteration degree.

The lava rocks show gray shades, darker or lighter, being predominantly represented by basalt, basanite, trachybasalt, trachyandesite and trachyte.

The pyroclastic flows, associated with very explosive eruptions of acid magma, show black to brown shades, and are predominantly represented by ignimbrite. O the other hand the pyroclastic rocks show a wide range of shades, the most common ones being: brown, black, red, and gray,

and are predominantly represented by lapilli tuff and breccia tuff.

Rock texture depends on the following factors: the number and how the larger minerals are distributed within a more or less abundant matrix constituted of smaller minerals; the small fissures; the alteration degree of minerals; and the number and size of pores and vacuoles.

In general, the effusive rocks show aphanitic or microcrystalline texture (Gomes & Silva, 1997), and the pyroclastic rocks show porous to very porous, vesicular to very vesicular, and rarely amygdalar and breccioid texture.

4 PETROGRAPHIC, MINERALOGICAL AND CHEMICAL CHARACTERIZATION

4.1 *Analytical methods*

The petrographic, mineralogical, and chemical characterization of the studied natural stones was carried out at the Departamento de Geociências da Universidade de Aveiro, and at the LNEG (Laboratório Nacional de Energia e Geologia).

The petrographic analysis was carried out using an optical microscope and polarized transmitted light and the observation of thin sections of the studied natural stones. Petrographic analysis provided information very useful for the interpretation of the physical-mechanical behavior of the natural stone, and for the identification of certain features or defects that could penalize natural stone quality.

The mineralogical analysis was carried out using X-Ray Diffraction (XRD) whenever due to minerals alteration they could not be identified by petrographic analysis.

The chemical analysis was carried out using X-Ray Fluorescence (XRF).

4.2 *Petrographic and mineralogical analysis*

Some types of the studied natural stones show great similarities, both in petrographic and in mineralogical terms:

i. Pedra do Porto Formoso (R1) and Pedra da Povoação (R2) from the island of São Miguel are characterized by their light gray and brownish black colours and by their small pores due to the release of gas; they show porphyritic texture, and fenocrysts of leucite, K-feldspar, Na-Ca feldspar, pyroxene and olivine have been identified; both rocks could be classified as leucite-basanites.

ii. Pedra do Chão das Gatas (R3) and Pedra do Dâmaso (R5), both from the island of São Miguel, correespond to extrusive mafic igneous rocks showing ligth gray colour; they show porphyritic texture and numerous pores; within the phenocrysts, K-feldspar, Na-Ca feldspar and pyroxene have been identified; both rocks could be classified as trachytes.

iii. The so-called Basalto do Bacharel (R4), in the island of São Miguel, is an extrusive volcanic rock of dark gray colour and pores of medium size; it shows porphyritic texture, and within their phenocrysts Na-Ca feldspar, pyroxene and olivine have been identified; this rock could be classified as basalt.

iv. Pedra da Vila do Porto (6) and Pedra do Anal (R7), both from the island of Santa Maria, are pyroclastic rocks of reddish brown and greenish gray colour, respectively; they show porous and vesicular texture, and the clasts have dimensions within the range 2–64 mm; Na-Ca feldspar, Ca-Mg pyroxene, olivine and analcite were the main minerals which have been identified; they were classified as picrite basalts.

4.3 *Chemical analysis*

Table 2 shows the results of the chemical analyses carried out on the studied natural stones.

The most comprehensive classification of volcanic materials being used as ornamental natural stone in the archipelago of the Azores had been established using the Francis´classification (1993) that is based on the relationship ($Na_2O + K_2O$) content *versus* (SiO_2) content, expressed weight %.

The representation of the data shown in Table 2 allowed the classification into seven typologies of the studied natural stones:

Table 2. Results obtained FRX the studied volcanic stone samples.

% Oxides	R1	R2	R3	R4	R5	R6	R7
	References of the natural stones						
SiO_2	60.11	61.41	55.92	45.26	56.10	44.75	45.08
TiO_2	1.01	1.03	1.81	3.78	1.83	3.18	3.14
Al_2O_3	18.24	17.92	17.66	14.34	17.68	16.99	16.75
Fe_2O_3 T	4.48	3.99	6.32	12,87	6.34	12.74	12.56
MnO	0.17	0.15	0.12	0.18	0.12	0.26	0.19
MgO	0.96	0.84	2.35	7.61	2.34	3.62	3.99
CaO	2.57	1.94	4.68	9.14	4.69	7.26	7.92
Na_2O	6.91	6.03	5.38	4.20	5.16	4.98	5.52
K_2O	4.93	6.36	5.25	1.92	5.25	0.99	0.75
P_2O_5	0.21	0.14	0.35	0.64	0.35	0.63	0.59
P.R.	0.40	0.20	0.17	0.05	0.14	4.60	3.50

i. Pedra do Porto Formoso (R1) and Pedra da Povoação (R2) from the island of São Miguel could be classified as trachytes.

ii. Basalto do Chão das Gatas (R3) and Pedra do Dâmaso (R5) from the island of São Miguel could be classified as trachybasalts.

iii. Basalto do Bacharel (R4) from the island of São Miguel could be classified as hawaiite.

iv. Pedra do Anal (R7) from the island of Santa Maria and Pedra da Vila do Porto (R6) from the island of São Miguel could be classified as picrite basalt.

Also, taking into account the data contained in Table 3 and the chemical classification of Le Bas *et al.*, (1986), the studied ornamental natural stones from the Azores archipelago, in terms of SiO_2 content, range from basic (45% > SiO_2 <52%) to intermediate (52% > SiO_2 < 63%).

5 PHYSICAL AND MECHANICAL TESTS

The properties that were determined on studied volcanic stone samples are listed bellow, as well as the European Standards (EN) used to perform the tests. Properties were grouped according to their purpose and a short procedure description was also included.

5.1 *Identification properties*

5.1.1 *Uniaxial compressive strength (EN 1926)*
Six cubic specimens were laid and centred on the plate of an ELE International ADR 200 compressive testing machine. It was applied a uniformly distributed and continuously increased load until specimen failure occurred. Compressive strength was calculated as the ratio between the failure load and the surface area submitted to compression.

5.1.2 *Flexural strength under concentrated load (EN 12372)*
Ten parallelepiped specimens were placed on two supporting rollers of an ELE International Tritest 50 Digital flexural testing machine. The distance between these rollers (span) was calculated according to the specimen thickness and previously adjusted. A force was applied by a third roller placed on the middle of the specimen. The load was progressively increased until failure occurred. Flexural strength was calculated by the following equation 1 below:

$$\text{Flexural strength} = \frac{3 \times \text{failure load} \times \text{span}}{2 \times \text{width} \times \text{thickness}^2} \quad (1)$$

5.1.3 *Apparent density and open porosity (EN 1936)*

Six cubic specimens were dried to constant mass and placed in a vessel to absorb distilled water while submitted to a standard vacuum. After an immersion period of time, also standard, saturated and immersed weightings were made. Apparent density and open porosity were calculated according to the following equations 2 and 3, respectively:

$$\frac{dried\ mass}{saturated\ mass - immersed\ mass} \times water\ density \tag{2}$$

$$\frac{saturated\ mass - dried\ mass}{saturated\ mass - immersed\ mass} \times 100 \tag{3}$$

5.1.4 *Water absorption at atmospheric pressure (EN 13755)*

For this test, six cubic specimens were also used. The test method is quite similar to the one used to determine apparent density and open porosity. The main differences were that vacuum was not used, the immersion was staged into 3 steps and specimens were left immersed in water until constant mass was reached. Water absorption was calculated by the ratio between the mass of absorbed water and the mass of the specimen.

5.2 *Performance after application properties*

5.2.1 *Rupture energy (EN 14158)*
Six specimen slabs were used to perform the test. After placing and levelling the slab on a silica sand bed of standard grain size, a steel ball of both standard mass and diameter was dropped on the upper face centre of the tested slab. On the first test, the ball was dropped from an initial height of 10 cm and the height was continuously increased, 5 cm each time, until the specimen broke. On the following five tests the procedure was the same, except for the initial height of the ball dropping: 15 cm bellow the result obtained on the first test. *Rupture energy* was calculated as the ball work when dropped from the height which produced specimen failure, according to the following equation 4:

$$Work = Ball\ mass \times g \times dropping\ height \tag{4}$$

where g is the value of gravity acceleration.

5.3 *Durability properties*

5.3.1 *Abrasion resistance (EN 14157)*
The faces exposed in use of six specimen slabs were abraded by means of a Capon wearing machine

Table 3. Results of CS and FS for the studied volcanic natural stones.

Volcanic stones	CS MPa	FS MPa
S. Miguel Island		
Pedra do Porto Formoso	81	13.6
Pedra da Povoação	6	1.4
Basalto da Chã das Gatas	32	7.8
Basalto do Bacharel	53	9.5
Pedra do Dâmaso	86	7.7
Sta. Maria Island		
Pedra da Vila do Porto	7	3.1
Pedra do Anal	8	4.7

Table 4. AD, OP and WA results obtained for the studied volcanic natural stones.

Volcanic stones	AD (kg/m^3)	OP (%)	WA (%)
S. Miguel Island			
Pedra do Porto Formoso	2220	9.4	4.7
Pedra da Povoação	1430	34.6	27.4
Basalto da Chã das Gatas	1930	6.5	3.7
Basalto do Bacharel	2160	7.0	3.7
Pedra do Dâmaso	2190	5.2	2.9
Sta. Maria Island			
Pedra da Vila do Porto	1430	26.0	21.1
Pedra do Anal	1630	24.5	16.1

Table 5. Results obtained for IR and AR for the studied volcanic natural stones.

Volcanic stones	IR Joules	AR mm
S. Miguel Island		
Pedra do Porto Formoso	4	23.0
Pedra da Povoação	2	42.5
Basalto da Chã das Gatas	4	22.0
Basalto do Bacharel	6	20.0
Pedra do Dâmaso	4	20.0
Sta. Maria Island		
Pedra da Vila do Porto	3	26.0
Pedra do Anal	4	25.5

(Tecnilab 440C). In this machine type, abrasion is obtained by using both a rotating abrasion wheel (with standard dimensions and made of a standard steel) and a standard abrasive powder (white fused alumina). A Certified Reference Material (CRM)—"Marbre du Boulonnais"—was used to calibrate the testing machine, but also to correct test results. The test was carried out under standard conditions and at the end a groove was obtained in the tested face. The abrasion resistance value is the chord dimension measured on the middle of the groove after correction according to the value obtained for CRM.

5.4 Physical and mechanical properties

The average results of *uniaxial compressive strength* (CS) and *flexural strength under concentrated load* (FS) obtained for the studied volcanic natural stones are listed in Table 3.

The average results obtained for *apparent density* (AD), *open porosity* (OP) and *water absorption at atmospheric pressure* (WA) are listed in Table 4.

Rupture energy, also known as impact resistance (IR) and abrasion resistance (AR) average results obtained for the studied volcanic stone samples are listed in Table 5.

6 DISCUSSION OF THE RESULTS OF THE PHYSICAL-MECHANICAL TESTS

Analyzing the results obtained for *compressive strength* on the studied volcanic natural stones, three of them from S. Miguel Island (Pedra do Porto Formoso, Basalto do Bacharel and Pedra do Dâmaso) present values above 40 MPa. Any values below the referred to value usually are considered low. Stones *Flexural strength* commonly has identical behaviour to the *compressive strength*. Stones that present low values of compressive strength also present low values of flexural strength and vice-versa. On the studied volcanic natural stones, the only exception to this statement is Pedra do Dâmaso that presents the highest value of *compressive strength* (86 MPa), but does not has the highest value for *flexural strength* that belongs to Pedra do Porto Formoso (13.6 MPa). All *flexural strength* values under 4 MPa usually are considered low; only two stones present values below this limit (Pedra da Povoação and Pedra da Vila do Porto).

The results obtained for the physical properties (presented in Table 2) are compatible with the types of natural stone being tested. Their open porosity is high, and consequently their *apparent density* is low. Concerning *open porosity*, values above 10% are considered very high and three of the studied volcanic natural stones exceed this limit (Pedra da Povoação and the two stones from Sta. Maria Island). These three stones are also the only ones which present *water absorption* values above 6%, value above which water absorption is considered high. As expected, physical properties have great influence on the mechanical behavior. That is why stones having higher apparent density and both lower open porosity and water absorption generally present higher values of compressive and flexural strength.

Analyzing the results of the test *resistance to impact*, only Pedra da Povoação presents a value of 2 Joules, which is considered very low, and Pedra da Vila do Porto is the only one that presents a value which is considered low (3 Joules).

In terms of *resistance to abrasion*, Pedra da Povoação presents the highest value (42.5 mm), and is the only stone that should not be used for paving. The two volcanic natural stones from Santa Maria Island should only be used for paving on low traffic conditions (for private applications). The remaining volcanic natural stones can be used for collective applications. Pedra do Porto Formoso and Basalto do Chão das Gatas only could be used for moderate traffic conditions. On the other hand, Basalto do Bacharel and Pedra do Dâmaso could be used for both moderate and strong traffic conditions.

7 CONCLUSIONS

The research being carried out on volcanic natural stones of different typology from the Azores archipelago clearly has shown how much the physical-mechanical properties are dependent upon rock lithology and texture. Such has been also found on similar studies carried out by Gomes & Silva (1997), Silva et al., 2002, and Silva & Gomes (2004), on volcanic natural stones from the Madeira archipelago.

Compiling the information obtained for the physical and mechanical properties of the studied volcanic natural stones from the Azores, it can be concluded that all stones are more suited for internal applications, due to the fact that the required water absorption value for external applications is usually ≤0.5%.

Pedra da Povoação from S. Miguel Island is not suitable for paving, not only because of its high abrasion resistance value (42.5 mm), but also because of its low impact resistance (2 Joules). In fact, for most of the applications, the selection of this stone and also of the two stones from Sta. Maria Island has to be deeply considered to avoid unsuccessful results.

In Portugal and in many European countries there are not national specifications which regulate property values that stones should satisfy to be used as construction materials. As a consequence, the selection of an adequate stone for a specific application depends mainly upon the prescriber expertise.

This study, in addition to the characterization of several volcanic natural stones exploited in the Azores archipelago, also put forward some considerations about stone applications.

Until the eighties the exploitation and processing of natural stone was performed using manual processes and the stone was used almost exclusively in religious, civil and military architecture. Since then new extractive methods have been introduced that allowed to increase production, to improve the quality and to increase the variety of products to be used in civil construction, public works and funerary art. In general, the applications given over time to the natural stone of the Azores are in accordance with the results obtained in the physical-mechanical tests being carried out.

The lithological varieties corresponding to the so-called "pedra de lavoura" and "welded pyroclasts" impede any type of polishing due to several factors: the low hardness of many of their minerals, the existence of glass in some stones, the porosity exhibited by others, and the alteration degree as well as the weak cohesion between minerals shown in some others.

Despite the existing textural (porosity, size of the volcanic fragments, and xenoliths) and compositional differences shown by the studied lithologies, the obtained results are very similar. Such findings are important to assure both quality and certification of the manufactured products and the consumer confidence. However, it is recommended the consultation of the harmonized standards corresponding to the different products to be manufactured, as well as the periodic control of raw materials, of products dimension and of the processing equipments.

The waste from the processing of natural stone in the Azores is almost totally utilizable. In what concerns the so-called "pedra de lavoura" (hawaiite, trachybasalt and trachyte) the waste is utilized, either in the production of pre-fabricated, or sold as the so-called "lajetas" applied to face walls and pavements, whereas the waste from the "welded pyroclasts" (tuff and lapilli tuff) after being ground is utilized both as soil for greenhouses, and as inert material in the manufacture of bricks for civil construction.

ACKNOWLEDGMENTS

Thanks are due to the enterprises Herdeiros de Agostinho Ferreira de Medeiros Lda, José Dâmaso e Filhas Lda, and Mário Batista for preparing and making available the set of natural stone samples required to perform the physico-mechanical tests as well as the petrographic, mineralogical and chemical analyses.

REFERENCES

Costa, L. Rodrigues, Leite, M.R. Machado & Moura, A. Casal (1995). O futuro da indústria das Rochas Ornamentais. Bol. Minas, v. 32, nº. 1, 1–13.

Caetano, S.D. (2007). Prospecção de Recursos Minerais: Modelo Integrador de Valores Ambientais e de Ordenamento do Território. Tese de Mestrado em Ordenamento de Território e Planeamento Ambiental. Universidade dos Açores, Ponta Delgada, Portugal.

Caetano, S.D., Lima, E.A., Medeiros, S., & Nunes, J.C. (2008). Caracterização Geotécnica de Agregados Naturais dos Açores. XI Congresso Nacional de Geotecnia, IV Congresso Luso-Brasileiro de Geotecnia. Coimbra, Portugal, 8p.

França, Z., Cruz, J.V., Nunes, J.C., Forjaz, V.H. & Borges, P. (2003). Geologia dos Açores: Uma Perspectiva Actual. Revista Açoreana. Sociedade Afonso Chaves. Ponta Delgada.

Gomes, C.S.F. & Silva, J.B.P. (1997). Pedra natural do Arquipélago da Madeira. Importância social, cultural e económica. Madeira Rochas-Divulgações Científicas e Culturais, Câmara de Lobos, 173p.

Francis, P. (1993). Volcanoes: A planetary perspective. Oxford University Press (Edition), 443p.

Moura, A. Casal (1991). Rochas Ornamentais: Características das rochas ornamentais portuguesas e a importância do seu conhecimento actual. Geonovas, nº. 2 (Especial), 123–136.

Moura et al. (1983/4/5 e 1995). Catálogo de Rochas Ornamentais Portuguesas. Edição Instituto Geológico e Mineiro, Lisboa, v. I, II, III, IV.

Nunes, João C. (2002). Novos conceitos em vulcanologia: erupções, produtos e paisagens vulcânicas. Geonovas, nº. 16, 5–22.

Le Bas, M.J., Le Maitre, R.W., Streckeisen, A. & Zanettin, B. (1986)—A chemical classification of volcanic rocks based on the total alkalis-silica diagram. Journal of Petrology, v. 27, 3, 745–750.

Silva, J., Ferraz, E., Moura, C., Grade, J. & Gomes, C. (2002). Natural stone of the Madeira Archipelago: commercial types and properties. Eurock 2002, Funchal, Workshop on Volcanic Rocks, 115–124.

Silva, J.B.P. & Gomes, C.S.F. (2004). Tipologias de pedra natural utilizadas no Simpósio Internacional de Escultura em Pedra (SINEP, 2004). Câmara Municipal de Câmara de Lobos e Madeira Rochas—Divulgações Científicas e Culturais, 77–98.

2   *Instabilities in volcanic islands: Slope stability, large landslides and collapse phenomena*

Volcanic Rock Mechanics – Olalla et al. (eds)
© *2010 Taylor & Francis Group, London, ISBN 978-0-415-58478-4*

Stability of the cone and foundation of Arenal volcano, Costa Rica

G.E. Alvarado, S. Carboni, M. Cordero, E. Avilés & M. Valverde
San José, Costa Rica

ABSTRACT: Arenal volcano is deforming the basement under ~20×10^3 kPa, and affects it for several kilometers below the surface and about 5 km around the volcano base. The total settlement below the present (1968–2009) lava field (0.75 km²; 0.6 km³) is 2 m o more, but it represent at the moment only 20% of the consolidation, so its deformation will be continuous for years. The volcano grew up on the top of weathered volcanic rocks (weak and plastic portion) conditions that are ideal for deforming the basement (subsidence, folding or faulting) and generate instability on the cone according to structural and volcanic models. The results of numerical models show that Arenal is at an incipient deformation stage by spreading of the basement. The overall effect generates stability at the interior of the volcano and its foundation. The twin edifice (cones C and D) can generate rock slides (cold or hot) as well as debris avalanches (0.03–0.75 km³).

1 INTRODUCTION

The subsidence due to overloads from infrastructure has been studied for thousands of years. However, the deformation produced by a natural overload on land is not as well known. Examples of this are the subsidence due the emplacement of lava flow or the growth of a volcanic cone, which can take from months to millions of years. As an illustration, a skyscraper transmits approximately 500 kPa to the ground, meanwhile, the load generated by a volcano varies between 20×10^3 kPa (Arenal Volcano) and 78.2×10^3 kPa (Olimpus Volcano, Mars).

The Arenal volcano, active since 1968 with a great lava effusion (approximately 0.6 km³), suits well to study the stress' effects and overload capacity (deformation and stability) generated on the soil underneath it. Studies from the past three decades (1985 to 2000) indicate, through the analysis of measurements at dry inclinometers that the foundation where the lava erupted has settled since 1968, are suffering a continuous deformation and its effects are detected several kilometers away from the emitter focus. Since Arenal is a geologically well known volcano, and its volume and distribution for the past 7000 years can be calculated, it helps understand the effects on the substrate or local basement (foundation), in this case on lateritic soils, ash soils and lava talus.

In our case, we want to model the growth of a volcano or cone (edifice or structure) on any surrounding land, which works as a substrate or

Figure 1. Scale comparison of the Arenal volcanic edifice with some great structures.

basement (foundation) of the eruptive edifice. To accomplish this, we must establish precisely its geotechnical conditions and characteristics. As well as understanding when the ladders of a volcano become unstable, are very important for the correct evaluation of the volcanic hazards.

This paper pretends, through different laboratory and field tests, and using different geological and geotechnical methodologies, the geological and geotechnical characterization of the Arenal volcano and its foundation, in order to estimate its deformation and stability conditions, as well as its geological danger. It also pretends to calibrate the results from the inclinometers in the volcano with the results from the geotechnical parameters from the laboratory tests. Finally, the possible risk generated by a slide (block and ash flow, rockslide, debris avalanche), which could affect any infrastructure or nearby population.

Past studies on the deformation due to the weight of a volcano and instability-dispersion of a cone

Deformations on volcanic media can be classified in four groups: a) subsidence, b) lifting, c) lateral movements (dispersion or spreading) and d) ladder instability. The process of subsidence might occur due to the deflation of the magmatic chamber, cessation of the magma's ascend, collapse of a lava tunnel or due to the lateral magmatic stresses, which can be transform at the surface in settlements, depressions, or in a larger scale into the formation of cauldron like volcanic features like calderas. The ground uplift might be caused by the intrusion of a magmatic body (cryptodome) or the inflation of the magmatic chamber due to presence of new magma or fluids.

The alternation of tephra layers and lava flows, usually deposited with a critical stability angle, combined with its fast alteration (weathering and hydrothermalism), produce impermeable clay silt soils (mechanically incompetent) and permeable fractured media in the lavas, with favorable conditions for sliding. We can also add the effects of the hydrothermal alteration, seismic movements, structural weaknesses, etc. These and many other factors are responsible of the sectorial collapses (craters and calderas) of some of the flanks that originate mega-slides (23, 27, 29). The work of Siebert et al (40), resumes the great volcanic avalanches in Central America (prehistoric and historic).

The deformation of the edifice-foundation in a volcanic structure was not evaluated quantitatively until a short time ago, even though Milne (32) was the first one to suggest that the shape of a volcano is related with its deformation due to its own weight during its building process, and Van Bemmelen (46) as well as Susuki (45) comment that a volcano's growing process and its deformation style depend on the type of basement where it develops. It is not until the last 30 years, that the topic was again approach, as a model that could explain in a different way the shape and form of volcanoes, the substrates function and the reason of existence of many tectonic-volcanical structures (10–12, 15, 31, 47, 48).

The keys to the volcanic dispersion are (9): a) the existence of a weak basal layer, and b) a sufficiently high mass and magma influx to drive the process. When a volcano grows, it goes through 5 steps (that can be repeated, omitted and superimposed): 1) building, 2) compressing, 3) trusting at the foot, 4) intrusion due to stress relaxation, and 5) spreading (lateral movement) that creates an extensional field in the superior flanks of the volcano. The geological result is the formation of reverse faults and propagation pleats. This is why the identification of the geological conditions for the dispersion of a volcanic cone and its geological symptoms can help control and identify the dangerous sectors for the collapse of the volcano.

The studies on volcanic avalanches or volcanic instability under geotechnical bases (theoretical as well as data from the field and laboratory), are very few (51, 52). Recently, there have been published some new papers that share similar aspects with the present investigation (2, 5, 16, 17, 56). In the work of Alvarado and coworkers (1, 2, 5), there were summarize several results of their investigation, however, due to the lack of diffusion, this study is presented to the geotechnical community, in a revised, summarized and updated format.

2 GEOLOGICAL FRAMEWORK

The Arenal volcano grew on the tectonic valley on altered volcanic rocks (>1 million years old) displacing the Arenal river to its actual position. The oldest rocks of its foundation include tuffs, pumice and breccias and lava flows, hydrothermally altered and wetheared in different degrees.

The volcano is at least 7000 years old. Its eruptive activity has been known for the presence of explosive eruptions alternated with lava flows. The historical activity is represented by an initial closed conduit eruption with the formation of three new craters (A, B and C), that along with the antecedent (crater D), constitute a fissure system E-W originated on July 29–31, 1968. From September of that year to 1973, there was effusive activity in crater A or inferior, that generated a basaltic andesitic lava field (with a maximum thickness of 150 m). In 1974, the effusive activity migrated to crater C, where there have been hundreds of relatively small lava flows until now. The total lava field is 7.5 km² for an actual volume of 0.6 km³ (46, 54).

This lead to the transformation of the Arenal cone from a conic shape with a cuspidal crater and ladders with a 33°–34° inclination in its half superior part to a double cone that has been growing (C cone) on top of the antecedent cone (D cone), disturbing the volcano's symmetry. The eastern slope is steeper and shorter than the western slope, because of the prevailing wind directions, that tend to settle the pyroclastic deposits further west, softening the slopes in that direction.

From the structural point of view, there is a semi curvilinear geoform in the west flank of the Arenal volcano, observed in aerial photographs taken before 1970, given that it was later covered by lava flows. Although it has been interpreted in various ways, if we assume that the Danta fault (N-S) has a certain normal-dextral component, according to the tectonic context (29), we may reinterpret the generation of a low angle reverse fault in response to its movement under the conical body. The faults with Holocene displacements are normal faults, NE-SW in the southern slope, that could affect the cone theoretically, even though this cannot be seen in the aerial photographs, since it is all covered with recent lava flows. There have been observed extensive open fractures (decimetric) cutting the middle section of the D cone, with apparent NNE and NNW tectonic trend.

The geophysical profiles made on the east and west flanks of the Arenal volcano suggest a superficial stratigraphy of 5 geophysical layers. The local basement appears to be related to ancient volcanic lateritic soils associated to the superior layer of the Aguacate group or to the Monteverde Formation: lavas and pyroclastics of the Pliocen- Lower-Pleistocene (28). Using this information, we can calculate the Poisson relation and the dynamic elastic modulus and the shear modulus. In the case of lava, this values are relatively low, considering the high values of the primary seismic wave, which implicates that the lavas are very good geomechanically. For igneous rock, the value of the elasticity modulus varies between 7×10^6 and 117×10^6 kPa, being the usual value between 40×10^6 and 62×10^6 kPa, for a Poisson relation between 0.01 and 0.40, in general in the range 0.15–0.22 (20).

The hydrogeological studies are limited to the geochemistry of the hot and cold springs (30).

Figure 2. Geological map of Arenal-Chato volcanic systema. uPPv, undivided Pliocen-Pleistocen volcanics; LC, lower Chato; CH; Chatito (and La Espina to northeast); mC, middle Chato lava field, uC, upper Chato; LA, lower Arenal; A4, A4 lava field; A3, A3 lava field; A2$_l$, A2 low lava field; A2$_h$, A2 high lava field; A1$_l$, A1 low lava field; A1$_h$, A1 high lava field; utAC, undivided tephra of Arenal and Chato; utAC$_{1235}$, Quebrada Guilermina pyroclastic flow; utAC$_{1968}$, 1968 pyroclastic flows; utAC$_{1975}$, 1975 pyroclastic flow; sd, sedimentary deposits; sd$_a$, alluvium; sd$_m$, mud flow deposits; sd$_t$, talus slope deposits.

These springs appear to be control by faults, recent lava fields or the contact between the volcanic massif and the ancient volcanic basement. We can conclude, in general terms, that the phreatic surface emerges in certain areas at 400 m.a.s.l in the eastern flank, at 600 m.a.s.l in the northern flank, at 500 m in the north-western side and at 550 m in the western slope; in the south flank of the Arenal volcano, there are not any important springs, since it is higher than the 600–700 meters, because of its connection to the Chato volcano ladders. Also, there are not any water sources at altitudes greater than 650 m in the Arenal cone, except for insignificant drips or leaks and only in the rainy season. Everything shows that most of the cone is in an unsaturated condition, due to the high primary permeability in the lavas, as well as the lack of impermeable levels inside the cone's nucleus. Five gravity permeability tests were made in a project called Aportes Fortuna, in the lavas from the Chato volcano (similar to ones from Arenal), whose values vary between 3.6×10^{-4} cm/s and 7.3×10^{-4} cm/s (34).

3 GEOTECHNICAL MODEL

To be able to apply the analytical models through formulations and computer software, it is necessary to establish a simplified geological-geotechnical model of the soils and rocks that constitute the volcano and its foundation. The results of studies based on the mechanic of soils, once checked, compared and selected, are shown in Table 2. If the reader is interested in the lab results or any other details, he can refer to the work of Alvarado (2) and Alvarado et al. (5).

Based on all the geological-geotechnical information, we raise a model of the rock massif from the foundation and the edifice (Fig. 3 a,b). We establish four layers or geotechnical levels (I to IV), conveniently assuming for the analysis that they behave as homogeneous materials, due to common geological-geotechnical features that support the previous simplification, even though there is a high degree of heterogeneity between the layers.

The Arenal cone (Layer I) is analyzed as a fractured rock mass, which at the same time can be study in two different ways: a) stratified, very fractured and drained, and b) as a mass, where the lavas and lava auto-breccias are analyzed as a whole and the pyroclastic rocks are subordinated. This was the case of study selected by simplification, as the macro behavior of the friction soils (c = 0, $\Phi \neq 0$). Parallel to this, we have Layer II, composed mainly by tuffs, as well as epiclastic deposits (transported soils), all moderately compacted, healthy, and 3000 years old or younger. Usually drained and

137

saturated near to the volcano and has an aquifer at its foot.

The foundation is well defined by the geotechnical levels III and IV. The upper section is constituted by old tuffs (>3000 years old), with a higher grade of alteration, along with the presence lateritic soils that make up Layer III. The water table is

Table 1. Synthesis of the geometrical parameters of the groups of soils, mainly from the foundation (the values between parenthesis are average values).

SUCS Classification	ML
Liquid limit	25–88 (65.5)
Plasticity index	24–52 (41.6)
Specific gravity Gs	2.43–2.88 (2.65)
Humid density (kN/m³)	14.4–19.1 (16.0)
Void ratio e	0.82–2.6 (1.75)
Dry density (kN/m³)	9.6
Saturation percentage	30.7–98.1 (68. 0)
Triaxial shear test CD/c', kPa	75–110
Internal friction angle/Ø'	11.5–13 (10–42°)
Triaxial shear test UU/Cu, kPa	80–190
Permeability (m/s)	10^{-5}–10^{-7}
Coefficient of consolidation (Cv)	0.17–6 m²/año
Coefficient of compressibility (Cc)	0.2–0.67
Pre-consolidation load (P'c), kPa	260–4000
Coefficient of recompression (Cr)	0.03–0.076 (0.05)

usually found in some level of this unit, and given its relatively reduced thickness of 10–15 m, it rises as a takeoff surface for the generation of landslides. The Layer III overlies the rocks with a high lateral heterogeneity as well as vertical (Layer IV), compose of rocks with a moderate permeability and with a significant and variable grade of fracturation in the lava flows, mesobreccias and tuffs. Table 3 summarizes some of these characteristics. From the practical point of view, layers I and IV differ only in their age, degree of compaction and tectonism, while layers II and III are very similar (soils), increasing their degree of compaction and fines with their age, and for the analysis, layers II and III are taken as a single layer. The angle $\Phi \neq 0$, was modeled as a small value, according to the results from the tests and to be on the side of safety. All of this information provides the base to develop the analysis of stability of the huge natural edifice of Arenal through different models, for example, instability of the crater C, of the cone or of the entire volcano and its foundation, or applying methods, for example: Buckingham, geotechnical, etc. The mechanical properties of the materials reported on Table 3 for the different geotechnical layers, are according to the ones reported for other volcanoes (14, 17, 51, 52, 56).

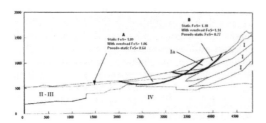

Figure 3.a. Results of the stability analysis of the occidental flank of the Arenal Volcano. A: critical fault surfaces found for the talus global stability. B: critical fault surfaces for the cone stability. The lowest factor of safety is given for the global fault with pseudo-static loads.

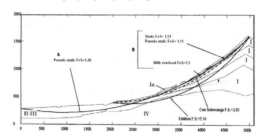

Figure 3.b. Results of the stability analysis of the east flank of the Arenal volcano. A: critical fault surfaces found for the talus global stability. B: critical fault surfaces for the cone stability. The lowest factor of safety is given for the global fault with pseudo-static loads.

Table 2. Basic rock parameters at the Arenal Volcano.

Lithology condition	Tuffs		Lavas and breccias	
	Healthy	Altered	Healthy	Altered or fractured
Unit weight	14–16.8		17.54–28.34	
GSSS			2.33–2.71	
G_s			2.22–2.64	
G_b			2.39–2.72	
Absorption			0.08–48	
Ø	10°–42°	10°–37°	33°–47°	33°–38°
V_p	0.5–0.9	0.3–0.6	2.1–3.9	1–2.5
Poisson's ratio	0.33–0.42	0.33–0.40	0.26–0.42	0.26–0.37
Uniaxial compressive strength (kPa)	8000–15 000	250–8000	40 000–300 000	10 000–250 000
Permeability (m/s)	10^{-5}	10^{-6}–10^{-7}	10^{-5}	10^{-5}–10^{-6}

Table 3. Geotechnical model for the case of the Arenal volcano.

Layers	Description	Unit weight γ kN/m³	Cohesion kPa c	Cohesion kPa c'	Friction ϕ	Friction ϕ'	Poisson	Static deformation modulus E 10⁶ kPa	Compress-ibility index Cc
I	Mainly lavas and autobreccias	20–25	–	–	40–44°	–	0.28	2.5	–
Ia	Lava Talus, pyroclastic deposits with alternate lava flows	15–20	–	–	33–36°	–	–	–	–
II	Alternation of pyroclasts and recent epiclasts	14	145	92	–	12° (10–42°)	0.37	1.042	0.6
III	Old pyroclastic rocks and laterites	14	90	75	10°	12°	0.33	2.238	0.67
IV	Rocks (lavas and old pyroclastic rocks)	17	200–400	–	40–45°	38°–42°	0.33–0.42	13.91	–

4 STABILITY ANALYSES

4.1 Talus analysis

Is the Arenal Volcano susceptible to a creeping process? To test this, a series of analysis that involve from analytical solutions to the computer programs SIGMA/W and STABL 5M was established.

The seismic movements with M ≥ 4.0 (26) could generate landslides in an active seismic zone like Arenal. The topographic effect of the cone of the Arenal volcano would amplify the seismic signals. The intense degree of fracture in the lavas and their auto-breccias, combine with the alternation of tephra layers makes the rock massif behave as a unique unit of soil with big blocks. Because of this, the massif would probably fail through poorly defined stratification surfaces a) Fault in crater C, b) Faults in cones C or D and c) Fault in the cone's foundation, instead of failing through the joints.

The safety factor is affected by different conditions with a wide variety of alternatives, depending on the media and available time for investigations. It is also particularly sensitive to judgment errors like the anisotropy of the media, wrong selection of the volumetric weight, calculation methods, etc. Also, the factor of safety does not reflect the probability of failure, because it does not involve any information about the used values of stress and resistance. In the present investigation, a safety factor of 1.3 is used as critical.

4.1.1 Stability analysis of the cone C
The cone C with its active crater has partially failed, generating a collapse of its faces combined with the overflow of the lava pool, producing pyroclastic flows in 1975 and 1993; and small ones in 1998, 1999, 2000 and 2001 (3). An analysis of this instability problem can be treated with an approach based on the mechanic of rocks.

– Case of a fracture filled with a liquid (i.e. lava pool): In this case, the rupture could be model as a plane fault in a block with a pressure generated by the presence of a lava pool, which behaves as a huge fracture in our case filled with lava and due to geometrical issues, its shape is similar to a square, instead of circular or tabular. Since the fluid is lava, a non-Newtonian and very viscous liquid with an elevated temperature, gas and crystal content, the pore pressure in the base of the failing surface must be cero (without water), but another pressure generated by the gases will exist. Some parameters have to be taken based on studies made of similar deposits. Also, the slopes and weights are estimated based on the indirect measures taken on field (slid rock mass), based on observations made by Alvarado and Soto (3). The total volume of slid material during the eruption of August 28th, 1993 was $2.3 \pm 0.8 \times 10^6$ m³, assuming a depth of the lava pool of 80 m, for a $\gamma Lv = 25$ kN/m³ (due to the porosity between blocks). The equivalent volume of dense rock or young non vesiculated magma was $1.35 \pm 0.4 \times 10^6$ m³ (= VLv).

For the instability analysis, an approach of stability of plane fault with crown fracture can be used (Fig. 4). The used nomenclature was the same used in Hoek & Bray (21). The parameters were: H = 150 m, Z = Zw = 80 m, $\Psi f = 36°$, $\Psi p = 25°$, $\gamma Lv = 25$ kN/m³, c = 600 kPa (obtained through rugosity) and $\Phi = 36°$. Seismic activity and sub-pressures in the base of the slid block were not

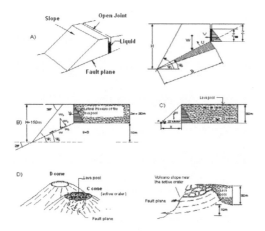

Figure 4. Simplified model of slope stability analysis of the C cone and its lava pool.

considered. For these parameters, the factor of safety is 1.3. From the stability analysis of the model, it was determined that with only an increase in the slope of the fault plane of 1.5° (upgrading Ψp), the block fails (factor of safety = 0.99). This may be due to gases or intrusion (dike or crypto-dome). It is interesting to note that 1.5° is trans-lated to distance B, right in the internal lava lake with the wall, with an opening of 4.5 m near the crater that would need only 6000 kPa to cause the crater wall collapse.

In addition to the parameters exposed here, we have to contemplate the force generated by the daily lava discharge (21 600–43 200 m³/day) and its associated pressure on the sides (lava and gas pressure), combined with the explosive erup-tions (that in the case of the La Soufriére Volcano had pressures between 3000 and 15 000 kPa; (19)) that weaken the structure, decreasing its cohesion gradually.

– Dam (i.e. crater wall) under the weight of a fluid by the reservoir (lava pool): Once again, we can model that one of the crater walls (the slid one) behaved as a retaining wall or dam holding the fluid in the reservoir (in our case a lava pool). The volume of slid material during the eruption of 1993 was 2.3 ± 0.8 × 10⁶ m³ (3). Assuming that the depth of the lava pool was 80 m, for γLv = 25 kN/m³ and a lava wall of 22–25 kN/m³ (due to the porosity between blocks). The equiva-lent volume of dense rock was 1.35 ± 0.4 × 10⁶ m³ (= VLv), therefore, we can calculate the weight of the lava pool and wall or slid wedge that had a volume of 0.55 × 10⁶ m³. To keep the lava pool stable, it must be contained between the walls with a determined or minimum thickness (and

stability). Therefore, the base of the crater wall that supports these fluids should equilibrate the overturning moment caused by the lava thrust: the horizontal hydrostatic force from the lava pool was bigger than the gravitational force contained in the lava wall. The sum of moments in point A (Fig. 4c), indicates that the width of the theoretical base to cause pivoting, would be around 120 m. As a matter of fact, based on field observations and oblique photographs, we can estimate that the crater wall in 1993 had a thickness of at least 60 m, so it was unstable due to its relatively thin thickness.

4.1.2 General stability analysis (cone and foundation)

Non-dimensional parameters according to the Buckingham Π Theorem

Merle & Borgia (31) and van Wyk de Vries & Matela (49) made experiments in the laboratory to study the spreading during the building phase of volcanoes. Alvarado (1, 2, 5) applied this method to the Arenal Volcano. The experiments deter-mined multiple non-dimensional relations named Π which predict if the spreading (material dis-placement at its base) may or may not take place, therefore, there are a series of referred variables to Π relations (Table 4) and at least three dimensional data (gravitational, inertial and viscous force), whose relations are needed to maintain the same value in nature and the experiments.

Therefore, the main geometrical variables are: height of the volcano (H), radius of the volcanic cone (R), the slope angle (α), the volcano's stratifi-cation thickness (E), the brittle of the substratum (D) and the thickness of the ductile susbstrate (I), the depth of the sectorial collapse (DA), the layer thickness and the displacement of the fault (F). Within the variable properties of the materials we have the density of the volcano (ρv) and the sub-strate (ρs), the time spam for deformation (T), the plastic substrate viscosity (μ), the angle of internal friction (Φ), the cohesion (cv) and the unit density (γ). The gravity acceleration (g) is the responsible force of the destabilization process. Therefore, a mode of analysis, based on laboratory information and its comparison with real cases is the one that considers the non dimensional values shown in Table 4. Of particular importance, we can find the relation between the thickness of the brittle layer (foundation) and the height of the volcano (Π_2), as well as the ratio between the thickness of the brit-tle layer and the thickness of the weak substrate (Π_3). Π_2 must be big enough to cause the substrate failure before the dispersion occurs. Meanwhile, Π_3 controls the deformation style. If $\Pi_3 \gg 1$, the volcano will not deform, however it twists with the basement; if $\Pi_3 = 1-2$, one single concentric ridge

Table 4. Average geometry and mechanical variables of volcanoes in general (31, 37, 49) and the case of the Arenal volcano.

Variable	Definition	Value Volcanoes	Arenal
H (m)	Height of the volcano	$0.6–12 \times 10^3$	1.1×10^3
R (m)	Radius of volcanic cone	$4–600 \times 10^3$	4×10^3
D (m)	Thickness of brittle substratum (Foundation)	$0–40 \times 10^3$	$1–2 \times 10^2$
I (m)	Thickness of the weak substratum (Foundation)	$0.5–100 \times 10^3$	$20–125$
ρv (kg m^{-3})	Density of volcanic cone	$2.5–2.8 \times 10^3$	2.8×10^3
ρs (kg m^{-3})	Density of substratum	$2.0–2.5 \times 10^3$	2.65×10^3
T (s)	Time of span for deformation	$5 \times 10^9–1.5 \times 10^{12}$	2.2×10^{11}
μ (Pa · s)	Viscosity of the weak substratum	$10^{17}–10^{22}$	10^{19}
Ø	Angle of internal friction	$15°–40°$	$33°$
α	Slope angle	$2°–35°$	$30°–34°$
G (ms^{-2})	Gravity acceleration	9.8	9.81
C (Pa)	Cohesion	$10^4–10^7$	10^5
γ (kNm^{-3})	Unit density (volumetric weight)	$19–24$	24

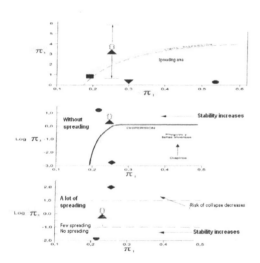

Figure 5. Π graphic that shows the limits between the stability domains.

develops around the spreading volcano: folding and thrusting. For $\Pi_3 \ll 1$, a fold belt develops around the spreading cone (Fig. 5).

The values of Π_3, Π_4 and Π_a suggest that the Arenal Volcano is in an incipient and critical stage of spreading due to dispersion (material displacement on its base) and cone instability based on the results of Π_1 and Π_2. However, obvious evidences that support dispersion are not shown; unless the dry tilt meters indicate the phenomenon in a very slow way (for example. inclinometer E). We might think that the dispersion tends to occur in a direction towards the regional slope, for example towards La Fortuna, combined with the fact that the new cone C generates extra weight on the old cone D, even though, the asymmetry in the pyroclastic thicknesses in the western part of the cone might also be a conditioning factor for the dispersion towards the Arenal lake reservoir.

The Arenal Volcano is in a construction stage, of at least 7000 years, and it has formed a new cone (C) since 1974, that has increased at least 270 m from 1460 m to approximately 1730 m.a.s.l. Also, between crater A (the inferior one formed in 1968) and crater C, there have been erupted 0.6 km³ of lava since 1968 to 2009, that are deforming the ground due to its weight. On the other hand, the volcano is basically constitute of a huge lava flow body, which was build on a basement made of volcanic rocks, the oldest ones deeply fractured lavas, old pyroclastic rocks with a paleo-surface of laterites.

Based on the model of Lagmay et al. (27) and the cases quoted on that paper, we could hypothesize that the occidental flank of the volcano (new cone C) is the most prone to the formation of a partial collapse through the formation of a debris avalanche or rockslide. However, if we assume the sigmoid model, with the thrust fault (occidental pseudo-border of the caldera), where the faults NNW have a dextral component and the Danta fault is dip towards the east and of normal type (2, 5, 6), then, we have that the volcano is unstable to the formation of a sectorial collapse leaded NE.

Also, the surrounding parts of the craters of C and D cones posses fumarolic manifestations that chemically affect the rocks, combined with its high grade of fracture (also see (15)), particularly in cone D (Fig. 6). Many of the gravitational slides that unchain debris avalanches generally occur in dilated sectors rich in fluids (52).

– Limit Equilibrium Analysis

Table 5. Non-dimensional Π numbers of volcanoes in general (15, 31, 48, 49) and the case of the Arenal Volcano.

Dimensional ratios	Volcanoes (Arenal)	Comments	Interpretation
$\Pi_1 = H/R$	0.15–0.55 (0.27)	The relation between Π_1 and Π_3 controls the dynamic evolution of the volcano. Π_1 ends to decrease with time, however its increase becomes a factor of instability	Its intermediate value suggests some cone instability. Since the volcano is still growing, its instability increases with a constant radio.
$\Pi_2 = D/H$	0–1.5 (0.09–0.18)	For a given thickness, it provides the critical height where the weight impulses the deformation: < 0.4. Si $\Pi_2 > 0.4$, the cone is stable	Since $\Pi_2 < 0.4$, there is some dispersion or cone instability
$\Pi_3 = D/I$	0–20 (0.8–10) (4.3)	Determines the type of deformation and grabens number. If < 4, the collapse begins. If $\Pi_3 \cong 10$, no spreading is presented even if $\Pi_2 < 0,4$. If Π_3 is between 0,1 and 1, the number of grabens should be 4 and if it is 3, there would be 6 or 7 grabens.	Since the Arenal cone is young, the number of grabens are not still develop and the spreading on its base is incipient.
$\Pi_4 = \rho v/\rho s$	1–1.5 (1.05)	Helps estimate the volcano's weight on the substrate. The increase of Π_4 (from 1 to 1,5) has the same effect than the decrease of Π_3	There could be some spreading at its base as a fold belt.
$\Pi_5 = \rho v \cdot g\,H \cdot t/\mu$	0.01–790 (0.66)	Since volcanoes tend to grow, an increase of Π_5 transforms the cone from a stable type to spreading type.	The Arenal volcano is in an incipient spreading stage.
$\Pi a = R/I$	0.3–50 (32–200)	If the ductile layer is too thin ($\Pi a > 1$), spreading occurs, while if it is too thick ($\Pi a < 1$), flexion occurs	Since the weak layer is very thin compare to the volcano ($\Pi_a > 1$), spreading occurs
* $\Pi b = \mu \cdot I / c \cdot D \cdot t$	166–1 000 000 (45.4–568) (247.1)		

The most common way to evaluate a talus stability, is through the determination of a factor of safety for the critical fault surface. This is can be done using limit equilibrium techniques that require information on soil resistance, but not on the relation between stress and deformation. Inside a homogeneous cone with an specific slope, the location of the sliding surface is controlled by the non dimensional relation:

$$\lambda = c/ (\gamma\,H\,\tan\Phi) \qquad (1)$$

where H is the height of the cone, c is the cohesion, γ is the unit density and Φ is the internal angle friction. Using the information on Table 6 and based on the work by Reid et al. (37), we have that $\lambda = 0.0065$ for the Arenal volcano. With this data, we can estimate a curvilinear fault surface of a maximum depth of 200 m for a volume approximately 5% (around 0.6 km^3) of the total volcanic edifice (cf. (37), p. 6048).

– STABL 5M Program

The slope stability analysis was made for two profiles of the Arenal cone, using the STABL 5M Program (Fig. 3 and Table 6). It analyzes the stabil-

Figure 6. Oriental flank of the D cone (highly fractured).

ity through the Janbu and Bishop methods. In this paper, the Bishop method was the only one used (8), which apart from being very extensive on its use, it can be apply to intensely fractured rock massif, for example in cones C and D at Arenal, or for extremely altered soils like the ones in its foundation. The factors of safety that were determined with the Bishop method differ in approximately 5% from more accurate solutions (41), this is why in many cases, we use the following method that is easier to handle.

We assumed that the fault might be a: a) Global fault of the massif or b) Cone fault. The first one has a deep fault surface and can go through the volcano's foot or deeper. The second case has a superficial fault surface and it does not regularly get to the volcanic cone's foot.

The analysis was made for three conditions:

– Static condition: Models the normal stage of the massif.
– Static condition with an overload: Models the situation that can be presented with a copious lava eruption and with the presence of an overload due to the material accumulation on the massif. For this case, a uniformly distributed load of 900 kPa was used (60 m of lava). It was modeled for the occidental flanks were the lava effusions are frequent, but also for the oriental flank to see how critical the situation is if the cone C erupts towards this section.
– Pseudo static condition: Models the situation in case of an earthquake. According to regional probabilistic estimations, there was obtained,

that for a return period of 500 years, the values of peak acceleration, where the volcano is located, are around 0.34 g and the intensities are MM VII to VIII (A. Climent, com. verbal, 2003). The seismic forces are supposed to be pseudo static and proportional to the sliding mass weight, and the coefficients kh and kv that represent the acceleration produced by the seismic movement. Since it is a highly seismic zone and the materials to be analyzed are similar to an earth dam, we used, a horizontal acceleration of $kh = 0.15$ g and a vertical acceleration of $kv = 0$ g (25, 38, 55).

In both profiles, the regional water table is shown. The four geotechnical levels shown in Table 6 are used. We also did tests with different friction angles and cohesion values, recording variations in the factor of safety, which express the need of a better geomechanical calibration to refine the stability of these slopes that genetically unstable.

– Stability for the occidental flank: The results are shown on Fig. 3a and Table 7. The critical fault surfaces match for all the load states (static, with an overload and pseudo static), for the global stability analysis of the massif as well as for the cone stability. The cone stability analysis presents safety factor slightly higher than the global analysis, close to 1. The most critical case is in the pseudo static stage, where the factor of safety is around 0.7. The talus global stability presents the lowest factors of safety. For the static case, the safety factor for the global stability is close to 1.1, this is why it might be considered that this fault surface is closed to the critical stage. In case

Table 6. Geotechnical model used for the stability analysis.

Soil	Dry volumetric weight (kN/m³)	Humid volumetric weight (kN/ m³)	Cohesion (kPa)	Angle of friction Ø
I	15	20	0	40–44°
Ia	20	25	0	33°
II–III	14	18	90–92	12°
IV	17	22	200–400	44–45°

Table 7. Results of the stability analysis for both profiles with an overload of 900 kPa.

Flank	Stability condition	Seismic coefficient	Overload (kPa)	Minimum safety factor	Volume (km³)
Occidental	Global	–		1.09	0.05–0.1
		–	900	1.06	0.05–0.1
		0.15		0.64	0.05–0.1
	Cone	–		1.18	0.03–0.07
		–	900	1.14	0.03–0.07
		0.15		0.77	0.03–0.07
Oriental	Global	–		2.16	0.04–0.1
		–	900	2.03	0.01–0.03
		0.15		1.28	0.03–0.07
	Cone	–		1.51	3.5×10^{-3}
		–	900	1.11	1.1×10^{-3}
		0.15		1.5	3.5×10^{-3}

there are any overloads due to volcanic eruptions, the safety factor is not very altered, and it is equally close to 1,1. The presence of seismic movement is very important, since it makes the factor of safety lover than 1 (around 0.6), what indicates that the sliding of an important slope mass could be imminent. For all the cases above, the fault surfaces are very close to the foot of the volcanic edifice, between the surfaces Ia and II–III, indicating that the mechanical characteristics of these two layers are the ones that determined the threat of sliding for this flank.

– Stability of the oriental flank: The results are shown on Fig 3b and Table 7. The critical fault surfaces do not match for all the load states. The lowest safety factors are given for the stability analysis of the cone with relatively small fault surfaces, meanwhile, for the global stabilization, the fault surfaces are very extensive but shallow, which shows that the mechanical properties of the first layers are the ones that determine the risk of sliding for this volcanic cone.

In all cases, the factor of safety is above 1, this is why the risk of sliding for this flank is low compare to the factor of safety of the occidental flank. The lowest factor is 1.3, so we have to be careful, but the sliding is not imminent, unless the geotechnical conditions change (for example tectonism, excessive load due to the growth of cone C, vibrations due to the explosions).

5 ANALYSIS OF STRESS AND DEFORMATION

5.1 Stress Calculation

There are many methods to calculate geostatic stresses through the theory of elasticity (36). Stamotopoulos & Kotzias (41) clarify that even though the elastic solution provide an acceptable method to calculate the changes in the vertical stress, they are based on suppositions that simplify the work, thus the results are only reasonable estimations. Different cases that could be presented in the Arenal Volcano are shown in the following information.

– Case of an individual lava flow: For a flow that reaches the pyroclastic deposits at the foot of the volcano, with a 60 m front and a 16 m thickness, we can calculate the vertical stress at 14 m of depth. The result is 390 kPa. A lava flow of 10 m of thickness transmits a load of 250 kPa, while the admissible load for the Arenal soils are between 50 and 250 kPa (see (44), p. 507).

– Case of a lava field: The maximum vertical stresses generated at the Arenal lava field under a 150 m thickness would be 3625 kPa and 2014 kPa for the horizontal stress.

– Case of the Arenal cone: We can assume a simplified geometrical shape as a vertical structure of 1100 m (height), and where γ is the volumetric weight of the lava (25 kN/m³) for an overload of 27 500 kPa. It is clear that a bidimensional shape of triangular section or a three-dimensional shape of conic section would be closer to reality. The overweight that transmits the cone under its peak, and considering the relations presented by Hoek & Brown (22), to its foundation may vary between 8000 and 29 700 kPa (maximum value) and horizontally between 3127 and 55 350 kPa. However, generally these values are over estimated (particularly the vertical component of the stress), which can be even 5 times inferior, specially at depths between 500 and 1500 m, since the equation used is only the best fit curve. If we use a finite element method (SIGMA/W Program), we may observe that the vertical stresses under the cone's peak are around 19 875 kPa and the horizontal stresses are 10 426 kPa. At 5100 m of depth from the base, they would be >5000 and 1250 kPa, respectively. These increases are caused by the effect of the cone, without considering the weight of the foundation. In conclusion, the most realistic results of the stresses generated by the cone in contact with the foundation may be around 20 000 kPa for the vertical σ and 10 000 kPa for the horizontal σ.

The shear stress caused by a body with an ideal triangular section on the foundation, has a maximum value around the depth h (equal to a quarter of the width of the base, i.e. 2 km), whose shear stress ($\Delta\tau$) is close to 0.3p, where p represents the maximum pressure under the cone's peak (1,1 km height) by the unit weight (Fig 7). Thus, we may conclude that the shear stress τo, at depth h under the cone's axis, that under the foundation acts by symmetry to 45° from the base, is very close to the maximum value (41), p. 68), that is to say:

$$\tau o = 0.25\ \gamma h + 0.3p \qquad (2)$$

where γ is the volumetric weight of the lava, with an average value of 25 kN/m³ and a humid weight of the foundation of 22 kN/m³.

In the case of study,

$$\tau o = 0.25\ (22\ kN/m^3) \times 2000\ m + 0.3\ (25\ kN/m^3) \\ \times 1100\ m = 19\ 250\ kPa$$

If τo is less than the shear resistance, the soil stresses would be values close to ones determined by the elastic solutions, and the soil won't yield.

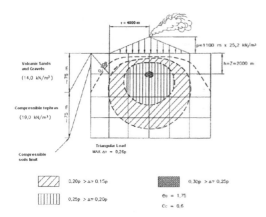

Figure 7. Shear stress by the load of the Arenal, in the triangular and conic section, obtained from (41). Actually, the compressible tephras, volcanic sands and tephras are linked.

5.2 Settlements based on the consolidation tests and the superimposed lava loads

5.2.1 Initial settlements

When there are loads applied on the field, it is subjected to an increase of stresses that produces an instant deformation and a long term deformation, with a decrease in the void ratio due to the water flow towards the exterior of the soil (consolidation). If the instant initial settlement wants to be analyzed in the case of a lava flow, the theory of elasticity provides a convenient formula (41):

$$\partial = \frac{0,07\,\rho H}{E} \qquad (3)$$

where ρ is the average pressure in the soil at the surface, and it is found multiplying the total weight of the lava flow (25.2 kN/m³) by its thickness (20 m); H is the thickness of the compressible unit (arbitrarily taken as 10 m in compressible soils) and E is the Young modulus (for pyroclastic rocks), value derived from geophysics (between 0.36 and 3.7×10^6 kN/m²) or E can also be estimated as 1000 Su to 100 Su (that is to say between 13 500 and 90 000 kN/m²), where Su = c (in our case 90 kN/m²) in a UU test (see 41). The initial instant settlements may vary from 1 m to 4 cm, given that a lot of values of modulus are considered. The total cone settlements were not analyzed, since its growth has taken several millions of years.

Geology and geophysics suggest that en some areas, volcanic soils can reach thicknesses of 150 m, particularly under erupted lava between 1968 an 1973. Thereby, the initial estimated settlement may vary between 30 cm and 4.1 m, considering

2 m as an acceptable value when an E = 150 Su is assumed, by recommendation of Stamatopoulos & Kotzias (41). If we use values of E provided by geophysics (Table 4), the settlements would be between 11 and 100 cm. An average value of 1.3 m may be taken as a representative value for this case of analysis.

The vertical stress at the foundation in Arenal under the vortex of a conic load under the volcano summit, at the same depth of 2 km and using the relations provided by Poulos & Davis (36), is σ = 13 750 kPa. The increase of stresses is still significant (influence zone; (7)) at a depth of 24 km (about the crust/upper mantle limit).

5.2.2 Immediate-primary compression using the compressibility index

When a lava flow runs on a granular soil (thick ashes to lapilli), the excess in pore pressure (if there is any) dissipates in a short time and the consolidation would be fast. In low permeability silt and clay soils, the water flow will be very slow and the dissipation process will take a considerable amount of time. In these cases, we may need to predict: a) The total settlement of the lava flow or lava field and b) The velocity at which it occurs.

The settlement calculation under the actual lava field (once the process ends) would be around 7.8–15 m using the compressibility index (Cc = 0.6–0.67) according to the methodology explained by Stamatopoulos & Kotzias ((41), p. 124–125). In a larger scale, how the settlement generated by the growth of the volcanic edifice in the last 7000 years is, using a simplified model shown in Fig. 7. The final result is a total settlement of 20.0 to 29.7 m.

Through the consolidation coefficient (Cv), we may obtain the time that takes a settlement on the field, given a layer with a drainage longitude Hf. A lava flow would generate a consolidating effect on the pyroclastic package of 10 m just in 10 years, lapse necessary to finish the theory of consolidation due to the effects of overweight. Since the tephra or pyroclastics thickness in the west flank of the volcano has been estimated as 150 m, from which 50% can be assumed as thick non cohesive-granular tephras. If we analyze the whole lava field of the Arenal Volcano (1968–2009), for a 50% and 90% consolidation of a tephra package of 75 m of thickness and whose drainage conditions are not very clear, it takes 132 and 568 years, respectively. From another point of view, the actual lava field, in the last 41 years, has only reached approximately 18% or more of the consolidation process, so we can expect the detection of more deformations by the vigilance inclinometers for one or many centuries.

5.3 Displacement and stresses based on elastic and elasto-plastic models

Six different geological-geotechnical models were analyzed: a) the cone (Layer I) with the foundation (Layer IV) with a constitutive elastic model, b) the cone with the foundation and an intermediate layer majorly formed by pyroclastics (Layers II and III) in an elastic way, c)the effect of the cone on its foundation in a elasto-plastic way, d) the cone (Layer I) developed on a pyroclastic level (Layer II and III) in a elasto-plastic way, e) the growth of the cone by stages and f) the effect of the recent lava field (particularly develop between 1968 and 1973) on the foundation (see Table 8 and Fig. 8).

5.4 Deformation based on inclinometry

The detected deformations at the dry inclinometry stations of the Arenal volcano, for volcanic

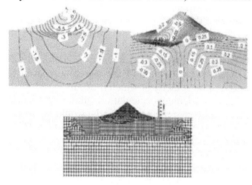

monitoring, are observed still at distances of 5 km from de axial principal area of the flows (maximum thickness of the lava field), 41 years after its effusion. Based on the data of the precise leveling campaigns at the Sakurajima Volcano in Japan (comparisons on Table 9), through the Showa lava field of 1946, it was evident that a local subsidence was superimposed to the regional deformation at the bottom of the volcano, still 56 years after the emplacement (24, 42). The same phenomenon has been observed at the Etna (Italy), in recent investigations on leveling and inclinometry. In the Sakurajima case, the lava thickness in the deformation areas is around 50 m, and there is not any local subsidence beyond 1.7 km from the lava axis. At the Etna Volcano, the subsidence is still active, more than three decades after its emplacement, in compound flows that reach thickness between 70 and 75 m and its effects are perceptible 200–300 m away from the border of the lava flows and more than 1 km from its axis. The deformation rate at a distance of 1 km from the flows axis is in the order of 40 µrad/year. In this case, the lava field is wider than the one at the Sakurijama Volcano and

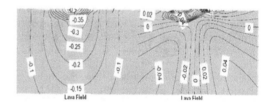

Figure 8a. Finite element analysis of the Arenal Cone: Vertical (left) and horizontal (right) deformation in meters.

Figure 8b. Elastic deformation (a: vertical; b: horizontal in cm) for a lava field of 150 m of thickness.

Table 8. Comparison of movements and stresses under different cases and models.

Case of study	Constitutive model	Maximum Vertical (V) and Horizontal (H) Displacements	Stresses (kPa)	Comments
Cone + Foundation	Elastic	V = 4.95 m H = 0.39 m	V = 16 157 H = 7500	
Cone (Layer I) on pyroclasts (Layers II and III) + Foundation (Layer IV)	Elastic	V = 7 m H = 0.42 m north y 1 m south	V = 19 875 H = 10 425	Horizontal deformation influence zone extends to 4800 m
Cone + Foundation	Elasto-plastic	V = 4.98 m H = 0.39 m		Horizontal deformation influence zone extends to 4830 m
Cone + Pyroclastic Layer	Elasto-plastic	V = 7.16 m		Plasticized sectors in the flanks, part of the volcano, foundation and surrounding setting.
Growth of the volcano in 3 stages	Elasto-plastic	V = 6.35 H = 0.89 m south y 0.26 m north, near the foundation		The plasticized sectors appear during the first stage (300 m) to manifest in the next two stages, at 700–900 m
Lava field (150 m thickness) on Layers III and IV	Elastic	V = 72 cm H = 4–11 cm	V = 362.53 H = 2014.4	Consolidation process is not considered

Table 9. Comparative analysis of the deformations produced by lava fields in different volcanoes.

Volcano	Year	Lava field parameters			Deformation		Detection (km) from area with maximum thickness	Source
		Thickness (m)	A (km²)	V (km³)	μrad/a at 1 km from the axis	mm/year		
Sakurajima (Japan)	1946	50		0.18	14	14	1.7	(24)
Pitón de la Fournaise	1976					20 (1986–89)		(18)
	1976					5 (1986–89)		
	1986					10 (1986–89)		
Arenal (Costa Rica)	1968–present	150 (compound thickness)	7.5	0.75	30–60	60 (1968–2003)	5	(5)
Etna (Italy)	1981, 1983	75						(35)
	1986–1987	15	6.5	0.06		22		(14)
	1989	10 (compound 40)	3.3	0.0033		47		

Table 10. Settlements determined under different methods and media (in meters).

Type and thickness		Geological media of load		
		Lava (20 m)	Lava field (150 m)	Cone (900–1100 m)
Period		Days–Months	35 years	7000 years
1	Initial settlements	0.001–0.04	0.11–4.1 (1.3–2)	0.44–16 (5.2–8)
2	Immediate-primary settlements using Cc	0.48–0.53	7.8–15.0	20.0–29.7
3	Elastic methods through finite element analysis	–	–	4.98–7.16
4	Elasto-plastic methods through finite element analysis	–	0.72	4.95–7
5	Total settlements (1 + 2)	0.48–0.57	7.9–19.1	20.4–45.0
6	Settlements based on inclinometers (~18% of consolidation)	–	~2	–

the volume is at least the times (0.6 km³ against 0.18 km³), as well as its thickness than the one at the lava profile of the Sakurijama Volcano. It also has twice its thickness than the profiles of the Etna flows (around 75 m).

When we compare the deflation rates associated to the Sakurajima and Etna flows with the ones from the Arenal Volcano, we find that they are very similar. Considering the differences in the covered area, thickness and volume, 1 km away from the axes, we have 40 μrad/year at the Etna, 14 at the Sakurajima, and 60 and 30 at the B and C stations at the Arenal Volcano (total weight is 1012 kg), seems like a plausible and concrete explanation, according to the field observations and data from the inclinometers (2, 4, 33, 42, 53).

Theoretically, we can pass from angular measurements to settlements at the maximum thickness axial axis. In 41 years (1968–2009), we can argue that close to the maximum lava thickness and when a μrad has a value of a millimeter in the vertical direction (when d = 1 km), we have a slow deformation from 40–64 mm/year, that by extrapolation would be 1.41 m (according to Soto, 1991) and 2.03–2.24 m (5) of subsidence. A value of 2 m seems likely, since it counts with a larger amount of data.

6 CONCLUSIONS

– Since the Arenal volcano is made predominantly of a lava cone that grew up on a volcanic basement composed mostly of fractured (incompetent mechanical level), an overweight would generate the necessary geotechnical conditions for: 1) deform its foundation (subsidence, folding or failure due to an overload) and, 2) unsettling the cone.

- The initial settlements in the case of a lava flow of 20 m of thickness, may vary in a range of <0,3 cm to >3.92 cm. Geology and geophysics suggest that the total settlements can reach 19 m, once the effect of subsidence caused by the actual lavas is finished. In a larger scale, the settlement generated by the growth of the Arenal volcano during the last 7000 years, can be calculated at a height of 45 m.
- A lava flow would generate consolidation on a pyroclastic deposit of 10 m of thickness in 10 years. If we extrapolate this for the entire historic lava field of the Arenal Volcano on a package of consolidatable tephras of 75 m (thick), we will require between 132 (50% of consolidation) and 568 years (90% of consolidation), respectively.
- The deformations shown in the dry inclinometers can be observed 5 km away from the flow axis.
- The results and analysis of mechanic of rocks and soils indicate that the Arenal cone and its foundation have a poor to medium mechanical quality. Therefore, to analyze the cone and its foundation stability, we applied several methodologies. If we consider the one provided by the Buckingham Π Theorem, we have that the Arenal volcano is in an incipient stage of deformation by dispersion of its base and a clear cone instability, in continuous growth. In general, all these loads applied during a long time may be producing continuous deformations known as creep. The lateral deformations during consolidation represent around 15% of the vertical ones. Both processes (creep and vertical deformation) involve the cone spreading.
- When we use the SIGMA/W Program, plasticized sector (that is to say areas that already failed or moved their maximum resistance) in the flanks, volcano core and with a less impact on its foundation. However, since the cone is mainly made by loose fractured or fragile materials, it is not easy to see incipient sliding materials. However, it is interesting to see that the superior-medium slope of the volcano (cone D) is highly failed.
- We conclude that the new cone C of the Arenal volcano is prone to the formation of sectorial collapses with debris avalanches towards the occident, while the old cone D, given its high grade of fragmentation and the overweight provided by the new cone C could be prone to the formation of rockslides towards NE, an area consider with relatively low hazard until now. As a matter of fact, the occidental flank of the Arenal shows an unexpected situation, since the global stability of the massif (foundation + cone) is lower than the stability of the cone, even yet, the factor of safety for this one is the lowest one (0.6) for a pseudo-static condition, what indicates that in case of a telluric event with enough magnitude to induct important accelerations in the region, there is a risk of big slides (0.03–0.75 km^3) towards the occident. A debris avalanche of this type may reach distances between 5 and 10 km from the Arenal summit.
- On a lower scale, the stability estimations on cone C and its old lava pool showed that the width of the crater wall, the height of the lava lake and the continuous explosions are the main factors that control its stability. Nowadays, cone C (attached lava domes) is taller than its predecessor, cone D. Both domes and lava pools, given their instability, when they collapse, produce pyroclastic fluids. The continuous growth of the C cone over the D cone is increasing the overweight on its predecessor, as well as the possibility of pyroclastic flows and rockslides towards an area that not many years ago was consider relatively protected, that is to say the oriental flank, right were the commercial and housing activity increases every day.

ACKNOWLEDGEMENTS

To Mario Arias, Benjamín van Wyk de Wries, Julio Macías, Carolina Sigarán and Francisco Cervantes that contributed with valuable comments during the final study phase. The *Laboratorio Geotécnico del ICE* and the *Laboratorio Nacional de Materiales y Modelos Estructurales* (LANAMME) from the University of Costa Rica, helped with laboratory tests. The translation of this text from the Spanish version was done by Melisa Quirós Echeverría.

REFERENCES

1. Alvarado, G.E., (2002), "Análisis de la Estabilidad del Cono y Comportamiento de la Fundación del Volcán Arenal (Costa Rica) mediante el uso de parámetros adimensionales de acuerdo con el teorema Π de Buchigam", *VIII Seminario Nacional de Geotecnia, 3er Encuentro Centroamericano de Geotecnistas: "Geotecnia en la Prevención de Desastres en el Entorno"*. San José: 267–275.
2. Alvarado, G.E., (2003), "Diagnóstico de la estabilidad del cono y comportamiento de la fundación debido al crecimiento del edificio volcánico del Arenal (Costa Rica)", -xv + 138 pp. Univ. Costa Rica, *M.Sc Thesis*.
3. Alvarado, G.E. & Soto, G.J., (2002), "Pyroclastic flow generated by crater-wall collapse and outpouring of the lava pool of Arenal Volcano, Costa Rica", *Bull. Volcanol.*, 63: 557–568.

4. Alvarado, G.E., Argüeta, S. & Cordero, C., (1988), "Inter prentación preliminar de las deformaciones asociadas al volcán Arenal (Costa Rica)", *Bol. Obs. Vulc. Arenal*, 1(2): 26–43.

5. Alvarado, G.E., Carboni, S., Cordero, M., Avilés, E., Val verde, M. & Leandro, C., (2003), "Estabilidad del cono y comportamiento de la fundación del edificio volcánico del Arenal (Costa Rica)", *Bol. OSIVAM*, 14 (26): 21–73.

6. Alvarado, G.E., Soto, G.J., Schmincke, H.-U., Bolge, L.L. & Sumita, M., (2006), "The 1968 andesitic lateral blast eruption at Arenal volcano, Costa Rica", *J. Volcanol. Geotherm. Res.*, 157 (1–3): 9–33.

7. Berry, P.L. & Reid, D., (1987), "An Introduction to Soil Mechanics". Spanish translation: "Mecánica de Suelos", McGraw-Hill, xv + 415 pp., México, 1993.

8. Bishop, A.W., (1955) "The Use of Slip Circles in the Stability Analysis of Earth Slopes", *Geotechnique*, 5: 7–17.

9. Borgia, A., (1994), "Dynamic basis of volcanic spreading", *J. Geophys. Res.*, 99 (B 9): 1779–1780.

10. Borgia, A. & Treves, B., (1991), "Volcanic plates over rinding the oceanic crust: Structure and dynamics of Hawaiian volcanoes", *En: L.M. Parson (ed.): Ophiolites and their Modern Analogues. Geol. Soc. London, Spec. Publ.*, 60: 277–299.

11. Borgia, A., Burr, J., Montero, W., Morales, L.D. & Alva rado, G.E., (1990), "Fault Propagation folds Induced by Gravitational Failure and Slumping of the Central Costa Rica Volcanic Range: Implications for Large Terrestrial and Martian Volcanic Edificies", *J. Geophys. Res.*, 95 (B9): 14357–14382.

12. Borgia, A. & van Wyk de Vries, (2003), "The volcano-tectonic evolution of Concepción, Nicaragua", *Bull. Volcanol.*, 65: 248–266.

13. Borgia, A., Poore, C., Carr, M.J., Melson, W.G. & Alvara do, G.E., (1988), "Structural, stratigraphic and petrologic aspects of the Arenal-Chato volcanic system, Costa Rica: Evolution of a young stratovolcanic complex", *Bull. Volcanol*, 50: 86–105.

14. Briole, P., Massonnet, D. & Delacourt, C., (1997), "Post eruptive deformation associated with the 1986–87 and 1989 lava flows of Etna detected by radar interferometry, *Geophys. Res. Lett.*, 24 (1): 37–40.

15. Cecchi, E., van Why de Vries, B., Lavest, J.-M., (2005), "Flank spreading and collapse of weak-cored volcanoes", *Bull. Volcanol.*, 67: 72–91.

15. Choncha-Dimas, A., (2004), "Numerical Modeling in un derstanding catastrophic collapse at Pico de Orizaba, México", 341 pp., *Tesis de Ph.D.*, Univ. Nevada, Reno.

16. Concha-Dimas, A. & Watters, R.J., (2003), "Preliminary Evaluation of Volcanic Flank Stability Using Finite Difference Modeling: Citlaltépetl Volcano, Mexico", *In: P.J. Cullingan, H.H. Einstein & A.J. Whittle: Soil & Rock America 2003, 12th Panamerican Conference on Soil Mechanics and Geotechnical Engineering*, VGE, Proceeding 1.

17. Delorme, H., (1994), "Apport des deformations à la compréhension des mécanismes éruptifs: le Piton de la Fournaise", *PhD. Thesis*, Univ. Paris VII, 449–467.

18. Druitt, T.H., Young, S.R., Baptie, B., Bonadonna, C., Calder, E.S., Clarke, A.B., Cole, P.D., Harford, C.L., Herd, R.A., Luckett, R., Ryan, G. & Voight, B., (2002), "Episodes of cyclic Vulcanian explosive activity with fountain collapse at Soufriére Hills Volcano, Motserrat", *En: T.H. Druitt, & B.P. Kokelaar (eds.): The Eruption of Soufriére Hills Volcano, Montserrat, from 1995 to 1999.* Geol. Soc., London, Memoirs, 21: 281–306.

19. Franklin, J.A. & Dusseault, M.B., (1989), "Rock Engin nering". x + 600 pp., McGraw-Hill Publis. Co., New York.

20. Hoek, E. & Bray, J.W., (1981), "Rock slope Engineering" [3era ed.]. 358 pp. Inst. Mining and Metallurgy, London.

21. Hoek, E. & Brown, E.T., (1980), "Underground excava tion in rock", 641 pp. McGraw Hill.

22. Hürlimann, M. & Ledesma, A., (1999), "The influence of Residual Soil on the Stability of Volcano Flanks. Application of the Large Landslides Ocurred on the Northern Slopes of Tenerife, Canary Islands", *IX Panamerican Conference on Soil Mechanics & Geothecnical Engineering*, 4: 1305–1311.

23. Ishirara, K., Takayama, T., Tanaka, Y & Hirabayashi, J.,(1981), "Lava flows at Sajurajima Volcano (I)"— volume of the historical lava flows-, *Ann Dis. Prev. Res. Inst.*, Kyoto Univ., 24 (B-1): 1–10 (in Japanese).

24. Japanese National Committee of Large Dams, (1988), "Earthquake Resistant Design Features of Dams", *En: Earthquake Resistant Design for Civil Enginnering Structures in Japan.* Japanese Society of Civil Engineereers, Tokyo, pp. 1–30.

25. Keefer, D.K., (1984), "Landsides caused by earthquakes", *Geol. Soc. Amer. Bull.*, 95 (2): 406–421.

26. Lagmay, A.M.F., van Wyk de Vries, B., Kerle, N. & Pyle, D.M., (2000), "Volcano instability induced by strike-slip faulting", *Bull Volcanol.*, 62: 331–346.

27. Leandro, C. & Alvarado, G.E., (1999), "Estudio Geológi co-Geofísico de una sección Oriental y Occidental en el volcán Arenal". *Bol. OSIVAM*, 11 (21–22): 50–60, 1998; San José.

28. López, A., (1999), "Neo- and Paleostress partitioning in the SW corner of the Caribbean Plate and its fault reactivation potential". Tübinger Geowissenchaftliche Arbeiten, 53: xi + 294 pp.

29. López, D.L., Bundschuh, J., Soto, G.J., Fernández, J.F. & Alvarado, G.E., (2006), "Chemical evolution of thermal springs at Arenal Volcano, Costa Rica: Effect of volcanic activity, precipitation, seismic activity, and Earth tides", *J. Volcanol. Geotherm. Res.*, 157 (1–3). 166–181.

30. Merle, O. & Borgia, A., (1996), "Scaled experiments of volcanic spreading". *J. Geophys. Res.*, 101 (B6): 13805–13817.

31. Milne, J.F.G.S., (1878), "On the form of volcanoes". *The Geological Magazine*, II (VIII): 337–345.

32. Mora, M., (2003), "Étude de la structure superficielle et de l'activite sismique du volcan Arenal, Costa Rica", 155 pp + 4 anexos, Universidad de Saboya [PhD. Thesis].

33. Morera, J.F., (1990), "Informe de apoyo geológico-geotécnico para la construcción de la presa del proyecto Fortuna", *Inf. Interno ICE*, Geology Department, september, 20 pp + maps.

34. Murray, J.B., (1988), "The influence of loading by lavas on the siting of volcanic eruptions vents on Mt Etna", *J. Volc. Geotherm. Res.*, 35: 121–139.

35. Poulos, H.G. & Davis, E.H., (1973), "Elastic solutions for soil and rocks mechanics". 411 pp., Wiley, Nueva York.

36. Reid M.E., Christian, S.B. & Brien, D.L., (2000), "Gravitational stability of three-dimensional stratovolcano edifices", *J. Geophys. Res.*, 105: 6043–6056.

37. Seed, H.B., (1979), "Considerations in the Earthquake-Resistant Design of Earth and Rockfill Dams", *Géotechnique*, 29 (3): 215–263.

38. Sieber, L., (1984), "Large volcanic debris avalanches: cha racteristics of source areas, deposits, and associated eruptions", *J. Volcanol. Geotherm. Res.*, 22: 163–197.

39. Siebert, L., Alvarado, G.E., Vallance, J.W., and van Wyk de Vries, B., (2006), "Large-volume volcanic eficice failures in Central America and associated hazards", *In: W.I. Rose, G.J.S. Bluth, M.J. Carr, J. Ewert, L.C. Patino, and J. Vallance. (eds.): Volcanic hazards in Central America. Geol. Soc. Amer. Sp. Paper*, 412: 1–26.

40. Stamatopoulos, A.C. & Kotzias, P.C., (1985), "Soil Im provement by preloading". John Wiley & Sons, Inc. Spanish Translation: "Mejoramiento de suelos por precarga" [1ª edición, 1990], Ed. Limusa, México, 296 pp.

41. Soto, G., (1991), "Análisis de inclinometría seca en el volcán Arenal, 1988–90", *Bol. Obs. Vulc. Arenal*, 4 (7): 33–61, San José.

42. Soto, G.J. & Alvarado, G.E., (2006), "Eruptive history of Arenal Volcano, Costa Rica, 7 ka to present", *J. Volcanol. Geotherm. Res.*, 157: 254–269.

43. Sowers, G.B. & Sowers, G.F., (1970), "Introductory foil Mechanics and Foundations" [3era ed.]. The Macmillan Company. Spanish Translation: Introducción a la Mecánica de Suelos y cimentaciones [2ª reimpresión 1978], Ed. Limusa, México, 677 pp.

44. Suzuki, T., (1968), "Settlement of volcanic cones", *Bull Volcanol. Soc. Japan*, 13: 95–108.

45. van Bemmelen, R.W., (1949), "The Geology of Indonesia. General Geology of Indonesia and Adjacent Archipielagos. 1ª". *Government Printing Office*, The Hague.

46. van Wyk de Vries, B. & Francis, P., (1997), "Catastrophic collapse at stratovolcanoes induced by gradual volcano spreading", *Nature*, 387: 387–390.

47. van Wyk de Vries, B. & Borgia, A., (1996), "The role of basement in volcano deformation", *In: W.J. Jones & J. Neuberg (eds.) Volcano Instability on the Earth and Other Planets. Geol. Soc. Sp. Publ.*, 110: 95–110.

48. van Wyk de Vries, B. & Matela, R., (1998), "Styles of volcano induced deformation: numerical models of subs tratum flexure, spreading and extrusion", *J. Volcanol. Geotherm. Res.*, 81: 1–18.

49. van Wyk de Vries, B. & Merle, O., (1998), "Extension in duced by volcanic loading in regional strike-slipe zones", *Geology*, 26 (11): 983–986.

50. Voight, B. & Elsworth, D., (1997), "Failure of volcano slopes", *Géotechnique*, 47: 1–31.

51. Voight, B., Janda, J., Glicken, H. & Douglas, P.M., (1983), "Nature and mechanics of the Mount St. Helen rockslide-avalanche of 18 May, 1980", *Géotechnique* 33: 243–273.

52. Wadge, G., (1984), "The magma budget of Volcan Arenal, Costa Rica, from 1968 to 1986", *J. Volcanol. Geotherm. Res.*, 19: 281–302.

53. Wadge, G., Oramas, D. & Cole, P.D., (2006), "The mag ma budget of Volcán Arenal, Costa Rica from 1980 to 2004", *J. Volcanol. Geotherm. Res.*, 157 (1–3): 60–74.

54. Wang, J.G.Z.Q. & Law, K.T., (1994), "Siting in Earth quake Zones". 115 pp., A.A. Balkema, Rotterdam.

55. Zimbelman, D., Watters, R.J., Bowman, S. & Firth, I., (2003), "Quantifying Hazard and Risk Assessment at Active Volcanoes", *EOS, Trans., Amer. Geophys. Union*, 84 (23): 213, 216–217.

Volcanic Rock Mechanics – Olalla et al. (eds)
© 2010 Taylor & Francis Group, London, ISBN 978-0-415-58478-4

Etna flank dynamics: A sensitivity analysis by numerical modelling

T. Apuani
Dipartimento di Scienze della Terra "A. Desio", Università degli Studi di Milano, Milan, Italy

C. Corazzato
Dipartimento di Scienze Geologiche e Geotecnologie, Università degli Studi di Milano-Bicocca, Milan, Italy

ABSTRACT: The present work investigates the flank deep instability dynamics of Mount Etna volcano by means of bi-dimensional numerical modelling, comparing finite element and finite difference methods, and by limit equilibrium analysis. The complicated conceptual model was first simplified and progressively implemented with a sensitivity analysis to evaluate the effect of topography, geometry and rheological behaviour of the structural units. The model is then implemented considering the presence of magma pressure along the feeding system. The results are expressed in terms of stress-strain field, displacement pattern, plasticity states and shear strain increments, or factor of safety.

1 INTRODUCTION

Most active volcanoes show clear evidence of flank instability resulting from several interacting causes including gravity force, magma ascent along the feeding system, and local and/or regional tectonic activity (Voight *et al.*, 1981, 1983; Elsworth & Voight, 1995; Iverson, 1995; McGuire, 1996; Voight & Elsworth, 1997; Elsworth & Day, 1999; Voight, 2000). The complexity of such dynamics is still an open subject of research.

1.1 *Geological-structural setting*

The present work focuses on the dynamics of Mount Etna, the largest active European stratovolcano, located on the eastern coast of Sicily (southern Italy, Fig. 1). It rises above a complex regional tectonic framework, dominated by a N-S trending direction of maximum compression, due to the Eurasia-Africa plates collision, where the Hyblean plateau is subducted beneath the Apennine-Maghrebian thrust chain, and a related E-W trending direction of maximum extension, associated with the development of the Malta Escarpment, the possible surface expression of a tear in the subducting Ionian slab (Lentini *et al.*, 1996). On the eastern flank of Etna, the Pleistocene subetnean clay unit is interposed between the Apennine-Maghrebian flysh units and the volcanic edifice products (Di Stefano & Branca, 2002).

Several volcanotectonic elements, represented by both rift zones and faults, influence the volcano dynamics (Fig. 1). A deep-seated instability affects the eastern and south-eastern flank (Borgia *et al.*, 1992; Tibaldi & Groppelli, 2002; Corazzato & Tibaldi, 2006). To the north, the boundary of such unstable sector is represented by the E-W trending Pernicana Fault System (Azzaro *et al.*, 1998; Acocella & Neri, 2005; Neri & al., 2003; and references therein) extending from the NE Rift to the coastline, with a predominant left-lateral motion. Here the flank shows a predominant ESE slip. To the south, the slip of the flank appears less consistent, being directed towards SE and S, and controlled by several structures, with different geometry and kinematics (Rasà *et al.*, 1982; Groppelli & Tibaldi, 1996, 1999; Monaco *et al.*, 1997; Azzaro *et al.*, 1998).

The instability is long-duration and apparently steady-state, although with documented accelerations related to magma pressure and with differential movements within the unstable mass (Rust & Neri 1996; Acocella *et al.*, 2003; Acocella & Neri, 2005; Bonforte & Puglisi, 2006).

1.2 *The approach*

As a contribute to understand the Etna eastern flank dynamics, a stability analysis was performed, aimed first at evaluating the role of basement geometry and rheological characters, in terms of constitutive law and the associated strength and deformability properties, and then at exploring the effect of magma pressure.

The problem was first approached by limit equilibrium analyses (LEM), and then by numerical

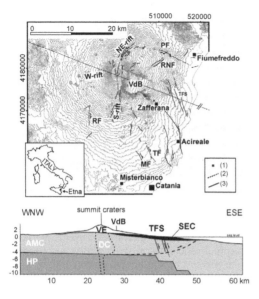

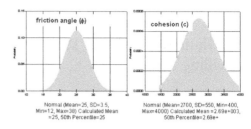

Figure 2. Frequency distribution of cohesion (c) and friction angle (φ).

Figure 1. Structural sketch map and cross section of Etna volcano. VdB–Valle del Bove, RF–Ragalna Fault, MF–Mascalucia Fault, TF–Trecastagni fault, TFS–Timpe Fault System, PF–Pernicana Fault, RNF–Ripe della Naca Faults. Inset: location of Mt. Etna. (1) rock-mass characterization site; (2) section trace; (3) main structural elements. VE–volcanics, A-MC–Apennine-Maghrebian Chain, HP–Hyblean Plateau, SEC–Subetnean clays, DC–dyke complex.

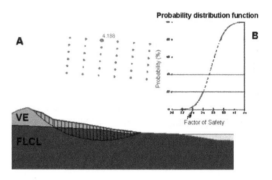

Figure 3. A. Analysed sliding surface and wedge sub-division. B. Factor of Safety Probability Distribution Function (saturated case).

modelling, performed by means of finite difference (FDM) and finite element (FEM) codes.

To support a reliable input dataset for the analysis, a dedicated structural-mechanical field survey was carried out along the eastern and southeastern volcano flank, aimed at: i) verifying the main structural framework controlling the flank instability, ii) characterizing the rock mass geomechanical quality at outcrop conditions, mainly in lava and flysh units, and iii) extrapolating average values representative for deep conditions. Some direct geotechnical tests were carried out at sample scale, to contribute to the parameterization (Koor *et al.*, 2010).

2 LIMIT EQUILIBRIUM ANALYSIS

A preliminary limit equilibrium analysis (LEM) was carried out in order to determine the Factor of Safety (FoS) related with rotational sliding occurring in the eastern flank of the volcano. The involved units are represented by the volcanic edifice (VE) built-up on the Apennine-Maghrebian flysh (FLCL). The Mohr-Coulomb properties are reported in Table 1.

To take into account the uncertainties in the shear strength parameters of the FLCL unit,

which are the most affecting and most doubtful values, a probabilistic analysis was performed assuming a gaussian frequency distribution of cohesion (c) and friction angle (φ) and applying a Monte Carlo sampling procedure to each of the analysed surfaces. The stability analysis considered two cases: i) a dry slope; ii) a hypothetical water table in the volcanic edifice and connecting to the sea level.

The analyses carried out to evaluate possible rotational sliding phenomena along surfaces several hundred metres deep show FoS>>1, even taking into account poorer geotechnical properties and the effect of a hypothetical deep aquifer (Fig. 3). Considering sliding surfaces crossing the subetnean clays unit (CL), and assuming for them a frequency distribution of cohesion (c) and friction angle (φ), the instability is never reached (FoS>>1) if gravity is the only perturbation, in both dry and saturated cases.

3 NUMERICAL MODELLING

The numerical modelling was aimed at the reconstruction of the stress pattern and at a stress-strain analysis in order to evidence plasticization areas in conditions of increasing complexity of the model.

The bi-dimensional finite difference modelling (FDM) was performed by FLAC 5.0 code (Itasca), the finite element (FEM) by SIGMA/W code (GeoSlope International), along the WNW-ESE section, crossing the volcano summit and the Valle del Bove depression, with a horizontal extent of 61 km and to the depth of 10 km (Fig. 1).

3.1 FDM

3.1.1 Conceptual model and sensitivity analysis

The conceptual model was first simplified and progressively implemented to analyse the effect of topography, geometry and rheological behaviour of the structural units. Five main geological and lithotechnical units were considered:

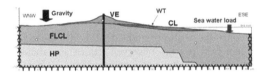

Figure 4. Geological-technical conceptual model. The vertical black line indicates the location of the interface for the application of magma pressures. At the sides, boundary conditions (fixed x- or y-velocities) are imposed. The white dots indicate the average depth at which equivalent Mohr-Coulomb parameters are calculated as a function of the confining stress.

1) volcanic edifice (VE), 2) subetnean clays (CL), 3) Apennine-Maghrebian flysh (FLCL), 4) Hyblean plateau (HP), 5) intrusive complex (DC). The boundary conditions were imposed by fixing x- and y-velocities equal to zero respectively at the base and side boundaries of the model. Sea water load was taken into account, and the hydrogeological conditions assumed considering first a dry model and then with FLCL and HP completely saturated with a static water table (Fig. 4). The model meshing has an average resolution of 100 m, and it is adjusted to fit the topography and the main units.

A first phase of analysis consisted in the initialization of the stress field in elastic conditions under the effect of gravity alone.

Then, a second phase consisted in an elastic-plastic equilibrium under the effect of gravity. In this phase the lithotechnical units were assigned a Mohr-Coulomb constitutive law and the associated strength and deformability properties in selected value ranges. Hoek-Brown properties were also considered (Tab. 1).

A sensitivity analysis (Fig. 5) was performed in order to evaluate the effect of the main assumptions concerning: i) topographic complexity; ii) geometry and asymmetry of the model; iii) role of the distance of boundary condition from the area of interest; iv) geometry of the contact between HP and FLCL (Fig. 5); v) rheology and constitutive laws attributed to the lithotechnical units (Mohr-Coulomb, Hoek-Brown; Table 1);

Table 1. Material properties assigned in the analyses. If multiple values are indicated, sensitivity analyses were performed.

	Lithotechnical units			
	VE	FL/FLCL	CL	HP
Elastic rock mass properties				
Dry unit weight, γ (kN/m^3)	2500	2600	2300	2700
Elastic modulus, E (GPa)	4; 25	6.6; 16	1.9; 10	25
Poisson ratio, υ	0.3	0.28	0.28	0.28
Mohr-Coulomb properties calculated at the specified minimum confining stress σ_3				
σ_3 (MPa)	17	22	5	–
Cohesion, c (MPa)	3.55	5; 2.9; 1;	1	–
Friction angle, ϕ °	33	29; 30	29	–
Tensile strength, σ_t (MPa)	0.035	0.036	0.126	–
Dilation angle, δ °	0	0	0	–
Hoek-Brown rock mass properties				
UCS* (MPa)	65	100	50	–
s coefficient	0.0013	0.0067	0.0016	–
mb parameter	2.346	2.0046	0.6300	–
a coefficient	0.511	0.5040	0.5099	–
Disturbance factor D	0	0	0	–

*Uniaxial compressive strength.

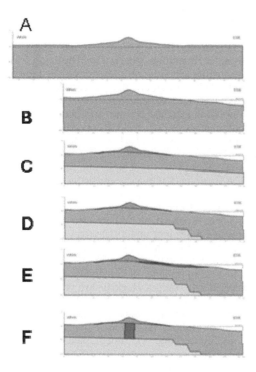

Figure 5. Sketch of the conceptual models considered in the sensitivity analysis. A. Symmetric topography, equal boundary distance. B. Real topography, different boundary distance, one unit (FLCL). C. Model with 3 units and flat HP/FLCL contact. D. Model with 3 units and stepped HP/FLCL contact. E. Model with 4 units and stepped HP/FLCL contact. F. Model including the dyke complex as a rigid elastic unit.

iv) presence of a unit representing the dyke complex as a rigid elastic body.

The results of this sensitivity analysis were expressed in terms of stress-strain field, displacement pattern, plasticity states and shear strain increments.

While a symmetrical model (Fig. 5A) showed symmetrical stress and strain fields, the choice of setting asymmetrical boundaries (Fig. 5B) resulted in asymmetrical fields. This assumption conditions all the following modelling results, but it seems to be an adequate tool to take into account the different tectonic regimes (buttressed by the Apennine-Maghrebian Chain to the NW, and transtensional to the SE). The attention was then focused to SE flank of the volcano, neglecting the strain field to the NW, which is affected by this boundary. The strain pattern shows a predisposition to SE-ward displacement, both when flat or stepped HP/FLCL contact models (Fig. 5C-D) are analysed. In the second case the tendency to develop subsidence and eastward movement is enhanced, especially

when low elastic modulus values are assumed for the FLCL unit.

The mean values assumed in the Mohr-Coulomb models well represent the rheological behaviour of the system. In fact, when the Hoek-Brown criterion, which enables to compute the equivalent Mohr-Coulomb c and φ as a function of the minimum stress σ_3 is applied, the strain field doesn't show any significant difference in relation to deep-seated instabilities.

These evidences support the choice of implementing the analysis using the asymmetrical model with stepped HP/FLCL contact and Mohr-Coulomb constitutive law for VE and FLCL.

The complexity of the system was then increased by introducing the CL unit (Fig. 5E), as well as the dyke complex (Fig. 5F). The presence of the dyke complex has a similar buttressing effect of making the boundary closer to the volcano centre, but the strain field in the area of interest is too conditioned.

Once these results enabled to choose the model which best fits the observed Etna flank dynamics, the third phase of analysis was the application of magma pressure along the feeding system. An interface was introduced in the model, in order to apply magmatic pressures. Such interface is characterized in terms of physical and mechanical properties (Normal and shear stiffness $K_n = K_s = 3.21e8$ Pa/m; c and φ equal to those of the host rock). Magma pressure is considered in terms of: i) magmastatic component (P_m); ii) overpressure component (P_0). The magmastatic component has a triangular distribution and is obtained as $P_m = \gamma_m z$, where z is the depth from the summit craters and γ_m is the magma bulk density, which ranges in values along the conduit in relation to gas content, exolution and vesiculation processes, from 0.02 g/cm³ to 2.57 g/cm³. The overpressure component is applied as a constant of 2 MPa all along the interface, according to Iverson (1995) and Apuani & Corazzato (2009), and references therein.

3.1.2 FDM main results
The main results of elasto-plastic model assuming Mohr-Coulomb properties are here presented.

The results of the analysis at elasto-plastic condition in the case of a 3 units-model, presented in Figure 6, are obtained using the worst values (in bold) of Table 1. The stress field, expressed in terms of horizontal and vertical stress contours, is conditioned by the rheological contrasts and by the stepped geometry of the FLCL/HP contact. The displacement field (Fig. 6A) is asymmetrical, with SE-ward directed displacement vectors also in the western portion of the model and maximum values of 8 m. The maximum shear strain increment

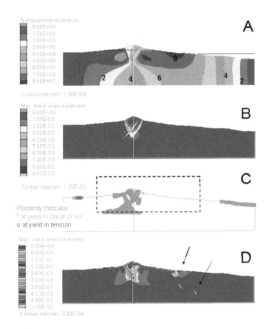

Figure 6. Results of the analysis considering three units (VE, FLCL, HP) and reasonably poor MC properties. A. displacement contours. B. Shear strain increments C. Plasticity indicators. D. Shear strain increments resulting from a simulation with less stiff VE and FLCL units (lower elastic moduli).

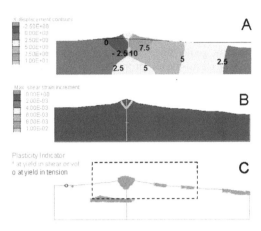

Figure 7. Results of the analysis when magma pressure is applied to the tree-units model with the relatively poor MC properties as in Figure 6. A. x-displacement contours related only to the application of magma pressure. B. shear strain increments. C. plasticity indicators.

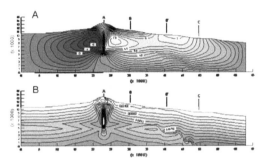

Figure 8. FEM Output of the case with 4 units and the application of magma pressure components. A. Displacement contours. B. Maximum shear strain contours.

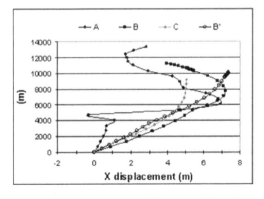

Figure 9. x-displacement profiles across the volcano (profile traces in Fig. 9A).

contours (Fig. 6B) and the plasticity indicators (Fig. 6C) are concentrated along high-angle critical zones below the volcano edifice, drawing the path for magma rising. No critical deep-seated sliding surfaces develop.

When rock mass elastic moduli are reduced, a more evident subsidence phenomenon is observed. In addition to the critical zone in the central part of the model, a zone of shear strain increment develops along the prosecution of the buried HP steps, cropping out in correspondence of the Timpe faults area (Fig 6D).

The introduction of the CL unit does not produce significant changes in the strain field if no additional perturbations are added.

Figure 7 shows the results adding the magmastatic pressure to the three-units model. The effect on the strain field is symmetrical even if the general cumulated pattern still presents an eastward direction. The high angle critical zones below the volcano edifice are better delineated.

The addition of a magma overpressure component favors the development of shallow instability on the eastern slope, especially when the CL unit is included in the model.

3.2 *FEM*

The finite element modelling was performed using the same geotechnical conceptual model and elastic and elasto-plastic properties as for the FDM analysis.

The results are generally in agreement with those of finite difference modelling, and show that the models with three or four units (including CL) are comparable as regards displacement and strain pattern. Displacements are concentrated along the eastern flank and both the topographic gradient and the geometry of HP control their direction (Fig. 8A). Maximum values of displacement and shear strain are recorded at the contact HP/FLCL, due to rheological contrasts (Fig. 8B, Fig. 9). No critical surfaces developed. The effect of magmatic pressures mostly affects the eastern flank, and the maximum horizontal displacements are recorded in the FLCL unit. The presence of the clay unit amplifies the effect of magma pressure. No sliding surfaces develop in the clay horizon neither at depth.

4 CONCLUSIONS

Allowing for the assumptions for the stress field initialization and the assigned elastic-plastic properties reasonable at depth in absence of direct investigations, the stress-strain analyses show that:

1. the stepped geometry of the contact between the Hyblean Plateau and the Apennine-Magrebian chain flysh conditions the stress field. Moreover, assuming low elastic moduli for the volcanic and flysh units and a very stiff Hyblean Plateau, it is possible to evidence a shear strain concentration along the prosecution of the steps in agreement with the Timpe structural lineaments.
2. gravity alone is not sufficient to develop deep-seated instabilities. Plasticity states are reached only in correspondence of high-angle surfaces below the crater zone, creating favourable conditions for magma rising.

Magmatic pressure emphasize this tendency, highlighting superficial displacements, but they do not seem to worsen the plasticity field at depth.

Comparing the results of deformation monitoring and the presented models, it can be argued that the deformation mechanisms of the eastern flank of Etna are surely controlled by magmatic activity, nevertheless it is not recognised so strong a cause-effect relationship to explain all the active deformation scenarios. It can be supposed that the instability process is controlled also by other factors, such as regional tectonics and seismic activity, that should be further investigated.

The results show that the geotechnical parameters describing the behaviour of the units, as well as their geometries in the model, can be considered realistic and reliable. A 3-D model is ongoing, taking into account the validity of these assumptions and the obtained results.

ACKNOWLEDGMENTS

This work was funded by Dipartimento Protezione Civile (DPC)—INGV grants to T. Apuani (project V4-FLANK, RU 02), and was performed in the framework of the ILP project "New tectonic causes of volcano failure and possible premonitory signals".

REFERENCES

Acocella, V., Neri, M., 2005. Structural features of an active strike-slip fault on the sliding flank of Mt. Etna (Italy). Journal of Structural Geology, 27, 343–355.

Acocella, V., Behncke, B., Neri, M., D'Amico, S., 2003. Link between major flank slip and 2002–2003 eruption at Mount Etna (Italy), Geophys. Res. Lett., 30(24), 2286, doi:10.1029/2003GL018642.

Azzaro, R., Ferreli, L., Michetti, A.M., Serva, L. & Vittori, E., 1998. Environmental hazard of capable faults: the case of the Pernicana fault (Mt. Etna, Sicily). Natural Hazards, 17, 147–162.

Apuani, T., Corazzato, C., 2009. Numerical model of the Stromboli volcano (Italy) including the effect of magma pressure in the dyke system. Rock Mechanics and Rock Engineering, 42, 53–72. doi: 10.1007/s00603-008-0163-1.

Bonforte, A., Puglisi, G., 2006. Dynamics of the eastern flank of Mt. Etna volcano (Italy) investigated by a dense GPS network. Journal of Volcanology and Geothermal Research 153, 357–369.

Borgia, A., Ferrari, L., Pasquarè, G., 1992. Importance of gravitational spreading in the tectonic and volcanic evolution of Mount Etna. Nature, 357, 231–235.

Corazzato, C., Tibaldi, A., 2006. Fracture control on type, distribution, and morphology of parasitic volcanic cones: an example from Mt. Etna, Italy. In: Tibaldi A. and Lagmay M. (eds.), "Interaction between Volcanoes and their Basement", J. Volcanology and Geothermal Research, 158 (1–2), 177–194. DOI: 10.1016/j.volgeores.2006.04.018.

Di Stefano, A., Branca, S., 2002. Long-term uplift of the Etna volcano basement (southern Italy) based on biochronological data from Pliestocene sediments. Terra Nova, Vol. 14, No. 1, 61–68.

Elsworth, D., Voight, B., 1996. Evaluation of volcano flank instability triggered by dyke intrusion. In: McGuire B (ed) Volcano Instability on the Earth and other planets. Geol Soc Spec Publ 110, pp 45–53.

Elsworth, D., Day, S. 1999. Flank collapse triggered by intrusion: the Canarian and Cape Verde Archipelagoes. J Volcanol Geoth Res 94(1–4): 323–340.

Groppelli, G., Tibaldi, A., 1996. Nuovi dati per un modello sullo sviluppo della porzione orientale della faglia della Pernicana, Monte Etna. Proceedings 1996 Congress of Gruppo Nazionale per la Vulcanologia, CNR, 3–5 March 1996, Rome, 180–181.

Groppelli, G., Tibaldi, A., 1999. Control of rock rheology on deformation style and slip-rate along the active Pernicana Fault, Mt. Etna, Italy. Tectonophysics, 305, 521–537.

Iverson, R.M.1995. Can magma injection and ground-water forces cause massive landslides on Hawaiian volcanoes? J Volcanol Geoth Res 66(1–4): 295–308.

McGuire, W.L., 1996. Volcano instability: a review of contemporary themes. In: McGuire WJ, Jones AP, Neuberg J (eds) Volcano instability on the earth and other planets. Geol Soc London Spec Publ 110, pp 1–23.

Monaco, C., Tapponnier, P., Tortorici, L., Gillot, P.Y., 1997. Late Quaternary slip rates on the Acireale-Piedimonte normal faults and tectonic origin of Mt. Etna (Sicily). Earth Planetary Science Letters, 147, 125–139.

Koor, N., Rust, D., Apuani, T., Corazzato, C., 2010. Mineralogical and geotechnical characterization of a clay unit that underlies the unstable flanks of Mount Etna - Sicily. Proceedings of the Geologically Active - 11th IAEG Congress, Auckland, New Zealand, 5–10 September 2010. Accepted.

Neri, M., Acocella, V., Behncke, B., 2003. The role of the Pernicana Fault system in the spreading of Mt. Etna (Italy) during the 2002–2003 eruption, Bull. Volcanol., 66, doi:10.1007/s00445–003–0322-x.

Rasà, R., Romano, R., Lo Giudice, E., 1982. A structural survey of M. Etna on a morphologic basis, with a morphotectonic map at 1:100,000 scale. Memorie Società Geologica Italiana, 23, 117–124.

Rust, D., Neri, M., 1996. The boundaries of large-scale collapse on the flanks of Mount Etna, Sicily. In: McGuire, W.J., Jones, A.P., Neuberg, J. (Eds.), Volcano Instability on the Earth and Other Planets, Geological Society Special Publication, 110, pp. 193–208.

Tibaldi, A., Groppelli, G., 2002. Volcano-tectonic activity along structures of the unstable NE flank of Mt. Etna (Italy) and their possible origin. Journal of Volcanology and Geothermal Research 115 (2002) 277–302.

Voight, B., 2000. Structural stability of andesite volcanoes and lava domes Phil. Trans R. Soc. London 358: 1663–1703.

Voight, B., Elsworth, D., 1997. Failure of volcano slopes. Geotechnique 47(1): 1–31.

Voight, B, Glicken, H, Janda, RJ, Douglas, P.M., 1981. Catastrophic rockslide avalanche of May 18. In: Lipman PW, Mullineux DR (eds) The 1980 eruption of Mount St. Helens. U.S. Geol. Survey Prof. Paper, 1250, pp 347–377.

Voight, B., Janda, R.J., Glicken, H., Douglass, P.M., 1983. Nature and mechanics of the Mount St. Helens rockslide-avalanche of 18 May 1980. Geotechnique 33: 243–273.

Volcanic Rock Mechanics – Olalla et al. (eds)
© 2010 Taylor & Francis Group, London, ISBN 978-0-415-58478-4

The origin and geotechnical properties of volcanic soils and their role in developing flank and sector collapses

R. del Potro
Costa Rican Volcano Observatory (OVSICORI-UNA), Universidad Nacional, Costa Rica
Department of Earth Sciences, University of Bristol, United Kingdom

M. Hürlimann
Technical University of Catalonia (UPC), Barcelona, Spain

ABSTRACT: Giant volcanic landslides are one of the most hazardous geological processes. Still, the mechanisms that trigger them remain unresolved. Recent studies suggest that the presence of weak volcanic materials is likely to play an important role. Herein, we present a study of the weakening effect of weathering and hydrothermal alteration of phonolitic lavas, pyroclasts and ignimbrites from Tenerife. A comprehensive geotechnical characterisation of these materials reveals that, from weathering, the weakest units are porous, sandy-silty, non-plastic soils (SM) that are cohesionless, with high peak strengths and significantly lower residual strengths. In the case of hydrothermal alteration, the weakest units are porous, silty, clay-rich, medium plasticity soils (MH) with low cohesion values and varying angles of internal friction (17–45°). Secondary mineralogy produced by alteration, mainly halloysites and the presence of bonding in weathered soils and kaolinites or alunites in hydrothermally altered soils, appears to control the behaviour of the soils.

1 INTRODUCTION

Large volcanic landslides are common phenomena at most volcanoes worldwide. The deposits and morphologies that characterise these events have been identified in subaerial and submarine environments worldwide (e.g. Siebert, 1996). The importance of studying large volcanic landslides is highlighted by their potential to cause significant damage to areas up to tens of kilometres from the volcano. There is also the potential for damage on a larger scale when landslides enter the sea and generate a tsunami. Volcanic landslides are generally more voluminous than non-volcanic events. The largest volcanic landslides, which can reach volumes in excess of hundreds of cubic kilometres, occur as volcanic islands flank collapses (e.g. Masson et al., 2002; Moore et al., 1989), Similar but relatively smaller events, with volumes in the range 0.1–20 km³, have been observed at stratovolcanoes worldwide (e.g. Siebert, 1996; Vallance et al., 1995). In many cases, collapses are recurrent phenomena at one same volcanic edifice.

Despite a great recent effort to understand the mobility and emplacement mechanisms of the large debris avalanches little work has been devoted to understanding the mechanism of initiation of such huge landslides. For most landslides, destabilising agents include morphological, geological, tectonic and climatic factors, but for volcanic slopes additional elements such as magmatic intrusions apply. Unfortunately, our poor knowledge on the strength and mechanical behaviour of volcanic material has limited modelling efforts to quantify the different hypothesis. In this way, a number of complex, but invalid, models have been published. For example, Voight et al. (1983), who were the first to study large volcanic landslides, assumed that an angle of friction of 40° and negligible cohesion was representative of the internal strength at Mount St. Helens when deep seated landslides developed in May 1980. Subsequent field characterisation and measurements of volcanic materials confirmed that volcanic rocks had strengths in excess of those assumed by Voight et al. (op. cit.), and future workers were forced to include unjustified empirical correlations in order to reduce them (i.e. disturbance factor D > 0 in the Hoek and Brown criterion (Hoek et al., 2002)). In the last decade a number of scientific studies have addressed this issue and focussed on the geotechnical properties of volcanic rock masses (Hernández et al., 2006; Moon et al., 2005; Serrano et al., 2002a; Watters et al., 2000). The findings of these studies provided a first order approach to quantifying the mechanical properties of volcanic rock masses. A normalised compilation is provided by del Potro and Hürlimann (2008).

A few recent studies have used these values to test the hypothesis of the different landslide driving/triggering mechanisms (e.g. Hürlimann et al., 2000; Moon et al., 2005). Their preliminary results show that a combination of increase in pore fluid pressures and transient stress (regional seismicity or large volcano tectonic events; i.e. caldera formation) could cause flank or sector collapses. However, it appears that volcanic rock masses are too strong. Herein, it has been suggested that the presence of weak layers such as 'destructured' pyroclastics (Serrano et al., 2002b), residual soils (Hürlimann et al., 2001), hyaloclastites (Schiffman et al., 2006) or hot olivine cumulates (Clague and Denlinger, 1994) can be very significant in reducing the stability of a slope, and these have been detected at many volcanoes which have experienced instability. An alternative weakening mechanism is hydrothermal alteration that can reduce lavas to clay-rich soils (Watters and Delahaut, 1995) and is thought to have been critical in the generation of many volcanic landslides (e.g. López and Williams, 1993; Opfergelt et al., 2006). In this study we present a geotechnical study of the weakening effect of weathering of phonolitic ignimbrites and that of hydrothermal alteration of phonolitic lavas and pyroclasts.

2 VOLCANIC RESIDUAL SOILS

Two common mechanisms for the genesis of volcanic residual soils are climatic weathering and hydrothermal alteration. In order to attempt quantifying the expected decrease in shear strength, relative to the parent material, we have collected several samples of volcanic residual soils in the island of Tenerife. Hydrothermally altered soils are the weakest materials exposed on Teide volcano—the most recently active volcano in the central volcanic complex. Three hydrothermally altered soil unit (SA, SE and SK) have been collected from inside the small summit crater (Fig. 1). These soils are hydrothermally altered products from phonolitic lava flows of the last summit eruption (~1 ka ago). Persistent hydrothermal activity has changed lavas to clay-rich soils by sulphate leaching. Units SA and SE are light coloured soils exposed along the summit fumarole field and are both found as extensive units comprising a significant volume of ~5 cm sulphur crystals as well as ~20 cm clasts from the parent rock. Unit SK, on the other hand, is a red, more homogeneous unit found on the southern inner rim of the summit crater. Undisturbed samples of SA and SE (SA1, SE1a and SE1b) were extracted from a depth of between 10 and 40 cm using a 12 cm-wide metal tube. In addition, remoulded samples from SK

(SK1) and two from the finest sectors of SA and SK (SA2 and SK2 respectively) were also collected. Both samples of SA were collected from the extrusion location of a prominent fumarole.

The field survey on Tenerife revealed that, away from the most recent crater of Teide volcano, widespread residual soils (also called paleosols) are the weakest materials. Red residual soil units are found located directly above ignimbrites—pyroclastic deposits produced by explosive eruptions of phonolitic magmas. On Tenerife, these deposits have a thickness of several decimetres and are found as part of the central volcanic complex, exposed by erosion on the scarps of large island flank collapses. Weathering processes change the phonolitic pyroclastic deposit into a residual soil. This soil is then eventually covered by a lava flow or pyroclastic deposits which further modify the soil through thermal processes. Hence, such residual soils are characterised by a double cementation or bonding that is caused first by the lithification processes during the deposition of the hot pyroclastic material and secondly by the thermal alteration ('baking') induced during the emplacement of the hot unit overlaying the soil layer. Two residual soil samples (RS1 and RS3) were collected at various locations in the Tigaiga massif (Fig. 1), a remnant of the northern flank of the central Las Cañadas Edifice. Samples RS1 was cut from the western lateral landslide scarp of the La Orotava near a site called 'Piedra de los Pastores'. The outcrop is exposed along a forestry road and it is

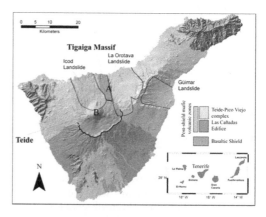

Figure 1. Simplified geological map of the central part of Tenerife Island. Inset shows the location of the island of Tenerife within the Canary Archipelago. The thick black line denotes the position of the Las Cañadas Caldera wall and the dotted lines the sub-aerial extend of the three most recent island flank collapses. The location of the weathered residual soils is indicated by letter A and the location of the hydrothermally altered soils is indicated by letter B.

red-coloured and about 1 m of residual soil has formed at the top of a thick phonolitic pyroclastic unit. The soil horizon is overlaid by a thick lava flow. Sample RS3 is a red soil that was taken from the top of the Tigaiga massif at a site called 'Mirador de Sergio' The RS3 unit exposes several fault planes that dip eastwards and black slickensides are clearly visible. The emplacement of both pyroclastic units predates the large Orotava landslide, dated at ~0.56 Ma BP.

3 GEOTECHNICAL CHARACTERISATION

3.1 Methodology

To fully characterise the properties of the residual soils collected we have performed a series of geochemical and geotechnical tests and measurements.

X-ray refraction (XRD) was performed on all samples to characterise the main mineral assemblages present in each case. Furthermore, a sample of unaltered phonolite was also analysed to provide a baseline measurement. Grain size distributions and plasticity tests were also performed for all soil samples, as was a general soil characterisation. Consolidation tests were performed on two different settings. Weathered soil RS1 was tested in an oedometer with normal loads of up to 1600 kPa. Consolidation of the hydrothermally altered soils (SA1, SE1 and SK1) was tested during the loading of the sample probes prior to direct shear tests. In this way only the initial load is carried out on an undisturbed sample, and further tests are on sheared samples. All soil units except RS3 were tested in strain-controlled, consolidated, drained, direct shear and ring shear tests. For both tests normal loads were in the range 50–600 kPa, and strain rates were: ~0.005 and ~0.089 mm/min for the direct shear and ring shear tests respectively. Weathered unit RS1 was further tested in a high-normal load shear box where normal loads ranged 5.2–15.59 MPa.

In addition to the tests described above, RS1 was also tested in triaxial conditions. Two different apparatus were used. A standard GDS stress-path system including a Bishop-Wesley hydraulic cell for pressures up to 1.7 MPa, and an in-house built larger apparatus for pressures up to 5 MPa. Two different tests have been carried out: consolidated-undrained tests (hereafter CU-tests) and consolidated-drained tests (hereafter CD-tests). The former provides undrained shear strength of the soil alongside the increase in pore water pressure in the probe. The latter provide drained shear strength as well as volumetric strain behaviour.

The results of the basic testing of the residual soils have already been published: Hürlimann et al. (2001) analyse the weathered ignimbrite while del Potro and Hürlimann (2009) present the study of the hydrothermally altered summit soils. Here we present a summary and compare both soils as well as the results of further, more complex geotechnical tests.

3.2 Results

3.2.1 General characterization

General soil properties are summarised in Table 1. X-ray difractograms of the fine fraction of the tested samples show the presence of secondary clay mineralisation in the residual soils (see Table 1). Soil particle distributions in Figure 2 show a significant clay fraction only for hydrothermally altered soils. Plasticity values are given in Table 1. Finally, the soil was classified according to the Unified Soil Classification System, USCS: residual soils from the weather ignimbrites are medium plasticity sands (SP) and soils from hydrothermal alteration are medium plasticity silts (MH).

3.2.2 Consolidation analysis

The analysis of the consolidation curves obtained from the oedometer tests on sample RS1 has given

Table 1. General properties, plastici and main composition from XRD of the soil samples collected, respectively.

Soil	Natural density [g/cm3]	Dry density [%]	Void ratio [–]	Liquid limit [%]	Plasticity index [%]	XRD main component*
SA	1.69	0.99	1.78	64.37	28.03	Al–Ka
SE	1.56	1.23	1.22			Al–Ka
SK	1.56	0.91	1.97x	71.75	30.57	Felds–Ka
RS1	1.53	1.24	1.40	n/a	n/a	Felds–Ha

* Al: alunite, Ka: kaolinite; Felds: feldspars; Ha: Halloysite.

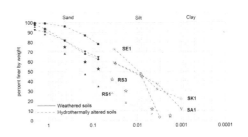

Figure 2. Soil particle distribution of the soils from Tenerife.

the following stress-strain parameters: compression index $c_c = 0.287$, average coefficient of compressibility $a_v = 0.087$ MPa^{-1} and average confined Young modulus $E_m = 27.8$ MPa. Preliminary results on the consolidation behaviour of the summit soil samples (SA, SE and SK) show no volume increase during the initial unloaded saturation phase. The consolidations of the undisturbed samples indicate total vertical strains ranging from about 4.5% for a normal load of 100 kPa up to 12.5% for a maximum normal load of ~450 kPa. In contrast, for remoulded samples, the measured absolute vertical strains range between 5 and 27% for the normal load range of 49–371 kPa. Consolidation curves indicate that primary compressions were achieved instantly after loading.

3.2.3 Shear tests

The results obtained from the drained direct shear tests can be used to produce Mohr-Coulomb envelopes in order to determine the strength parameters of the soil (Fig. 3). Hydrothermally altered soils, tested at low normal loads, show no difference between peak and residual shear strength. These soils have low-cohesion values (~100 kPa) and angles of internal friction in the range 16–34°. Weathered soil RS1 shows distinctive peak and residual behaviour as well as a clear decrease in strength relative to increasing confining pressure. At low normal loads sample RS1 has an angle of internal friction ~40°, clearly higher than the hydrothermally altered soils. However, for higher

normal loads the angle of internal friction of soil RS1 decreases to ~30° and this is very similar to the extrapolated values from the hydrothermally altered soil (Fig. 3b). Figure 4 shows the normalised shear strength and residual mobilised angles of internal friction of remoulded samples tested in ring shear conditions. In spite of the low and small range of normal loads allowed by the experimental settings, a relationship with increasing normal load is clearly visible, as is the difference in residual shear strength of the residual soils tested.

3.2.4 Triaxial tests

Triaxial test results fall into three main groups: permeability, volumetric behaviour, and stress-strain behaviour during failure. For the first, constant flow-rate permeability test, carried out prior to consolidation of the probes, provide a permeability range, $k = 1.04 \cdot 10^{-4} – 4.8 \cdot 10^{-4}$ cm/s. These values are in agreement with those from the oedometer tests. For the second set of results, volumetric changes have been collected during the confining phase of the tests. The results, shown in Figure 5, indicate the lack of an explicit yield strength or collapse. However, there is a clear contrast in behaviour above an effective confining stress of around 700 kPa. Beyond this value the e–$\log(p')$ curves are parallel and define a normal compression line. Again, these results are in good agreement with values from oedometer tests, and the RS1 shows values close to those of bonded residual soils (Blight, 1997).

The analysis of the stress-strain behaviour starts with the undrained tests results are shown in Figure 6. Generally, the maximum shear stress increases rapidly up to failure, which occurs at

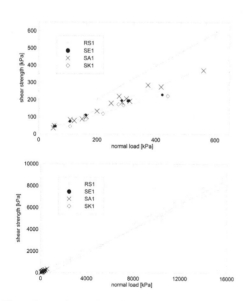

Figure 3.   Mohr envelopes from direct shear test results.

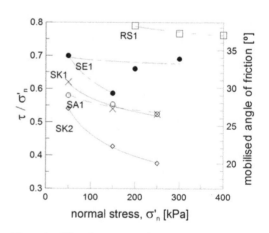

Figure 4.   Ring shear test results.

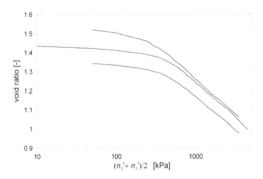

Figure 5. Void ratio versus effective confining stress plots during the confining phase.

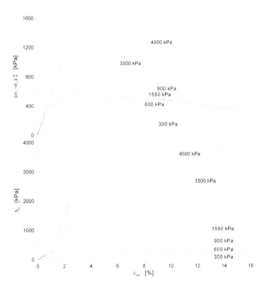

Figure 6. Undrained triaxial test results for soil RS1.

The results of the drained triaxial tests are illustrated in Figure 7. The slopes of the stress-strain curves are much lower than in the undrained tests and no clear peaks have developed. The initial tangent stiffness increases for higher effective confining pressures, as is to be expected for bonded residual soils (Maccarini, 1993). For the lower confinement pressures, the maximum shear stresses, at failure, are reached at axial strains higher than 16%. For an effective confining pressure of 3500 kPa, the drained triaxial test renders a stress-strain curve that increases continuously with a smooth slope, with no clear failure developing within an axial strain of 24%. In all cases volumetric strain decreases continuously and hence no dilatancy is observed, contrary to what would be expected for bonded residual soils (e.g. Aversa and Evangelista, 1998; Zhu and Anderson, 1998). The final volumetric strain values show an enormous reduction of volume during failure of between 8 and 13% for effective confining pressures ranging between 300 and 1550 kPa, reaching almost 14% for a confinement pressure of 3500 kPa.

Finally, effective stress path plots (Fig. 8) of undrained triaxial tests show the appearance of loops

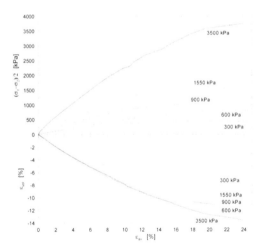

Figure 7. Drained triaxial test results for soil RS1.

an axial strain of between 2 and 5%, and then decrease slightly. The behaviour then further varies at an axial strain of around 12% and after this show a grater decrease for increasing axial strain. At this same point pore pressures show a small but continuous increase. The clear difference in the magnitudes of samples subject to effective confining stresses of 300 kPa may indicate a double mechanism of silty-sandy soils caused by the different particle sizes. At higher effective confining pressures the results show a large increase in pore water pressure due to undrained loading, which reach a maximum value of 3750 kPa for the maximum effective confining stress. In these cases, after failure, strength decreases but a further variation to the pore water pressure at higher strains, as seen at lower confining stresses, does not occur.

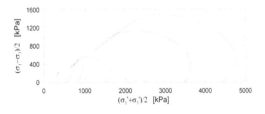

Figure 8. Stress-path plots of the undrained (solid line) and drained (dashed line) triaxial tests of soil RS1 at low and high confining pressures.

163

at the top in the case of low confining pressures. These loops are rare and characteristic of the failure of bonded soils (Maccarini, 1993), and diminish in the stress path for failures at higher confining pressures until they eventually disappear. Stress path plots also show an exceptional increase in pore pressures for high effective confining pressures, where after exceeding the yield strength they aims for the origin.

Peak and residual strengths values obtained from both drained and undrained tests are divided into four groups: peak strength from drained test (CD peak), residual strength from drained test (CD residual) and the same for undrained test—CU peak and CU residual respectively. Fitting Kf-lines through p-q(q') values of the four groups provides the angle of internal friction values in Table 2. These values indicate a decrease in the angle of internal friction with increasing confining stress.

Additionally in the case of undrained tests an apparent friction angle (φ'_{ap}) can be calculated as proposed by Sassa (1992):

$$\frac{q_{max}}{p'_{ini}} = \tan\phi'_{ap} \qquad (1)$$

where q_{max} is the maximum shear strength and p'_{ini} is the initial effective stress of the undrained triaxial test. The results are given in Table 3. In this way the decrease in angle if internal friction for increasing confining stresses is accentuated. Another parameter that picks up the variation dependant on the confinement stress is the pore pressure parameter, A, of Lambe and Whitman (1979).

Table 2. Angle of internal friction from triaxial tests [°].

Confining stress [kPa]	CU		CD	
	Peak	Residual	Peak	Residual
Low (300–900)	38.66	33.57	33.59	32.58
High (1550–4500)	30.76	33.24	31.25	29.91

Table 3. Angle of internal friction from triaxial tests [°].

Effective confining stress [kPa]	φ'_{ap} [°]	A [–]
300	38.24	0.34
900	33.98	0.41
1550	20.68	0.97
4500	18.75	0.99

4 THE ROLE OF RESIDUAL SOILS IN FLANK AND SECTOR COLLAPSES

Low shear strength values have been measured for residual soil units present in the island of Tenerife. If these units have spatial continuity along the island or large volumes are present forming the core stratovolcanoes, this would have serious implications in the future for the generation of large-scale deformations and landslides.

X-ray difractograms of the tested samples show a clear evolution of the residual soils from the parent phonolitic material. There is a clay component from secondary mineralisation that shows the first significant difference between the weathered and the hydrothermaly altered soils. Samples of the former show halloysites and the latter alunite and kaolinite. While the chemical composition of the secondary clay minerals is the same, there is an important structural difference. Halloysites are tubular clays and hence have a well-evolved three-dimensional structure. Alunite crystals are pseudo-tetrahedral minerals formed by combinations of two trigonal pyramids that hence form a micro-three dimensional structure. Kaolinites, on the other hand, are platy, two-layered clays—a very weak structure. These observations are supported by shear tests that reveal decreasing shear strength with weaker clay structure. It therefore appears that the small fine fraction of the volcanic residual soils ultimately controls their shear strength properties.

Examination of normalised stress-strain curves, from shear tests, only shows a clear difference in the case of low normal loads for RS1. This is proof of the presence of bonding on the weathered soils. As expected, this bonding is created by the thermal alteration of the residual soil as the overlying layer (lava in this case) is emplaced. In spite of the clear stress-strain behaviour given by the bonding, our tests show that its yield strength is surpassed at the higher normal load end of our test settings, which corresponds to relatively shallow depths relative to the depth of large volcanic landslide decollments.

In the triaxial tests, the enormous reduction from the drained angle of friction of 30° to the apparent friction angle of less than 20° (Table 3) verifies the significance of the pore water pressure during undrained loading. Moreover, the large values in the pore water pressure parameter attained at the high end of the confining stress in the test set-up, approach one, and this can be associated with a loose structure that collapses upon the application of a load. This opens a window into understanding the impact this could have in the behaviour of the soil units during large volcanic landslides as they may act in this manner after fast, undrained loading, if the confining pressure is greater than about 1 MPa.

The plausibility of widespread residual soils is high if they develop at the top of large ignimbrite units, hence weak halloysitic residual soils such as the one tested here, appear to be perfect candidates for shear surfaces to develop through causing large flank island collapses. In the case of the hydrothermally altered soils from advanced argillic alteration, exposed at the summit region, it is harder to determine their extension at depth and hence their significance in sector and flank collaspes. Taking the hydrothermal conditions described for stratovolcanoes (e.g. Frank, 1995; Zimbelman et al., 2005), alunite minerals in association with kaolinite form in surface steam-heated environments, but there is also evidence to support that these minerals form in deep magmatic hydrothermal environments too. Frank (op. cit.) suggests that superficial zones of alteration go down possibly to a depth of 1000 m. Following more extensive hydrothermal models for volcanic hydrothermal systems as well as for epithermal and porphyry ore deposits (e.g. Sales and Meyer, 1948), it is also possible that areas of near-neutral, intermediate-argillic, montmorillionite-rich alteration are present surrounding and grading into the advanced argillic, alunite–kaolinite alteration of the central-system. These units appear not to be exposed on Teide, although this may be because recent lava fields have covered them.

CONCLUSIONS

The origin of residual volcanic soil units and their role in the development of large volcanic landslides has been studied through units of weather and hydrothermally altered soils exposed on Tenerife Island. These soils have been studied by carrying out standard laboratory tests. The results highlight the significance of the structure of the clay fraction in controlling the shear strength. They also highlight the presence of a bond in the weathered residual soils. In fact, geotechnical tests revealed that the mechanical behaviour of the residual soil changes greatly if this bonding is broken. However, because the bonded structure generally fails when the effective normal stress surpasses the yield strength of the bonding, in the case of large volcanic landslides with thicknesses up to several hundred meters, the high overburden easily exceeds this value. In the light of the phenomena studied here, weathered residual soils have a residual strength value. This work brings to light the necessity to test samples at high normal and confining stresses in order to better constrain their real behaviour within a volcanic edifice.

In any case, the low shear strength values of the soil units analysed reduces the overall strength of a volcanic edifice units such as the one analysed in this study are therefore perfect candidates for the potential failure surfaces of large volcanic landslides.

REFERENCES

Aversa, S. and Evangelista, A. 1998. The mechanical behaviour of a pyroclastic rock: yield strength and 'destruction' effects. Rock Mechanics and Rock Engineering, 31(1): 25–42.

Blight, G.E. 1997. Mechanics of residual soils. Balkema, Rotterdam, 237 pp.

Clague, D.A. and Denlinger, R.P. 1994. Role of olivine cumulates in destabilizing the flanks of Hawaiian volcanoes. Bulletin of Volcanology, 56: 425–434.

del Potro, R. and Hürlimann, M. 2008. Geotechnical classification and characterisation of materials for stability analyses of large volcanic slopes. Engineering Geology, 98(1–2): 1–17.

del Potro, R. and Hürlimann, M. 2009. The decrease in the shear strength of volcanic materials with argillic hydrothermal alteration, insights from the summit region of Teide stratovolcano, Tenerife. Engineering Geology, 104(1–2): 135–143.

Frank, D. 1995. Surficial extent and conceptual model of hydrothermal system at Mount Rainier, Washington. Journal of Volcanology and Geothermal Research, 65: 51–80.

Hernández, L., Rodríguez, J.A., Olalla, C., Perucho, A. and Serrano, A. 2006. Estudio de caracterización geotécnica de las rocas volcánicas de las Islas Canarias, AT-SGT-01. Regional Ministry of Works. Government of the Canary Islands.

Hoek, E., Carranza-Torres, C. and Corkum, B., 2002. Hoek-Brown criterion—2002 edition, NARMS-TAC Conference, Toronto, pp. 267–273.

Hürlimann, M., Ledesma, A. and Martí, J. 2001. Characterisation of a volcanic residual soil and its implications for large landslide phenomena: Application to Tenerife, Canary Islands. Engineering Geology, 59: 115–132.

Hürlimann, M., Martí, J. and Ledesma, A. 2000. Mechanical relationship between catastrophic volcanic landslides and caldera collapses. Geophysical Research Letter, 27(16): 2393–2396.

Lambe, T.W. and Whitman, R.V. 1979. Soil mechanics. Wiley, New York, 553 pp.

López, D.L. and Williams, S.N. 1993. Catastrophic volcanic collapse: Relation to hydrothermal processes. Science, 260: 1794–1796.

Maccarini, M. 1993. A comparison of direct shear box tests with triaxial compression tests for a residual soil. Geotechnical and Geological Engineering, 11: 69–80.

Masson, D.G. et al. 2002. Slope failures on the flanks of the western Canary Islands. Earth-Science Reviews, 57: 1–35.

Moon, V., Bradshaw, J., Smith, R. and de Lange, W. 2005. Geotechnical characterisation of stratocone crater wall sequences, White Island Volcano, New Zealand. Engineering Geology, 81(2): 146–178.

Moore, J.G. et al. 1989. Prodigious submarine landslides on the Hawaiian Ridge. Journal of Geophysical Research, 94: 17465–17484.

Opfergelt, S., Delmelle, P., Boivin, P. and Delvaux, B. 2006. The 1998 debris avalanche at Casita volcano, Nicaragua: Investigation of the role of hydrothermal smectite in promoting slope instability. Geophysical Research Letters, 33(15).

Sales, R.H. and Meyer, C. 1948. Wall Rock Alteration at Butte, Montana. Transactions of the American Institute of Mining and Metallurgical Engineers, 178: 9–35.

Sassa, K. 1992. Landslide volume—Apparent friction relationship in the case of rapid loading on alluvial deposits. Landslieds News, 6: 16–18.

Schiffman, P., Watters, R.J., Thompson, N. and Walton, A.W. 2006. Hyaloclastites and the slope stability of Hawaiian Volcanoes: Insights from the Hawaiian Scientific Drilling Project's 3-km drill core. Journal of Volcanology and Geothermal Research, 151(1–3): 217–228.

Serrano, A., Olalla, C. and Perucho, A., 2002a. Evaluation of non-linear strength laws for volcanic agglomerates. In: C. Dinis da Gama and L. Ribeiro e Sousa (Editors), Workshop on volcanic rocks, ISRM international symposium on rock engineering for mountainous regions, Eurock 2002, pp. 53–61.

Serrano, A., Olalla, C. and Perucho, A., 2002b. Mechanical collapsible rocks. In: C. Dinis da Gama and L. Ribeiro e Sousa (Editors), Workshop on volcanic rocks, ISRM international symposium on rock engineering for mountainous regions, Eurock 2002, pp. 105–113.

Siebert, L., 1996. Hazards of large volcanic debris avalanche and associated eruptive phenomena. In: S. Tilling (Editor), Monitoring and Mitigation of Volcano Hazards. Springer Berlin, pp. 541–572.

Vallance, J.W., Siebert, L., Rose Jr., W.I., Girón, J.R. and Banks, N.G. 1995. Edifice collapse and related hazards in Guatemala. Journal of Volcanology and Geothermal Research, 66: 337–355.

Voight, B., Janda, R.J., Glicken, H. and Douglass, P.M. 1983. Nature and mechanics of the Mount St Helens rockslide-avalanche of 18 May 1980. Géotechnique, 33(3): 243–273.

Watters, R.J. and Delahaut, W.D., 1995. Effect of argillic alteration on rock mass stability. In: W.C. Haneberg and S.A. Anderson (Editors), Clay and Shale Slope Instability. Reviews in Engineering Geology. Geological Society of America, Boulder, CO, pp. 139–150.

Watters, R.J., Zimbelman, D.R., Bowman, S.D. and Crowley, J.K. 2000. Rock mass strength assessment and significance to edifice stability, Mount Rainier and Mount Hood, Cascade Range volcanoes. Pure and Applied Geophysics, 157(6–8): 957–976.

Zhu, J.H. and Anderson, A. 1998. Determination of shear strength of Hawaiian residual soil subjected to rainfall-induced landslides. Géotechnique, 48(1): 73–82.

Zimbelman, D.R., Rye, R.O. and Breit, G.N. 2005. Origin of secondary sulfate minerals on active andesitic stratovolcanoes. Chemical Geology, 215(1–4): 37–60.

Volcanic Rock Mechanics – Olalla et al. (eds)
© *2010 Taylor & Francis Group, London, ISBN 978-0-415-58478-4*

The role of hyaloclastite rocks in the stability of the volcanic island flanks of Tenerife

M. Ferrer
Instituto Geológico y Minero de España, Madrid, Spain

J. Seisdedos
Prospección y Geotecnia S.L., Madrid, Spain

L.I. González de Vallejo
Universidad Complutense de Madrid, Spain

ABSTRACT: The failure mechanisms that could originate the mega paleo-rockslides of Güímar and La Orotava in Tenerife (Canary Islands) are analyzed, based on the geomechanical site investigations carried out on the pre-failure volcanic materials of Tenerife island flanks. Geological and geomorphological modelling and geomechanical characterization of the materials are presented. Hyaloclastites rocks are forming the submarine substratum of the island edifice presenting a highly deformable behaviour. Preliminary stability analyses have suggested potential failure surfaces in the hyloclastites rocks.

1 INTRODUCTION

Güímar and La Orotava valleys in Tenerife were originated by mega rockslides. The resulting slided masses, deposited on the ocean floor, cover areas of hundreds of square kilometers. These paleo-landslides have been considered as ones of the largest known in the world by their volume. Both constitute exceptional examples due to their geomorphological features and the fact that the slided deposits have been identified in the ocean floor and inside the galleries excavated in the island (Navarro & Coello, 1989). In spite of their importance, only few investigations have been carried out to analyze these processes under a geomechanical point of view. The authors are carrying out detailed studies on the geomechanical properties of the materials involved, including in situ testing and geophysical surveys, to evaluate the instability processes of the volcanic islands flanks (Ferrer *et al.*, 2007, 2008).

This paper presents the preliminary results of the site investigation carried out on the submarine materials formed by hyaloclastites and the role of these rocks on the stability of the volcanic edifice of Tenerife.

2 GÜÍMAR AND LA OROTAVA ROCKSLIDES

Güímar and La Orotava valleys, 9 and 12 km wide, present opposite orientations, ESE and NNW respectively. Their heads are located in the *Cordillera Dorsal*, main rift zone in the island with

NE direction and maximum heights between 1700 and 2200 m (Fig. 1).

The morphological characteristics of the valleys are singular, outstanding the symmetry and the important height of the lateral scarps (500 m), formed by pre-landslide volcanic materials with slope angles higher than 35°. The depressions formed were filled by post-landslide volcanic materials with slope angles lower than 15°.

The estimated volume of these rockslides is in the order of 30–50 km³.

The age of Güímar rockslide has been estimated approximately 1 Ma (Ferrer *et al.*, 2008). The age of La Orotava rockslide has been estimated between 0.54 to 0.69 Ma (Cantagrel *et al.*, 1999).

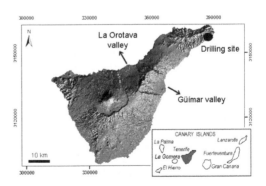

Figure 1. Güímar and La Orotava valleys (Tenerife) and drilling site location.

3 GEOLOGICAL AND GEOMECHANICAL CONDITIONS OF THE PRE-FAILURE VOLCANIC EDIFICE

In order to analyze the mechanical behavior of the flanks of the volcanic edifice, the geological, geomorphological and geomechanical representative modeling of the pre-landslide edifice have been prepared.

The different factors influencing volcanic landslide processes have been considered: morphology, lithology, geological structure, hydrogeological conditions, geomechanical properties, weathering and alteration and state of stress. Also the triggering factors on stability have been studied: volcanic and hydrothermal activity, dike intrusion and seismic activity.

The geometrical model before the occurrence of Güímar and La Orotava rockslides has been assessed considering paleomorphological data obtained from the slopes and lateral scarps of the volcanic edifice not affected by landsliding processes. The morphological features of the submarine slopes have been obtained from bathymetric data analysis.

It has been assumed that the ground water conditions for the pre-failure edifice could be similar to those encountered before intensive ground water exploitation of the island aquifers had taken place. A water table at 600–700 m below surface was estimated according with ground water records, with exception of the coastal areas. In the central part of the edifice due to the presence of a large number of dykes the ground water levels should be higher.

Geological and geotechnical data were recorded from field survey and from the extensive network of small diameter tunnels, with a total length over of 4000 km, excavated in the inland flanks for ground water supply purposes. Geotechnical properties of the volcanic materials of the emerged edifice have been also obtained from engineering geological surveys (Seisdedos, 2008; González de Vallejo et al., 2008).

With respect to the geological and geomechanical data of the submarine edifice, only morphological and tectonic data are available from oceanographic surveys. In the Easter corner of the island site investigations have been carried out where submarine rocks are outcropping (Fig. 1). Three boreholes have been drilled in hyaloclastites reaching one of them 200 m depth.

Hyaloclastites and basaltic lavas have been the rock materials core drilled in the three boreholes, although hyaloclastite has been the predominant lithology. Hyaloclastites are composed of clastic particles of irregular shape and sizes ranging from 0.5 to 3 cm, forming a green, grey or brown coloured breccia. This material is poorly consolidated and weakly cemented. Voids and vacuoles are occasionally present with sizes ranging from 1 to 3 cm. Secondary minerals are observed inside them. Fracture zones and slikenside surfaces have been identified.

Pressuremeter tests, borehole geophysics (sonic, acoustic, televiewer camera, calliper) and laboratory tests have been carried out. Laboratory and geophysical results are not yet available at time of writing this paper (2009, December).

The deformational properties of the hyaloclastites were obtained from 16 pressuremeter tests carried out at different depths in one of the boreholes (Table 1). The values for pressuremeter moduli ranged from 10 MPa to 3212 MPa. Most frequently intervals range from 50 to 80 MPa and from 125 to 135 MPa. A representative value of 129 MPa has been considered for modelling purposes.

The simplified geological model for the pre-failure edifice is shown in Figure 2. The materials were grouped in those corresponding to the emerged edifice (above sea level) and the submarine edifice (below sea level). The following lithological units have been distinguished as representative of the simplified geological model of Tenerife island flanks for geomechanical purposes:

Forming the flanks of the island:

1. Lava flows (60%) and autoclastic breccias (40%).
2. Lava flows (70%) and autoclastic breccias (30%).
3a. Altered lava flows (90%) and pyroclastic deposits (10%), below the water level.
3b. Altered lava flows (80%), pyroclastic deposits (10%) and dykes (10%), below the water level.

Forming the structural axis of the island:

4. Lava flows (30%), autoclastic breccias (20%), pyroclastic deposits (40%) and dykes (10%).
5. Altered lava flows (40%), pyroclastic deposits (30%) and dykes (30%), below the water level.

Forming the submarine edifice:

6a. Hyaloclastites (70%) and pillow lavas (30%).

Table 1. Hyaloclastites pressuremeter moduli (E_p).

Depth m	E_p MPa	Depth m	E_p MPa
23.5	159	95	1833
35	48	96	235
38.5	169	102	262
47.5	60	103	10
60	58	107	903
62	81	107.5	123
74	145	118	335
89	824	118.5	3212

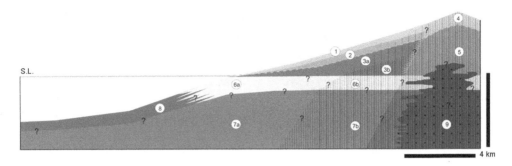

Figure 2. Simplified representative geological model of the flank of the volcanic edifice of Tenerife. Numbers represents the different lithological units considered. 1, 2, 3a, 3b, units forming the flanks: lava layers predominant; 4, 5, units forming the structural axis: pyroclastic deposits predominant; 6a, 6b, submarine rocks: hyaloclastites predominant; 7a, 7b, submarine rocks: pillow lavas predominant; 8, Fragmentary submarine deposits; 9, Plutonic complex: dykes predominant. S.L.: Sea level. Dashed line: water level. ?: uncertainties. The same scales horizontal and vertical.

6b. Hyaloclastites (65%), pillow lavas (25%) and dykes (10%).
7a. Pillow lavas (90%) and hyaloclastites (10%).
7b. Pillow lavas (85%), hyaloclastites (5%) and dykes (10%).
8. Fragmentary submarine deposits.
9. Dykes (90%) and pillow lavas (5%).

4 STABILITY CONDITIONS OF GÜÍMAR AND LA OROTAVA PRE-FAILURE EDIFICES

Stability analysis has been carried out in the pre-failure edifices of Güímar and La Orotava applying stress-strain methods. A first analysis has been carried out using rock mass parameters obtained from the Hoek-Brown criterion (Table 2). Figure 3 shows the results of this analysis showing a large deformation surface affecting the whole edifice. In this case a factor of safety of 1.34 was obtained applying c-ϕ reduction procedure.

A second stability analysis has been carried out using deformability values for the hyaloclastites. Figure 4 shows the results obtained. In this case, the distribution of the maximum deformations shows larger deformations affecting the hyaloclastites and a new failure surface.

In this second analysis values of 0.1 MPa for cohesion and 16° for angle of internal friction are needed to reach limit equilibrium.

Limit equilibrium methods have been also applied showing similar results, Figure 5 (Seisdedos, 2008).

Although these results are still preliminary they present significant potential failure surfaces that are in accordance with the geomorphological and geological features observed in Güímar and La

Table 2. Geomechanical properties (c, ϕ, E) obtained for the pre-failure edifice using Hoek-Brown criterion.

Unit	γ_d kN/m³	c MPa	ϕ °	E MPa	ν -
1	20.7	0.9	51	6756	0.30
2	21.7	1.7	47	8921	0.30
3a	20.2	2.3	34	4204	0.29
3b	18.0	3.1	25	2779	0.29
4	18.0	0.9	33	2299	0.26
5	18.9	2.8	22	2056	0.26
6a	22.6	1.0	23	1012	0.33
6b	23.0	2.5	17	1176	0.32
7a	26.8	8.0	36	12023	0.28
7b	27.2	11.4	34	13183	0.28
8	19.0	1.0	20	1000	0.30
9	27.6	13.2	33	10233	0.28

Figure 3. Shear strains results using Table 2 properties (maximum shear strain 4.3%). Same scale horizontal and vertical.

Orotava valleys, as well as with the geomechanical properties of the materials.

The importance of the hyaloclastites rocks has been also pointed out on the stability of the Hawaiian volcano flanks (Schiffman et al., 2006).

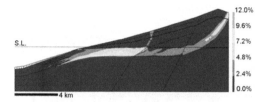

Figure 4. Shear strains results using pressuremeter moduli for hyaloclastites (maximum shear strain 11.24%). Same scale horizontal and vertical.

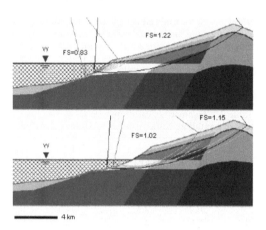

Figure 5. Limit equilibrium analysis results showing potential failure surfaces for the properties included in Table 1 and considering c = 0.1 MPa and ϕ = 14° for hyaloclastites. Same scale horizontal and vertical.

5 CONCLUSIONS

The highly deformable hyaloclastite rocks can play a primary factor in the destabilization process of the flanks of Tenerife. The preliminary results obtained have shown the geometry of the potential failure surfaces of Güímar and La Orotava rockslides. These results are in accordance with the geomechanical properties of the rocks, the surface and submarine geomorphological data and the geological processes involved. The results suggest that the large instability processes common of the volcanic islands flanks depend on the high deformability

properties of hyaloclastites, as well as the morphological conditions of the emerged volcanic edifice, mainly the height and the slope angle of their flanks. Other influencing or triggering factors such as dyke intrusion pressures and volcanic seismicity activity should be also considered.

ACKNOWLEDGEMENTS

This investigation has been carried out with the financial support of the Ministry of Science of Spain (CICYT) and the Geological Survey of Spain (IGME).

REFERENCES

Cantagrel, J.M., Arnaud, N.O., Ancochea, E., Fuster, J.M., Huertas, M.J., 1999. Repeated debris avalanches on Tenerife and genesis of las Cañadas caldera wall (Canary Islands). *Geology* 27 (8): 739–742.

Ferrer, M., Seisdedos, J., García, J.C., González de Vallejo, L.I., Coello, J.J., Casillas, R., Martín, C., Navarro, J.M., 2007. Volcanic mega-landslides in Tenerife (Canary Islands, Spain). In Malheiro & Nunes (eds.), *Volcanic Rocks*: 185–191. London: Taylor & Francis Group.

Ferrer, M. (coord.) & various authors (2008). Large rockslides hazards in Tenerife island. Geological analysis and geomechanical modelling of instability mechanisms ("GRANDETEN"), IGME-CICYT CGL2004–00899, internal report (unpublished).

González de Vallejo L.I., Hijazo, T., Ferrer, M., 2008. Engineering geological properties of the volcanic rocks and soils of the Canary Island. *Soils and Rocks* 31: 3–13.

Navarro, J.M., Coello, J., 1989. Depressions originated by landslide processes in Tenerife. *Proc. European Science Foundation Meeting on Canarian Volcanism, Lanzarote, Spain:* 150–152.

Schiffman, P., Watters, R.J., Thompson, N., Walton, A.W., 2006. Hyaloclastites and the slope stability of Hawaiian volcanoes: Insights from the Hawaiian Scientific Drilling Project's 3-km drill core. *Journal Volcanology and Geothermal Researches* 151: 217–228.

Seisdedos, J. 2008. Large paleo-rockslides of Güímar and La Orotava (Tenerife): Geological analysis, instability mechanisms and geomechanical modelling. PhD Thesis (UCM). Madrid: E-prints Complutense.

Volcanic Rock Mechanics – Olalla et al. (eds)
© 2010 Taylor & Francis Group, London, ISBN 978-0-415-58478-4

Passive anchors within retaining walls to stabilize volcanic rock slopes in road widening

M.A. Franesqui
Department of Civil Engineering, University of Las Palmas de Gran Canaria, Las Palmas, Spain

ABSTRACT: An economical and environmentally-friendly solution to stabilize jointed vertical rock slopes in works of improvement and cross section widening of a local road section in Gran Canaria Island (Spain) is presented in this paper. Due to the mountainous relief of this territory, this road cross over an extremely narrow section between two deep cliffs with vertical rock slopes on jointed phonolitic ignimbrites. The structural solution involves the construction of traditional gravity retaining walls with passive fully-grouted steel bar anchorages within its foundation. The rock mass nailing under the foundation of the retaining walls and even the adjacent rock slopes is also designed. This system combines traditional constructions of high simplicity with modern techniques of rock reinforcement.

1 INTRODUCTION

Infrastructure maintenance and alignment and roadbed improvement of local roads represent an especially significant component of the total actions carried out by local and regional road Agencies operating these road network levels.

The natural drainage in most of the volcanic territories with particularly abrupt topography is frequently shaped by a radial scheme with deep ravines divided by extremely narrow basin watersheds. This is a common relief pattern in Gran Canaria Island, where communications between coasts and central summits are fairly frequently performed along local roads with alignments roaming on these constricted watersheds.

These were the GC-503 local road conditions, with a sinuous non-homogeneous horizontal alignment and a particularly narrow cross section as a result of its historic origin as a rural bridleway and an extremely mountainous relief. This road is currently classified in the Gran Canaria road network as a minor way, but servicing to several residential areas with increasing population. Its winding layout is daily run by the aforementioned residents (Annual Average Daily Traffic [AADT] forecast for design life over 500 vehicles/day).

Thus, difficulties for cross section widening reside in the rugged ground conditions of the roadbed foundation, on an extremely narrow section between two deep cliffs with 15–20 m high vertical rock slopes on jointed rock masses.

Design specifications, dictated by the road Insular Administration Agency for roadbed widening solutions including structural system design and construction, specifically stated to design an especially simple construction procedure in order to be economical and systematically built with safety without needing skilled labour, owing to the complicated access to working site and high unevenness of the slopes. Furthermore, the maximum environmental and landscape protection was specified, given that 64% of the Canary territory is environmentally protected. In consequence, solutions involving expensive and complex cantilever structures to support the roadbed widening outside the existing ground had to be abandoned.

With this aim, this paper describes a combined stabilization system, designed to widen the road infrastructure and that includes numerous rock slope reinforcement measures, achieving the former requirements of minimum environmental impact and maximum building simplification.

2 GEOMECHANICAL BACKGROUND

Volcanic materials in the working area were formed during the Cycle I (Upper Miocene) and comprise welded phonolitic ignimbrites of scoriaceous nature, although highly fractured. Ignimbrites are originated from welded pyroclastic flow deposits whose coarse fragments are flattened and stretched. Their high welding level, one of the main characteristics of this type of material, is related to its deposition at temperatures higher than 400°C–550°C, and the foliation and joints are due to its cooling (González de Vallejo *et al.* 2007).

Figure 1. Detail of roadbed limits extremely close to the top of a vertical rock slope, restricting widening solutions.

The rock mass structure stratification had a dip angle around 8°. Two additional main joint sets can be distinguished. Rock blocks are subangular, of metric and decimetric size with big cavities at the slope toe. Modes of slope failure are mainly vertical movements as rock fall and collapse phenomena, in addition to block slips or rock wedge and planar slides created by jointing, in accordance with Corominas (2004).

In order to characterize the rock mass, 3 geomechanical stations have been implemented, in which rock matrix properties and main joint sets were studied according to the International Society of Rock Mechanics (ISRM 1981). These data were employed to classify the rock masses using the geomechanical classifications of Bieniawski (RMR) and Barton (Q index). Likewise, the aforementioned empirical classification criteria and performed laboratory tests have allowed to estimate some rock mass geotechnical properties for designing such us uniaxial compressive strength, cohesion, angle of internal friction, the m and s Hoek & Brown (1980) failure criterion parameters and deformation modulus. The most representative discontinuities could be grouped into three prevailing sets using a hierarchical cluster analysis and were represented on sterographic projection. The shear strength of rock joints has been analysed applying the Barton & Choubey (1977) failure criterion with the obtained values of JRC and JCS from Tilt test and Schmidt rebound hammer.

The rock mass degree of weathering, according to ISRM (1981), has been assessed as W_2 (slightly weathered), and was also determined using the United Alteration Index Classification (UAI) proposed by Kilic (1999) with a similar result.

The Rockslope Deterioration Assesment (RDA) index to evaluate the potential for deterioration of engineered slopes (Nicholson & Hencher 1997) was applied, with a result of class number 4 (high

Figure 2. GC-503 road on an extremely narrow cross section between two deep ravines with vertical jointed rock slopes on phonolitic ignimbrites.

sensitivity). Similarly, the Rockfall Hazard Rating System, RHRS (Pierson et al. 1990) and the Ontario Rockfall Hazard Rating System, RHRON (Franklin & Senior 1997) have been assessed, revealing a medium rock-fall hazard (RHRS = 464). Moreover, the rock mass was classified according to its rippability (Hadjigeorgiou & Scoble 1990) yielding IE = 50 (very difficult case of excavation).

3 STRUCTURAL SYSTEM FOR OVERALL STABILIZATION

3.1 General description

The designed structural system is composed of gravity retaining walls to support the backfill bearing the road cross section widening, and a rock mass reinforcement of the wall foundation by means of rock nailing with passive grouted permanent steel bar anchors. These bars are inserted in predrilled (in the rock) or precast (in the mass concrete, using corrugated steel tubular ducts) boreholes to the design depth and embedded in a cement grout.

Retaining and stabilization structural types had to be as standardized as possible with the purpose of achieving the most economical construction, considering the complicated access to site and working conditions and the extremely irregular and erratic rock slope geometry, with the presence of cavities in the rock mass. Hence, the structural system had to be of uncomplicated insertion on any irregular rock slope cross section and employing a systematic building procedure with reliability and, simultaneously, without needing highly skilled labour. For that reason, solutions including prestressed anchorages had to be discarded.

To facilitate the construction of the retaining structure, a typology of gravity retaining wall was selected, built with cast-in-place mass vibrated concrete and stair-shaped batter of the wall rear face (extrados) with steps vertically spaced at distances of 2.0 m (normally, total lift height should not exceed this distance according to concrete standard specifications, as known), which allows simple placement of concrete in series of horizontal layers and trouble-free shuttering. Stone masonry facing with local materials is utilized as non-recoverable formwork system, which provides a high quality surfacing and environmental integration.

The maximum design wall height at any road cross section is 10 m. At the foundation base, below the intersection of the wall rear plane with the existing rock slope profile, a systematic set of passive pressure fully-grouted bolts are installed, joining the mass concrete to the rock with corrugated high strength steel B500S GEWI® bars of diameter 25 mm, to form a strong mechanical and chemical reinforcement. This solution allows maximum reduction of wall base width, adapting its insertion to whichever existing rock profile. Figure 3 illustrates the main components of the stabilization system.

Additionally, with the intention of stabilizing the rock mass underneath the bottom of the wall base, for concentrated loads transferred by the retaining wall, a rock nailing reinforcement with similar steel tie rods has been designed and, as Figure 4 shows,

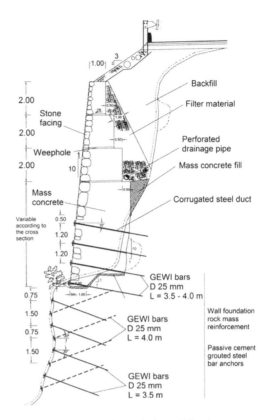

Figure 3. Cross section of the stabilization system. Retaining wall with bar anchors within its foundation and rock nailing of the rock slope beneath.

similarly even on the adjacent rock slopes where retaining walls are unnecessary to widen the road cross section.

The most significant advantages of the selected stabilization structural system can be summarized in:

– Lower volume of excavation for gravity wall foundations against the bigger base slabs for reinforced concrete cantilever retaining walls. In this case, this advantage is essential due to the confined space to lay the foundation and to minimize rock excavations. Furthermore, this base section reduction can be achieved through anchoring the wall foundation with passive steel tie bars, allowing an easier adjustment to a highly irregular rock slope.
– Excellent landscape integration as a result of reclaiming some excavation products (in the mass concrete and on the stone facing).
– Longer long-term integrity of the retaining wall owing to the non-existence of concrete steel reinforcing. Likewise, some studies have revealed

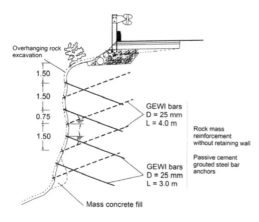

Overhanging rock excavation

1.50
1.50
0.75
1.50

GEWI bars
D = 25 mm
L = 4.0 m

Rock mass
reinforcement
without retaining wall

Passive cement
grouted steel bar
anchors

GEWI bars
D = 25 mm
L = 3.0 m

Mass concrete fill

Figure 4. Cross section of the stabilization system on the rock slope sections without retaining wall.

Table 1. Geomechanical design properties of welded phonolitic ignimbrites in the working area.

Geomechanical parameter		Unit	Design value
Moist unit weight	γ	(kN/m³)	20.21
Saturated unit weight	γ_{sat}	(kN/m³)	21.50
Basic joint friction angle	ϕ_{BJ}	(°)	57.2
Residual joint friction angle	ϕ_{RJ}	(°)	41.5
Rock mass friction angle	ϕ	(°)	46.6
Rock mass cohesion	c	(kPa)	35.0
Joint Roughness Coeff.	JRC_{20}		8
Joint wall Comp. Strength	JCS	(MPa)	12.0
UCS (intact rock)	σ_{ci}	(MPa)	43.0
Point Load Test Strength	Is	(MPa)	2.5
Hoek-Brown parameter	m	–	0.851
Hoek-Brown parameter	s	–	1.27×10^{-3}
Elastic modulus (laborat.) (intact rock)	E_i	(GPa)	9.261
Estimated elastic modulus (rock mass)	E	(GPa)	2.985
Rock Quality Designation	RQD	(%)	80
Rock Mass Rating	RMR		57 (Class IIIa)
Q Index	Q		4
Slope Mass Rating	SMR		53 (Class IIIa)

that mono-bar anchors have shown less suscepti-bility to corrosion than cable anchors as a result of groundwater attack (Shaqour 2006). On the other hand, the utilization of mass concrete facilitates the construction procedure, especially in a slope with complicated access to working site and high irregularity to systematize a steel reinforcement.
– Achievement of elevated resistant reaction force in order to guarantee a specified factor of safety.
– The batter of the wall with a rear face mean gradient of 1H/8V takes advantage of the verti-cal stabilizing force due to backfill. Designing a stair-shaped rear face simplifies the construction using horizontal lifts.
– The election of a cast-in-place mass concrete retaining structure, in preference to pre-cast con-crete, is due to difficulties with standardizing pre-cast elements because of rock unevenness, with presence of cavities in the rock mass, reduced foundation spaces and difficult excavations.

A previous excavation of overhanging rock and breached or detached blocks, as well as bush clear-ing (in order to avoid roots growing in rock cracks) and wall drainage and waterproofing with bitumi-nous products have been also prescribed.

3.2 Stability analysis and verifications

Design parameters of the welded phonolitic ign-imbrite rock mass were estimated from de geome-chanical and geotechnical study and are summarized in Table 1.

In project, stability checks for the anchored walls have been performed, assuming earth pressures of the backfill placed over tie back level and the traffic surcharge according to design approaches and load

combinations specified in Spanish Standards IAP (1998) and EHE (2008). Lateral earth pressures on the wall can be calculated (whether the mass rock is stable) applying "silo effect" earth pressure the-ory due to backfill confinement between the rock mass and the new wall structure. This assumption allows to consider more reduced active pressures than classical Rankine and Coulomb theories, being a more realistic approach in this particular case, because of the limited width of the confined backfill. Previously, conditions to utilize this meth-odology were checked.

The rock slope safety against overall and local rock wedge and planar instabilities created by jointing has been verified taking into account the reaction force introduced by the anchored gravity retaining wall. This force was calculated with the condition of assuring a specified global Factor of Safety (F_S = 1.5) with respect to overall and local slope failure, according to limit state design methods for persistent design situation and quasi-permanent load combination. Safety factors depend on the uncertainties and risks involved for the conditions encountered. The choice of this overdesign Factor, required for anchorage calculation, obeys to the great variability observed in respect of the geomechanical properties of the rock mass, being thus the upper value of the recommended interval ($F_s \geq 1.3$ to 1.5) according to the Spanish "Foundation Design Guide for Road Works" (Ministerio de Fomento. Dirección General de Carreteras 2003).

Achievement of the necessary structure reaction against the ground pressure requires a sufficient wall displacement, as a consequence of employing passive bolts. Its calculation has allowed to check the compatibility of this displacement with road operation.

Two different analysis methodologies has been applied to check the rock slope stability:

- When the rock mass presents defined discontinuity surfaces (according to stereographic projection of joint sets), determining the possibility of cinematic instabilities, analyses using Equilibrium Limit methods applied to planar or wedge rock slides have been performed. To calculate the shear strength of rough rock joints, the aforementioned Barton-Choubey criterion was utilized. Rock wedges are studied with the Hoek-Bray method.
- If the rock has not presented defined discontinuity surfaces, as in the event of extremely fractured rock masses with joints of all directions, failures have been assumed on cylindrical and planar slip surfaces, being analysed with Equilibrium Limit methods assuming the mass rock performance similar to a dense coarse-gravel soil, with a Mohr-Coulomb failure criterion. In that case, stability is studied both in drained and undrained conditions.

In every former situation, the slope Factor of Safety against all the actions is checked. A minimum overdesign Factor of 1.5 has been required for the stability. If the calculated F_s results lesser, a force must be applied against the sliding slope until this factor is equal or over. Precisely, this force coincides with the minimum reactive force that the retaining structural system must guarantee to stabilize the slope with the specified safety, keeping in consideration the reaction dip with respect to the horizontal.

It should be remembered that methods of total equilibrium do not consider the plasticization of the slices and consequently, species in the proximity of the re-enforcement interventions, localized breaks could realistically occur even when the retaining structure is able to offer a reaction sufficient to stabilize the slope: this since the ground may not be able to transmit the exerted pressure of the slope to the intervention due to poor geotechnical characteristics or reduced thickness in the upstream zone of the retaining structure. In the interaction between slope and retaining structure the analysis includes not only the sliding surface that without a retaining structure presents a minimum safety factor, but also includes all the potentially unstable sliding surfaces. The safety factor value is, in fact, a relative quantity while that of the retaining structure reaction is absolute, and it is for this reason that the force necessary to stabilize a slope along a sliding surface with a minimum safety factor may not be sufficient for another surface with an initially greater safety factor but with a unit weight, mass, and then force very much superior in play.

In addition, the rock slope nailing underneath the concrete retaining wall is analysed. Its safety, with respect to overall and local rock wedge and planar instabilities, has been also checked. Here, loads on the top of this slope are bigger because of the wall weigh concentrated load. The applied analysis methodologies and failure criteria are the same. Likewise, the stability of the rock slopes where retaining walls are unnecessary was verified, designing the necessary passive anchor system to assure the specified Factor of Safety (Fig. 4).

With the results of the required stabilizing reaction forces, whose maximum values are presented in Table 2, the anchorages were designed.

For the reason of simplifying the installation, taking into consideration the complicated access to site and working conditions, a typology of passive steel bar anchor was the ultimate choice. The system is constituted, as aforesaid, by passive corrugated steel B500S GEWI® bars of the same diameter, only varying their length depending upon each road cross section.

To form a mechanical and chemical reinforcement, the space between the bolt and the borehole

Table 2. Total required stabilizing pressure on the anchors system to assure a specified overdesign Factor ($F_s = 1.5$) against planar and wedge sliding.

Slope	Station	Height (m)	Required Reaction (kN/m²)	Factor of Safety F_s
Slope behind the retaining wall	5 + 000	7.17	64.00	1.50
Slope beneath the retaining wall	4 + 940	2.50	54.40	1.50
Slope with anchors (without retaining wall)	4 + 960	8.00	32.50	1.50

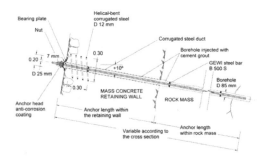

Figure 5. Passive fully grouted permanent mono-bar anchor inserted in a pre-drilled (in rock mass) and pre-cast (in mass concrete, using corrugated steel ducts) borehole.

is pressure fully-grouted after the insertion of each bar, employing a CEM I 42.5R cement laitance. The anchor category as a function of its design life, social and economical allowable risk if failure, and injection system result to be: category C5 (ACHE 2005), type 7 (DGC 2003b).

Optimum spacing between anchor heads and dip angle has been calculated, depending on the main joint set dip angle and joint friction, though within some practical construction limits.

Bar lengths and diameters were designed by analysing the individual performance of each anchor, checking the design steel tensile strength and the safety against pulling out (both with respect to grout-steel bar failure and grout-rock mass failure).

Some recommendations related to ultimate shear strength for ground anchorages, which can be consulted in Littlejohn (1995) and in the afore-mentioned Spanish standards (DGC 2003b), and our own experiences with volcanic massive rocks were applied for mechanical design. The section of the bars was standardized to 25 mm for simplifying, and anchorage total lengths of 3.0, 3.5 and 4.0 m have been calculated.

4 CONCLUSIONS

Due to environmental, economical and building reasons, solutions involving complex and high-priced cantilever structures, in order to provide the requested widening of the roadbed infrastructure, had to be discarded.

Therefore, a combined structural system for rock slope stabilization, that integrates numerous reinforcement measures joining together traditional retaining wall construction techniques and more modern rock nailing technologies, has been suggested.

This structural solution aims the combination of the highly-experienced local stonework construction procedures, traditionally utilized in the Canaries (especially masonry retaining walls), with a reinforcement consisting of passive steel bar anchorages, allowing an important reduction of wall foundation surface and achieving the minimum impact on landscape.

Unitary construction cost of the proposed slope stabilization structural system was assessed around 126 €/m² slope surface, including health and safety at work measures. Based on this result, it is concluded that the required investment expenditure is notably lower with respect to other possible structural solutions. The working period of time, for a road section of 340 m long, can be estimated around 2 months.

REFERENCES

Asociación Científico-técnica del Hormigón Estructural ACHE. 2005. *Recomendaciones para el proyecto, construcción y control de anclajes al terreno.* 3th ed. Madrid: Colegio de Ingenieros de Caminos, Canales y Puertos.

Barton, N & Choubey, V. 1977. The shear strength of rock joints in theory and practice. *Rock Mechanics* 10 (1).

Comisión Permanente del Hormigón. 2008. *Instrucción de Hormigón estructural.* EHE-08. Madrid: Ministerio de Fomento, Secretaría General Técnica.

Corominas, J. 2004. Tipos de roturas en laderas y taludes. In López Jimeno, C. *et al.* (eds), *Ingeoter* 4: 191–213. Madrid: U.D. Proyectos, E.T.S.I. Minas, UPM.

Franklin, J.A. & Senior, S.A. 1997. The Ontario Rockfall Hazard Rating System. *Engineering Geology and Environment.* Rotterdam: Balkema.

González de Vallejo, L.I., Hijazo, T., Ferrer, M. & Seisdedos, J. 2007. Geomechanical characterization of volcanic materials in Tenerife. In Malheiro & Nunes (eds.), *Volcanic Rocks.* London: Taylor & Francis Group.

Hadjigeorgiou, J. & Scoble, M. 1990. Ground characterization for assessment of case of excavation. *Mine planning and equipment selection.* Calgary.

Hoek, E. & Brown, T. 1980. Empirical strength criterion for rock masses. *J. Geotech. Eng. Div., ASCE* 106 (GT9): 1013–1035.

International Society of Rock Mechanics ISRM. 1981. *Suggested methods for rock characterization, testing and monitoring.* Oxford: E.T. Brown. Pergamon Press.

Kilic, R. 1999. The Unified Alteration Index (UAI) for Mafic rocks. *Environ. Eng. Geosci., AEG* (4): 475–483.

Littlejohn, S. 1995. Rock Anchorages. *News Journal. International Society for Rock Mechanics* 2 (3,4).

Ministerio de Fomento. Dirección General de Carreteras DGC. 1998. *Instrucción sobre las acciones a considerar* en el proyecto de puentes de carretera. IAP. Madrid: Centro de Publicaciones del Ministerio de Fomento.

Ministerio de Fomento. Dirección General de Carreteras DGC. 2003. *Guía de cimentaciones en obras de carretera*. Madrid: Centro de Publicaciones del Ministerio de Fomento.

Ministerio de Fomento. Dirección General de Carreteras DGC. 2003. *Guía para el diseño y la ejecución de anclajes al terreno en obras de carretera*. Madrid: Centro de Publicaciones del Ministerio de Fomento.

Nicholson, D.T. & Hencher, S. 1997. Assessing the potential for deterioration of engineered rockslopes. *Engineering Geology and Environment*. Atenas: Balkema.

Pierson, L.A., Davis, S.A., & Van Vickle, R. 1990. *The Rockfall Hazard Rating System: Implementation Manual*. Technical Report FHWA-OR-EG-90-01. US Department of Transportation.

Shaqour, F. 2006. Ground anchors in an aggressive hydro-environment. *Bull. Eng. Geol. Env.* 65: 43–56.

Volcanic Rock Mechanics – Olalla et al. (eds)
© 2010 Taylor & Francis Group, London, ISBN 978-0-415-58478-4

Detailed studies and stabilization methods of volcanic rocky slopes in coastal areas, Canary Islands, Spain

A. Lomoschitz & A. Cilleros
Departmento de Ingeniería Civil, Universidad de Las Palmas de Gran Canaria, Spain

R. García-Ferrera
JOFRAHESA, S.L., Las Palmas de Gran Canaria, Spain

ABSTRACT: Natural relief on mountainous volcanic islands has deep ravines, steep rocky slopes and high coastal cliffs. Volcanic formations have very heterogeneous rocks and soils. Besides, civil and building works sometimes include high cuts on the terrain and, as a result, many urban areas have been affected by rock falls and landslides. We show two case studies of the Canary Islands: (1) Rock fall hazard study and stabilization methods on Los Teques slope, Mogán, Southern Gran Canaria Island, and (2) Geologic-geotechnical study for a footpath project in Morro Jable coastal cliff, Pájara, Southern Fuerteventura Island. We conclude that sometimes classic methods of rock masses characterization are ineffective, while detailed geological studies are the best way to define and evaluate unstable zones on the slopes and to design the most convenient stabilization methods.

1 INTRODUCTION

Volcanic areas present an outstanding variety of rock and soil materials in a relatively random pattern. This wide diversity is a direct consequence of the genesis of the material by a wide range of volcanic activity and external factors which control eruption dynamics, transport, emplacement and reworking of the material being generated (del Potro & Hürlimann 2007).

Occasionally, simplified graphic profiles of volcanic formations have been used, showing an alternance of lava flows, of middle to high strength, and interbedded pyroclastic layers, of low strength (Serrano et al. 2008). Moreover, natural slopes of volcanic layers are variable and sometimes they have a conical spatial disposal (volcanic cones) around central points of eruption. Besides, other features have to be considered such as the existence of syngenetic cooling joints, which are previous to tectonic jointing (Blyth & de Freitas 1984, Schmincke 2004) and the influence of the strata thickness on the bearing capacity of the terrain and the stability of slopes (Lomoschitz, 1996).

In recent years, many works have been done on the geotechnical characterization and behaviour of volcanic rocks. They are about four main topics: (a) geomechanical characterization of intact rocks, from massive lava flows to pyroclasts, which are of a variety of geochemical compositions and textures (Rodríguez-Losada et al. 2007a,b); (b) field

and geomechanical classification of volcanic formations (e.g., González de Vallejo et al., 2007, del Potro & Hürlimann 2007); (c) strength and deformability models to interpret the behaviour of volcanic rocks (Kwasniewsky 2002, Serrano et al. 2002, 2007) and (d) case studies of engineering projects and works (e.g., Erichsen 2002, Simic 2007).

In this chapter, two case studies of volcanic rock slopes in coastal areas are shown (Fig. 1), which

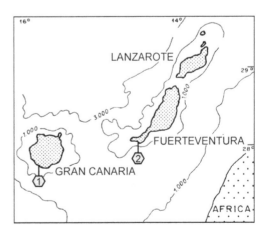

Figure 1. Map of the Eastern Canary Islands, 100 km off the northwest coast of Africa, and location (polygons 1 and 2) of the two study cases.

were affected by a variety of rock falls and slides. Specific fieldwork methods were necessary for an effective characterization of rocks and to determine the most convenient stabilization techniques.

2 CASE STUDIES

2.1 *Los Teques apartments and slope, Southern Gran Canaria Is.*

In December 2005 a 11.6 metric ton rock block fell down on the back corridor of the apartments and the structure of the building was damaged. For some days, other smaller rock falls occurred. Afterwards, by means of a detailed geological study of the whole slope, 105 m long and 30 m high, 9 zones of medium to high risk were found out. In the study there were considered the strength and weathering of the rocks, potential rock fall mechanisms and trajectories, and the superficial and underground waters influence.

In addition, many specific solutions were proposed for the slope stabilization. Five building companies were consulted and some of them proposed more general and expensive stabilization methods than required. Finally, a variety of techniques were used (rock bolts, cable nets and dowels, concrete walls, shotcrete, cable belts, cable barriers, etc.) which were applied on specific areas.

Fieldwork was focused on three main targets: (1) Geological characterization of rocks and soils; (2) identification of unstable rock blocks and loose debris, and (3) definition of the best construction methods to retain or avoid potential rock falls, taking also in account the cost and effectiveness of each method.

The geological materials of the slope are of three main types: (a) acid volcanic agglomerate of fragments with ash and pumice which is a soft rock also affected by a fresh water spring; (b) 2 to 3 lava flow layers, which are ignimbrite of medium to high strength, and (c) some superficial scree and debris deposits. All rocks are of trachyte-rhyolite composition and they correspond to the Miocene magmatic cycle of Gran Canaria Is.

The volcanic agglomerate outcrops at the base of the slope, just behind the apartments, with a thickness of 10–15 m. Its high slope angle (70–80°) and weathering had caused the undermining of the upper lava flow layers, which are 15–20 m thick. These layers had a number of unstable rock blocks with varied (wedge-, columnar-, cubic-) forms (Fig. 3).

When the unstable zones had been identified, one by one, a rock fall hazard map was done, considering three categories of risk: low, medium and high. Construction solutions were mainly focused on the stabilization of medium and high risk zones (Fig. 4).

Figure 2. General view of Los Teques apartments and rocky slope in Southern Gran Canaria Island.

Figure 3. Removal of the 11.6 tons rock block that fell down at the back of the building. Damages caused by this rock fall.

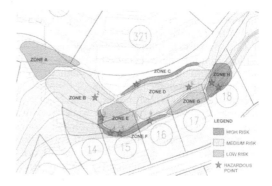

Figure 4. Risk map of Los Teques slope. Stabilization methods were focus on medium to high risk zones.

To evaluate the cost of construction works which were needed for the slope stabilization five building companies were consulted and some of them proposed more general and expensive stabilization methods than required. Finally, a variety of techniques were used (Fig. 5):

Figure 5. Stabilization methods used on Los Teques slope.

- Removal of loose stones and debris (450 m^2).
- Concrete buttress to support rock above cavities (18 m^2).
- 20 drain holes to reduce water pressure within slope.
- Wire mesh hung on vertical face of slope (260 m^2).
- Reinforced cable net and rock bolts on unstable rock blocks (180 m^2).
- Small fences (1 m high) to retain canalized debris flows (2 fences of 15 m).
- Cable belts around semi-unstable rock columns (3 zones).
- Tensioned rock bolts in rock blocks (10 units).

The final cost of the stabilization works was €45,000, less than a half of the budget cost of other more extensive and general solutions.

2.2 Morro Jable cliff and footpath, Southern Fuerteventura Is.

In 2006 a project for a coast footpath, 450 m long, was done. It would connect Morro Jable harbour and El Matorral beach, located in a main tourist area at Southern Fuerteventura Island. The path had to be designed on a narrow strip of land, along a rocky cliff 30 m high, with a lot of caverns, landslide and rock-fall zones and very heterogeneous volcanic formations (Fig. 6).

About 40 percent of the footpath would be supported by the natural terrain and the remaining 60 percent by a metal structure of piles and girders. A detailed geological and geotechnical study was needed for an adequate development of the project, because of the irregular morphology of the shore and the geomechanical features of the cliff. In the study there were included: (a) a geomorphologic

Figure 6. General view of Morro Jable Cliff in Fuerteventura Island. Differentiation of geological materials on the slope.

map of the shore, indicating cavern and slide zones, rocky platforms, gravel and sand beaches; (b) a geological section along the cliff with a description of the volcanic layers and dykes; (c) the location of unstable rocky wedges and blocks; (d) volume measurement of caverns and caves at the base of the slope; (e) proposal of protection methods against the coastal erosion and stabilization techniques for the unstable zones; and (f) evaluation of the soil or rock bearing capacity for each of the foundation points.

The geological study has been focussed on the two parts of the zone: the tidal shore platform and the rocky cliff.

The tidal shore platform is composed of basalt rocks, corresponding to intrusive bodies and tephra agglomerates, small patches of carbonate beachrock and boulder and gravel beaches.

Figure 7. Designed footpath structure.

The cliff includes three groups of materials: (a) Irregular layers of basaltic breccia of scoria and lapilli in the first one third part of the cliff wall; (b) basaltic lava flows, massive and extensive, which compound the upper two thirds of the cliff, and (c) a number of volcanic dikes, of basaltic and trachytic composition, that intrude the country rock. All rocks correspond to the Miocene volcanic edifice of Jandía.

It was done a geological and geomechanical description of every material, which included an evaluation of the bearing capacity at each support point of the structure (Fig. 7).

However, along the shore there are many hazardous zones for the project of the coast footpath:

– 2 zones affected by large landslides: the first is 60 m long and 20 m high and has had big rock topple and fall movements, and the second is 40 m long and 18 m high and has produced a huge debris flow deposit on the shore platform.
– 7 potentially unstable zones at different points of the cliff wall.
– 9 zones with caverns and caves at the base of the cliff.
– 3 zones on the cliff with a high degree of erodibility.

Every zone has been studied individually, having searched a building solution for each one. For instance: concrete buttresses and stone walls to support rock above caverns and caves; rock removal at some places; concrete injection inside open joints; tensioned rock bolts; reinforced cable net and bolts on unstable rock wedges and columns, etc.

3 CONCLUSION

In many occasions, slopes on volcanic formations do not follow homogeneous or simple patterns (in terms of geometry, thickness, jointing, weathering, etc.) In these cases, classic methods for rock masses characterization are ineffective, and detailed geological studies are the best way to define and evaluate unstable zones on the slopes and to design the most convenient stabilization methods.

REFERENCES

Blyth, F.G.H. & de Freitas, M.H. 1984. A Geology for Engineers. London: Edgard Arnold.
del Potro, R. & Hürlimann, M. 2007. Strength of volcanic rock masses in edifice instability: Insights from Teide, Tenerife. In A.M. Malheiro & J.C. Nunes (eds), Volcanic Rocks: 175–184. London: Taylor & Francis Group.
Erichsen, C. 2002. Geotechnical models or classification of the bedrock as a basis for the design and for the construction of rock structures: 15–26. In C. Dinis da Gama and L. Ribeiro e Sousa (eds), Workshop on volcanic rocks, Eurock 2002, Funchal, 27 November, 2002.
González de Vallejo, L.I., Hijazo, T., Ferrer, M. & Seisdedos, J. 2007. Geomechanical characterization of volcanic materials in Tenerife. In A.M. Malheiro & J.C. Nunes (eds), Volcanic Rocks: 21–28. London: Taylor & Francis Group.
Kwasniewsky 2002. A note on the triaxial strenght on volcanic rocks: 27–36. In C. Dinis da Gama and L. Ribeiro e Sousa (eds), Workshop on volcanic rocks, Eurock 2002, Funchal, 27 November, 2002.
Lomoschitz, A. 1996. Caracterización geotécnica del terreno, con ejemplos de Gran Canaria y Tenerife. Escuela Técnica Superior de Arquitectura, Universidad de Las Palmas de Gran Canaria.
Rodríguez-Losada, J.A. Hernández-Gutiérrez, L.E. Olalla, C. Perucho, A. Serrano, A. & del Potro, R. et al. 2007a. The volcanic rocks of the Canary Islands. Geotechnical properties. In A.M. Malheiro & J.C. Nunes (eds), Volcanic Rocks: 53–58. London: Taylor & Francis Group.
Rodríguez-Losada, J.A., Hernández-Gutiérrez, L.E. Lomoschitz, A. 2007b. Geotechnical features of the welded ignimbrites of the Canary Islands. In A.M. Malheiro & J.C. Nunes (eds), Volcanic Rocks: 29–34. London: Taylor & Francis Group.
Schmincke, H-U. 2004. Volcanism. Berlin: Springer.
Serrano, A., Olalla, C. & Perucho, A. 2002. Evaluation of non-linear strength laws for volcanic agglomerates: 53–60. In C. Dinis da Gama and L. Ribeiro e Sousa (eds), Workshop on volcanic rocks, EUROCK 2002, Funchal, 27 November, 2002.
Serrano, A., Olalla, C. & Perucho, A. 2008. Estabilidad de taludes en materiales heterogéneos. II Jornadas Canarias de Geotecnia, Tacoronte, Tenerife, 21–23 May 2008.
Serrano, A., Olalla, C., Perucho, A. & Hernández-Gutiérrez, L.E. 2007. Strength and deformability of low density pyroclasts: 35–44. In C. Dinis da Gama and L. Ribeiro e Sousa (eds), Workshop on volcanic rocks, EUROCK 2002, Funchal, 27 November, 2002.
Simic, D. 2007. Foundation of the "Los Tilos" arch bridge in La Palma Island: 113–122. In C. Dinis da Gama and L. Ribeiro e Sousa (eds), Workshop on volcanic rocks, EUROCK 2002, Funchal, 27 November, 2002.

Volcanic Rock Mechanics – Olalla et al. (eds)
© *2010 Taylor & Francis Group, London, ISBN 978-0-415-58478-4*

Slope stability in the Canary volcanoes based on geotechnical criteria

J.A. Rodríguez-Losada & A. Eff-Darwich
Department of Soil Science and Geology, University of La Laguna, Tenerife, Spain

L.E. Hernández
Regional Ministry of Works, Government of the Canary Islands, Spain

C. Olalla Marañón, A. Perucho & A. Serrano González
Centre for Civil Engineering Studies and Research, CEDEX, (Ministery of Civil Works, Spanish Government), Madrid, Spain

ABSTRACT: The stability of natural slopes in different areas of the Canary Islands have been analysed through the relation between cohesion, friction angle and slope height. The combination of estimates for geomechanical parameters of intact rocks of the Canary Archipelago, the geological strength index (GSI) and textural features were used to deduce geomechanical parameters of rock massifs. This paper discusses the changes in cohesion and friction angle as a function of the slopes height for different rock massifs and geological conditions expressed in the form of GSI. Such differences may define the threshold between stability and instability of slopes and have relevant implications in the volcanic hazards of certain areas of the Canary Islands which are discussed here.

1 INTRODUCTION

In recent years, the geomechanical characterization of volcanic rocks in the Canary Islands has gained more attention due to systematic works on the wide spectrum of volcanic rocks in the entire Canary Archipelago (Rodriguez-Losada *et al.*, 2007; Del Potro & Hurlimann, 2007). Thus, except for references in other volcanic areas (e.g. Chaplow, 1998; Apuani *et al.*, 2005; Davies and Moon *et al.*, 2005), previous works referred to geomechanical studies of volcanic rocks in the Canaries are scarce. Most of them were made by private construction and civil engineering companies without any later scientific relevance. In this sense, it is essential to be concerned about the role of the geomechanical characterization of rock masses as an essential tool to build approaches and models of ground stability in the context of geological hazard and particularly in volcanic hazard (Hürlimann, 1999; Hürlimann *et al.*, 1999; Hürlimann *et al.*, 2001; Thomas *et al.*, 2004; Gonzalez de Vallejo & Ferrer, 2006; Malheiro, 2006; Rodriguez-Losada *et al.*, 2007; among others).

Parameters such as strength of the rocks, cohesion, friction angles among others are essential to understand the stability of volcanic terrains. In the case of the Canary Archipelago, many recent works analysed the occurrence of giant landslides in the islands and the factors that increase the risk of lateral collapse of volcanic edifices (Cantagrel *et al.*, 1999; Carracedo, 1999; Day *et al.*, 1999; Elsworth & Day, 1999; Keating & McGuire, 2000; Ward & Day, 2001; Walter & Schmincke, 2002; Munn *et al.*, 2006; Marquez *et al.*, 2008; del Potro & Hurlimann 2008; del Potro *et al.*, 2009; Rodriguez-Losada *et al.*, 2009; del Potro and Hurlimann 2009). It is well known that height and steep slopes constitute an effective association to trigger giant landslides. High gradient slopes spread over large areas of the islands and two antagonistic processes are involved in their formation, namely erosion and the building up of lavas, scoria and pyroclastic layers. Simple erosion or mass wasting processes occur in previously unstable slopes. In this case the steeped areas are constrained to the edges and at the head of the resulting depressions and normally, progressive erosion in these areas is expected with no significant risk of slope collapse. These landscapes appear associated to the oldest basaltic outcrops of the islands (e.g. "Anden Verde" and deep valleys in Gran Canaria, Northern flank of "Jandia" massif in Fuerteventura, West flank and deeply eroded ravines in "Los Ajaches" massif in Lanzarote, deeply eroded ravines and arcuate head of "Valle de Aridane" at the "Cumbre Nueva" massif in La Palma, along the radial network of ravines in La Gomera and specially the Arguamul,

Tazo, Alojera areas at the northeast most part of the island, the head areas of "El Golfo" and "Las Playas" depressions in El Hierro and finally along the deep ravines and coast cliffs of the "Anaga" and "Teno" massifs, the wall of the "Cañadas" caldera, head and edges of the "Güimar" and "La Orotava" valleys in Tenerife Island. There, slope gradient vary from 65–50° to 30–26° and almost vertical in cliff areas. Most of these areas are the result of previous landslides processes and consequently, there is not expected an early occurrence of giant mass wasting processes. Rocky debris and small avalanches are expected to be the mainly subsequent erosive mechanisms.

By contrast, accumulation of volcanic lava flows and interbedded pyroclastic layer is the result of cycles of continuous volcanic eruptions that may evolve to the build-up of large steeped edifices with poorly stabilized slopes and high risk of landslides. The growth of such unstable volcanic edifices may take place over previously collapsed areas filling the consequent deeply eroded depressions with the consequent possibility of later landslides. It is assumed that the high gradient slopes originated by accumulation of lavas and pyroclasts, constitute zones of potential hazard of landslide and hence,

final discussion will be focused on them. Examples of these steep areas highlighted here are "El Julan" (SW flank of El Hierro island) which exhibit an average gradient slope of 25°, the volcanic rift of "Cumbre Vieja" (S of La Palma island) with an average gradient slope of up to 20° and "Teide stratovolcano", where the gradients of the flanks vary between 25° and 30° of slope (Fig. 1).

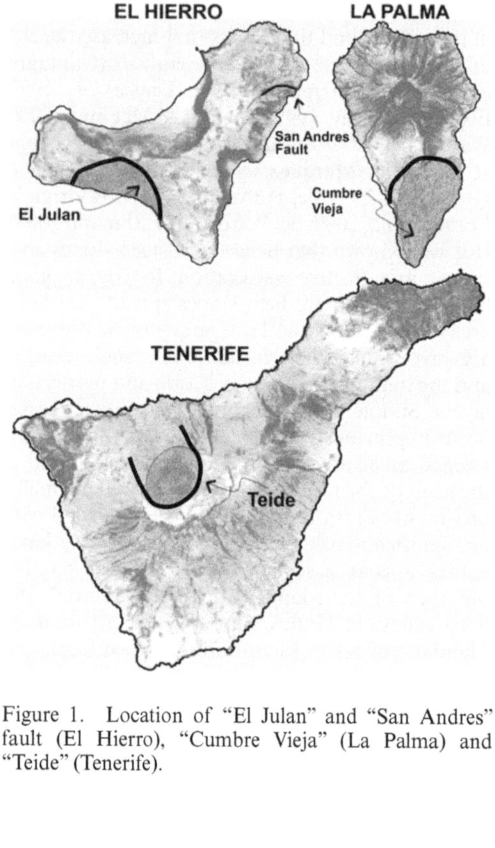

Resistance parameters of rock masses such as cohesion and friction angle may be deduced from the Hoek and Brown criterion (Serrano and Olalla, 1994) by means of the resistance-related properties of intact rock and the geological conditions, namely joints, fissures and alteration degree. Both cohesion and friction angle could be quantified through the Geological Strength Index or GSI (e.g. Hoek et al., 1992) that may be easily estimated through the method of Marinos & Hoek (2000) and Marinos et al. (2005) or the RMR (the Rock Mass Rating from Bieniawski, 1989), which depends on parameters such as the GSI among others.

The Mohr-Coulomb fits: cohesion and friction angle of the rock masses for different rocks were estimated by means of the Hoek and Brown failure criterion (Hoek et al., 2002) based on previous geomechanical data of intact rock (Rodriguez-Losada et al., 2007; Rodriguez-Losada et al., 2009).

2 THE ROLE OF THE LOW CEMENTED PYROCLASTS

Low cemented pyroclasts conform significant deposits outcropping in all the islands in the form of air fall salic deposits (pumice) or lipilli deposits, both being either slightly cemented or non-cemented fragments, that could be considered as typical examples of macroporous rocks. In this sense, special and complex geotechnical behaviour is found for these materials. At low stress levels pyroclasts behave like a rock with very high deformation moduli; however, under high stress levels the internal structure is destroyed and the rock deformability dramatically increases, behaving as a soil (Uriel & Serrano, 1973; 1976). This phenomenon is known as mechanical collapse and the involved materials as mechanically collapsible. Under isotropic compression, the pyroclastic deposits react in one of the two following ways: a) continuous increase of the specific weight and stiffness with no failure processes and b) continuous increase of the specific weight until it reaches a threshold value at which, a sudden collapse takes place.

Under laboratory experiences, it has been noted that intact pyroclastic samples may experience at least two consecutive collapses: a first one produced by breakage of particle contacts and a

Figure 1. Location of "El Julan" and "San Andres" fault (El Hierro), "Cumbre Vieja" (La Palma) and "Teide" (Tenerife).

second one at higher stresses by means of breakage of the own particles until the failure of the material as a soil is reached. In general terms a threshold of dry specific weight of around 15 kN/m³ separates the collapsible materials of lower specific weight from the non collapsible ones of higher specific weight than the threshold (Serrano *et al.*, 2007). However, not all low specific weight pyroclasts collapse and factors such as the structure of the samples in relation with their porosity could explain such anomalies. Pyroclastic samples have dry specific weight between 6.5 to 13.3 kN/m³, uniaxial compressive strength of 0.3 to 3 MPa, elastic moduli of 30 to 1140 MPa and isotropic pressure collapse from 0.3 to 5.4 MPa. In this sense, the isotropic collapse pressure may be estimated as equal to twice and half times the uniaxial compressive strength and the elastic modulus equal to a hundred eighty times the uniaxial compressive strength (Serrano *et al.*, 2007).

In rock masses, these parameters tend to decrease by reducing factors that depend on the geological conditions quantified by the GSI (Serrano *et al.*, 2002). Pyroclastic rock masses have a GSI that ranges from 25 to 60. In the low cemented ones this value is assumed to be 25 (Serrano *et al.*, 2002).

Taking into account these data, the estimated Mohr-Coulomb failure parameters in the low cemented pyroclasts have averages of cohesion around 0.012 MPa and friction angle of roughly 45°.

3 DISCUSSION

In order to estimate the stability of natural slopes in different areas of the Canary Islands, we have analysed the relation between cohesion, friction angle and slope height. Initially, for smaller slope heights, lowest values for cohesion and largest for friction angles were found. As slope height increases, cohesion also increases thus contributing to improve the stability of the massif, whereas friction angle decreases reducing the flank stability.

Based on the Hoek and Brown failure criterion (Hoek *et al.*, 2002), variations of cohesion and friction angle for different types of rock masses have been estimated. Such changes were estimated for a range of slope heights between a minimum of 5 m and a maximum of 2000 m (Figs. 2 & 3). According to the results, cohesion may be expressed as a function of the slope height in the following form: $c = kh^l$, where h(m) is the slope height, whereas k and l are constants that depend on the type and geological conditions of the rock mass. The parameter k ranges from 0.37 in fresh massifs to 0.04 in fairly weaken massifs, whereas l varies from 0.41 to 0.60. For both fresh and weaken massifs, respectively. In cemented ignimbrites, the values for k and l were

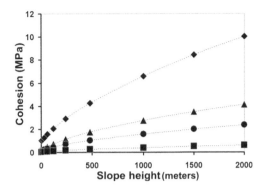

Figure 2. Variation of cohesion (MPa) vs. slope height (meters) for slope application on selected types of massifs. Rhomboids: Fresh massive lavas; triangles: Weathered massive lavas; circles: Ignimbrites; squares: Low cemented pyroclasts.

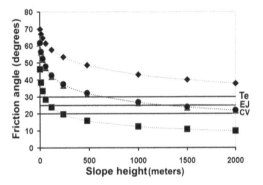

Figure 3. Variation of the friction angle (degrees) vs. slope height (meters) for slope application on selected types of massifs. Rhomboids: Fresh massive lavas; triangles: Weathered massive lavas; circles: Ignimbrites; squares: Low cemented pyroclasts. The horizontal solid lines intersect the vertical axis in the angular value corresponding to the average slope on the flanks of "Teide (Te)", "El Julan (EJ)" and "Cumbre Vieja (CV)".

0.05 and 0.5, respectively, whereas the values for k and l in low or non cemented pyroclasts were 0.007 and 0.6, respectively.

The relation between the internal friction angle (ϕ) and the slope height was logarithmic, namely $\phi = -k. Ln(h) + l$, being h(m) the slope height, whereas k and l are constants that, as in the case of cohesion, depend on the type and conditions of the rocky or fragmentary material. Respective values for k and l are 5.6 and 82 in unaltered rocks, 7 and 75 in weakened rocks or cemented ignimbrites and finally 6 and 54 in poorly or not cemented pyroclastic deposits.

Under the most unfavorable geological conditions (low values of GSI), cohesion and friction angle suffer significant decrease in the same type of rock massifs. That would imply a situation of instability for the areas highlighted in this paper (Figs. 4 & 5).

Equivalent reductions on the GSI may be assumed as a result of internal decomposition of the rock mass usually accompanied by generalised secondary fracturing.

If we assume the cohesion as nearly zero and the gravity as the main acting force, failure of natural slopes may occur if the slope gradient exceeds the friction angle of the rock mass. In the case of the Canary Archipelago, the most significant average slope gradients are found in "El Julan" flank, "Cumbre Vieja" rift zone and "El Teide" stratovolcano. As illustrated in the figure 3, friction angles for unaltered rock masses are always larger than the slope gradients and hence, lateral collapse would never take place. However, due to the internal corrosion of rocky massifs or due to the presence of dominant ignimbrite rock masses, friction angle may become smaller than slope gradient as the slope height rises to 700–1300 m. In this case and exclusively from a geotechnical point of view, if internal cohesion drops, conditions for potential lateral collapse would occur. Although the rock massifs are not only constituted by rocky (highly cohesive) or pyroclastic materials, it is expected that the presence of low cemented pyroclasts and certain alteration degree of the intact rock will reduce the cohesion and friction angle of the rock masses. In this sense, an increase of flank instability is expected and hence, the likelihood of sudden collapse of buried pyroclastic layers as isotropic-like stress reaches their threshold value. This is what it could take place at "El Teide", "El Julan" and "Cumbre Vieja" areas, since weathering of the intact rocks, joints affecting the rock massifs, unfavourable dip of layers and certain amount of weakly consolidated pyroclastic layers, reduce the internal cohesion (Fig. 4). Moreover, friction angles may drop to values below the slope gradient, creating the conditions for flank failure (Fig. 5). Additionally, flank collapse could be reinforced by external factors such as massive injection of volcanic fluids, massif distortion by magma injection, groundwater pressurization (Elsworth & Voight, 1995) or terrain fracturing induced by magma intrusion. Pre-existing fractured zones would also increase the risk of slope failures in certain locations such as the sector of the "San Andres" fault system in El Hierro island, considered as an aborted landslide (Day et al., 1997). Among the three sites where the geotechnical characterization of flank failure was carried out, "Cumbre Vieja" is the only one where previous works (Ward & Day, 2001) suggested a catastrophic scenario due to the failure of its west flank by the displacement of hundred of cubic kilometres of rock mass into the sea). This is a possibility that should not be mistreated if the volcanic system reactivates and a similar reflection should be kept in mind for the "El Julan" flank, "Teide stratovolcano" and the landscape affected by the "San Andres" fault system whose level of risk has been minimized. As concluding remarks it may be stated the following.

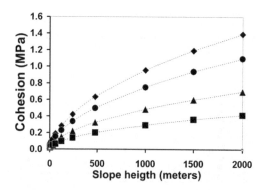

Figure 4. Variation of cohesion (MPa) vs. slope height (meters) for slope application on selected types of massifs for assumed GSI lower than 30. Rhomboids: Fresh massive lavas; triangles: Weathered massive lavas; circles: Ignimbrites; squares: Low cemented pyroclasts.

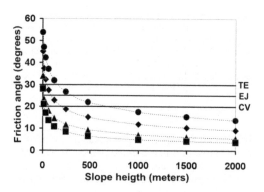

Figure 5. Variation of the friction angle (degrees) vs. slope height (meters) for slope application on selected types of massifs for assumed GSI lower than 30. Rhomboids: Fresh massive lavas; triangles: Weathered massive lavas; circles: Ignimbrites; squares: Low cemented pyroclasts. The horizontal solid lines intersect the vertical axis in the corresponding average value of slope angle on the flanks of "Teide (Te)", "El Julan (EJ)" and "Cumbre Vieja (CV)".

1. The combination of estimates for geomechanical parameters of intact rocks of the Canary Archipelago, the geological strength index (GSI) and textural features were used to infer geomechanical parameters of rock massifs.
2. Estimates for the Mohr Coulomb parameters, namely cohesion and internal friction angle, were calculated through the Hoek and Brown criterion.
4. Estimates for the friction angle of low cemented pyroclasts were low enough to cross down the average slope gradient at the three selected areas from a slope height of roughly 300 m. This will be reduded to less than 50 m in the case of more unfavorable geological conditions. That means that if low cemented pyroclasts were the only existing materials, the three selected areas would surpass the threshold value of slope height by which, slope gradient is higher than friction angle and then landslide processes will occur as significant drop of cohesion takes place.
5. As rock masses are constituted by a sequence of massive lava flows, scoriaceous and pyroclastic layers in uncertain proportions, it is expected that the equivalent Mohr-Coulomb parameters will correspond to intermediate values between the purely unweathered massive rocks and the purely pyroclastic massif. However, the relevance of the relative proportions between rock mass and pyroclastic, dramatically decreases when geological conditions deteriorate (as GSI decreases). Such intermediate cohesion and friction angle are doubtful but for significant slope gradients the risk of landslide dramatically increases. Concerning this fact, three areas of special relevance have been highlighted for their potential risk of catastrophic landslides in the Canary Archipelago: "Teide Stratovolcano" (Tenerife) with average slope gradient of 30°, "El Julan" (El Hierro) with average slope gradient of 25° and the "Cumbre Vieja Rift Zone" (La Palma) with average slope gradient of 20°. All of them with slope heights higher than 1000 m exhibit conditions of slope gradient that overcome the friction angle of their rock massifs and of low cohesion that became those areas as susceptible of suffering gigantic slides. Internal cohesion will drop by shallow magma injection, fluid injection or groundwater pressurization. Additional factors such as the pre-existing fractured zones will reinforce the risk of slides.
6. This last option was already discussed for the flank affected by the "San Andres" fault system where the risk of potential flank failure was probably minimized. Special emphasis must be made for the "Cumbre Vieja Rift Zone".

There, the question of a potential risk of slope failure was pointed out earlier (Ward & Day, 2001), especially if the occurrence of a significant fracture roughly parallel to the rift zone such as the 1949 fault break is combined with a shallow magma intrusion and a significant slope gradient.

REFERENCES

Apuani, T., Corazzato, C., Cancelli, A., Tibaldi, A., 2005. *Physical and mechanical properties of rock masses at Stromboli, a dataset for volcano instability evaluation.* Bull. Eng. Geol. Environ. 64 (4), 419–431.

Bieniawski, Z.T., 1989. *Engineering rock mass classifications.* John Wiley and Sons Inc. 272 p.

Cantagrel, J.M., Arnaud, N.O., Ancochea, E., Fuster, J.M., Huertas, M.J., 1999. *Repeated debris avalanches on Tenerife and genesis of Las Canadas caldera wall (Canary Islands).* Geology 27 (8), 739–742.

Carracedo, J.C., 1999. *Growth, structure, instability and collapse of Canarian volcanoes and comparisons with Hawaiian volcanoes.* J. Volcanol. Geotherm. Res. 94 (1–4), 1–19.

Davies, N., Chaplow, R., 1998. *Geotechnical characteristics of the Borrowdale Volcanic Group.* Proceedings of the Yorkshire Geological Society 52, 189–197 Part 2.

Day, S.J., Carracedo, J.C., Guillou, H., 1997. *Age and geometry of an aborted rift flank collapse, the San Andrés fault system, El Hierro, Canary Islands.* Geol. Mag., 134 (4), 523–537.

Day, S.J., Carracedo, J.C., Guillou, H., Gravestock, P., 1999. *Recent structural evolution of the Cumbre Vieja volcano, La Palma, Canary Islands, volcanic rift zone reconfiguration as a precursor to volcano flank instability.* J. Volcanol. Geotherm. Res. 94 (1–4), 135–167.

Del Potro, R., Murlimann, M., 2007. *Strength of volcanic rock masses in edifice instability: Insights from Teide, Tenerife.* Proceedings of the International Workshop on Volcanic Rocks W2. 11 ISRM Congress. Ponta Delgada (San Miguel, Azores). Session 1, Characterization of volcanic formations, 175–183.

Del Potro, R., Hurlimann, M., 2008. *Geotechnical classification and characterisation of materials for stability analyses of large volcanic slopes.* Eng. Geol. 98(1–2): 1–17. doi:10.1016/j.enggeo.2007.11.007.

Del Potro, R., Hurlimann, M., 2009. *The decrease in the shear strength of volcanic materials with argillic hydrothermal alteration, insights from the summit region of Teide stratovolcano, Tenerife.* Eng. Geol. 104(1–2): 135–143. doi:10.1016/j.enggeo.2008.09.005.

Del Potro, R., Pinkerton H., Hurlimann, M., 2009. *An analysis of the morphological, geological and structural features of Teide stratovolcano, Tenerife.* J. Volcanol. Geoth. Res. 181 (1–2): 89–105. doi:10.1016/j.jvolgeores.2008.12.013.

Elsworth, D., Day, S.J., 1999. *Flank collapse triggered by intrusion, the Canarian and Cape Verde Archipelagoes* J. Volcanol. Geotherm. Res. 94 (1–4), 323–340.

Elsworth, D., Voight, B., 1995. *Dike intrusion as a trigger for large earthquakes and the failure of volcano flanks.* J. Geophys. Res. 100 (B4), 6005–6024.

Gonzalez de Vallejo, L., Ferrer, M., 2006. *Caracterización geomecánica de los materiales volcánicos de Tenerife. Spanish IGME. Serie Medio Ambiente.* Riesgos Geológicos 8. 147 p.

Hoek, E., Wood, D., Shah, S., 1992. *A modified Hoek-Brown criterion for jointed rock masses.* In, Hudson, J.A., editor. Proceedings of the Rock Characterisation Symposium of ISRM, Eurorock 92. British Geotechnical Society, 209–214.

Hoek, E., Carranza-Torres, C.T., Corkum, B., 2002. *Hoek-Brown failure criterion–2002 edition.* In Proceedings of the Fifth North American Rock Mechanics Symposium, Toronto, Canada 1, 267–273.

Hürlimann, M., 1999. *Geotechnical analysis of large volcanic landslides. The La Orotava events on Tenerife, Canary Islands.* Ph.D. thesis. Technical Univ. of Catalonia, 220 p.

Hürlimann, M., Ledesma, A., Martí, J., 1999. *Conditions favouring catastrophic landslides on Tenerife (Canary Islands).* Terr. Nova 11 (2–3), 106.

Hürlimann, M., Ledesma, A., Martí, J., 2001. *Characterisation of a volcanic residual soil and its implications for large landslide phenomena, application to Tenerife, Canary Islands.* Eng. Geol. 59 (1–2), 115–132.

Keating, B.H., McGuire, W.J., 2000. *Island edifice failures and associated tsunami hazards.* P. Appl. Geophys. 157, 899–955.

Malheiro, A. 2006. Geological hazards in the Azores archipelago: Volcanic terrain instability and human vulnerability. J. Volcanol. Geoth. Res. 156 (1–2): 158–171. doi:10.1016/j.jvolgeores.2006.03.012.

Marinos, P., Hoek, E., 2000. *GSI—A geologically friendly tool for rock mass strength estimation.* Proc. GeoEng2000 Conference, Melbourne.

Marinos, V., Marinos, P., Hoek, E., 2005. *The geological strength index, applications and limitations.* Bull. Eng. Geol. Environ. 64 (1), 55–65.

Marquez A., Lopez I., Herrera R., Martin-Gonzalez F., Izquierdo T., Carreno F., 2008. *Spreading and potential instability of Teide volcano, Tenerife, Canary Islands.* Geophys. Res. Lett., 35, L05305, doi: 10.1029/2007GL032625.

Moon, V., Bradshaw, J., Smith, R., de Lange, W., 2005. *Geotechnical characterisation of stratocone crater wall sequences, White Island Volcano, New Zeal.* Eng. Geol. 81 (2), 146–178.

Munn, S., Walter, T.R., Klugel, A., 2006. *Gravitational spreading controls rift zones and flank instability on El Hierro, Canary Islands.* Geol. Mag. 143, 257–268.

Rodriguez-Losada, J.A., Hernandez-Gutierrez, L.E., Olalla, C., Perucho, A., Serrano, A., Rodrigo del Potro, 2007. *The volcanic rocks of the Canary Islands. Geotechnical properties.* Proceedings of the International Workshop on Volcanic Rocks W2. 11 ISRM Congress . Ponta Delgada (San Miguel, Azores). Session 1, Characterization of volcanic formations, 53–57.

Rodriguez-Losada, J.A., Hernandez-Gutierrez, L.E., Olalla, C., Perucho, A., Serrano, A., Eff-Darwich, A., 2009. *Geomechanical parameters of intact rocks and rock masses from the Canary Islands: Implications on their flank stability.* J. Volcanol. Geoth. Res. 182 (1–2):67–75. doi. 10.1016/j.jvolgeores.2009.01.032.

Serrano, A., Olalla, A., 1994. *Ultimate bearing capacity of rock masses.* Int. J. Rock Mech. Min. Sci. 31 (2), 93–106.

Serrano, A., Olalla, C., Perucho A., Hernández-Gutiérrez, L., 2007. *Strength and deformability of low density pyroclasts.* Proceedings of the International Workshop on Volcanic Rocks W2. 11 ISRM Congress. Ponta Delgada (San Miguel, Azores). Session 1, Characterization of volcanic formations, 35–43.

Thomas, M.E., Petford, N., Bromhead, E.N., 2004. *Volcanic rock-mass properties from Snowdonia and Tenerife, implications for volcano edifice strength.* J. Geol. Soc. 161, 939–946 Part 6.

Uriel, S., Serrano, A., 1973. *Geotechnical properties of two collapsible volcanic Soils of low bulk density at the site of two Dams in Canary Island (Spain).* 8th Congress of I.S.S.M.F.E. Moscú, vol. 1, 257–264.

Uriel, S., Serrano, A., 1976. *Propiedades geotécnicas de algunos aglomerados volcánicos en las islas Canarias.* Mem. Simp. Nac. de Rocas Blandas. Tomo 1. A-10. Madrid.

Walter, T., Schmincke, H.U., 2002. *Rifting, recurrent landsliding and Miocene structural reorganization on NW-Tenerife (Canary Islands).* Int. J. Earth Sci. 91 (4), 615–628.

Ward, S.N., Day, S.J., 2001. *Cumbre Vieja Volcano— Potential collapse and tsunami at La Palma, Canary Islands.* Geophys. Res. Lett., 28 (17), 3397–3400.

Volcanic Rock Mechanics – Olalla et al. (eds)
© *2010 Taylor & Francis Group, London, ISBN 978-0-415-58478-4*

Cliff stabilization solutions at the south coast of Madeira Island

V.C. Rodrigues, F.A. Sousa, S.P.P. Rosa & J.M. Brito
Cenorgeo, Engenharia Geotécnica, Lda., Lisboa, Portugal

ABSTRACT: The necessity to protect important infrastructures constructed, along the south coast of Madeira Island, at the top and at the bottom of high cliffs, against the collapse of isolated rock blocks or of significant masses of soil and rock debris, has lead to the development of complex geotechnical projects and to the execution of specific stabilization works. This paper describes, for some of those cliffs, the existing geological conditions and the main instabilization processes that affect them. Attending to the acquired experience from these works, some considerations are presented about the advantages and limitations of the stabilization solutions considered, especially in what concerns the required logistics for their execution.

1 GENERAL CHARACTERISTICS OF THE CLIFFS

Along Madeira Island south coast, between Arco da Calheta and Machico, there are several cliffs of significant height showing stability problems, threatening the safety of the infrastructure located at their top or bottom.

As a consequence, some of these infrastructures had their use or occupation conditioned or interdicted.

Over the last years, several cliffs have been identified as requiring stabilization works involving significant human and material resources.

Table 1 presents the characteristics of the most important cliffs that were already stabilized, namely its location, length, height and cost of the stabilization works.

2 GEOLOGICAL CHARACTERISTICS

The south coast of Madeira Island presents a very accentuated relief, formed by deep valleys and elongated mountains known as "lombos".

These elevations are, in most cases, cut by subvertical cliffs, usually over hundred metres height, originated in the Post-Miocene Volcanic Complex β^2 (Fig. 1).

This complex is constituted of alternating layers of basaltic lava flows and breccias. Interstratified in these formations also occur tuffs layers, with lenticular and irregular contours, generally of minor expression (Serviços Geológicos de Portugal, 1974).

The basaltic lava flows can be either compact (βC) or fractured (βF) exhibiting, mainly, subvertical contractional joints, showing continuity and

Table 1. Main characteristics of the most important cliffs stabilized along Madeira Island south coast.

Location	Description	Length (m)	Maximum height (m)	Cost (€)
Arco da Calheta	Cliff above the ER101 at Sítio do Massapez	150	50	2.0×10^6
Ponta do Sol	Cliff above the ER222 at Sítio da Vargem	40	40	0.7×10^6
Ponta do Sol	Upper section of the cliff above the Lugar de Baixo Seaport	200	60	7.2×10^6
Ribeira Brava	Cliff above the ER222 at Sítio do Calvário	50	20	0.3×10^6
Machico	Cliffs beneath the Saint João Baptista Fort	45 to 50	10 to 20	0.2×10^6

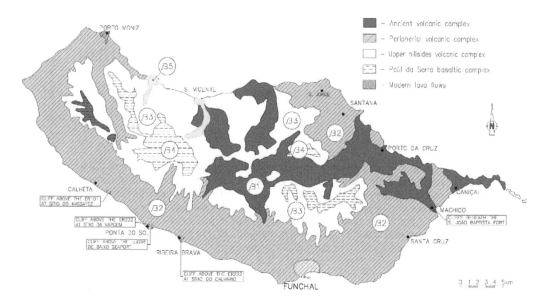

Figure 1.   Geological map of Madeira Island.

spacing from a few centimetres to a metre, that lead to the formation of polygonal columns.

In addition to this joint set, there is also a sub-horizontal one due, either to successive flows deposition, or to cooling conditions, or to superficial decompression. The combined action of both joint sets leads to block individualization.

The breccia (Br) and tuff (T) layers are less resistant to the atmospheric agents than the basalts, reason why they are generally significantly eroded. These formations may be found predominantly compact (BrC and T) or disaggregated (BrD and TD).

3   CLIFF GEOMORPHOLOGICAL EVOLUTION AND INSTABILITY CAUSES

The geomorphological evolution of the cliffs essentially depends on their stratigraphic succession, namely the sequence of very resistant, though often fractured rock layers, intercalated with friable layers.

In the volcanic complex, the cliffs' morphological evolution is due to the erosion of the breccia and tuff layers, gradually leaving the overlying basalt layers overhanging originating, on occasions, undercut cliffs up to 2 m wide.

This evolutionary process, allied to the fractures that affect the rocky layers, leads to the collapse, either of individualized blocks, or of rocky and terrigenous masses of significant volume.

This decompression process is often aggravated by superficial water flow that infiltrates the bedrock through the fractures in the basalts and through the breccias and sandy tuffs layers originating the formation of unconfined water levels delimited at the base by silty or clay tuff layers. The consequent gradual softening of these formations origins surfaces with lower mechanical resistance. Rock mass failure processes are, thereby, more frequent during stronger rainfall periods. Superficial weathering deposits (DC) occur on the top and on the terrace levels of the cliffs. Slope deposits or talus piles (DV) are due to the accumulation of rock and soil that falls from the cliffs' surface, at its base. Both these deposits reach, many times, significant thickness and are found in precarious stability conditions.

4   STABILIZATION SOLUTIONS ADOPTED

Considering that the definition of most appropriate solutions for cliff stabilization depends strongly, both on the occurring geological and geotechnical conditions and on the acquired experience, either on design and on monitoring similar works, it is critical to have a plain understanding of the evolutionary processes that affect the cliffs.

However, there are generally many obstacles to a full characterization of the cliffs, due, on one hand, to the lack of topographical data that accurately represents their geometry, and, on the other hand, to the impossibility of performing detailed geological surveys.

Table 2 summarises the adopted stabilization solutions used for the presented cliffs.

Table 2. Stabilization solutions adopted on the cliffs.

Stabilization solutions	Geology				
	Sítio do Massapez BrD/β	Sítio da Vargem βF + BrD + T	Lugar de Baixo βF + Br + T	Sítio do Calvário β + BrD	Saint João Batista Fort T + βF
Excavation	X	X	X		
Shotcrete associated to bolts	X	X	X		
Shotcrete				X	X
Locally applied bolts	X	X	X	X	
Joint sealing				X	X
Cave filling with cyclopean concrete					X

In general, when the cliffs are affected by a considerable rock mass volume instabilization, the stabilization works consist on the removal of the unstabilized mass and on the protection of the resulting slope with shotcrete and bolts.

In other cases, when superficial cliff instability situation are identified, lighter interventions are recommended, such as shotcrete, locally applied bolts, joint sealing and cave filling with cyclopean concrete.

After the execution of the designed solutions, it was also necessary to pay attention to landscape impact. Colour additives matching the surrounding geological formations were applied in the shotcrete and sealing grout.

5 STABILIZATION WORKS EXECUTED

5.1 Cliff above the ER101 at Sítio do Massapez

In February of 2006 a crack about 13 m long, 2.5 m deep and 1 to 2 m wide was identified at the top of the cliff at elevation 240.00 m.

After a detailed survey it was possible to confirm that this crack represented the top of an unstable mass of about 30 000 m³, with a sliding surface 60 m wide and 40 m high, located 12 m to the inside of the cliff.

This cliff is crossed by the ER101 road at elevation 104.00 m, below which Fajã village is located.

The cliffs comprehended essentially thick layers of disaggregated breccias and fractured basalts, overlaying a thick layer of volcanic tuffs. Slope deposits were identified on the terraces that existed along the cliff and at its base (Cenorgeo, 2006a).

The adopted solution was to excavate the unstable mass from elevation 240.00 m to 210.00 m, along 2V/1H slopes, with 3 m wide and 10 m vertically spaced berms.

To deal with the essentially breccious nature of the rock mass, a systematic shotcrete lining of the

a. General view of the slope after the stabilization works

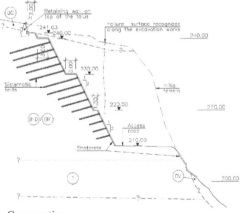

b. Cross-section

Figure 2. Cliff above the ER101 at Sítio do Massapez.

slope associated to bolts 4 to 12 m long was carried out. Wherever possible, in the basalt layers this lining was substituted by locally applied bolts (Fig. 2).

In this case, the aim was to assure that the final slope was located behind the unstable surface. That was accomplished by a continuous topographical survey during the excavation.

During the excavation, a strip about 2 m wide and 1 to 2 m high was maintained at the outside edge of the cliff, as a safety measure to prevent workers and equipments to fall to the bottom of the cliff during stabilization works at that level, being only removed in a secondary phase, when the works were completed.

5.2 Cliff above the ER222 at Sítio da Vargem

On June 18th 2008 an estimated volume of 350 to 400 m^3 of basaltic blocks fell off the cliff at Sítio da Vargem, onto a section of the ER222 road. This rockfall completely blocked the road and the access to an adjacent house.

The unstable mass, about 40 m in length and 40 m in height, was bounded at the top by a set of terraces used for agricultural purposes, supported by drystone packed walls.

The cliff was formed, at its base, by a layer of relatively compact and resistant tuffs, on top of which occurred disaggregated breccias. Basaltic flows overlayed these two layers, where several columns and very large blocks were identified as unstable. Disaggregated breccias also occurred on the inside of the cliff, behind the basalts. On top of this flow, there was yet another level of fairly compact breccias (Cenorgeo, 2008a).

The adopted solution was to progressively excavate the basalts, along a 3V/1H slope, with berms with a minimum width of 1.5 m and vertically spaced by 12 m.

The excavation solution appeared to be the most appropriate solution, based on the survey conducted on the top of the cliff (shortly after the rockfall) that allowed the identification of several important cracks, bounding a possibly failure surface.

A shotcrete lining associated to 6 to 12 m length bolts was used on the disaggregated breccias and on the fractured basalts.

The stabilization of the compact basalt blocks and columns was assured by locally applied bolts (Fig. 3).

It is noteworthy that immediately after the rockfall an access road to the top of the cliff was created, to allow the beginning of the stabilization works.

Attending to the urgency on resolving this situation, namely due to inconvenience of the road interdiction, the works began to be conducted with a daily monitoring by the design team, which was permanently adapting the design to the geological and topographical conditions encountered as the works were being executed.

It should also be noted that due to the lack of space, it was not possible to maintain the road

a. General view of the slope after the stabilization works.

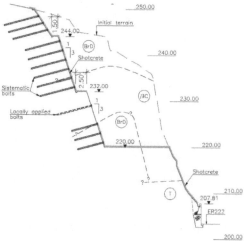

b. Cross-section

Figure 3.   Cliff above the ER222 at Sítio da Vargem.

access along the successive excavation levels, inhibiting the circulation of transport vehicles of the excavated materials for dumping.

Therefore, it was necessary to create a reception basin on the ER222, through the construction of a delimiting barrier in its exterior alignment (to prevent the material to overlap the basin), being the excavated material then carried for dumping using this road.

For the same reason the equipment that performed all the excavation, could only be removed after the completion of the stabilization works.

5.3 Upper section of cliff above Lugar de Baixo Seaport

The Lugar de Baixo Seaport was constructed at elevation 4.00 m, at the base of a cliff with about 200 m height. A parking building, a swimming pool complex and several service buildings were constructed at the base of the cliff (Fig. 4). The distance between these infrastructures and the cliff was between 3 and 10 m.

When the seaport was constructed, stabilization works consisting on bolted metallic wire meshes were executed from the base up to elevations 85.00 to 90.00 m, where it essentially presented vertical inclination and was constituted by basalt lava flows interbeded with breccious levels.

A flexible barrier was placed at a platform located at these elevations, with the purpose of retaining any material that might fall from the upper section of the cliff, which had a less aggravated inclination and consisted of thick layers of compact tuffs and of basaltic flows. Layers of intercalated volcanic breccias were also identified within the basaltic flows.

a. General view of the cliff after the stabilization works

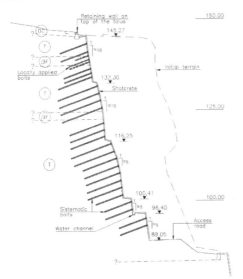

b. Cross-section of the upper section of the cliff

Figure 4.   Cliff above the Lugar de Baixo Seaport.

Over time it became clear that this barrier did not provide sufficient protection against rockfall from the upper section, which led to the interdiction of the parking building, the swimming pool complex and the service buildings (Cenorgeo, 2006b).

Hence, a stabilization solution was developed consisting on the excavation of the upper section, from the top of the cliff, at elevations 90.00 m to 150.00 m, to the platform located at elevations 85.00 to 90.00 m.

This work was carried out along a length of about 200 m, along slopes 5V/1H at the base to 10V/1H at the top.

Given the nature of the rock mass, consisting essentially of tuffs, a shotcrete lining was applied on the excavated slope along with the execution of 8 to 12 m long bolts.

It is noteworthy that the resulting slope was designed to guarantee a minimum working platform width of 6 m, to assure a safe circulation of the equipments.

5.4   Cliff above the ER222 at Sítio do Calvário

This cliff has a length of about 50 m and an approximate maximum height of 50 m.

It is limited, at the top, by a sub-horizontal platform at elevation 175.00 m, where a house was built, and, at the base, by the ER222 road, at elevation 130.00 m.

It is constituted, at the base, by a first slope about 15 m height, excavated with a sub-vertical inclination and it is formed essentially by relatively compact and resistant breccias that are, nevertheless, disaggregatable.

There is a narrow platform on the top of this slope, about 1 m width, which widens until reaching about 5 m.

Overlaying this platform there's another slope, about 20 m height, also sub-vertical, consisting of a thick and compact, but fractured, basaltic flow, where it is possible to observe several polygonal columns and large rock blocks that could fall in the short term and affect the road (Cenorgeo, 2008b).

The solutions adopted for the stabilization of this cliff consisted on the removal of all the unstable rock blocks, open joints sealing and the bolting of the unstable blocks that were not possible to remove.

The disaggregated breccias found in the lower slope, underlying the basaltic bedrock, were protected with a shotcrete lining (Fig. 5).

To carry out these works it was necessary to temporarily close the road, introducing a significant impact on the local population.

In order to cut the time of execution of the stabilization works to a minimum and to ensure that the time scales established at the outset were accomplished, a permanent monitoring was assured by the project designer, so that the predicted solutions could be quickly adjusted to real conditions.

It is noteworthy that, after the vegetation removal from the surface of the cliff, which was practically covered, it was possible to complete the work design, particularly the definition of the rockbolts location.

5.5 Cliffs below the Fort of Saint João Baptista in Machico

At the NE top of the Machico Bay coastline road, there is a headland at the top of which the Saint João Baptista Fort was built.

The headland is bordered by two cliffs, one to the south, with about 45 m width and 10 to 15 m height, and another to the east, with about 45 to 50 m length and 15 to 20 m height.

Both these cliffs have an approximately rectilinear configuration though, on the east side, there is also a small cove, possibly associated with an existing cave.

This cave has about 10 m width, a height between 1.5 and 4 m and a depth of about 6 to 8 m. Two other caves have also been identified in the southern cliff, although smaller in size and located above the sea level.

There appeared to be a certain predominance of breccious formations over basaltic formations. The basalts are located essentially at the base level,

a. General view of the cliff after the stabilization works

a. View of the southern cliff after the stabilization works

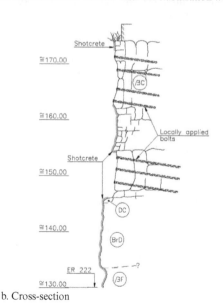

b. Cross-section

Figure 5.   Cliff above the ER222 at Sítio do Calvário.

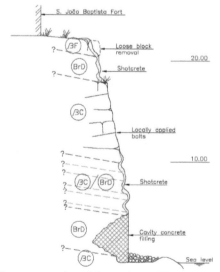

b. Transverse section of the southern cliff

Figure 6.   Cliffs underlying the Saint João Baptista Fort.

while the breccia are located above the basalts and show significant erosion signs, evidenced by the formation of the already mentioned caves (Cenorgeo, 2007).

The stabilization works carried out aimed to protect the rock mass against erosion at the Fort foundation level which could put its stability at risk.

The stabilization works of the basaltic layers involved unstable rock blocks removal and joint sealing.

The disaggregated breccias and tuffs were protected with shotcrete lining with fiber reinforcement.

In the west top of the southern cliff, above the pier, a very resistant basalt layer was identified overhanging at the top. This 5 m overhang of about 15 m in length was partially excavated (Fig. 6).

The caves were filled with cyclopean concrete with a final exterior shotcrete lining, providing the cliffs a homogeneous appearance (Fig. 7).

a. General view of the southern cliff after the stabilization works

b. General view of the eastern cliff after the stabilization works

Figure 7. Cliffs underlying the Saint João Baptista Fort.

It is noteworthy that the base of the Fort wall was protected with a shotcrete lining in a strip about 0.50 m high.

6 CONCLUSIONS

The construction of important infrastructures, such as roads, ports and buildings, at the edge of cliffs undergoing constant evolution has increased the need to carry out stabilization works to provide protection and enable continuous safe use of these infrastructures.

These types of interventions have led to the need to carry out increasingly detailed studies, meaning identifying the main problems at each location, with the aim of designing the stabilization solutions most appropriate to each case.

A recommendation arising from the experience acquired in designing and monitoring these types of works is that, besides the design phase, the execution phase of the works should be monitored by experts with geotechnical training, preferably those who developed the project phase of the studies.

Only this way it is possible to make the necessary adaptations in good time when faced with unexpected situations during the design, due, mainly, to the lack of direct access to the surface of the cliffs and the dense vegetation that often covers them.

This permanent monitoring translates into significant savings, particularly in the scheduling, the type and the development of the works to be done and in the optimisation and flexibility of the solutions implemented as a result of better understanding of the problems, allowing a maximum use of the resistance of the volcanic bedrock and the local conditions occurring, which contributes, in turn, to a reduction in costs and an increase in the reliability of the works.

ACKNOWLEDGEMENTS

The authors wish to extend their thanks to the Regional Directorate of Infrastructures and Equipment (DRIE) and to the Estradas da Madeira (RAMEDM) of the Regional Government of Madeira (GRM), for the permission provided for the production of this document.

REFERENCES

Cenorgeo. 2006a. *Consolidation of the slope above the ER101 at Sitio do Massapez*. Final design. GRM. DRIE (in Portuguese).

Cenorgeo. 2006b. *Stabilization of the cliff above the Lugar de Baixo Seaport*. Preliminary design and final design (in Portuguese).

Cenorgeo. 2007. *Stabilization of the cliff underlying the Saint João Baptista Fort in Machico*. Final design. GRM. DRIE. (in Portuguese).

Cenorgeo. 2008a. *Consolidation of the slope above the ER222 at Sítio da Vargem*. Final design. GRM. RAMEDM. (in Portuguese).

Cenorgeo. 2008b. *Stabilization of the slope above the ER222 at Sítio do Calvário*. GRM. RAMEDM. (in Portuguese).

Serviços Geológicos de Portugal. 1974. *Madeira Island Geological Map* (in Portuguese).

Volcanic Rock Mechanics – Olalla et al. (eds)

Shear behaviour of Stromboli volcaniclastic saturated materials and its influence on submarine landslides

P. Tommasi
Institute for Geo-Engineering and Environmental Geology, National Research Council, Rome, Italy

F. Wang
Research Centre on Landslides, Disaster Prevention Research Institute, Kyoto University, Kyoto, Japan

D. Boldini
Department of Civil, Environmental and Material Engineering, University of Bologna, Bologna, Italy

T. Rotonda & A. Amati
Department of Structural and Geotechnical Engineering, Sapienza University, Rome, Italy

ABSTRACT: On 30 December 2002 a submarine landslide generated high tsunami waves and destabilized the subaerial slope of the NW flank of the Stromboli volcanic island. The volcano flank is a large subaerial and subaqueous scar filled by loose volcaniclastic materials. Their susceptibility to undrained shear failure is investigated through stress- and displacement-controlled large-scale ring shear tests (LRST), conducted at DPRI-Kyoto University at different hydraulic boundary conditions. Results are presented in the form of stress paths and time-histories of shear resistance and pore pressures and are discussed with reference to the different teting conditions. Finally shear band formed in LRST are analyzed in terms of changes in porosity and grain size distribution in order to investigate the development of grain crushing at failure and at large displacement.

1 INTRODUCTION

In subaerial volcanic edifices, large flank instabilities occur essentially in the form of sector collapses (McGuire 1996), which in many cases are directly driven by volcanic activity through stress changes induced by magma intrusions or/and fluid pressurization (Voight & Elsworth 1997).

More subtle instability phenomena may occur on the submerged part of volcano flanks formed by loose volcaniclastic deposits. In fact the resistance of the saturated deposit can drops if it is subjected to loading or deformations as fast as to establish undrained conditions. This circumstance can occur if porosity and grain size changes accompany the shear process as a consequence of grain crushing.

High deformation/loading rates are believed to have been experienced by the submerged part of the NW flank of the Stromboli Volcano (Sicily, Italy) during the 2002–2003 eruption, when a rapid submarine failure, 6×10^6 m^3 in volume, produced tsunami waves that run up the inhabited coasts of the Stromboli Island (Chiocci et al. 2008). Failure was preceded by deformations of the subaerial and of the nearshore submerged slope, which sharply

accelerated a couple of hours before the submarine failure.

The NW flank of the Stromboli Island is a large sector collapse scar (more than 200 m deep), filled by the products of the continuous explosive activity of the volcano, which are incessantly re-mobilized by slides over the whole subaerial and submarine slope. This process forms a sequence of volcaniclastic layers with interspersed blocks and thin lava flows. Volcaniclastic layers are predominant and have large continuity over the submerged and subaerial slope thus controlling the mechanical behaviour of the whole deposit.

In order to individuate mechanisms governing a sudden submarine failure, investigations were focused on the behaviour of the volcaniclastic saturated material. The large initial dimensions of grains and the strong dependence of grain size distribution and void ratio from strain magnitude due to grain crushing, have addressed experimental activity towards tests that involve large specimens and encompass a wide range of strains/displacements.

In this respect a first series of large scale ring shear tests were performed with the scope of assessing the susceptibility of volcaniclastic material to static liquefaction (Boldini *et al.* 2009). Further tests, presented in this paper, were conducted at DPRI of Kyoto University using different drainage and shearing conditions, measuring void ratio and grain size changes. The aim is to link shear failure to the evolution of state parameters and soil texture within the shear zones, in the perspective of defining mode of slope evolution leading to failure.

2 THE MATERIAL

The material used in all tests comes from layers of sandy gravels (curve A in Fig. 1) sampled from horizons of the volcaniclastic sequence outcropping at the foot of the Sciara slope. To ensure a convenient ratio between the maximum grain size and the minimum specimen dimension, tests were performed on the fractions passing the 8 mm sieve. Therefore the in situ grain size distribution curve was shifted parallel to the grain size axis obtaining a reference soil (curve SG in Fig. 1) classified as sand and gravel, with uniformity coefficient of 9.6. In situ void ratio, calculated from the in-situ density of a volcaniclastic layer with a particle size distribution similar to that of the material used in laboratory tests, is 0.83.

Grains are mostly basaltic scoriae and scoria fragments with high roughness and low sphericity. Average porosity, created by widespread pores often tortuous, is 17.5%. Previous tests (Tommasi *et al.* 2005) indicate that grains exhibit relevant crushing occurring as collapse and rupture of surface asperities. This behaviour induced to evaluate the strength of the grains, which was investigated through standard tests on cylindrical specimens cored from blocks

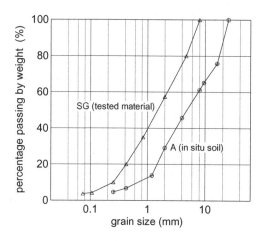

Figure 1. Particle-size distribution curves of the in situ volcaniclastic soil (curve A) and tested material (SG).

of scoriaceous lava (autoclastic breccia) having the same porosity of the grains (Rotonda *et al.* 2010). This material, taken at the lava flow fronts, represents a major source for the productions of volcaniclastic grains. The uniaxial compressive strength and the tensile strength from Brazil tests, determined on cylindrical specimens 42 mm in diameter, are characterised by a mean value of 34.5 MPa and 4.01 MPa respectively which account for crushing at moderate stresses.

3 PREVIOUS TESTS

Results of a first set of tests performed using the large-scale ring-shear apparatus DPRI-6, developed at the DPRI, Kyoto University (Sassa 1997) are described in detail by Boldini *et al.* (2009). The DPRI-6 device is characterised by an inner diameter of 250 mm, an outer diameter of 350 mm and a maximum sample height of 150 mm.

Tests can be conducted in both drained and undrained conditions, thanks to a water-leakage tightness system formed by O-rings on the upper loading platen, and bonding rubber edges on the two confining rings of the lower rotary pair (Sassa *et al.* 2004). Pore pressure is measured by two transducers located 2 mm above the shear surface.

All the specimens were prepared using the dry pluviation method with the aim of obtaining a very loose initial configuration, similar to that produced by the deposition process at Sciara del Fuoco. In all tests, saturation was obtained by exposing the sample to a CO_2 flux followed by a de-aired water flux for a time (generally more than 12 hours) sufficient to obtain a value of Skempton's pore pressure parameter B higher than 0.95.

Torque-controlled tests were performed in undrained conditions and with free drainage at the sample top. During the latter procedure excess pore pressure can generate depending on material behaviour and loading rate.

In both cases, samples were consolidated at a normal stress of 230 kPa and a shear stress of 122 kPa. These values, calculated assuming that the material is normally consolidated (Boldini et al 2009), correspond to the lithostatic stresses along the reconstructed slip surface of the submarine slide, located at the average depth and dip of 37 m and 28°, respectively. At the end of consolidation, the dry unit weight ranged between 16.6 and 17.2 kN/m^3. The shearing stage was carried out by applying a shear torque with a loading rate of 55 Pa/s.

In undrained tests, the volcaniclastic material experienced liquefaction, evidenced by an abrupt increase in pore-water pressure up to values higher than 200 kPa and a corresponding drop of the shear resistance down to a few kilopascals. The apparent friction angle at the final stage (given by the arctangent of the ratio between the mobilised shear resistance and the total normal stress) was about 3°. Evident crushing was observed in all specimens under the applied testing conditions.

In the second type of tests conducted with drainages open at the specimen top, shear resistance initially increased gradually up to a maximum value, corresponding to a peak friction angle of about 44°. The linear increase in shear indicates that shearing is likely to occur under drained conditions. After peak, the shear resistance temporarily dropped to a value corresponding to an apparent instantaneous friction angle of 18° and thereafter increased up to a stable value, yielding a friction angle at large displacements of roughly 34°. The drop in shear strength (below the dynamic residual friction angle) is likely to be a consequence of local excess pore-water pressures within the shear zone, which can temporarily build-up due to the high crushability of the material.

4 NEW TESTS

The new campaign of LSRST tests carried out at the DPRI consists of 6 tests on fully saturated specimens subjected to different shearing and drainage conditions. Only four tests reached the incipient liquefaction stage or large displacement conditions stage, which are of interest for the analysis of instability mechanisms. Specimens were prepared and consolidated according to the same procedure described by Boldini et al. (2009) and reported in the previous section. Conditions adopted in the different tests are reported in Table 1 together with values of the void ratio determined at different stages of each test.

4.1 Stress controlled test

Test #2 was carried out in stress controlled conditions. Test #2 was performed by applying a shear torque with a loading rate of 200 Pa/s, four times

higher than the analogous test by Boldini et al. (2009). In the same hydraulic conditions and at higher loading rate, an incipient liquefaction was again observed.

Figure 2 reports time-histories of total normal stress, shear resistance, pore pressure and shear displacement, δ_s, after the consolidation phase. The test was stopped at the onset of liquefaction in order to measure void ratio. Before this stage, as expected, pore pressure initially increases and then gradually decreases down to around −30 kPa. Correspondingly a decrease in the effective normal stress followed by an inversion, due to material dilation, can be observed (Figs. 2–3). When shear displacement reaches 15 mm, in correspondence with the minimum pore pressure, the sign of pore pressure increments inverts even though it cannot be appreciated in Figure 2 due to the plot scales. Imminent liquefaction can be argued from the drop of shear resistance.

The instantaneous friction angle during the dilative phase (i.e. when pore pressures are negative) was found to be 43° (Fig. 3), higher than those (39°) obtained in the past testing campaign.

The void ratio was accurately measured in the shear zone. Its thickness, h_s, was estimated to be 26 mm. This value was calculated adopting the ratio between h_s and the specimen height, h, observed in other ring shear tests on crushable materials (Coop et al. 2004). The specimen was ideally subdivided into three layers (upper layer, shear zone, lower layer) which were separately recovered. The material of each layer was weighted after oven-drying at 105°C for 25 h and void ratio, e, was calculated as

$$e = \frac{V - V_s}{V_s} \tag{1}$$

The solid volume V_s derives from the dry weight and the solid matrix density, whilst V was calculated from h, h_s and from the diameter of the inner and outer rings. In the shear zone a void ratio equal to 0.39, greatly lower than that obtained on the whole, was measured (Table 1).

Table 1. Test conditions and void ratio measured at different stages of the test.

Test	Control type	Rate	Drainages	e_{0i}	e_c	e_f	Notes
#2	stress	200 Pa/s	closed	0.79	0.69	0.69 (0.39)	e_f at $\delta_s = 0.015$ m
#4	displacement	5 mm/s	open	0.79	0.69	0.57 (0.36)	e_f at $\delta_s = 20$ m
#5	displacement	100 mm/s	open	0.79	0.71	0.52 (0.36)	e_f at $\delta_s = 1.6$ m
#6	displacement	5 mm/s	open	0.83	0.72	0.59 (0.25)	e_f at $\delta_s = 20$ m

* the bracketed value refers to the shear zone.

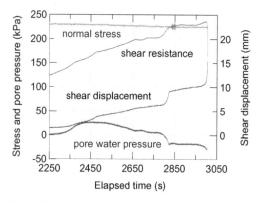

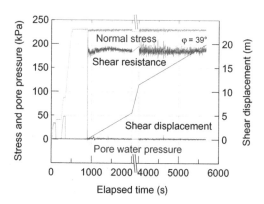

Figure 2. Time-histories of total normal stress, shear resistance, pore pressure and shear displacements recorded in test #2 (undrained conditions).

Figure 4. Time-histories of normal stress, shear resistance, pore pressure and shear displacement recorded in test #4. Note that the x-axis is interrupted at 2000 seconds.

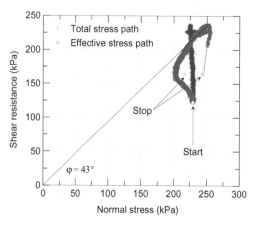

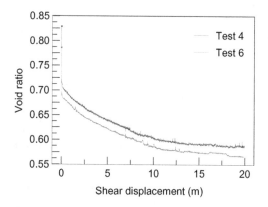

Figure 3. Effective and total stress paths in test #2 (undrained conditions).

Figure 5. Void ratio decrement during tests #4 and #6.

4.2 Displacement controlled tests

Three additional tests were performed controlling the shear displacement rate with open drainages. In open-drainage tests, measured pore pressures reflect only the actual values that establish near the transducers (i.e. in the upper ring). As a consequence excess pore pressure rising far from the transducers could not be detected.

In tests #4 and #6 shear displacement was applied at a constant rate of 5 mm/s until a total displacement of 20 m was reached. A displacement rate of 100 mm/s was instead applied in test #5, which was stopped at 1.6 m of total displacement (Table 1).

Figure 4 shows the time-histories of normal stress, shear resistance, pore pressure and shear displacement during test #4. Soon after the application of shear displacement increment, the mobilised

shear resistance attained a constant level ranging between 180 and 190 kPa, corresponding to an angle of shear resistance equal to about 39°. The absence of evident drops of shear resistance indicates that drained conditions maintained during the test. Similar results were obtained in test #6, carried out at the same shear displacement rate.

Values of void ratio versus shear displacement for tests #4 and #6, including the initial consolidation stage, are reported in Figure 5. The two curves are parallel and only differ for the initial value of void ratio, e_0, equal to 0.79 and 0.83 for specimen #4 and #6, respectively. After consolidation, e gradually decreased until it tended to an asymptotic value indicating that steady-state conditions were reached. At the end of the test ($\delta_s = 20$ m) e_f was equal to 0.57 and 0.59 for test #4 and #6, respectively (Table 1).

After the removal of the upper part of the ring shear apparatus, a thorough observation of the tested material revealed the presence of an intensively crushed inner shear zone about 20 mm thick and an outer shear zone characterised by a less severe crushing (Fig. 6). The total thickness of the shear zone was estimated to be about 30 mm.

Since a significant grain crushing was expected during such a prolonged shearing, an attempt to measure the void ratio directly within the shear zone was performed after the test #4. Miniaturised samplers, 30 mm in height and about 20 mm in diameter, were driven into the shear zone (Fig. 7) and immediately weighted. Successively, they were oven-dried at 105°C in order to determine their water content. The inner shear zone and the outer shear zones can be clearly individuated on the extruded oven-dried sample (Fig. 7).

The void ratio was calculated adopting the following relationship

$$e = \frac{V_w}{V_s} \qquad (2)$$

where V_w and V_s are the water and solid volumes, respectively. Eq. (1) is not recommended in this case because the specimen is irregular and its volume is considerably lower than that of the sampler. A value of 0.36 was determined for the whole shear zone; it could be slightly overestimated because some swelling (about 4 mm) occurred when the apparatus was opened at the end of the test.

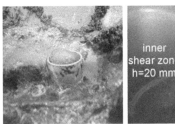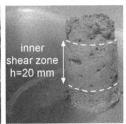

Figure 7. Miniaturised sampler driven into the shear zone after test #4 (left) and material extruded from the sampler after oven-drying (right).

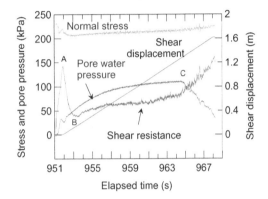

Figure 8. Time-histories of normal stress, shear resistance, pore pressure and shear displacement recorded in test #5.

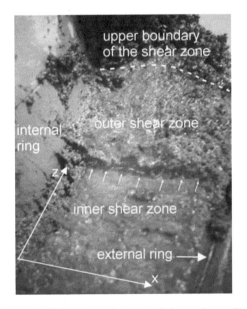

Figure 6. Sub-horizontal section of the specimen after test #4. X- and Z-axes are directed diametrally (outward) and tangentially. Small arrows indicate the step marking the passage between the inner and outer shear zone.

On the specimen of test #6 the void ratio was determined using the procedure described in section 4.1 (miniaturised samplers were not used); the obtained value of 0.25 (Table 1) is different from that found in test #4 and it is deemed to be unlikely low.

In the last test (test #5) a much higher shear displacement rate was adopted, in order to investigate the influence of movement rate on liquefaction triggering. Figure 8 shows the time-histories of normal stress, shear resistance, pore water pressure and shear displacement, while the effective and total stress paths are plotted in Figure 9.

After an initial increment up to 141 kPa (point A), shear resistance rapidly drops to 38 kPa (point B). This is due to the contemporary build-up of pore pressure, which increases until point C, where it attains the maximum value of 110 kPa. The mobilised friction angle, calculated from a regression of the data comprised between C and the end of the test, is equal to 39° (Fig. 9), the same obtained in test #4.

This response can be interpreted as a phenomenon of limited liquefaction, related to the local

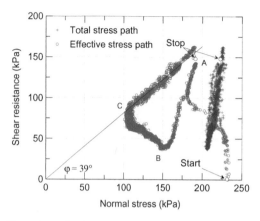

Figure 9.   Effective and total stress paths during test #5.

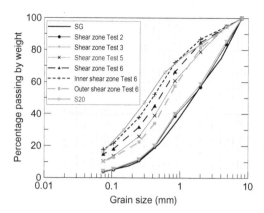

Figure 11.   Grain size distributions determined on the original material (SG) and on the material forming the shear zones of specimens from different tests.

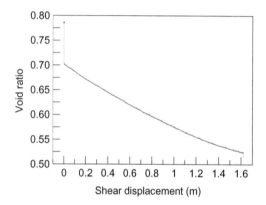

Figure 10.   Void ratio decrease recorded during test #5.

build-up of pore water pressure as a consequence of intense grain crushing in the shear zone. Evidence of this is the significant decrease in void ratio recorded during the test (Fig. 10). The final void ratio determined for the shear zone is equal to 0.36 (Table 1), calculated through the procedure reported in section 4.1 disregarding swelling of the specimen (about 2 mm), following the opening of the apparatus.

4.3   Grain size distribution after shearing

The grain size distribution of the shear zone material was determined for different tests (Fig. 11). The grain size distribution of the inner and outer shear zone of test #6 specimen is also provided.

Figure 11 shows that the material from the shear zone of tests #2 was subjected to a very limited crushing and the corresponding curves almost match that of the original material SG.

Conversely, the material forming the inner shear zone of test #6, subjected to a shear displacement

of 20 m, experienced severe crushing. Its grain size distribution is very similar to that determined by Boldini et al. (2009), denoted as S20, in stress-controlled ring shear tests carried out to values of shear displacement larger than 30 m.

Finally, the grain size distribution of the shear zone in test #5, despite the limited shear displacements reached at the end of the experiment, indicates that a significant crushing occurred, which helps in explaining the limited liquefaction phenomenon observed soon after the increase in shear displacement.

5   CONCLUSIVE REMARKS

The new LRST testing programme confirmed previous results and gave new insight into the texture modification experienced by the volcaniclastic material during shearing.

The high susceptibility of the Stromboli volcaniclastic soil to static liquefaction, observed in previous and new tests conducted under stress-controlled undrained conditions, was confirmed by displacement-controlled tests performed with open drainages. Even though a threshold for the insurgence of static liquefaction was neither established, nor could be easily extended to the real slope, a specimen tested at high displacement rate (100 mm/s) temporarily experienced shear failure, whilst specimens sheared at 5 mm/s did not exhibit any significant drop in shear resistance. This behaviour could be assimilated to the limited liquefaction described by Castro (1969) in undrained triaxial tests. Alternatively, since drainages are open, the strength recovery could be simply due to the dissipation of the excess pore pressure.

Given that all tests yield similar values of the void ratio after consolidation (Tab. 1), the void ratio of the shear zone measured at incipient liquefaction in the test #2 results highly reduced ($e_f = 0.39$). Such a decrease indicates that even though the shear displacement is low, crushing occurs and grains are strongly rearranged within the shear zone. This consideration is supported by the low reduction of void ratio (e passed from 0.69 to 0.65) which was instead experienced by a further test (test #3) stopped fairly before liquefaction (not described in the previous sections). Liquefaction at higher displacements (test #5) does not induce a further reduction of void ratio.

The shift of grain size curve of tests #2 (Figure 11) indicates that grain size modifications are appreciable. An identical curve was obtained in the stress-controlled test #3, carried out with open drainages. The grain size reduction is quite higher in the specimen of test #5, which attains liquefaction (limited or local) at larger shear displacements.

The larger influence that liquefaction conditions have on the void ratio induces two considerations: i) at relatively low displacements grain crushing seems to manifest itself as ruptures at grain surfaces, which improve particle packing; ii) void ratio well before liquefaction experiences moderate changes due to shearing.

The measurements of void ratio during large-displacement tests conducted with open drainages suggest that a critical void ratio (at steady state conditions) can be individuated and that it depends on the initial void ratio. This indication could be useful for establishing, in the case of crushable soils, the ultimate resistance of the material. In this respect the increment of total displacement from 20 to 30 m (curves "inner shear zone Test 6" and "S20" in Fig. 11) produces minor changes in the grain size distribution within the inner part of the shear zone.

ACKNOWLEDGEMENTS

Authors thank E. Tempesta, P. Millozzi and L. Passeri of CNR-IGAG and Eng. G. Caldarini for sample preparation.

REFERENCES

Boldini, D., Wang, F., Sassa, K. & Tommasi, P. 2009. Application of large-scale ring shear tests to the analysis of tsunamigenic landslides at the Stromboli volcano, Italy. *Landslides* 6: 231–240.

Castro, G. 1969. Liquefaction of sands. *Harvard Soil Mechanics Series 87*, Harvard University, Cambridge, Massachusetts.

Chiocci, F.L., Romagnoli, C., Tommasi, P. & Bosman, A. 2008. The Stromboli 2002 tsunamigenic submarine slide: Characteristics and possible failure mechanisms. *Journal of Geophysical Research* 113(B10102): 11 p.

McGuire, W.J. 1996. Volcano instability: a review of contemporary themes. In McGuire, W.J., Jones, A.P. & Neuberg, J. (eds), *Volcano Instability on the Earth and Other Planets, Geological Society Special Publication No. 110*: 1–23.

Rotonda, T., Tommasi, P. & Boldini, D. 2010. Geomechanical characterization of the volcaniclastic material involved in the 2002 landslides at Stromboli. *Journal of Geotechnical and Geoenvironmental Engineering* 136(2): 389–401.

Sassa, K. 1997. A new intelligent type dynamic loading ring-shear apparatus. *Landslide News* 10: 1–33.

Sassa, K., Fukuoka, H., Wang, G. & Ishikawa, H. 2004b. Undrained dynamic-loading ring-shear apparatus and its application to landslide dynamics. *Landslides* 1: 9–17.

Tommasi, P., Boldini, D. & Rotonda, T. 2005. Preliminary characterization of the volcaniclastic material involved in the 2002 landslides at Stromboli. In Bilsel & Nalbantoglu (eds.), *Problematic Soils GEO-PROB 2005; Proc. int. conf., Famagust*, 3: 1093–1101, Famagusta: Eastern Mediterranean University Press.

Voight, B. & Elsworth, D. 1997. Failure of volcano slopes. *Geotechnique* 47: 13–16.

3   *Geoengineering and infrastructures in volcanic environments*

Volcanic Rock Mechanics – Olalla et al. (eds)

The suitability of volcanic tuff from the Ethiopian plateau for earth dam construction and foundation

E.E. Alonso & E. Romero

Department of Geotechnical Engineering and Geosciences, UPC, Barcelona, Spain

ABSTRACT: The intended site for the dam is located in a wide valley of the Gilgel Abbay River with an overall elevation difference between the upper plateau and the river level of about 85 m. The river has eroded a series of volcanic rock units of early Tertiary origin, with alternative levels of basaltic lava flows and ash or tuff deposits. The lava flows result in hard to medium rocks, jointed and fractured. The two volcanic units mentioned (lava flows and tuffs) are approximately laid in a horizontal manner. Tuffs and ash deposits, which have a very similar appearance, are white in colour and they are clearly identified when exposed. They constitute the substratum of a dominant proportion of the dam foundation. They tend to produce gentle slopes and, in the lower cultivated plots they are covered by alluvial clays. The exposed tuff is not cultivated. The exposed tuff is eroded by running waters and it shows erosion patterns similar to other soft clayey rocks. The resulting erosion forms tend to be rounded. However no firm evidence of piping was observed. The paper reports laboratory experiments on specimens of the intact tuff material. The suitability of the tuff, once compacted, as a core material for a zoned-earth dam was also investigated.

1 INTRODUCTION

Current projects to develop the Lake Tana District in Ethiopia involve the construction of dams for water supply and irrigation purposes. This paper refers to the studies carried out during the design of an earth and rockfill dam in the area (Figure 1). The project is located in the Central volcanic plateau of the country. Mantles of Tertiary and Pleistocene volcanic rocks and soils have been eroded by the river network, which eventually discharge into the lake Tana.

The valleys were excavated in a sequence of basaltic and rhyolitic rocks and volcanic ash and tuffs (Figure 2). This paper describes the geotechnical investigations performed to characterize the volcanic tuffs in view of its suitability as foundation soils for dams. In addition, the possibility of using tuffs as a compacted material in the dam core or shoulders was also investigated.

The main aspects addressed in this paper refer to the strength, permeability and dispersion susceptibility of these materials. Intact material properties were investigated in rock cores recovered during the site investigation for the intended dam. Compacted materials were manufactured by crushing cores and passing #4 ASTM sieve, mixing material them with water and compacting samples in the usual manner (repeated tamping or static compaction; Standard Proctor was the reference compaction energy). Most of the tests reported here were performed at the Geotechnical Laboratory of the Civil Engineering School of Barcelona.

The two volcanic units mentioned (lava flows and tuffs) are approximately laid in a horizontal manner. Tuffs (this terms will be used also for the ash deposits which, have a very similar appearance) are white in colour and they are clearly identified when exposed (Fig. 2).

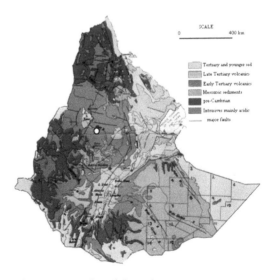

Figure 1. Location of the project.

They constitute the substratum of a dominant proportion of the dam foundation. They tend to produce gentle slopes and, in the lower cultivated plots they are covered by alluvial clays. The exposed tuff is not cultivated probably due to the high soil acidity.

When exposed in cuts the tuff appears as a soft massive rock. Unconfined compression strength, determined in cores lie in the range 1–10 MPa. In general the SPT sampler does not penetrate in the tuff. Exposed contacts between tuff and basaltic flows are apparently tight. No open cracks or discontinuities have been observed. Recovered cores provide also information on the nature of contacts. A review of the cores in transitions indicated also a tight transition between tuff and basalt (Fig. 3).

Tuff is exposed in some locations by intermittent creeks which erode the surface mantle

Figure 4.   Erosion patterns of tuff by running water.

Figure 5.   Remoulded tuff.

Figure 2.   Exposed volcanic tuff covered by a thin mantle of residual soils. In the background a cultivated valley (alluvial soils).

Figure 3.   Tight transition between tuff and basalt.

of colluvial soils. The exposed tuff is eroded by running waters and it shows erosion patterns similar to other soft clayey rocks (Fig. 4). The resulting erosion forms tend to be rounded. However no firm evidence of piping was observed.

The tuff has a low unit weight but it is very heterogeneous. Unit weights in the range 18.3–15.5 kN/m^3 have been reported. Porosity is difficult to determine because no reliable indication of the tuff minerals is available and, therefore, the solid unit weight is not available. When wetted it displays a soft, clay like, slippery touch, a consequence of the present minerals, described later (Fig. 5).

2   INTACT TUFF ROCK TESTS ON CORES

2.1   *Background*

Tests were performed on core pieces recovered in borings. Cores were kept in dry conditions, "in situ", exposed to the atmosphere and later transported to the Laboratory. Some of the determinations were performed also under "dry" conditions at the RH prevailing at the laboratory in Barcelona

Table 1. Initial (natural) conditions of intact core samples.

Sample	Water content %	Water content at RH = 50%	Natural density Mg/m³	Dry density Mg/m³	Void ratio	Degree of saturation %
NB-4	7.3 to 9.4	11.7	1.31–1.40	1.22–1.28	0.88–0.98	18–26
NB-7	12.3	12.3	1.47	1.31	0.84	35

(around 50%). However, some of the fundamental tests, specifically the shear strength of the material "in situ" were determined under saturated or close-to-saturated conditions. In general the objective of tests was to evaluate the suitability of the tuff as a foundation material. Therefore, in addition to strength, there was an interest in determining the tuff permeability and its susceptibility to dispersion through the pin-hole test.

2.2 Basic identification

The X-ray diffraction plot was difficult to interpret because the powdered sample was fundamentally a mixture of amorphous (i.e. non-crystalline) phases. The expansive clay mineral montmorillonite was also identified. In addition, plagioclase (calcium/sodium feldspar) was distinguished as a minority phase. The soil investigated displays other crystalline phases, which are difficult to identify.

The initial conditions of two intact core samples tested are summarized in Table 1. The density of solids was measured on sample NB-4, giving a value of $\rho_s = 2.41$ Mg/m³. The same value was used on NB-7 to estimate void ratio and degree of saturation. The relative humidity (RH) prevailing at laboratory conditions is around 50%. Since core samples were not properly isolated to prevent water content changes, the water content was affected by this condition (especially on sample NB-4).

Core samples were crushed for particle size study and afterwards remoulded for consistency limit determination. Table 2 summarizes the main results. Figure 6 shows the consistency limits plotted in the plasticity chart (inorganic silt MH). Figure 7 presents the particle size distribution curves obtained by sieve analysis.

A mercury intrusion porosimetry (MIP) test provided information on the multiple porosity network of the intact sample NB-7. This information is useful to improve the understanding of the hydro-mechanical behaviour of the intact material and to compare the pore network with that obtained using a compaction procedure. MIP tests attained maximum intrusion pressures of 220 MPa (entrance pore size of 6 nm according to Washburn equation). A freeze drying process was used to dry the samples and to preserve the pore network before the injection of mercury.

Table 2. Consistency limits and content of fines (<75 μm).

Sample	Liquid limit	Plastic limit	Plasticity index	<75 μm
NB-4	56%	45%	11%	53%
NB-7	76%	45%	31%	57%

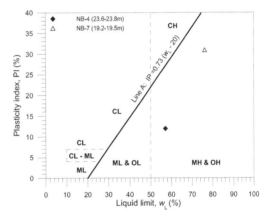

Figure 6. Consistency limits plotted in the plasticity chart.

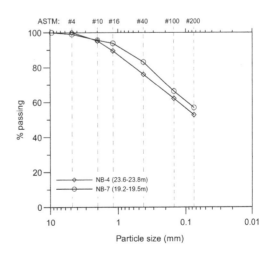

Figure 7. Particle size distribution.

209

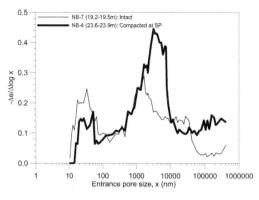

Figure 8. Pore-size distribution curve of compacted (Standard Effort, NB-4) and intact (NB-7) samples.

Figure 8 shows the pore-size distribution curve in terms of the log differential intruded void ratio as a function of the entrance pore size, which aids in the visual detection of the dominant pore modes. As observed, the sample displayed a double-porosity network with two dominant modes, one formed by large macro-pores at 1–2 μm and a small micropores range at 20–40 nm.

2.3 Strength

Tests were carried out in a direct shear cell with a box 60 mm in diameter and 25 mm high. The following steps were performed: a) loading at constant (initial) water content; b) soaking at approximately 1 min (this saturation stage was maintained for 24 hours); and c) shearing during 1 day at constant horizontal displacement rate of approximately 5 μm/min up to a maximum of 8 to 9 mm. The shearing rate was low to ensure drained conditions. The vertical stresses applied were 52, 104 and 208 kPa. As observed in Figure 8, the natural sample displayed some swelling on soaking at low stresses.

The evolution of shear stresses at different vertical effective stresses during direct shear tests is presented in Figure 9. Figure 10 plots the shear strength envelopes for different conditions (peak, just after peak and ultimate conditions), along with the shear strength parameters.

Figure 11 indicates that the tuff has a marked brittle behaviour. Peak shear strength parameters are high ($\phi' = 44°$; $c' = 176$ kPa). Post-peak behaviour is characterized by ($\phi' = 37°$; $c' = 32$ kPa), which is judged suitable for foundation conditions. The residual shear strength, measured on remoulded samples by means of a ring shear apparatus provides $\phi'_{res} = 29°$. This value is high and it is not explained by the high plasticity measured in samples. However, the correlations plasticity-residual friction established for sedimentary/regular

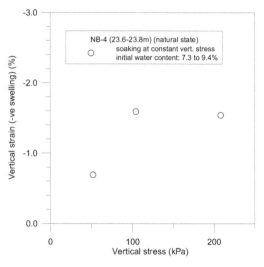

Figure 9. Evolution of vertical strain (swelling) during soaking at different vertical stresses. Intact core sample NB-4.

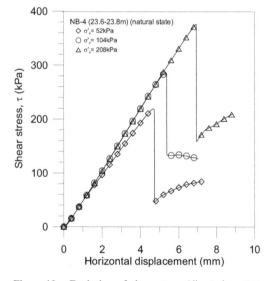

Figure 10. Evolution of shear stress (direct shear test on intact core sample NB-4).

clay soils are of doubtful validity here. The fact is that the direct determination of the residual friction provided a high friction angle in the specimen tested. The amorphous mineral content may explain this result.

2.4 Hydraulic related properties

A water permeability test was run on intact NB-7 sample using controlled-gradient conditions in an

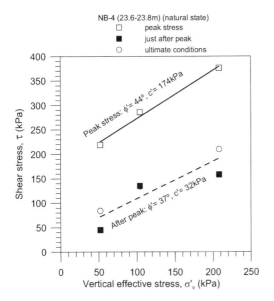

Figure 11. Shear strength envelopes for different conditions (peak, just after peak and ultimate conditions). Intact core sample NB-4. Shear strength parameters.

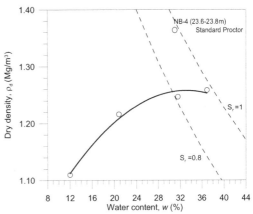

Figure 12. Standard effort compaction curve (NB-4).

oedometer cell under loaded conditions (50 mm diameter and 20 mm high specimen). The vertical total stress applied was 75 kPa. A water pressure of 30 kPa was applied at the bottom of the sample by a GDS Instruments advanced pressure/volume controller with a pressure resolution of 1 kPa and a volume resolution of 0.5 mm^3/step. Pressure at the top of the sample is maintained under atmospheric conditions. The saturated water permeability is determined under steady-state conditions.

Saturated permeability is low (3.5×10^{-10} m/s) (in just one determination). This is a value representative of "matrix" conditions. In the absence of fractures/joints tuff permeability is low. This low value is significantly lower than some field determinations in boreholes.

The pin-hole test performed identified a non-dispersive material (ND-1).

2.5 Compaction

Core samples were crushed and dynamically compacted in one lift using Standard effort: 600 kJ/m³. Maximum dry density is 1.26 Mg/m³ (void ratio of 0.920) at water content between 32 and 37% and degree of saturation between 84 and 97% (refer to Figure 12).

The pore-size distribution curve of the compacted material NB-4 obtained by mercury intrusion porosimetry is also presented in Figure 8 together with the intact material for comparison. The compaction process led also to a double-porosity

network formed by large inter-aggregate pores between aggregations or grains of approximately 3–8 μm, and smaller intra-aggregate pores inside aggregates or grains between 20–50 nm. Although the compaction at Standard effort allowed easily reaching dry densities similar to the natural condition, it is important to remark that this density was achieved with dominant macro-pore sizes and macro-porosity larger than those found for the natural condition (refer to Fig. 8). Based on this fact, a slightly larger water permeability and compressibility on loading and soaking (collapsibility) of the compacted material compared to the intact one is expected.

2.6 Wetting under load behaviour and strength tests

These tests were carried out in a direct shear cell. The same steps were applied as for the intact specimens: a) loading at constant (initial) water content; b) soaking at approximately 1 min (this saturation stage was maintained for 24 hours); and c) shearing during 1 day at constant horizontal displacement rate of approximately 5 μm/min up to a maximum of 8 to 9 mm. The vertical stresses applied were 52, 104, 208 and 364 kPa. During soaking at constant vertical stress, the compacted sample underwent some compression (collapse), which increased with vertical stress, as shown in Figure 13. The larger compressibility on loading and soaking (collapse) is consistent with the larger macro-pore sizes and macro-porosity found by mercury intrusion porosimetry on the compacted material.

The evolution of shear stresses and vertical displacements at different vertical effective stresses during direct shear tests is presented in Figure 14. The specimen is now essentially ductile. Dilatancy is recorded for the low stress range (<100 kPa of

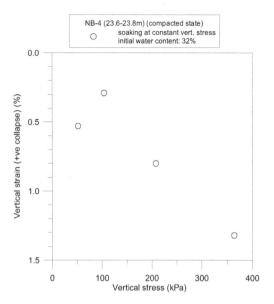

Figure 13. Evolution of vertical strain (collapse) during soaking at different vertical stresses. Compacted sample NB-4.

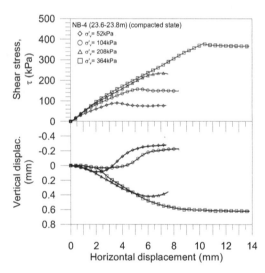

Figure 14. Evolution of shear stresses and vertical displacements during direct shear test on compacted sample NB-4.

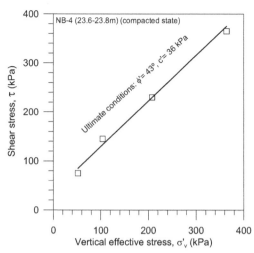

Figure 15. Shear strength envelope for ultimate conditions. Compacted sample NB-4. Shear strength parameters.

Figure 16. Compacted tuff sample. Saturated sample after pinhole test. The 1 mm hole did not erode.

compression stress). Figure 15 shows the shear strength envelopes for ultimate conditions, and the shear strength parameters.

Compacted tuff exhibits a high drained friction angle (43°) and a ductile behaviour. However, when remoulded and tested for residual conditions the friction reduced to 29°. The question of the appropriate strength parameters to be used in practice is dependent on the amount of "damage" or remoulding induced by excavation and compaction procedure.

2.7 Hydraulic-related behaviour

A water permeability test was also run on compacted sample NB-4 using controlled-gradient conditions in an oedometer cell under loaded conditions (50 mm diameter and 20 mm high specimen). The vertical total stress applied was 75 kPa. Water pressure of 30 kPa was applied at the bottom of the sample. Pressure at the top of the sample was maintained under atmospheric conditions. A saturated permeability of $k = 3.5 \times 10^{-10}$ m/s under steady state conditions was measured.

Table 3. Ethiopian tuff. Properties of saturated cores and compacted specimens (Standard Proctor optimum).

Type of sample	Structure	Wetting under stress of 100 kPa	Peak strength	Post peak strength	Residual strength	Saturated permeability	Dispersivity. Pinhole
Saturated core (soft rock)	Double porosity (Fig. 8)	Expansion $\varepsilon_v = -1.6\%$ (from RH = 50%)	$c' = 176$ kPa $\phi' = 44°$	$c' = 32$ kPa $\phi' = 37°$	$c' = 0$ $\phi' = 29°$	3.5×10^{-10} m/s	Non-dispersive ND-1
Compacted to optimum standard proctor	Double porosity (Fig. 8) (more marked)	Collapse $\varepsilon_v = 0.4\%$ (from $S_r = 0.80$)	$c' = 36$ kPa $\phi' = 43°$	$c' = 36$ kPa $\phi' = 43°$	$c' = 0$ $\phi' = 29°$	3.5×10^{-10} m/s	Non-dispersive ND-1

This is a low value, which makes the compacted tuff a suitable material to build impervious barriers.

In the pinhole dispersion test (colloidal erodibility) performed on the compacted NB-4 sample water run through a 1 mm hole drilled in the compacted sample for a maximum of a10 minute period at increments of water head of 85, 285, 590 and 1560 mm. No effluent turbidity was detected and the hole did not erode, indicating an erosion resistant material (non-dispersive ND1) according to the pinhole test (Figure. 16).

3 DISCUSSION AND CONCLUSIONS

The "in situ" tuff is a hard soil of high porosity (void ratio approaching 1). A main mineral component identified is montmorillonite. The remoulded soil is identified as a high plasticity clay (CH) (liquid limits of 55–75%). This implies a low permeability, which is a positive feature. Possible foundation preferential paths for the water include the contacts between different volcanic units and the presence of discontinuities in rhyolitic and basalt levels. However the tuff matrix is an impervious mass (saturated permeability of 3.5×10^{-10} m/s). Its stiffness is high and it should not pose settlement problems. The presence of montmorillonite could imply, a priori, reduced friction angles, especially under conditions leading to remoulding or weathering of the rock. However, the tests performed reveal that the tuff is a frictional material. Especially surprising is the high residual friction (29°) which is probably a consequence of the amorphous mineral content and its non-platy nature.

The intact material displays a marked brittle behaviour. After peak states it maintains a significant friction, however. The natural tuff exhibits a significant heterogeneity. Other tests, not discussed here, show that strength parameters depend markedly on the tuff density. Tuff density seems to vary strongly from point to point although the effects of sampling (coring) disturbance cannot be minimized. However, the presence of montmorillonite in the fine fraction of the tuff forces to be cautious.

The tuff appears to be non-dispersive in the pinhole tests performed. Some volcanic tuffs around the world are known to be dispersive and erodible, but in general these phenomena are present when the plasticity of the material is low.

Specimens of crushed tuff and compacted at Standard Proctor energy reached dry densities similar to the "in situ" intact soft rock. It is then interesting to compare the properties of both materials. This is done in Table 3.

There are two main differences when the table is examined. The first one is the high peak strength of the intact material and its brittle behaviour. This behaviour is interpreted as an effect of the natural structure and cementation. Interestingly, immediately after the peak strength is reached the intact specimen adopts strength values similar to the compacted specimen at the same dry density. The second difference concerns the volumetric behaviour when wetted under equivalent stresses. The intact material tends to expand (but measured strains are moderate) whereas the compacted specimen collapses. Collapse increases with confining stress in this case (up to a vertical stress of 400 kPa), a typical behaviour of compacted clayey soils. The interesting observation is that the smectite content is not able to counteract the collapse behaviour, which is explained by the significant macroporosity of the compacted specimen.

The remaining properties examined here are similar for the two types of material.

It appears that the compacted tuff may be considered for dam construction as a preliminary conclusion. Its properties as foundation material seem to be also adequate. A word of caution is necessary because of the heterogeneity of these formations and the presence of smectite.

Volcanic Rock Mechanics – Olalla et al. (eds)
© 2010 Taylor & Francis Group, London, ISBN 978-0-415-58478-4

Geotechnical map and foundation solutions of Santa Cruz de Tenerife (Spain)

S. Álvarez Camacho & F. Lamas Fernández
Department of Civil Engineering, University of Granada, ETSICCP, Granada, Spain

L.E. Hernández
Regional Ministry of Public Works of the Government of the Canary Island, Spain

ABSTRACT: The town of Santa Cruz de Tenerife, NE Tenerife, Canary Islands, Spain, spreads on young volcanic materials which show a great heterogeneity. Very often this means a hard problem to overcome for building projects. This paper describes the geotechnical features of volcanic formations of Santa Cruz de Tenerife on data from building geotechnical studies, geotechnical maps, and rock geotechnical properties database of the Regional Ministry of Works of the Government of the Canary Islands. An inventory of different types of building foundations has been made in order to establish a correlation between volcanic terrains and foundation solutions. As a conclusion, a geotechnical zoning map of Santa Cruz de Tenerife is proposed as preliminary foundation solutions in building projects.

1 INTRODUCTION

The Canary archipelago is situated between 27° 37' and 29° 25' north latitude and 13° 20' and 18° 10' west latitude and consists of different geological units that forms seven volcanic buildings. (Ancochea et al., 1990).

By routine geotechnical tests similar to those made by companies operating officially in public works at the Spanish state's level, including: size, PLT, Brazilian test, Normal and Modified Proctor, direct cutting, etc... for soils and for rocks try to set their mechanical behavior. (UNE, 1999); (NLT, 1991).

This chapter studies the materials of the urban area of Santa Cruz de Tenerife in the geotechnical aspect, correlating with their mineralogical characteristics. With the results obtained are characterized and classified the materials described above, mapped in different areas of mechanical properties across the environment investigated. This allows formulating different models of ideal foundation for each urban area providing, for finally, a geotechnical mapping of possible foundations.

2 METHODOLOGY

To establish our findings we have made some cards that collect significant data on the types of lands in the municipality to subsequently correlate with foundations that run, frequently, on the island.

As a basis for classification we use the model proposed by R. Potro and M. Hürlimann (2009). Geotechnical studies were conducted in each of the examined areas, being made a total of 47 samples that we have characterized both geotechnical and geomechanical soil and rock.

The values presented below are from the collection of data obtained from the various

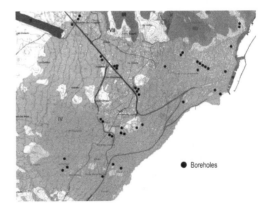

Figure 1. Location of boreholes.

terrains studied (Hausen 1956). The origin of the information is very diverse. Because of the different purpose of each borehole do not have all the data that characterize a soil, but the most significant was required to know.

The treatment we have given to the information is to gather data with similar characteristics, eliminating and establishing a mean to achieve generality.

The classification of terrains begins with an abbreviation that includes the different characteristics. The initial classification will be an R or S depending on whether volcanic rock or volcanic soil.

Then, and only for volcanic rocks, they continue with other letter according to R. Del Potro and Marcel Hürliman (2009), correspond to different volcanic terrains. L is for lavas, WPR for pyroclastics, IPR to slag and AB for gaps (not considered in our analysis). For the case of soils we create a set of letter that differentiate be AR for sandy soil and LIM for loamy soil. The third and final sets of acronyms denote the composition of rock samples. BOPM and BOPV indicate a massive olivine-pyroxene basalt or vacuolar olivine-pyroxene basalt. BES refers to scoriaceous basalt with certain degree of welding, unlike EBSS that indicates poorly welded slag. BAFM and BAFV show aphanitic basalt massive or vacuolar and TRQB reveals trashy existence.

While the writing above is present in the earth's surface as lava flows, pyroclastics also have other abbreviations like IGN refers no welded ignimbrites and LPS for basaltic pyroclastics.

As soils are also EBSS or PSC, they are slag or fall pyroclastics and, connecting with the sandy-silty nature of the samples was attached TUFF or BAS, initials as derived from basalt or tuff (see tables 1 and 2).

Parallel has developed a compendium of common foundation in this area of the island and was restrained to be predominant shoes and foundation slabs to bridge the heterogeneity of the terrain. However, the foundation type chosen depends on the type of building constructed, making it, sometimes, requires the execution of deep foundations (Justo et al., 2006).

In case that requires, micropiles are drilled, as the constructive approach of the piles cannot be carried out in volcanic terrain due to the hardness of it. A peculiarity of the area is to build on basaltic pyroclastics by making a hole to fill later with selected and compacted material (Gonzalez de Vallejo, 2002; Serrano, 1996).

It is also necessary to comment that the town's commercial activity concentrated in the port has implied the need to earn ground to the sea through the implementation of landfill, on which buildings

Table 1. Characteristics of rocks.

	ρ (Tn/m³)	σ·ρ*	σc (MPa)	σ·σc	σt(MPa)	σ·σt	Ultra-sound waves (m/s)	σ·Ultra-sound waves	c (KPa)	σ·c	φ	σ·φ	RMR	E (MPa)	σ·E
RLBOPM	2.76	0.10	101.3	57.03	38.84	17.36	4808.6	712.6	222	0.43	30.68	3.81	65.6	48720	23
RLBOPV	2.18	0.27	43.50	54.95	31.33	11.10	4731.5	859.7	250	0	27.5	0	49.2	28390	18.1
RLBES	2.42	0.05	21.15	46.84	31	0	2459.8	0	250	0	27.5	0	46.4	22940	19
RLEBSS	0	0	2.0	0	0	0	0	0	175	0	35.0	0	48.0	12600	0
RWPRIGN	1.15	0	4.8	0	2	0	0	0	150	0	20.0	0	27.1	6540	0
RLBAFV	2.56	0	43.95	0	0	0	5459.7	0	0	0	0	0	0	49600	0
RLBAFM	2.62	0.34	62.82	23.79	26.04	0	3976.3	774.2	0	0	0	0	0	26960	12.4
RLTRQB	2.57	0	48.36	46.75	46.75	0	0	0	0	0	0	0	0	49790	0

* σ corresponds to the standard deviation.

Table 2. Characteristics of soils.

	c (KPa)	σ·c	φ	σ·φ	ρ(Tn/m³)	E (KPa)	σ·E	Nspt	σ· Nspt
SLEBSS	5	0.05	34	1	1.85	30400	107	15	8.49
SARTUFF	0	0	32.5	4.95	1.73	35000	155	20.5	19.1
SLIM	40	0.14	25.5	10.6	1.70	30000	0	47.5	3.53
SARBAS	0	0	36	0	1.90	45000	0	0	0
SIPRPSC	0	0	32	0	1.00	32000	0	18	0
SIPRLPS	140	0	35	0	1.30	30000	0	11	0

have also been constructed. The most noteworthy, Tenerife Auditorium by S. Calatrava has requires large diameter piles and length.

3 STUDY AREA

The study focuses on the island of Tenerife, specifically in the provincial capital where there are 221, 956 inhabitants, INE (2009).

However our study will focus on the urban area of the municipality, where the population density is much higher than in rural areas. The city covers a total of 22.06% of the inhabitants of the province, of which 48.14% are men and 52.86% women. Of all of them a 60.17% are part of the workforce leaving the rest as pensioners, students and unemployed (the latter percentage is high due to the economic crisis we are experiencing) INE (2009).

Although the construction sector has demanded large numbers of staff in recent years, most of the island's workers are engaged in work under the services sector because the island's economic engine is tourism. The abandonment of the traditionally productive work such as agriculture and livestock, even industry, gives more importance to the construction. This, in addition to generating wealth, should give the public efficient infrastructure to meet the needs of citizens.

The island of Tenerife, with an area of 2058 km², is the largest of the Canary Islands, which occupies a central position. It rises about 8 km above the abyssal plain ocean and reaches a height above sea level of 3,718 meters, Teide Mountain, (Ridley, 1971).

The growth of the island has taken place by the accumulation of volcanic materials through a process that lasted several tens of millions of years and has continued to the present. The ascent and emission of magma has focused on bands, known as structural axes or ridges, which converge in the center of the island at angles of 120°. The erosive

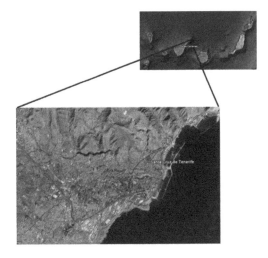

Figure 2. Situation of the study area.

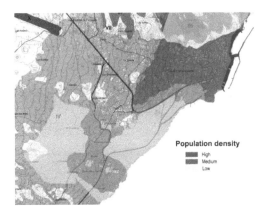

Figure 3. Population density.

unconformities created allow to group volcanic materials in units called Series, which correspond roughly whit large volcanic polygenic edifices (Ridley, 1971).

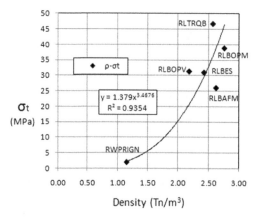

Figure 4. Density and tensile strength in rocks.

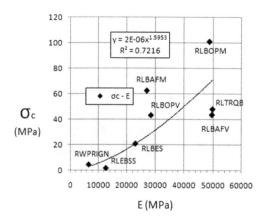

Figure 5. Compressive strength and deformability in rocks.

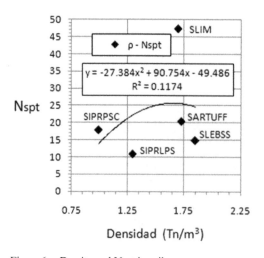

Figure 6. Density and Nspt in soils.

Our study area belongs to the Anaga's building, in the NE corner of the island. This building, along with Teno and other outcrops distributed over the island, are related to the oldest subaerial materials (Series I) covered today by most recent releases (Series III). IGME (1995).

4 EXPERIMENTAL DATA

4.1 Variability of some properties of rocks

4.1.1 Density—tensile strength
As shown in the graph, the different types of rocks have very different values on both properties. This variability can be compared to the potential function shown in Figure 4. The values closest to the function, coincide about the aspect more disintegrated of materials, however, the rest of values come from more compact materials.

4.1.2 Compressive strength—deformability
As in the previous graph, the values best adapted to the function are RWPRIGN and RLBES. The rocks are better suited to potential functions although, in this case, the values are further than in the previous function. It is possible to establish a limit of deformability module from which we can differentiate the terrains of granular and massive appearance.

4.2 Variability of some properties of soils

4.2.1 Density—Nspt
The graph correlates two values that determine the compactness of a soil. Unlike rocks, soils are more similar to a polynomial function. In this case tends to a second order polynomial function but if we increase the degree of the function will move away of the variability.

4.2.2 Deformation—angle of friction

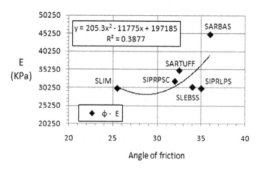

Figure 7. Deformation and angle of friction in soils.

5 RESULTS AND DISCUSSION

As explained in the methodology, the results obtained from the above tests were as follows.

In Tables 1 and 2 shown above, are collected the data more representative of the existing land in Santa Cruz de Tenerife. On these data we have relied to establish the most suitable type of foundation for the terrain and the type of construction. The values listed are extremely variable, which is characteristic of volcanic soils.

Despite having made a separation of terrain according to their lithotypes they cannot be considered individually. These soils and rocks are embedded in the land and their response to the solicitations is a mixture of both.

For this reason the municipality is divided into several zones, which are distinguished by their morphology, generated by eruptive processes that have created this sector of the island. Each morphology is associated with a type of rock or soil characteristic, shown in Table 3. This makes it easier to classify which enshrine different types of foundations.

Figure 8 shows the zoning of the urban area based on their morphology. Zone 1 covers Anaga's Residential, founded on ancient lava (Series I) and Zone 2 includes Toscal district, which as its name suggests is situated tuff or on non-welded ignimbrites.

Anthropic fillers are groups as Zone 3 and, like the Zone 5 (formed by basaltic pyroclastics), have a singular behavior that hinders the construction of large buildings. Most of the city of Santa Cruz is set up on "aa" and "pahoehoe" lavas and there are represented as Zone 4 and Zone 6.

"Aa" flows are characterized by being exposed to an alternation of lava and slag, which creates a problem to choose the most appropriate type of

Figure 8. Zoning of Santa Cruz de Tenerife.

foundation for each building. "Pahoehoe" lavas have a peculiarity of having large and small cavities, which should be evaluated by the geotechnical study to give them the appropriate treatment.

The basaltic Pyroclastics are located forming independent units like Taco's Mountain. Buildings in these areas are low-rise and require treatment of the terrain. The same applies to the buildings on landfills in the area of Puerto de Tenerife, in Zone 3 (Alcibiades Serrano, 1996).

As there are many constraints on each of the sectors, has prepared a chart that summarizes and clarifies the use of the foundations (see Table 4).

6 CONCLUSIONS

The terrain that forms the Canary Islands is very diverse, and presents very particular problems. The rocks forming the basement have suitable properties for construction of foundations because their resistance values are high; nevertheless, the problem is that it no appear alone in the environment.

The rocks outcrop surrounded by other less resistant materials (slag), and after several cycles of erosion, are also involved in volcanic soils of different origin. This plurality generates a decline in property values and requires an analysis to quantify the consequences.

The study seeks to encompass the solutions over time, based on the investigation and experimentation developed until now.

This aims to carry out a guide to facilitate and expedite future constructions and it can be extrapolated to other parts of the world with similar terrain to those found in this province.

Table 3. Correspondence between areas and terrains.

	Soils	Rocks
ZONE 1	SLEBSS	RLBOPM
ZONE 2	SARTUFF	RWPRIGN
		RLBOPM
ZONE 3	_(1)	_(1)
ZONE 4	SLEBSS	RLBOPV
		RLBAFV
		RLBES
ZONE 5	SIPRLPS	RLTRQB
ZONE 6	SLEBSS	RLBOPM
	SARBAS	RLBAFM
	SLIM	RLEBSS
	SIPRPRS	RLBES
	SARTUFF	

*(1) Stuffed anthropic.

Table 4. Foundation according to the area and the type of work.

	C-0	C-1	C-2	C-3	C-4
ZONE 1	15% Restrained shoes	20% Restrained shoes	65% Restrained shoes	–	–
ZONE 2	15% Restrained shoes	25% Restrained shoes	58% Restrained shoes	2% Foundation slab	–
ZONE 3	5% Foundation slab	70% Foundation slab	15% Piles	–	–
ZONE 4	5% Foundation Footing on Concrete of fills	10%	24% Foundation Slab Concrete of fills	41% Foundation slab on concrete of fills	20% Foundation slab or micropiles filling calvities
ZONE 5	30% Foundation slab above compacted fill material selected	55% Foundation slab above compacted fill material selected	25% Foundation slab above compacted fill material selected	– 25%	– 8% Micropiles
ZONE 6	16% Restrained shoes	17% Restrained shoes	34% Restrained shoes		

Therefore, this work is intended as a starting point, without losing the geotechnical studies for each case.

ACKNOWLEDGMENTS

This work would not have been possible without the information provided by the Regional Ministry of Works of the Canary Islands Government, and without the data given by the company Estudios del Terreno S.L.

REFERENCES

Ancochea, E., Fuster, J.M., Ibarrola, E., Cendrero, A., Coello, J., Hernan, F., Cantagrel, J.M., Jamond, C. 1990. "Volcanic evolution of the island of Tenerife (Canary Island) in the light of new K-Ar data". J. Volcanologist Geothermal Res 1990; 44:231–49.

González de Vallejo L. et al., 2002, "Ingeniería Geológica". Ed. Pearson educación 2002.

Hausen H., 1956, "Contribution to the geology of Tenerife". Soc. Sci. Fenmica, Com. Page 18 of 1247.

IGME, 1995, "Mapa geológico Las Galletas (Santa Cruz de Tenerife)", escala 1:50.000, hoja 1107, Plan MAGNA 50 (serie 2°), servicio de publicaciones.

INE, 2009, "Indicadores demográficos básicos; serie 1975–2008", Public Services of National Institute of Statistic of Spain.

Justo J.L., Justo E., Durand P., Azañon J.M. 2006, "The foundation of a 40-storey tower in jointed basalt", International Journal of Rock Mechanics & Mining Sciences 43, pages 267–281.

NLT-250/91. 1991. "Determinación de la resistencia a compresión simple de Probetas de roca". Normas de Ensayo del Laboratorio del Transporte y Mecánica del Suelo José Luis Escario. Cedex, Madrid, 1991.

Ridley, W.I., 1971. "The field relations of the Cañadas volcanoes, Tenerife, Canary Islands". Bull. Volcanol. 35, 318–335.

Rodrigo del Potro and Marcel Hürlimann, 2009. "The decrease in the shear strength of volcanic materials with argillic hydrothermal alteration, insights from the summit region of Teide stratovolcano, Tenerife". Engineering Geology, Vol. 104, pages 135–143.

Serrano A. 1996. "Algunos problemas geotécnicos en las construcciones en terrenos volcánicos". Ponencia, Sesión V, Jornadas Nacionales sobre Suelos y materiales pétreos en carreteras. Febrero 1996.

Serrano González A. & Olalla Marañón C. 1996. "Propiedades geotécnicas de materiales canarios y problemas de cimentaciones y estabilidad de laderas en obras viarias", Ponencia 16, Sesión V, Jornadas Nacionales sobre Suelos y materiales pétreos en carreteras. Unas dificultades crecientes. Febrero 1996.

UNE 22950-3:1990. 1999. "Propiedades mecánicas de las rocas. Ensayos para la determinación de la resistencia. Parte 3: Determinación del módulo de elasticidad (Young) y del coeficiente de Poisson". Geotecnia. Ensayos de Campo y de Laboratorio. AENOR, Madrid.

Volcanic Rock Mechanics – Olalla et al. (eds)
© *2010 Taylor & Francis Group, London, ISBN 978-0-415-58478-4*

Design and construction of the Machico-Caniçal expressway tunnels

C.J.O. Baião, J.M. Brito, A.R.J. Freitas, S.P.P. Rosa & M.F.M. Conceição
Cenorgeo—Engenharia Geotécnica Lda., Lisboa, Portugal

ABSTRACT: The Machico-Caniçal Expressway, 8 km in length, is part of the Madeira Island's new development program, which connects Funchal city to Caniçal village via a continuous expressway along the island's southern coast. This expressway crosses an extremely mountainous region affected by the existence of two very heterogeneous geological volcanic complexes, generally covered by unstable slope deposits or by thick alluvial deposits, leading to the construction of a wide range of civil engineering works. This paper presents the main aspects related with the design and construction of six double tunnels along the referred expressway, as well as a number of safety aspects related to these tunnels.

1 INTRODUCTION OVERALL CONSIDERATIONS

The geotechnical works associated with the new roadways of the Madeira Island, in particular the tunnels, are substantially affected by both the extensive heterogeneity, the structural and lithological complexity of the volcanic formations and the rugged terrain. In addition to this, a dense occupation of the land is often found along the defined road profiles (Brito et al. 2005). The combination of these factors led to the conception of a varied range of structural solutions with specific characteristics, at times with a degree of originality in order to achieve economic solutions adapted to the topographic and geological conditions, based on the latest technologies and most evolved construction processes.

Major technological development and knowledge attained within the domain of engineering have enabled the constricting results of nature's modelling to be overcome. We would point out the use of powerful means of earth removal with impressive production rates, the sprayed concrete and soil nailing applied as temporary tunnel linings, the laying of large amounts of concrete as definitive tunnel linings using large-scale metallic formworks and the use of the jet grouting technique at affordable costs, complemented with an improved knowledge of volcanic formations resulting from geotechnical investigation and an accumulated experience. It is worth pointing out the extreme importance of having, as soon as the conception phase of the route itself, the most in-depth knowledge possible of geomorphologic and geological-geotechnical conditions, in order to

make the best use of the rock-mass resistance and the existing local conditions, hence avoiding early on (or minimising the effects), and as far as possible, the more unfavourable zones, at times substantially conditioned by the existence of slope and alluvial deposits, and seams associated with the presence of water.

Our focus will be Machico's region, where the expressway closely follows the right bank of Machico's stream mid-slope until turning eastwards where it crosses this stream. On this initial stretch, with an extension of 2050 m, are inserted the Piquinho and Fazenda double tunnels, as well as the Queimada II tunnel (Fig. 1). The Machico and Natal valleys on the Caniçal side are reached via the double tunnel of Caniçal, a tunnel with 2100 m which crosses the mountain bedrock with more than 300 m covering. On the final stretch on the Caniçal side, with approximately 2145 m,

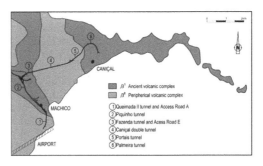

Figure 1. Volcanic complexes of Machico-Caniçal Expressway.

Table 1. Tunnel characteristics.

Tunnel designation	Width (m)	Section (m²)	Length (m)	Max. cover (m)	Geological conditions*
Queimada II	9.0	75	745	95	VCβ²
Access Road A	7.0	57	202	70	VCβ²
Piquinho (double)	9.0	75	2 × 449	35	VCβ¹ & VCβ²
Fazenda (double)	9.0	75	2 × 163	30	VCβ²
Access Road E	7.0	57	185	40	VCβ²
Caniçal Tunnel	9.0	75	2 × 2100	>300	VCβ¹ & VCβ²
Portais (double)	9.0	75	2 × 682	100	VCβ²
Palmeira (double)	9.6	83	2 × 915	60	VCβ¹ & VCβ²

* VC – Volcanic complexes.

Table 2. Relationship between the length of tunnels and the Machico-Caniçal Expressway (2 × 2 carriageways).

Total length of expressway (m)	7 770
Total number of tunnels (including double tunnels and access roads in tunnels)	14
Total development of tunnels (including double tunnels and access roads in tunnels) (m)	10 493
Total development of tunnel layout (including double tunnels and access roads in tunnels) (m)	4 991
Relationship between the development of tunnel layout and total development of expressway (%)	64

are the double tunnels of Portais and Palmeira (Table 1) (Cenor 1991a,b 1994; Cenorgeo 2002; Cenor-Grid 2003a,b).

Hence, a very significant part of the short stretch of expressway between Machico and Caniçal corresponds to tunnels. In fact, as can be seen in Table 2, 64% of the route corresponds to underground construction.

The current transverse cross-section of the single direction tunnels has a working width between walls of 9 m, comprising two 3.5 m road lanes and clearance of 5 m. The section is comprised of a semi-circular arch in the vault, with walls providing vertical extension (Fig. 2). Figure 3 depicts the Western portal of Fazenda Tunnel and Acess Road E.

2 GEOLOGICAL-GEOTECHNICAL SURVEY

The occurring geological-geotechnical conditions are described in specific detail in Rodrigues et al. (2009). Nevertheless, we would point out that the geotechnical surveying carried out was based on detailed geological surveys of the surface together with surface prospecting using trenches, and in-depth prospecting comprised basically of rotation boring. In order to survey the Machico-Caniçal Expressway 146 bores were taken to survey around 8 km of the route, comprising a total of 3071 m of drilling, which corresponds to 384 m of drilling for each kilometre of the expressway.

Of all the drilling made around 40% corresponds to drilling to survey tunnels whilst the remaining 60% is associated with the route layout itself, retaining works and bridges. Of the prospecting completed for the tunnels, this corresponds to an average of around 233 m per tunnel (144 m per km). Bearing in mind the costs of prospecting, each kilometre corresponds to on average €100,000 in prospecting work and €40,000 for each tunnel. These costs correspond to around 0.5 to 1% of the price of the construction work.

3 CHARACTERISATION & DESIGN

For volcanic formations, being characterised by substantial heterogeneity and lithological and structural complexity, the use geomechanical

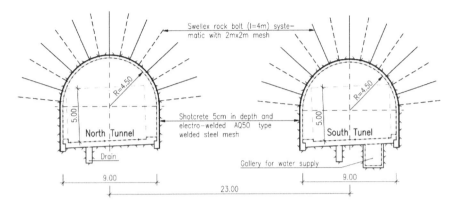

Figure 2. Current cross-section of tunnels with indication of primary lining within geotechnical zone ZG2.

Figure 3. Overall view of Western portal of Fazenda Tunnel and Access Road E.

classifications in tunnel design must be pondered with considerable reservations, since the estimation of the required parameters for these materials, especially for breccia and pyroclastic tuffs, presents a great number of difficulties.

On the other hand, the parameters used in these classifications are seldom those which best reflect the behaviour of the rock mass and, thus, appropriate to characterize some specific volcanic formations. It is also worth emphasising that the situations which may occur are so varied that these can hardly be entirely covered and described using these classifications.

The very peculiar characteristics of volcanic formations, which are reflected in the complexity of the representative parameters that control their behaviour, allied with the limitations of empirical methods, results that experience takes on an essential role, during both the design and construction phases.

4 PRIMARY & SECONDARY LININGS

The substantial lateral variation in the type and characteristics of the volcanic formations both along the length of the tunnels and their transverse cross-section, prevented precise zoning of the respective bedrock or the corresponding zones for the application of each type of support to be determined to a sufficient degree during the project phase.

Hence, based on the lithological, structural and mechanical characteristics of the formations crossed by the tunnels, materialised in their respective longitudinal and transverse geological-geotechnical profiles, geotechnical zoning of the bedrock was carried out comprising four zones, from ZG1 to ZG4, including one or several types of lithology which in average terms display either good, reasonable, poor or very poor behaviour.

As an example, in general, one refers to the geotechnical zone ZG2 as comprised of resistant rocks, faulted basalts and consolidated breccia, with shallow layers of disaggregated breccia or tuffs.

In general, it behaves as a formation of reasonable quality which, due to its degree of faulting, as a primary lining in the current tunnels, implied a systematic application on the vault of 4 m length Swellex rock bolts spaced at 2 m apart in quincunx, an AQ50 type electro-welded steel mesh and a 5 to 10 cm shotcrete layer (Fig. 2).

Again in general, rock bolts were not used on the walls. In sections with a wider span, such as in the lay-by sections, 6 m length rock bolts were adopted with a 1.5 m × 2m mesh and a 15 to 30 cm layer of shotcrete. The excavation and application of the primary lining were mainly carried out in full section or in a partial section with a lower and an upper half.

Secondary lining was generally comprised of reinforced concrete in a layer varying between 0.25 m in the ZG1 zone to 0.4 to 0.5 m in the ZG4 zone.

All the tunnels have internal drainage and impermeabilization systems, the latter comprising a PVC geo-membranes.

During the execution of the tunnels the rate of excavation and application of the primary lining was of the order of 5 m/day in the ZG2 geotechnical zone and 2 m/day in the ZG3 geotechnical zone. The rate of concreting of the secondary lining with a metallic frame-work was of around 10 m/day.

Next, the most relevant characteristics of the Caniçal Double Tunnel are presented. As a result of the experience acquired by the authors during the design and technical assistance phases of various projects, it is considered that this is one of the most representative and emblematic tunnels constructed along the Machico-Caniçal Expressway and whose location is indicated in Figure 1.

5   CANIÇAL DOUBLE TUNNEL

It will now be presented the most relevant characteristics of the Caniçal Double Tunnel which figures as the most representative and emblematic constructed along the Machico-Caniçal Expressway.

The Caniçal Double Tunnel is part of the Machico/Caniçal Expressway between km 2 + 518 and km 3 + 700 in an east/west direction.

The tunnel comprises two galleries 15 m apart in a straight direction with 2100 m long and a constant inclination of 3.5% in the northern gallery and 3.65% in the southern gallery, from level 70 on the Machico side and up to level 140 on the Caniçal side, comprising the largest double tunnel in Portugal on the date it was completed. Maximum covering rises up to 300 m and on average totals 140 m.

The current transverse cross-section of each gallery, with a working width between walls of 9.0 m and a maximum height of 7.75 m, ensures a minimum clearance of 5.0 m and construction of a transverse roadway profile comprised of a two-lane traffic carriageway, an outer hard shoulder of 0.75 m in width and an inner hard shoulder of 0.25 m in width plus two sidewalks, an outer of 1.0 m and inner of 0.50 m which form the limits of the platform (Fig. 4).

At the western end of its stretch, in order to enable construction of an additional traffic lane which comprises the exit and entry lanes to the Caniçal road junction, the tunnel galleries have a widened section whose working width between walls at the start of the stretch varies from 9.0 m to 12.25 m and a constant 12.25 m at the end of the stretch, hence accommodating 3 traffic lanes of 3.25 m.

Given its large extension, 3 emergency parking lay-bys have been constructed in each gallery, 1 connecting gallery between the two central lay-bys (Figs. 5, 6) which enables light vehicles to reverse in the event of an accident and 8 pedestrian evacuation galleries. The emergency parking lay-bys have a transverse cross-section of 12.0 m in working width, also ensuring a minimum clearance of 5.0 m, and an extension of 16.0 m and 32.0 m, for light vehicles and light/heavy vehicles, respectively.

From west to east the tunnel crosses two of the oldest volcanic formations on Madeira Island. On the Machico side, corresponding to the western portal, the tunnel crosses the Mio-Plicoenic volcanic complex (β^1); on the Caniçal side, where the eastern portal is constructed, the tunnel crosses the Post-Miocenic volcanic complex (β^2) (Baião et al. 2003).

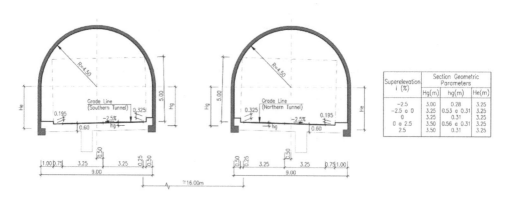

Figure 4.   Transverse type cross-section of tunnel galleries.

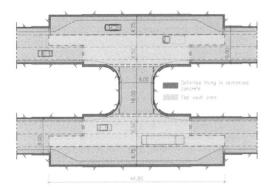

Figure 5.   Layout of central emergency parking lay-bys and connecting gallery for vehicles.

Figure 7.   Overall view of western tunnel portal.

Figure 6.   Overall view of heavy vehicle parking lay-by and connecting gallery for vehicles during impermeabilization phase and after completion of secondary lining.

The β^1 volcanic complex is formed of a chaotic pile of coarse materials resulting from angular block projections, volcanic bombs and scoriae, covered in a more or less developed fraction of fine pyroclastic material. Occasionally this complex is intercalated with highly altered basaltic lava. The β^2 volcanic complex is formed of alternating layers of basaltic lava and irregularly interstratified pyroclastic breccia materials which are also intercalated with episodic layers of volcanic tuff.

Based on the lithological and structural characteristics of the crossed formations obtained during the course of prospecting and characterisation works, although performed only in the portal zones given the high covering and length of the galleries, as well as on the experience acquired monitoring other civil works in formations with similar characteristics, the geotechnical zoning of the bedrock was carried out. Zones were defined covering all foreseeable situations with each geotechnical zone corresponding to a level of bedrock quality and behaviour. This zoning covered 4 geotechnical zones referred to as ZG1, ZG2, ZG3 and ZG4.

Geotechnical zone ZG1 was essentially comprised of compact basalts (βC) and compact breccia (BrC), very resistant and homogeneous formations, corresponding to a zone of good quality bedrock. Geotechnical zone ZG2 had a more heterogeneous structure, comprised of resistant rocks, faulted basalts (βF), occasionally intercalated with materials of poorer characteristics but of a shallow depth, behaving as a whole as a reasonable quality formation. In geotechnical zone ZG3 friable and weak cohesive formations predominate such as rocks of weak resistance and compacted or loosely consolidated soils, in other words, disaggregated breccia (DBr), compact tuff (CT) and compact tuffs with volcanic bombs (TVB), comprising formations of poorer quality. Geotechnical zone ZG4 encompasses all the formations of geotechnical zone ZG3 but associated with stretches of shallow covering and an important presence of water.

To calculate primary lining, empirical methods adjusted to the volcanic formations crossed were used, such as the geomechanical classifications of rocky bedrock for tunnels by Bieniawski, Manuel Rocha, Terzaghi and AFTES recommendations (Brito et al. 2002).

Based on the geological-geotechnical characterisation obtained and taking into account the primary lithological types present in each of the geotechnical zones determined, three distinct construction phases were defined for the current section, each one corresponding to a specific type of primary and secondary lining.

The main difference between the construction phases adopted resulted from the specified rate of excavation and subsequent application of primary lining.

Hence, in geotechnical zone ZG1 excavation was made in full cross-section at a maximum construction of 6 m and application of primary lining comprised of 4.0 m Swellex rock bolts associated with

5 cm of shotcrete applied only on faulted bedrock zones and/or those more substantially altered.

In tunnel stretches crossing formations belonging to geotechnical zones ZG2 and ZG3, excavation was carried out in phases, upper and lower mid-sections, progressing a maximum of 2 m and 1 m at a time respectively, ensuring a minimum difference between phases of 20 m.

The primary lining applied also differed, lighter for geotechnical zone ZG2, which consisted of applying 4.0 m Swellex rock bolts at 2 m intervals in the vault and 5 to 10 cm of shotcrete with metallic fibres, whilst application of shotcrete with metallic fibres was limited to 5 cm at walls in the zones of poorer quality bedrock. In geotechnical zone ZG3 heavier primary lining was applied comprising HEB160 steel ribs at 1 m intervals, AQ50 electrowelded steel mesh and 10cm of systematic shotcrete in the vault and walls (Fig. 8). Geotechnical zone ZG4 differed from ZG3 merely with regard to the spacing adopted for steel ribs which was 0.70 m.

Construction of the 3 parking lay-bys of the tunnel was adjusted during the construction phase itself using borings at the face in order to locate zones of bedrock with the best characteristics. Excavation was recommended in phases, beginning with the excavation of a central gallery located at the keystone of the vault in order to enable a detailed geological study of the bedrock during advance of the face, then proceeding with the dismantling of the lateral areas. After the upper mid-section had been excavated, excavation continued on the lower part, beginning with the dismantling of the central section followed by the lateral areas. Primary lining recommended for lay-bys consisted of the systematic application of 6 m Super Swellex rock bolts in a 1.5 m × 1.5 m mesh, AQ50 welded steel mesh and 15 cm of shotcrete.

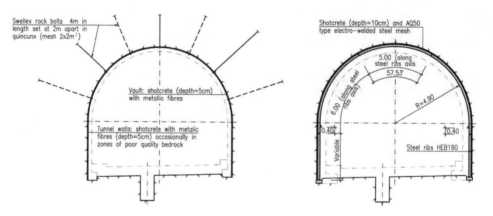

Figure 8. Primary linings defined for geotechnical zones ZG2 and ZG3.

Figure 9. Overall view of eastern tunnel portal.

For the widened section, excavated in formations of alternate bands of basalt and disaggregated breccia, the excavation was carried out using a methodology similar to the one defined for the parking lay-bys. Yet, due to the proximity of the tunnel portal, reinforcement for the primary lining was studied. This consisted of the application of HEB180 steel ribs with 1 m spacing, associated with 15 cm of shotcrete incorporating AQ50 welded steel mesh.

In the current section, the secondary lining in reinforced concrete had a constant thickness of 0.25 m and 0.40 m in accordance with the respective geotechnical zone which the tunnel was crossing. Figures 7 & 9 depict the final aspect of the tunnel portals.

Auscultation consisted of measuring the following installed instrumentation: i) precision topographic marks set in profiles transverse to the galleries ii) rod extensometers inserted in some of the profiles, with each one providing two reading points located at 2.0 m and 5.0 m from the top of the vault; iii) targets placed around the excavation edge in sections at distances of 25 m or 50 m for converging readings; iv) sealed inclinometer tubes in bore holes.

Great emphasis was given to safety concerns and an evolved system was implemented which comprised the installation of the following control and safety equipments: power supply; lighting; ventilation; signs; communications; control systems CCTV; auscultation; fire detection and fire extinguisher systems; fire hydrants; emergency parking lay-bys, connecting gallery and light vehicle reversing and pedestrian evacuation galleries; and access walkways for service personnel and emergency personnel.

6   CONCLUSIONS

The main aspects related with the geological-geotechnical survey, the design and the construction of six double tunnels inserted in the Machico-Caniçal Expressway were presented.

In the referred expressway, the percentage of tunnel length is unique, reaching 64% of the total length, mainly due to vigorous relief of the region.

Emphasis was given to the Caniçal Double Tunnel, which figures as the most representative and emblematic constructed along the Machico-Caniçal Expressway.

Concerning the design, although considering the significant limitations associated with the nature of the volcanic rock mass, it was possible to improve the geological recognition in order to obtain the necessary guidelines to define the geotechnical zoning.

It was also possible to estimate the temporary support applying empirical methods and define the soil characteristics required to the calculation models, consisting either by the finite element method, or by the hyperstatic reactions method.

This was complemented with a thorough observation by a continuous geologic survey of the rock mass, adapted to the development of the excavations and with complementary boreholes, as well as with measurements of displacements inside the tunnels, at the surface or inside the surrounding rock mass.

Therefore, it was possible to adapt the supports, in effective time, to the effective conditions of the rock mass. The observation design methodology and supervision of the works proved to be rather adequate and allowed a rational and safe adaptation of the support, thus enabling a significant economy during the construction stage.

Finally, reference was made to the integrated control and safety system installed in the tunnels.

REFERENCES

Baiao, Carlos J.O., Costa, P., Sousa, Luis R., Rosa, Sergio P.P. 2003. Caniçal Tunnel, Island of Madeira Geotechnical Characterisation & Observation. Jornadas HispanoLusas "Obras Subterraneas, Relevancia de la Prospeccion y Observacion Geotecnicas". Madrid.
Brito, Jose A.M., Costa, P., Monteiro, A., Oliveira, R., Gaiao, R., Sousa, Luis R. 2002. Geotechnical Analysis of the Caniçal Tunnel, Island of Madeira. International Symposium ISRM. Eurock 2002. Madeira.
Brito, Jose A.M., Baiao, Carlos J.O., Rosa, Sergio P.P. 2005. Modernisation of the Roadway Network of the Island of Madeira—Part 3: Tunnels Engineering & Life Magazine. No.9.
Cenor. 1991a. Variant to 101 Expressway in Machico. Geological-Geotechnical Study.
Cenor. 1991b. Tamega S/A—Somague A/S Construction Consortium Machico-Caniçal Expressway—1st Phase Queimada Stretch Tunnel to Caniçal Basic Project. Implementation Project.

Cenor, 1994. Machico-Caniçal Expressway—2nd Phase Preliminary Study.

Cenorgeo. 2002. SRESA/DRE. Machico-Caniçal Expressway Caniçal Double Tunnel Implementation Project.

Cenor-Grid. 2003a. Machico-Caniçal Expressway Caniçal Junction/Caniçal Roundabout Stretch Implementation Project.

Cenor-Grid. 2003b. Machico-Caniçal Expressway Queimada Tunnel/Caniçal Junction Stretch Implementation Project.

Rodrigues, Vitoria C., Rosa, Sergio P. Parada, Brito, Jose A.M., Baiao, Carlos J.O. 2009. Geological and Geotechnical Conditions of the Machico-Caniçal Expressway. 16th World Highway Convention Lisbon.

Volcanic Rock Mechanics – Olalla et al. (eds)
© *2010 Taylor & Francis Group, London, ISBN 978-0-415-58478-4*

Innovative aspects in the execution of the strengthening and stabilizing of the volcanic cavern of "Los Jameos del Agua", Lanzarote, Canary Islands, Spain

A. Cárdenas
Cabildo Insular de Lanzarote, Canary Islands, Spain

C. Olalla & A. Serrano
Escuela de Ingenieros de Caminos, Madrid, Spain

A. Gonzalo
HCC, Madrid, Spain

ABSTRACT: The rehabilitation of the volcanic natural cave that constitute "Los Jameos del Agua" Auditorium, was a challenge in the development of new technologies, materials and even machinery, specifically designed for strengthening and conditioning this impressive natural area. As a consequence that this is the biggest natural Auditorium in Europe, and a specially protected space, both, the project and its execution have required to contemplate the double condition of securing more than 800,000 people who use it every year, while maintaining the original aesthetics.

1 BACKGROUND

Los Jameos del Agua is located in the north of Lanzarote, one of the seven islands of the Canary Islands and is a part of the volcanic system of the Atlantic Cave, formed by the eruptions of the volcano La Corona three thousand years ago. Jameo is an aboriginal word that designates the part of a volcanic tube that has collapsed its ceiling, generating a circular hollow open to the light.

The Auditorium is the product of a geological volcanic pipe, which is currently in progressive degradation. The area consists of a natural cave with 73 meters in length, about 20 in width and 10 of maximum height. Since 1976 it has been fit out as an auditorium by the architect from Lanzarote, César Manrique being the biggest natural auditorium in Europe.

Several instabilities observed in the natural terrain that forms the vault caused the preventive closure to the public of the Auditorium in 2003. In 2005 the Authorities of the Cabildo of Lanzarote decided to entrust to a commission of experts the study of the problem and the issuance of a technical report on its status and predictable evolution from the point of view of security to the public use of the Auditorium.

This committee issued a report concluding that the Auditorium had various problems affecting its stability. So for opening to the public the

Figure 1. Overview of the auditorium.

Auditorium again in acceptable safety conditions, several actions should be carried out.

The high patrimonial value of the place imposed that these actions should be defined and run not only under structural criteria, but also of aesthetics, preserving the original appearance of the natural environment in which it is located.

2 PATHOLOGIES DETECTED

The pathologies detected are classified into four sections:

- Structural instability
- Local instability of large blocks of up to 4000 kg weight
- Local instability of small blocks of up to 160 kg weight
- General instability of small particles, type sand or gravel

Previous anthropic activities, as a mesh of steel protection, retention of loose rocks with conventional mortars, etc, were removed.

3 OBJECTIVES AND DETERMINING FACTORS OF THE WORKS

The essential objective of the intervention has been to secure the installation. At the same time the comfort has been improved by removing the dust and small stones that could fall from the ceiling.

Obviously being a specially protected area, constraints from the aesthetic point of view have been extremely important. Actually the work of restoration, have not only had to relate to previous anthropic activities, they have conditioned both rehabilitation and restoration solutions.

4 DEVELOPMENT OF THE WORK

4.1 Overview

To achieve the first objective, which was to strengthen the cave and avoid falls of large blocks of stone, it was decided to create a one

meter thick vault, to act as a monolithic arch. For that purpose, a first generalized short bolting was carried out. After that stage, a long bolting was carried out to assure the contribution of the higher strata to the stability of the cave. The medium size stones fall risk was solved through individualized fine bolts. A chemical sealing product served for consolidating partially the surface of the cave, preventing the fall of small rocks. Finally, a mineralizing product applied widely was the procedure chosen for the general consolidation of the surface.

The Works performed are described below.

4.2 Hydrodemolition

In some parts of the cave it had been placed an unsightly triple twist steel mesh to protect both the public and workers, from the fall of small blocks and gravel. Protective measures had to be taken to protect the staff that should work in the cave. Hydrodemolition techniques were applied with water pressure up to 3000 bar, forcing selectively to fall the rocks that were in worst conditions.

The application of this high-pressure water jet also eliminated the numerous efflorescences that appeared in the surface of the cave.

Figure 3.   Hydrodemolition techniques.

Figure 2.   Mesh of steel protection.

4.3 Short bolts

It was intended to create a consolidated arch of approximately one meter thick. In order to achieve this arch, meter long and 22 mm diameter boreholes were drilled. Fiberglass and polyester bolts, with high resistance to traction and inert to saline aggression, were introduced in the boreholes. The structural characteristics of the bolts are: deformation module 47 GPa, resistance to traction 1040 MPa and maximum elongation 2.3%. The approximate grid was 1 × 1 meter. Once each bolt was introduced the borehole was sealed and a medium viscosity epoxy resin (30,000 cP) was injected to form a consolidated stone block.

Since it was impossible to find in the market suitable drilling machinery for these works, very light and powerful pneumatic drilling machinery were designed. A total of 2000 short bolts were specially manufactured for this job.

4.4 Long bolts

Once this monolithic rock layer was achieved it was necessary to tie this first meter to the higher strata, and somehow "hang" that first consolidated meter from higher strata. To do so, 46 mm diameter boreholes were drilled, at least 4 m long. The grid was 2 × 2 m. It was planned to place in each of these holes a long bolt, fiberglass and polyester made, four meters long and 25 mm diameter, that later, as in the case of short bolts should be injected with medium viscosity epoxy resin. Before this step an inspection of the boreholes was carried out with a TV camera, founding out that there were numerous large hollows. This forced to unplanned supplementary treatment, described below.

Once the boreholes were drilled they were injected with expansive polyurethane resins, able to increase up to 20 times its original volume without applying pressure when swelling. By doing so, the hollows were filled allowing drilling the boreholes again, placing the fiberglass bolts and their subsequent injection with epoxy resins. More than 5000 liters of these resins were consumed.

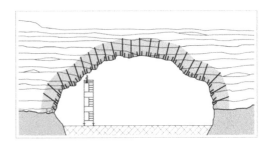

Figure 5. Consolidation of the 4 meters thick arch.

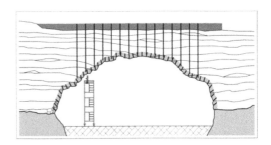

Figure 6. Lightweight concrete slab.

4.5 Lightweight concrete slab

In specific areas, higher strata, above the first meter consolidated arch, were considered insufficient to carry the long bolting. For this reason, an artificial stratum was created to anchor the long bolts, using a lightweight reinforced concrete slab.

Once lightened concrete had hardened, a grid of boreholes of 46 mm diameter was drilled from the concrete slab. The TV camera was introduced in every borehole, discovering numerous large hollows that decreased dangerously the remainder structural resistance to the entire area. It was decided to fill these gaps by injecting a grout of low-density, high strength, excellent adhesion to the rock and null retraction.

After that the boreholes were redrilled introducing a 25 mm fiberglass and polyester bolt in each of them, diameter, which was injected with resin epoxy resin.

4.6 Micro bolts

In specific areas, with the aim of holding loose rocks, small reinforcements of micro bolts of 8 mm diameter and variable length up to 50 cm long, and capable of supporting up to 160 kg were made with from fiberglass resin polyester. The annular space between rock and bolts was also injected with resin epoxy. A total amount of 1200 items were made.

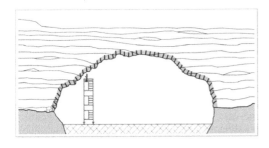

Figure 4. Consolidation of the one meter thick arch.

4.7 Chemical welding

Also, in specific areas, those rocks that had risk of falling, and were not large enough to stand a micro bolt, were sealed between them with a powerful epoxy adhesive. Thus large superficial areas of the vault were tied together.

4.8 Aesthetic recovery

Several works were carried out to return or even increase the aesthetic beauty of the cave. Firstly

Figure 7.   Setting in place of small bolts.

Figure 8.   Chemical welding.

Figure 9.   Final aspect of the rock.

the sealing of the ends of the bolts were carried out with a stone mortar of the same color of the environment, elaborated with natural stone from the place and epoxy resin. Then a mastic with suitable color was prepared and implemented to hide the chemical welding. Also other aesthetic treatments aimed to minimize the impact of previous anthropic action.

4.9 Generalized treatment of mineralization

Once earlier works were finished, a mineralizing product was applied with airless. Its mission was to improve the surface characteristics of the rock to avoid dust from the ceiling falling.

This product penetrates into the pores and capillaries of the element to be treated reacting with terrain oxides and carbonates resulting in the formation of insoluble crystals.

5 CONTROLS DURING THE EXECUTION

Before the beginning of the work, and during its development, a strict control of materials and execution was carried out. Perhaps highlights its importance the breaking test of the bolts. By doing them the values supplied by the manufacturer were checked and at the same time the adhesion force achieved in field was checked. Both values exceeded the design values.

As a precaution, during the work, a topographic detailed control was done.

During the remove of the falsework of the stratum improved with the lightweight concrete slab, a control in surface was carried out by 3 clinometers next to the area of major overhang. The interior of the cave was controlled, with the same accuracy, possible movements with topographic equipments, not detecting any movement.

Figure 10.   Bolts tests.

Figure 11.   General overview of the scaffolding system.

6   MATERIALS AND WORK EQUIPEMENT

6.1   *Scaffolding*

In order to have easy and safe access to all points of the cave, both to intervene and to inspect, after studying several possibilities, the placement of fixed scaffolding was decided. In the area of the lightweight reinforced concrete slab the scaffolding served as a falsework A total of 6000 m³ of European tubular scaffold were used.

6.2   *Epoxy resins*

Two components epoxy resins were used, able to harden, adhere and acquire their mechanical characteristics even under water. The reason for the use of this material, specially formulated, was the large amount of water introduced into the rock during drilling. Indeed, more than 4500 meters of boreholes were drilled to rotation, with diamond crowns, requiring water for cooling.

The essential features of the resins used for injection, were:

Resistance to compression: 700 kg/cm²
Resistance to traction: 300 kg/cm²
Elastic modulus: 20.000 kg/cm²
Pot-life: 45 min

6.3   *Injection pump*

Given the high viscosity of the resins to inject and the need to pump them at very high pressures, it required a pump design itself, capable of up to 600 bar pressure pump output. Also a few special injectors were designed to fill the holes where hosted the bolts from the bottom to the surface, avoiding air bubbles or poorly injected areas.

6.4   *Bolts*

With the maximum loads to stand by the bolts supplied by the designer over 4,500 fiberglass and polyester bolts were manufactured for this work.

Fiberglass and polyester bolts were preferred to metal bars because resistance to saline environment. It should not be forgotten that Los Jameos lies is situated near the Atlantic coast. Also this type of bolts is rather lighter than conventional.

7   CONCLUSION

The restoration of Los Jameos del Agua is the first important work of this type in such a volcanic environment. The success of the work has been completed with some magnificent, both structural and aesthetic results, with a perfect restoration to the natural environment. The specific conditions of the works resulted in the development of new materials and machinery, some of which have been the subject of patents.

8   STATISTICS

Total amount	3.000.000 €
Epoxy resins injected	43.000 kg
Expansive resins	5.000 kg
Short bolts	2.000 items
Long bolts	600 items
Micro-bolts	1.200 items
Grout	130.000 kg
Total bolts length	4.500 m

Volcanic Rock Mechanics – Olalla et al. (eds)
© 2010 Taylor & Francis Group, London, ISBN 978-0-415-58478-4

Geological risk at world class astronomical observatories

Antonio Eff-Darwich
Departamento Edafología y Geología, Universidad de La Laguna, Tenerife, Spain

Begona García-Lorenzo
Instituto de Astrofísica de Canarias, La Laguna, Tenerife, Spain

Jose A. Rodríguez-Losada
Departamento Edafología y Geología, Universidad de La Laguna, Tenerife, Spain

Luis E. Hernández
Area de Laboratorios y Calidad de la Construcción, Consejería de Obras Públicas y Transportes, Tenerife, Spain

Julio de la Nuez
Departamento Edafología y Geología, Universidad de La Laguna, Tenerife, Spain

Carmen Romero-Ruiz
Departamento de Geografía, Universidad de La Laguna, Tenerife, Spain

ABSTRACT: Future large and extremely large ground-based telescopes will demand stable geological settings. The world class astronomical observatories of El Teide (Tenerife, Canary Islands), Roque de los Muchachos (La Palma, Canary Islands), Mauna Kea (Hawaii) and Paranal (Chile) are in or closer to volcanic environments, and hence the impact of volcanic activity has to be studied in detail. In this sense, seismic activity, the extent of lava flows, eruptive clouds and ground deformation associated to volcanic/tectonic activity have studied in terms of probabilistic risk analysis. This information might be essential in ranking astronomical sites for emplacing future large telescope infrastructures.

1   INTRODUCTION

Some of the best astrophysical observatories in the world, namely the Canarian, Chilean and Hawaiian observatories, are located within active geological regions. This is not a coincidence, since the sky transparency that defines good astronomical sites is the result of the combination of factors, directly or indirectly related to geological activity, such as altitude, local topography and/or atmospheric stabilization induced by the presence of water bodies (*e.g.* ocean). Increasingly larger telescopes (10 and 40 metres classes) demand stable structures and buildings, and hence geological activity becomes an important parameter to take into account in astronomical site ranking. The structures of large telescopes have and will have to withstand the effects associated to seismic and/or volcanic activity, but they also have to minimize the loss of operational time, recalling the extreme precision in the alignment of mechanical and optical components (Salmon 2007).

Astronomical observatories are vulnerable to geological activity, hence it is necessary to consider the impact on the telescope facilities, but also on supporting facilities at the observatories and on local/regional communication and infrastructures. In this work, only direct hazards to telescopes have been considered, in particular those affecting the structural design of telescopes, namely lava flows, volcanic ashfall and seismicity. The analysis was carried out at four world-class observatories, namely Roque de Los Muchachos (La Palma, Canary Islands, 28.75°N, 17.89°E, 2400 m.a.s.l.), El Teide (Tenerife, Canary Islands, 28.3°N, 16.51°E, 2380 m.a.s.l.), Paranal (Chile, 24.62°S, 70.4°E, 2620 m.a.s.l.) and Mauna Kea (Hawaii, 19.82°N, 155.47°E, 4130 m.a.s.l.) and the candidate site of Cerro Ventarrones (Chile, 24.35°Ss, 70.2°E, 2200 m.a.s.l.). A common methodology was used to characterize the geological hazard, expressed in terms of probabilities of occurrence in the next 50 years, recalling that this period of time corresponds to the expected lifetime of a large telescope.

2 GEOLOGICAL CONTEXT

2.1 *Tenerife*

Tenerife is the largest island of the Canarian Archipelago and one of the largest volcanic islands in the world. It is located between latitudes 28–29°N and longitudes 16–17°W, 280 km distant from the African coast. The morphology of Tenerife (see Fig. 1) is the result of a complex geological evolution: the emerged part of the island was originally constructed by fissural eruptions that occurred between 12 and 3.3 Ma (Ancochea et al. 1990). In the central part of the island, where the observatory is located, the emission of basalts and differentiated volcanics gave rise to a large central volcanic complex, the Las Cañadas Edi-fice, that culminated in the formation of a large elliptical depression measuring 16×9 km^2, known as Las Cañadas Caldera (Martí et al. 1994). In the northern sector of the caldera, the Teide-Pico Viejo complex was constructed as the product of the most recent phase of central volcanism. Teide-Pico Viejo is a large stratovolcano that has grown during the last 175 Ka. The basaltic activity, which overlaps the Las Cañadas Edifice, is mainly found on two ridges (NE and NW), which

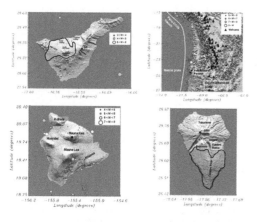

Figure 1. Simplified shaded-relief map of Central Chile and the islands of Tenerife, La Palma and Hawaii, indicating the most important geological features (see text for details). The filled circles of different sizes and colours indicate the location, magnitude and depth of the earthquakes registered during the period 1973–2008. The location of the observatories are indicated by black filled circles. In Tenerife, the region surrounded by the black solid line and the dark areas indicate the regions affected by recent (last 10 ka) and historical (last 500 years) volcanic activity, respectively. In La Palma, the region surrounded by the thick solid line indicates the region affected by recent and historical activity of Cumbre Vieja volcano. In Chile, the location of the Peru-Chile trench is indicated by a white solid line, whereas the volcanic arc is shown as filled black triangles.

converge on the central part of the island. Eruptive activity at Mount Izaña, where El Teide observatory is located, ceased more than 300 ka (Edgar et al. 2007), whereas recorded eruptive activity in Tener-ife has consisted of six strombolian eruptions, namely Siete Fuentes (1704), Fasnia (1705), Arafo (1705), Arenas Negras (1706), Chahorra (1798) and Chinyero (1909). The last three eruptions occurred at the NW ridge, the most active area of the island, together with El Teide-Pico Viejo complex, for the last 50,000 years (Carracedo et al. 2007).

2.2 *La Palma*

The island of La Palma is the fifth in extension (706 km^2) of the Canary Islands and the second in elevation (2426 m.a.s.l.) after Tenerife (Fig. 1). The two main stages of the development of oceanic volcanoes, the submarine and subaerial stages, outcrop in La Palma, since the submarine basement or seamount that was built during the Pliocene (4 to 2 Ma) were uplifted up to 1.3 km above the present sea level (Ancochea et al. 1994). The subaerial edifice is conformed by a series of overlapping volcanoes (Navarro & Coello 1993): (1) the Northern Shield of the island includes the Garafía shield volcano (1.7 to 1.2 Ma), the Taburi-ente shield volcano (1.1 to 0.4 Ma) and the Bejenado edifice (0.55 to 0.49 Ma); the Caldera de Taburiente, in the center of the old shield, and Cumbre Nueva ridge formed by a combination of large landslides and erosion; (2) The Cumbre Vieja Volcano (0.4 Ma to present) in the southern half of the island. The Cumbre Vieja ridge is interpreted to be a volcanic rift zone because of the prominent north–south alignment of vents, fissures and faults. Historic eruptions on La Palma have lasted between 24 and 84 days and were recorded in 1470/1492 (uncertain), 1585, 1646, 1677–78, 1712, 1949 and 1971 (Romero-Ruiz 1991). The Roque de los Muchachos observatory is located atop the Taburiente Edifice, at the rim of the Caldera de Taburiente, and hence eruptive activity in the proximity of the observatory ceased more than 0.4 Ma ago.

2.3 *Hawaii*

The island of Hawaii is the youngest island in a chain of volcanoes that stretches about 5600 kilometres across the northern Pacific Ocean. The island chain results from a magma source that originates deep beneath the crust. The ocean crust and lithosphere above the magma source, within the Pacific tectonic plate, move to the Northwest with respect to the deep stationary magma source. Over a span of about 70 million years, new island volcanoes are formed and older volcanoes are carried away from the magma source, erode, and eventually subside beneath sea level (Lockwood & Lipman 1987). The island of Hawaii, the largest of the entire Hawaiian chain, consists of five main volcanoes: Kilauea, Mauna Loa, Mauna Kea, Kohala and Hualalai (Fig. 1). Kilauea and Mauna Loa volcanoes are in the shield stage and erupt frequently, Hualalai and Mauna Kea volcanoes are in the post-shield stage and erupt every few hundred to few thousand years, and Kohala is dormant, having passed through the post-shield stage. Kohala last erupted about 120 Ka, Mauna Kea about 3.6 Ka, and Hualalai in 1800–1801 (Frey et al. 1984).

2.4 Central Andes, Chile

The Andes is a mountain chain which extends along the western edge of the South American plate. In this region, the Nazca Plate thrusts beneath South America at a rate of approximately 64 and 79 mm/ year (Angermann et al. 1999) in an east-north-east direction, forming the Peru-Chile trench in the ocean's floor (Fig. 1). The Nazca plate is divided into several segments, whose main difference are convergence direction and dipping angle of the slab (Barazangi & Isacks 1976). In general, normal or steep dipping slab segments have angles around 30°, whereas in shallow segments this angle does not exceed 15°. The angle of the segments also define the presence of volcanic activity, since in shallow segments there have not been volcanic activity at least since the Miocene. Both Paranal and Cerro Ventarrones are located in the Central Andes region (15°S to 27.5°S), above a steep dipping segment of the Nazca plate, hence volcanic activity is present, although this takes place more than 200 km eastwards of the sites. In this sense, volcanic activity ceased in the area of the observatories more than 20 million years ago. Unlike the low viscosity-high temperature magmas of Hawaii, the high viscosity and low temperature of the magmas in the Central Andes generate explosive eruptions that eject large eruptive columns.

3 SEISMIC HAZARD ANALYSIS

Seismic hazard is defined as the probabilistic measure of ground shaking associated to the recurrence of earthquakes. Hazard assessment for buildings and other structures commonly specifies 10% chance of exceedance of some ground motion parameter for an exposure time of 50 years, corresponding to a return period of 475 years. The operational and survival conditions of telescope structures will depend on the ground shaking level (seismic hazard) and the seismic design of the structure. Thus, two sites with different levels of seismic

hazard will need different seismic designs for the structures to reach the same survival conditions. In this sense, larger seismic hazard means increased building costs. In order to assign a common methodology to infer the probabilistic seismic hazard to the different sites, the data from the Global Seismic Hazard Assessment Program (GSHAP) were analyzed (Giardini et al. 1999). GSHAP hazard maps depicts Peak-Ground-Acceleration (PGA) with 10% chance of exceedance in 50 years, as illustrated in Fig. 2 and summarized in Table 1. Although all the sites are located within active

Table 1. Seismic hazard expressed in terms of the Peak Ground Acceleration (PGA) with 10% chance of exceedance for an exposure time of 50 years.

Observatory	PGA (g)
Mauna Kea	0.5
Paranal	0.47
Ventarrones	0.42
Roque de Los Muchachos	0.05
El Teide	0.06

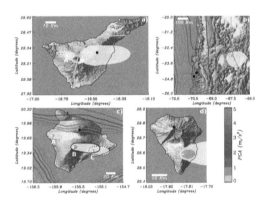

Figure 2. The contour lines represent the probabilistic seismic hazard (peak ground acceleration with 10% chance of being exceeded in 50 years), obtained from the GSHAP project. The spatial distribution of ashfall (at least 1 cm) after the explosive eruptions of the vents represented by filled white circles. The observatories are marked as black filled squares. *Panel a)* The yellow semitransparent areas represent the extent of tephra fall after a VEI = 2 eruption (A) and a VEI = 4 Montaña Blanca type eruption (B) in Tenerife, Canary Islands. *Panel b)* The yellow and orange semi-transparent areas represent the extent of tephra fall after a VEI = 4 and a VEI = 6 eruption of Lascar volcano, in Chile. *Panel c)* The yellow and orange semi-transparent areas represent the extent of tephra fall after a VEI = 2 and a VEI = 3 eruptions of Hualalai (A), Mauna Loa (B) and Kilauea (C) volcanoes, in Hawaii. *Panel d)* The yellow and orange semi-transparent areas represent the extent of tephra fall after a VEI = 2 and a VEI = 3 eruptions of Cumbre Vieja volcano, in La Palma, Canary Islands.

volcanic/tectonic regions, seismic activity strongly depends on the geological setting and thus, significant differences in seismicity are expected. This is clearly illustrated in the distribution of epicenters recorded by the National Earthquake Information Center (NEIC), http://neic.usgs.gov, for the period 1973 to 2008 (Fig. 1).

Seismic hazard for the Canary Islands is slightly higher than that reported by the GSHAP in the basis of recent results. In this sense, recent analysis of historical records revealed that earthquakes with intensities of VIII MSK took place during the dikefed volcanic eruptions in Cumbre Vieja, La Palma and the eruptions of 1704–1705 closer to Mount Izaña in Tenerife (Romero-Ruiz 1991). Also, observational seismicity from volcanic rift zones worldwide suggests the maximum magnitudes of dike-induced earthquakes are M = 3.8 ± 0.8 (1996). The focal mechanism of the M = 5.2 earthquake recorded in 1989 May 9 (the largest earthquake registered in the Canaries) and the aftershock distribution helped to identify a submarine fault parallel to the eastern coast of Tenerife (González de Vallejo et al. 2006). Hence, the activity of this fault and the effect of volcanic activity should be included in the hazard calculation, increasing the PGA from 0.15 m/s² to 0.56 m/s² (0.06 g) for eastern Tenerife (including El Teide observatory) and 0.5 m/s² (0.05 g) for the rest of the islands. In any case, following the convention of the GSHAP, seismic hazard at the Canary Islands remains at the lowest level.

The high rate of earthquake occurrence, coupled with the potential for large events, places Hawaii among the areas of highest seismic hazard (Figs. 1 and 2). Three types of seismicity are found (Wyss 1988). On one hand, particularly on the island of Hawaii, earthquakes are the result of the crustal stresses imparted by the volcanic activity, in particular in Kilauea and Mauna Loa volcanoes. Secondly, the largest earthquakes in Hawaii have occurred beneath the flanks of the Kilauea, Mauna Loa, and Hualalai volcanoes. The flanks of these volcanoes adjust to the intrusions of magma into their adjacent rift zones by storing compressive stresses and occasionally releasing it in crustal earthquakes. Largest earthquakes recorded in Hawaiian history (the 1975 M = 7.2 Kalapana earthquake beneath Kilauea's south flank, and the 1868 M = 7.9 Great Kau earthquake beneath Mauna Loa's southeast flank) are examples of these crustal earthquakes. The third type of earthquakes corresponds to deeper mantle earthquakes at approximately 30–40 km depth resulting from flexural fracture of the underlying lithosphere in long-term geologic response to the load of the island mass. Examples of these mantle earthquakes are the 2006 October 15 Kiholo Bay earthquake and the 1973 M = 6.2 Honomu earthquake (on the northeast coast of Hawaii Island).

Seismic hazard is very high on the Chilean sites of Paranal and Cerro Ventarrones (Figs. 1 and 2) due to the intense seismic activity. Seismicity results from the release of stresses generated by the subduction of the oceanic Nazca plate beneath the South American plate. The focal mechanisms of these earthquakes indicate subduction-related thrusting, likely on the interface between these two plates, that is located between 35 to 50 kilometres beneath the observatory sites.

4 VOLCANIC HAZARD ANALYSIS

Two different volcanic phenomena have been considered in the analysis of the effect of near future eruptions on astronomical sites, namely ashfall and lava flows, since they are the most likely volcanic hazards affecting all sites, either at short (years) or medium (decades) terms.

4.1 Lava flow hazard analysis

In the case of the Chilean sites of Paranal and Cerro Ventarrones, the closest volcanoes of the Central Andes are at least 200 km distant, being significative longer than the typical length of lava flows in the region (30 km). Moreover, volcanic activity at the observatory sites ceased more than 20 million years ago. Hence, hazard due to lava flows is negligible at Paranal and Cerro Ventarrones.

The Canarian site of Roque de Los Muchachos is emplaced atop the extinct Taburiente volcano, being approximately 1000 metres above and more than 15 kilometres distant from Montaña Quemada, the closest eruptive vent of Cumbre Vieja volcano (Romero-Ruiz 1991). In this sense, hazard due to lava flows is also negligible at Roque de los Muchachos observatory. El Teide observatory is emplaced atop Mount Izaña, where volcanic activity ceased more than 300 ka. However, the observatory is closer to El Teide-Pico Viejo complex, where there has been an intense volcanic activity in the last 150 ka, whereas several effusive basaltic eruptions took place about 35 ka and in the XVIII century within 2 and 10 kilometres from the observatory, respectively (Fig. 1). In any case, the caldera walls protect the observatory from lava flows originated within the caldera area, while local topography and the altitude of the observatory makes the hazard to lava flows from the NE ridge system negligible.

There is not lava flow hazard at Mauna Kea observatory due to the eruptive activity of other volcanoes such as Mauna Loa, Kilauea or Hualalai volcanoes, as the result of the topographic protection given by the altitude the observatory is located, above 4100 metres above sea level. However, Mauna Kea is still an active volcano which last erupted

3600 years ago and hence, lava flow hazard is low but larger than in Chile and the Canaries.

Following the convention defined by the United States Geological Survey (USGS), the level of hazard for lava flows has been scaled from 1 to 9, being 1 the most hazardous (Mullineaux & Peterson). The USGS assigns a hazard level 7 to the area of Mauna Kea observatory, meaning that at least 20% of this area has been covered by lavas in the last 10000 years (see Table 2). Hazard level at El Teide observatory was set to 8, since the area covered by lava flows in the last 10000 years is significantly lower than 20%, Both the Chilean sites and Roque de los Muchachos observatory in the island of La Palma have a hazard level of 9, since there have not been volcanic activity in the last 60 ka.

4.2 *Volcanic ash hazard analysis*

Ash/tephra fall poses the widest-ranging direct hazard from volcanic eruptions. Vast areas (10^4 to 10^5 km^2) have been covered by more than 1 cm of tephra during some large eruptions, whereas fine ash can be carried out over areas of continental size (Gardeweg 1996). Burial by tephra could collapse roofs, break power and communication lines, whereas suspension of fine-grained particles in air affects visibility, could damage unprotected machinery, cause short circuits in electrical facilities and affect communications.

The spatial extend of tephra fall depends on two factors, namely the strength and direction of the wind and the explosivity of the eruption (height of the eruptive column). Wind speeds at different heights above the selected sites were collected from the (NCEP/NCAR) Reanalysis database of the National Center for Environmental Prediction/ National Center for Atmospheric Research. The second major factor affecting the extent of ashfall is the explosivity of the eruption, expressed as the Volcanic Explosivity Index (VEI). The Explosivity Index for eruptions in the Central Andes that have been measured in historical times ranges from the VEI = 4 sub-plinian eruption of Lascar volcano in 1967 or 1993 (Gardeweg 1996) to the VEI = 6 eruption of Huaynaputina volcano in February 1600, corresponding to the largest volcanic explosion in South America in historic times. The

Table 2. Level of hazard by lava flows that was calculated following the USGS convention.

Observatory	Zone	Last eruption
Mauna Kea	7	3.6 ka
Paranal and Ventarrones	9	>1 Ma
Roque de Los Muchachos	9	400 ka
Teide	8	300 ka

Explosivity Index for eruptions in the island of La Palma has been low (VEI < 3), although eruptive columns up to 4 km high have been recorded in historical times when the uprising magma encountered groundwater, producing phreatomagmatic explosions (Romero-Ruiz 1991). In the case of the island of Tenerife, explosive eruptions might occur in the central part of the island, associated to El Teide-Pico Viejo strato-volcano and its peripheral vents. Last explosive event took place 2 ka and corresponded to the VEI = 4 sub-plinian eruption of montaña Blanca volcano (Ablay et al. 1995). Due to the complex nature of the central volcanism in Tenerife, other types of explosive eruptions have occurred and might occur in the future, namely phreatomagmatic and/or violent strombolian eruptions. Low intensity (VEI < 3) eruptions have occurred in the vicinity of El Teide observatory, and hence similar future eruptions should be considered in the hazard analysis associated to tephra fall.

The Explosivity Index for Hawaiian eruptions has also been low (VEI < 3), but with chances of phreatomagmatic eruptions like those reported for the Kilauea volcano in historical times, namely 600 years ago, in 1790 and May 1924 (McPhie et al. 1990) and in prehistoric times (2700, 2100 and 1100 years ago). Moreover, other widespread ash units point to larger phreatomagmatic eruptions at Kilauea and Mauna Kea, likely VEI = 3 (Mullineaux et al. 1987).

Unlike the hazard analysis for lava flows, there is not established convention to define the hazard due to ash/tephra fall. Hence, a numerical simulator of volcanic ashfall, TEPHRA2 (Connor et al. 2001), was used to analyze the extent of ash deposits at each site, considering the typical wind conditions and the VEI of the volcanoes located in the surroundings of the sites. In this sense, we supposed two different eruptive events (VEI = 4 and 6) for the Lascar volcano in Chile, the closest volcano to the Chilean sites (Fig. 2). Two events were also considered in Tenerife, a VEI = 4 eruption in Montaña Blanca and a VEI = 2 eruption on the NW ridge. In the case of La Palma, two cases were modeled, a VEI = 2 and a VEI = 3 eruptions located in Cumbre Vieja volcano. Finally, six events were considered in the island of Hawaii, namely VEI = 2,3 eruptions at Hualalai, Mauna Loa and Kilauea volcanoes.

The results of TEPHRA2 (Fig. 2) show that El Teide observatory might be affected by deposition of ash (exceeding 1 cm) from a VEI = 4 eruption of El Teide-Pico Viejo complex due to the prevailing wind conditions. A Montaña Blanca type eruption affecting the observatory would be followed by years of contamination of the site by reworked volcanic fine ash blown about by the wind (Mullineaux et al. 1987). A VEI = 2 eruption would deposit significant amounts of ash if the wind were blowing in the same direction than the vent-observatory alignment and the distance from the vent to the observatory did not exceed 10–15 kilometres. Several historical and recent (last 35 ka) eruptions have taken place within that distance from the observatory (Fig. 1).

The Chilean observatory of Paranal and the candidate site of Cerro Ventarrones could be affected by a VEI = 6 eruption if the wind blowed westwards, a condition that happens less than 0.5% of the time. In the case of La Palma and Hawaii, the chances to have significant depositions of ash at the observatory site are negligible for a VEI = 2 eruption. Probabilities remain low for VEI = 3 eruptions in both islands, considering that the directions from the possible vents to the observatories significantly differ from the prevailing wind directions. An eruption of Mauna Kea volcano in Hawaii would certainly be catastrophic, since ash deposition could exceed several meters in thickness in the proximity of the vent, besides the expected strong seismicity and the likelihood of lava inundation (Mullineaux et al. 1987).

5 CONCLUSIONS

The lowest geological hazard in both seismic and volcanic activity was found at Roque de los Muchachos observatory, in the island of La Palma. Seismic hazard is also low at the other Canarian site, El Teide observatory, since seismic activity in the Canary Islands is low in both number and magnitude of earthquakes. On the contrary, seismic hazard is very high in Paranal and Ventarrones (Chile) and in Mauna Kea (Hawaii). Hazard associated to lava flows during a volcanic eruption is not significant at any site, as the result of low volcanic activity in the regions where the sites are emplaced, topographical protection or distance to the eruptive vents. Hazard associated to volcanic ashfall is negligible at Mauna Kea, Roque de los Muchachos and Chile and low to moderate in Tenerife, depending on the prevailing winds and the still poorly known explosive volcanic activity.

ACKNOWLEDGEMENTS

Partially funded by the Spanish MICINN under the Consolider-Ingenio 2010 Program grant CSD2006-00070: First Science with the GTC (http://www.iac.es/consolider-ingenio-gtc). This work has made use of the NCEP Reanalysis data provided by the National Oceanic and Atmospheric Administration/Cooperative Institute for Research in Environmental Sciences (NOAA/ CIRES) Climate Diagnostics Center, Boulder, Colorado, USA, from their web site at http://www. cdc.noaa.gov/.

REFERENCES

Ablay, G.J., Ernst, G.G.J., Martí, J., and Sparks, R.S.J., 1995, Bull. Volcanol., 57, 337.

Ancochea, E., Fuster, J.M., Ibarrola, E., Cendrero, A., Coello, J., Hernán, F., Cantagrel, J.M., and Jamond, A., 1990, J. Volcanol. Geotherm. Res., 44, 231.

Ancochea, E., Hernán, F., Cendrero, A., Cantagrel, J.M., Fúster, J.M., Ibarrola, E., and Coello, J., 1994, J. Volcanol. Geotherm. Res., 60, 243.

Angermann, D., Klotz, J., and Reigber, C., 1999, Earth and Planetary Science Letters, 171, 329.

Barazangi, M., Isacks, B., 1976, Geology, 4, 686.

Carracedo, J.C., Rodriguez-Badiola, E., Guillou, H., Paterne, M., Scaillet, S., Pérez-Torrado, F.J., Paris, R., Fra-Paleo, U., and Hansen, A., 2007, GSA Bulletin, 119, 1027.

Connor, C.B., B.E. Hill, B. Winfrey, N.M. Franklin, and P.C. LaFemina, 2001, Natural Hazards Review, 2, 33.

Edgar, C.J., Wolff, J.A, Olin, P.H., Nichols, H.J., Pittari, A., Cas, R.A.F., Reiners, P.W., Spell, T.L., and Martí, J., 2007, J. Volcanol. Geotherm. Res., 160, 5985.

Frey, F.A., Kennedy, A., Garcia, M.O., West, H.B., Wise, W.S., and Kwon, S.T., 1984, Eos, Transations, American Geophysical Union 65, 300.

Gardeweg, M., 1996, MMA Site East of San Pedro De Atacama North Chile Volcanic Hazards Assessment and Geologic Setting, MMA Memo 251.

Giardini, D., Grünthal, G., Shedlock, K.M., and Zhang, P., 1999, Annali de Geofisica, 42, 1225.

Gonzalez de Vallejo L., García-Mayordomo J., Insua J.M., 2006, Bull. Seism. Soc. Am. 84, 2040.

Lockwood, J.P., and Lipman, P.W., 1987, Volcanism in Hawaii: U.S. Geological Survey Professional Paper 1350, 509.

Martí, J., Mitjavila, J., and Araña, V., 1994, Geol. Mag., 131, 715.

Masson, D.G.; Watts, A.B., Gee, M.J.R., Urgeles, R., Mitchell, N.C., Le Bas, T.P., and Canals, M., 2002, Earth Sci. Rev., 57, 1.

McPhie, J., Walker, G.P.L. and Christiansen, R.L., 1990, Bulletin of Volcanology 52, 334.

Mullineaux, D., and Peterson, D., 1974, U.S. Geological Survey Open File Report, 72–239.

Mullineaux, D., Peterson, D., and Crandell, D., 1987, U.S. Geological Survey Professional Paper 1350.

Navarro, J.M., and Coello, J., 1993, Mapa geológico del Parque Nacional de la Caldera de Taburiente. ICONA. Ministerio de Agricultura, Pesca y Alimentación.

Romero-Ruiz, C., 1991, Las manifestaciones volcánicas históricas del Archipiélago Canario, Consejería de Política Territorial, Gobierno de Canarias.

Salmon, D., 2007, http://www.gemini.edu/.

Smith, R.P., Jackson, S.M., and Hackett, W.R., 1996, J. Geophys. Res., 101(B3), 6277.

Wyss, M., 1988, Bulletin of the Seismological Society of America 78, 1450.

Volcanic Rock Mechanics – Olalla et al. (eds)
© 2010 Taylor & Francis Group, London, ISBN 978-0-415-58478-4

Construction experiences with volcanic unbound aggregates in road pavements

M.A. Franesqui
Department of Civil Engineeing, University of Las Palmas de Gran Canaria, Las Palmas, Spain

F. Castelo Branco
Department of Earth Sciences, University of Coimbra, Coimbra, Portugal

M.C. Azevedo
Civil Engineering Faculty, Catholic University of Portugal, Lisbon, Portugal

P. Moita
Marques SA, Ribeira Grande, Azores, Portugal

ABSTRACT: This experimental research discusses the performance of crushed granular materials resulting from volcanic rocks employed in the construction of subgrades and continuous grading unbound aggregate road base courses. Some frequent doubts related to their suitability for the aforementioned use, to their observance of the technical specifications of different countries, and to the methodology and criteria to control field compaction and bearing capacity are intended to be clarified. For this purpose some experiences of several Atlantic islands (Azores, Canary Islands, Cape Verde, Iceland) are compiled and additional recommendations are contributed for volcanic aggregates as those of Canaries or Azores. The experimental results reveal that it is possible to reach a good load-carrying capacity with these granular materials, and also that conventional tests can be used. The in-situ study has allowed us to obtain the rate of compaction as a function of the Effective Modulus of the base course foundation.

1 INTRODUCTION

Continuous grading aggregates (CGA), from volcanic pyroclastic deposits or from mechanically or natural crushed volcanic rocks such as basalts or phonolites, are typically applied to the construction of road subgrades and unbound granular subbase and base courses in volcanic regions. This work focuses on an experimental study carried out in a local road in Azores (Portugal) related to the construction of an unbound base course with a basaltic crushed all-in aggregate.

The construction project planned to build the untreated aggregate base on two different foundations along the road alignment:

– On a new subgrade constructed with *lapilli* (natural granular material from low density basaltic pyroclasts). This is also a characteristic porous granular material used in these volcanic regions, composed of a particle-size range from 2 mm to 64 mm. According to the project in these road sections, it was planned to construct a new roadbed structure including native ground excavation, fills and subgrade construction with these

pyroclastic materials, and finally spreading the unbound crushed aggregate as base course for the bituminous pavement.
– Directly on an existing pavement (after scarification and re-compaction) composed of half-penetration macadam (an obsolete traditional technique) on a subgrade formed of lapilli.

Initially, 12 trial road sections were built with the same type and origin of crushed all-in aggregate, varying the densification procedure (number of roller passes, layer thickness, base course foundation conditions and compaction equipment) in order to establish the most suitable construction procedure.

The regional road Administration had questioned the observance of Portuguese and Spanish road standard specifications for the local basaltic crushed materials, related to particle size distribution and to the utilization of the Modified Proctor test as the reference test to control field compaction, due to the possibility of not being strictly applied to volcanic aggregates such as basalts because of their greater insensitivity to moisture content variations. In addition, during the construction of the

241

Figure 1. Stockpiling of volcanic continuous grading aggregate (CGA) after moistening and mixing.

Figure 2. Trial road section: field compaction of unbound granular base course with lift thickness of 20 cm on a compacted red lapilli subgrade.

first test sections a tendency towards a more difficult densification (decreased unit weights) when compacting the granular base course on the existing pavement was verified.

In attempting to validate the utilization of these volcanic granular materials in untreated road base courses and the requirements and tests for field compaction, the object of this work has been to study the influence of foundation conditions and construction procedures on the rate of compaction.

2 EXPERIMENTAL METHODOLOGY

Two new experimental test sections of about 100 m each were constructed, where a detailed surveying program was performed with in-situ testing and sampling for laboratory tests. In both, conditions in relation to granular material origin and compaction equipment and procedures have been unmodified, and therefore only roadbed conditions have differed.

2.1 Test road sections

One experimental road section (Station 0 + 600 to 0 + 700) included spreading aggregate base on a compacted lapilli subgrade (Fig. 2). The unbound crushed stone base has been compacted with lift thickness of 20 cm and 4, 8, 12, 16, 20, 24, 28, 32 and 36 roller simple passes with a 18.5 T vibrating roller. The other 100 m long trial section (Station 3 + 600 to 3 + 700) has been constructed with a similar procedure with the difference of compacting the aggregate untreated base directly on the aged asphalt pavement after scarification and re-densification with 16 vibrating roller simple passes.

2.2 Sampling and laboratory tests

With the aim of verifying the characteristics of basaltic local material related to field compaction and to evaluate the effects of densification on its properties, especially on particle shape and grading, 7 aggregate samples have been recovered from exploration pits: one from stockpile after moistening and mixing, and 3 from each trial road section after densification.

Laboratory testing have included sieve analysis, flakiness and elongation indexes of coarse aggregate, particle density and water absorption (fractions: 0/4, 4/20 and >20 mm) and Modified Proctor test with coarse particle correction, according to Portuguese and Spanish Standards.

2.3 In-situ testing

Deflection dynamic response tests with a Falling Weight Deflectometer (FWD) Carl BRO Pri 2100 have been applied to determine differences of structural performance in the base course support. 20 FWD tests on each experimental road section had been carried out with load plate diameter of 450 mm, 20 and 40 kN of applied impulse loads, and 9 deflection sensors (geophones) on two parallel alignments coinciding with the road lane wheelpaths.

3 RESULTS AND DISCUSSION

3.1 Granular material characterization and comparison with Standard specifications

Geotechnical properties of volcanic aggregates may suffer frequent and important variations (including gradation, physical, chemical and mechanical prop-

Table 1. Laboratory test results of CGA samples obtained from stockpile and from trial road sections.

		From stock-piling	Compacted on existing pavement	Compacted on subgrade of lapilli
Flakiness index (%)		23	19	19
Elongation index (%)		29	25	20
Specific	# >20 mm	–	26.19	26.29
gravity of	# 4/20 mm	–	26.49	26.59
solids	# <4 mm	–	27.08	27.08
(kN/m^3)	Mean	–	26.78	26.78
Water	# >20 mm	–	2.1	2.0
absorption	# 4/20 mm	–	1.9	1.8
(%)	# <4 mm	–	1.2	1.8
	Mean	–	1.6	1.8
Modified	$\gamma_{d\,(max.)}$ (kN/m^3)	–	22.76	22.76
Proctor				
laboratory	$\omega_{optim.}$ (%)	–	7.5	7.5
test	$\gamma_{d\,(max.)\text{-}c}$ * (kN/m^3)	–	23.05	23.05
	$\omega_{optim.\text{-}c}$ * (%)	–	7.0	7.0

* Maximum dry unit weight and optimum moisture content with coarse-particle correction, according to ASTM D-4718-87.

Figure 3. Particle-size distribution curves of CGA samples and Portuguese and Spanish grading envelopes of road specifications.

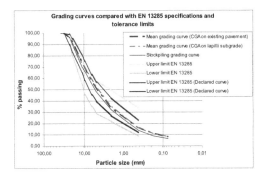

Figure 4. Particle-size distribution curves of CGA samples and allowable tolerance (with respect to manufacturer declared grading curve) and grading envelope specified in EN 13285.

erties, owing to their heterogeneous nature) caused by raw material alterations, aggregate manufacturing process (crushing, grinding and sieving), and stock-piling manipulation conditions (Franesqui 2009).

Table 1 summarizes laboratory test results. The mechanically-crushed local material has resulted slightly lacking in particle size over 9.5 mm, according to 0/31.5 grading envelope of Portuguese specifications (EPE-98), and over 2 mm in accordance with Spanish Standard PG-3 for ZA25. However, the Uniformity Coefficient and Coefficient of Curvature (Cu = 62; Cc = 1.9–2.0) indicate an easily compactable material. In the same way, analysed CGA samples are acceptably adjusted to 0/20 gradation according to French specifications (NF P-98-129).

Particle-size distributions have revealed a grading modification caused by field compaction with respect to stockpiling material, although not clearly significant (fine aggregate fraction increased by 1.5% and improvement of coarse aggregate particle shape).

The European Standard EN 13285 related to unbound mixtures states requirements regarding regularity of grading properties, and defines the tolerance limits with respect to the manufacturer declared grading curve. Figure 4 illustrates, similarly, that tested CGA samples were found to satisfy these allowable tolerances and the grading envelope established in the aforementioned Standard for the most severe categories (OC_{90} and G_A), confirming a good quality manufacturing and stockpiling control.

3.2 Control of field compaction: Criteria and tests

Field compaction trials were performed in both experimental road sections, using the same compaction equipment and only varying the number of vibratory roller passes. In Figure 5 is represented the evolution of relative compaction with the compactive effort, showing a substantial difference depending upon the granular base foundation conditions.

The minimum allowable field relative compaction (98% of M.P. maximum dry unit weight), according to EPE-98) has not been achieved when the unbound aggregate is compacted directly on the existing pavement. Consequently, experimental results demonstrate that the same volcanic material responds differently, and that the efficiency of the densification process is determined by a new factor affecting field compaction: the mechanical properties of the layer foundation.

As previously mentioned, the regional road Administration had raised objections, based on its previous experience, to the utilization of the generalized laboratory compaction tests as specified tests to control field compaction of volcanic aggregates due to their greater insensitivity to moisture content variation, obtaining quasi-horizontal compaction curves. This finding is attributable to high permeability of volcanic crushed aggregates, and therefore moisture content directly measured on the compacted lift underestimates the effectively participant water in the compaction process.

Nevertheless, Modified Proctor compaction test curves on aggregate samples recovered from stockpile and from each test road section after densification reveal that the volcanic CGA employed for base course responds to moisture variations, as can be seen in Figure 6. This figure compares water content-dry density curves from all-in aggregates of different petrographic natures (volcanic and non-volcanic), suggesting that the basaltic aggregates from Azores even achieve higher maximum dry unit weights than the others represented on the graphic.

These volcanic crushed mineral aggregates are habitually lacking in fine-grained sizes (even from continental regions), have high porosity and mean water absorption values around 3%.

On the same Figure 6, compaction curves for other basaltic crushed granular materials from some recently built roads in the Canaries and Cape Verde have been plotted. As it can be seen, all of them provide evidence of a normal performance in compaction tests.

The observation of this result not only has provided some support to the assumption that Proctor tests can be applied to volcanic materials, but also is consistent with the European Standard EN 13285

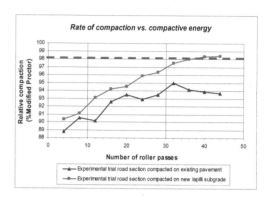

Figure 5. Variation of relative compaction with compaction effort depending on foundation conditions.

which specifies the Modified Proctor laboratory test as relative compaction test for unbound mixtures, independently of the petrologic nature of the soil or aggregate.

3.2.1 Control methodology in the Canary Islands

Volcanic granular materials utilized in the Canaries for unbound road bases are obtained from crushing basaltic and phonolitic rocks, since at the present time sedimentary deposits of gravels present a very restricted quarrying due to environmental reasons, and low density pyroclastic aggregates (lapilli) are only employed to form subgrades as they do not observe Spanish road standard specifications (PG-3) as for fracture strength, gradation and Sand Equivalent.

In Spain, the general employment requirements for unbound continuous grading aggregates (CGA) used in bases and subbases are founded on road construction Standard PG-3 (Article 510. "Zahorras"). This granular material has to observe some general conditions with respect to reduced plasticity, particle fracture strength, shape and angularity, being spread with compacted lift thickness not greater than 30 cm. End-product control criteria are based upon:

- Rate of compaction control: for heavy traffic flow >200 heavy-vehicles/day, a level of relative compaction of 100% (of P.M. maximum dry unit weight) in the construction is required. For lower traffic categories and on road shoulders, a relative compaction not less than 98% is accepted.
- Bearing capacity control by means of in-situ plate loading tests: The modulus of vertical reaction measured in the second cycle of plate load test (E_{v2}) is specified to be over 80 to 180 MPa depending on the heavy traffic conditions, and not to be lesser than the modulus of the

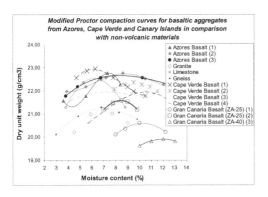

Figure 6. Modified Proctor compaction curves for some volcanic granular aggregates from different origins, compared with non-volcanic materials.

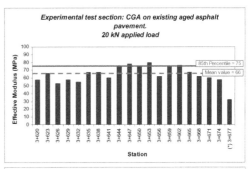

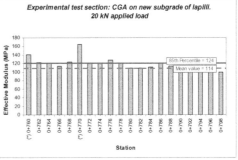

Figure 7. Effective Moduli of the CGA base course foundation, backcalculated from FWD tests (20 kN applied load). [Values marked with (*) have been excluded from percentile calculations for being unusually extreme].

layer below multiplied by 1.3. Likewise, the rate $K = E_{V2} / E_{V1}$ (E_{V1} = modulus measured in the first cycle of load) is restricted to be lower than 2.2.

Experience with volcanic materials from Canary Islands indicates that, in general, ordinary control criteria applied to granular soils and aggregates can be also implemented to volcanic unbound materials. However, the standard requirement related to the allowable limit of the coefficient $K = 2.2$ results achievable with some difficult according to previous experience, even with the lift rightly compacted. This can be explained by the lack of fine-size particles in crushed volcanic aggregates. Thus we suggest a less restrictive value of $K = 2.5$ with excellent structural performance of granular layers.

3.2.2 Control methodology in Iceland

In accordance with the Icelandic Building Research Institute (IBRI), the field compaction control is founded on attaining a minimum value of 98% of Modified Proctor maximum dry density for base courses and 95% for subbases. It is also habitual to specify the K-coefficient to be lower than 2.5 for the heaviest traffic, what provides support and is consistent with our experience with Canary materials and suggested value of K.

3.2.3 Control methodology in Hawaii

Standard specifications for aggregate road base courses of Hawaii Department of Transportation (HDT 2005) based on relative density and moisture content are also end-product specifications. Article 304.03 "Aggregate base course. Construction" states to achieve at least 95% of relative compaction (based on Modified Proctor maximum dry unit weight) and to obtain water content within 2% above or below optimum. It should be noted that maximum compacted thickness of one lift shall be 6" (15 cm) according to this Standard.

3.3 Characterization of the granular course foundation

Assuming that all significant factors affecting compaction (granular material characteristics and origin, moister content, lift thickness, compaction equipment type and compaction energy) have been unchanged in both in-situ trial road sections, and only the base course foundation conditions have varied, the differences between rates of compaction observed suggest that this latter factor has determined the compaction process. Hence, FWD tests have been carried out with the purpose of characterizing the supporting subgrades and to directly evaluate their dynamic response.

In Figures 7 and 8, the measured Effective Moduli (for 20 and 40 kN of applied load) of the CAG base course foundation are illustrated, comparing both trial road sections. The Effective Modulus is a measure of the effective or combined resilient stiffness (i.e. ratio of the applied cyclic stress to the recoverable or elastic strain after many cycles of repeated loading) of all layers below the tested surface, backcalculated from FWD defections.

The resulting values show that Moduli of the CAG base course foundation are notably different in both situations. Stiffness obtained on existing

245

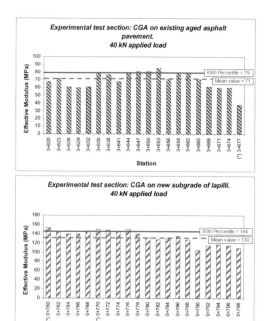

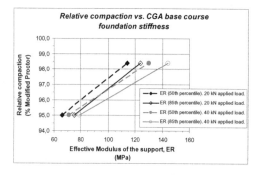

Figure 8. Effective Moduli of the CGA base course foundation, backcalculated from FWD tests (40 kN applied load).

Figure 9. Empirical relationship of Effective Modulus (of the foundation, obtained from FWD) versus granular base relative compaction.

aged asphalt pavement has resulted significantly reduced compared with that on a new roadbed structure (including native ground excavation, fills and subgrade construction with lapilli), with reductions up to 45%. The interpretation for this could probably be the advanced distress of the existing pavement structure, causing dissipation of the compaction energy without increasing the rate of densification of the new granular base course. Consequently, the observed compaction

inefficiency when the CGA layer is spread on the existing pavement is related to the reduced bearing-capacity of the latter due to the ageing.

As a result, the importance of achieving adequate stiffness of every down-up compacted granular course has been reflected in this study. When a granular layer is built on an existing pavement, previous testing (plate load test, FWD) should be performed in order to assure sufficient stiffness of its support. Figure 9 summarizes the empirical correlation between effective stiffness of the support (from FWD) and relative compaction of the granular base, found in this field experimental research.

4 CONCLUSIONS

Variability of geotechnical properties of granular aggregates resulting from volcanic rocks is generally more significant than with non-volcanic materials, thus manufacturing and reception control should be more intensive. However, laboratory test results have revealed that it is possible to attain with them satisfactory adjustment to grading and allowable tolerances of different specifications in Europe, provided that an optimum regularity of the product can be assured with the production and stockpiling control. In this way, they can be employed in road structural courses even with heavy traffic conditions.

Moreover, field experiences have confirmed that, in general, a good densification can be achieved with them if a continuous grading curve is obtained (Maximum dry unit weight: 21.6 to 23.0 kN/m³; Optimum moisture content: 6 to 11%). Grading modification caused by field compaction has not produced increasing of fine aggregate fraction upper to 1.5%, with improvement in flakiness and elongation indexes of coarse aggregate.

Despite of some previous experiences in relation to their greater insensitivity to water content variation during compaction, laboratory test results and data compiling of similar materials from several Atlantic volcanic regions have confirmed that conventional compaction tests and criteria can be also applied to volcanic materials. It is interesting to note, however, that moisture directly measured on the compacted lift underestimates the effectively participant water content in the compaction process owing to their high permeability.

Plate loading tests can contribute to a better control of the in-situ performance as they allow verifying the stiffness of the compacted layer. Experience with volcanic materials from Canary Islands indicates that, generally, the Spanish specification related to the allowable limit of the plate load test K-coefficient ($K = E_{V2}/E_{V1}$) results achievable with some difficulty, so a less restrictive value of K = 2.5 has been suggested, with excellent granular layer structural performance.

Experimental field testing has clearly identified that the efficiency of the densification process is influenced by the mechanical properties of the supporting layer. Compaction on insufficient foundation stiffness produces dissipation of the compaction energy into the underlying courses without increasing the rate of densification of the new unbound granular base layer. In this field experimental research a empirical correlation between Effective Modulus of the support (obtained from FWD tests) and rate of compaction of the employed volcanic material has been suggested, verifying that standard specifications generally required by technical Codes (relative compaction >98% of P.M. maximum dry density) only can be achieved for greater subgrade stiffness than 110–120 MPa (E2 subgrade quality, according to currently Spanish Pavement Standard 6.1-IC).

ACKNOWLEDGEMENTS

The authors thank to the Contractor Marques, SA, for their permission to use information from various unpublished reports. Special thanks to Dr. Petur Petursson for the information about Iceland Specifications and to Dr. Letizia de Lannoy Kobayashi for the information about Hawaii Specifications.

REFERENCES

Asociación Española de Normalización AENOR. 2003. *Áridos para capas granulares y capas tratadas con conglomerantes hidráulicos para su uso en capas estructurales de firmes.* UNE-EN 13242:2003. Madrid: AENOR.

Comité Français pour les Techniques Routièrs. 1994. *Assises de chaussées. Graves non traitées. Définition. Composition. Classification.* NF P 98-129:1994. Paris: CFTR.

Comité Français pour les Techniques Routièrs. 2006. *Mise en application de la nouvelle norme grave non traitée.* NF EN 13285: 2006. Paris: CFTR.

European Committee for Standardization CEN. 2003. *Unbound mixtures. Specification.* EN 13285:2003. Bruxelles: CEN.

Estradas de Portugal. 1998. *Cuaderno de Encargos 5-03 Pavimentação.* EPE-98. Lisboa: EP.

Franesqui, M.A. & Castelo, F. 2009. Áridos volcánicos en capas granulares no tratadas: control de la compactación e influencia de la rigidez del cimiento. *Carreteras* 164: 34–47.

Hawaii Department of Tansportation HDT. 2005. *Standard Specifications. 304-Aggregate base course.*

Instituto Português da Qualidade IPQ. *Agregados para materiais não ligados ou tratados com ligantes hidráulicos utilizados em trabalhos de engenharia civil e na construção rodoviária.* NP EN 13242:2005. Lisboa: IPQ.

Ministére des Transports. Direction des routes. 1984. *Mémento des spécifications francaises. Chaussées.* Paris: MT.

Ministerio de Fomento. Dirección General de Carreteras. 2004. *Pliego de Prescripciones Técnicas Generales para Obras de Carreteras y Puentes. 510-Zahorras.* Orden Ministerial FOM/891/2004.

Volcanic Rock Mechanics – Olalla et al. (eds)
© 2010 Taylor & Francis Group, London, ISBN 978-0-415-58478-4

Geotechnical investigation guide for building in volcanic environments

L.E. Hernández
Regional Ministry of Works, Government of the Canary Islands, Spain

J.A. Rodríguez-Losada
Department of Soil Science and Geology, University of La Laguna, Tenerife, Spain

C. Olalla
E.T.S.I.C.C.P., Universidad Politécnica de Madrid, Spain

J. Garrido-Manrique
Department of Civil Engineering, Universidad de Granada, Spain

ABSTRACT: The recent emergence of the Spanish Building Technical Code set a regulatory landscape where geotechnical studies become mandatory for construction in Spain. This code provides a classification for building and terrain types, depending on which performs the geotechnical research planning. It is therefore necessary to identify and classify the terrain as one of the three types defined in the code. The Government of the Canary Islands has developed a guide that will allow code enforcement to volcanic terrains of the Canary Islands. In this paper, the geotechnical units of the Canaries as well as their classification according to the code are defined. In addition, the number and type of minimum geotechnical surveys carried out in each geotechnical unit is specified as a function of the planned building. Since the Canary Islands consist of a wide range of volcanic products, this guide can be applied to any other volcanic region.

1 INTRODUCTION

The Canary Islands are located in the Atlantic Ocean, close to the northwest coast of Africa, in front of Western Sahara and Morocco. All of islands are volcanic in origin and form the Macaronesia ecoregion with the Azores, Madeira, Cape Verde and the Savage Isles.

Canary Island's territory covers an area of 7.447 km² and populations reached 2 million people in 2009. The economy is based primarily on tourism, but in recent decades, due to tourism development and population growth, the construction sector has been very important in the economy of the islands.

Due the increasing demand for quality in the buildings by the society, the Government of Spain recently published the Spanish Building Technical Code, "Código Técnico de la Edificación" (CTE), 2006). This code creates a situation where the geotechnical studies for building are needed and prescriptive.

According to the CTE, geotechnical surveys will be carried out depending on the type of building (Table 1) and the type of terrain (Table 2).

Canaries terrains of volcanic origin are very different to the Spanish mainland, so that implementation of the CTE in the islands has had some difficulties.

In the past decade canarian government has done several studies about geomechanical properties of volcanic materials. Results has created the first database with geomechanical and geochemical properties, and correlations between various parameters studied, which have been published and accepted by the scientific community (Rodriguez-Losada et al., 2006, 2007, Hernandez-Gutierrez et al., 2007). In addition, the government has developed geotechnical mapping of the Canary Islands.

Table 1. Type of building according to the CTE.

Type of building	Description
C-0	Buildings of less than 4 levels and built area less than 300 m²
C-1	Other buildings of less than 4 levels
C-2	Buildings from 4 to 10 levels
C-3	Buildings from 11 to 20 levels
C-4	Monumental, unique, or more than 20 levels

The Department of Public Works and Transportation of the Canarian Government has drafted a guide to assist the correct interpretation and implementation of national code in the Canaries: The Geotechnical Investigation Guide for Building Projects in the Canary Islands (GETCAN-10).

The guide, presented in this article, articulates the basic appropriate methodology for the planning of geotechnical prospecting in building projects and for conducting geotechnical studies concerned, in accordance with current regulations and Canarian terrains.

Logically, the guide is a very extensive document and in this paper only aspects relating to the planning of geotechnical survey are presented.

2 GEOTECHNICAL UNITS

The geology around the Canary Islands is dominated almost entirely by a succession of volcanic materials and structures.

Geotechnical mapping of the entire archipelago has been made as first step of the guide project.

A geotechnical units classification of the Canarian terrain was necessary in order to establish areas and groups of terrains with similar geomechanical behaviour (Table 3). These units have been further classified into groups of terrains that define the CTE (Table 2).

Geotechnical areas of behavior more or less homogeneous have been considered, with the

Table 2. Groups of terrains according to the CTE.

Group	Description
T-1	*Favourable terrains*: those with little variability, and where the usual practice in the area is direct foundation by isolated elements.
T-2	*Intermediate terrains*: those who show variability, or that the area does not always happen the same foundation solution, or where it can be assumed to have some relevance anthropic filled, although probable it not exceed 3.0 m.
T-3	*Unfavourable terrain*: those who do not qualify in any of the previous types. It is especially this group will be considered in the following areas:
	a) Expansive soils
	b) Collapsible soils
	c) Soft or loose soils
	d) Karstic terrains in gypsum or limestone
	e) Variable terrains in composition and condition
	f) Anthropic filled with more than 3 m thick
	g) Terrains in susceptible areas to landslides
	h) Volcanic rocks in thin layers or with cavities
	i) Land with slope greater than 15°
	j) Residual soils
	k) Land of marshes

Table 3. Geotechnical units of the Canary Islands and their classification according to CTE terrains.

Units	Subunits	CTE terrains
Unit I: Basal complex		T3e
Unit II: Salic lava flows and salic massif		T1
Unit III: Altered basaltic massif		T3h
Unit IV: Fresh lava flows	IVa: "aa" lava flows little scoriaceous	T1
	IVb: "Pahoehoe" lava flows and "aa" lava flows very scoriaceous	T3e
Unit V: Pyroclastic deposits	Va: Ignimbrites y tuffs	T2
	Vb: Loose or weakly cemented pyroclastic deposits	T3b
Unit VI: Volcanic breccias		T2
Unit VII: Alluvial-colluvial deposits		T3c
Unit VIII: Coastal sands		T3c
Unit IX: Clay and silt soils		T3j
Unit X: Anthropic filled		T3f

limitations that the scale and nature of the materials allow. These areas have a similar treatment when planning geotechnical surveys.

2.1 Unit I: Basal complex

The basal complex of the Canary Islands is represented by Cretaceous sediments, submarine lavas and plutonic rocks (gabbros and syenites). This set is traversed by numerous dykes intrusion with a density so high that often leave no trace of the rock sticking. Typically, with a high degree of alteration as rock materials are very slippery and difficult to recognition. This gives them features of soft rock RMR_b values were below 40 and can occasionally reach up to 60. They are considered as T3e terrain.

2.2 Unit II: Salic lava flows and salic massif

This unit consists of highly resistant rock materials. There are two forms of outcrop: 1) As very thick lava flows, usually with horizontal arrangement or as thick tabular packages with slopes not too steep and large horizontal extension. Sometimes these packages may consist of very compact breccias with Salic fragments and 2) As domes, like a large rock massif. In any case, for practical purposes, the geotechnical features of both types of upwelling are considered similar and should therefore be considered as the same geotechnical unit. They are massifs of trachytic-phonolitic composition, generally moderate to high bearing capacity, characteristics of hard rock and RmR_b values from 80 to 90. They are considered as T1.

2.3 Unit III: Altered basaltic massif

Composed of basaltic lava flows of small thickness (around 1 m or less) and moderate to high disturbance. The remarkable peculiarity of these basaltic lavas is a vertical alternation of compact basaltic levels (basalt rock) and scoria levels (granular material).

Pyroclastic mantles and burned paleosols may appear interspersed in these massifs.

The presence of interspersed scoriaceous levels produces a high heterogeneity, due alternation both vertically and horizontally. Overall, scoriaceous levels tend to behave like a granular soil, little or nothing compact. But these features fade in Unit III materials due to the advanced state of alteration. So, these materials show problems as expansiveness, high deformability and slope instabilities. In addition, there may be caves due to water circulation and low compaction.

They are generally soft rocks and spread mainly in the areas of outcrop of the Antique Series

(phases of formation of large shield volcanoes in the early subaerial volcanism of the Canary Islands). Usually they present in surface RmR_b values from 50 to 60. They are considered as T3h terrain.

2.4 Unit IV: Fresh lava flows

In this unit lists the basaltic lava flows that retain their original structure due to their low state of alteration, so it can distinguish the types "pahoehoe" and "aa".

"Pahoehoe" lavas are characterized by a smooth and undulating surface, although in detail they are formed by interlocking corrugated ropes. Internally is to highlight the presence of large numbers of small vacuoles or spherical voids that give them high porosity. However, the more remarkable internal detail is the presence of volcanic tunnels or tubes that can reach kilometers in length and several meters in diameter. Often during surveys those tunnels or tubes are not detected, which does not mean that they none exist.

"Aa" lavas or scoriaceous lavas have an extremely rough or spiny surface. Vertical section consists of a central band of dense rock crossed by a network of joints, limited below and above by two irregular scoriaceous bands.

The basalt rock massif levels in general have high bearing capacity, RmR_b valued between 60 and 85. However, scoriaceous levels may show low bearing capacity and high deformation, if scorias are loose and without matrix. Besides, they show moderate bearing capacity and low deformability, if they are welded or with a weakly cementation degree.

So for geotechnical surveys purposes, the Guide considers that this unit recognizes two subunits.

2.4.1 Subunit IVa

"Aa" lavas with compact basalt thicknesses equal to or greater than 2 m, while retaining its lateral continuity across the plot; with less than 0.5 m scoriaceous levels, absence of cavities and a slope of field less than 15°. They are considered as T1 terrain.

2.4.2 Subunit IVb

It includes "pahoehoe" lavas and "aa" lavas with compact basalt thicknesses less than 2 m, interspersed scoriaceous levels and/or presence of cavities. They are considered as T3e.

2.5 Unit V: Pyroclastic deposits

It consists of areas of undifferentiated pyroclastic deposits. Thickness and dip depend on the topography on which they were deposited at the time of the eruption. They can be subdivided into.

2.5.1 Subunit Va

Ignimbrites and tuffs: hard or medium hard rock corresponding to highly compact pumice or cinder pyroclastic deposits. This variety of materials occurs when a mass of pyroclastic products are transported in the form of gas dispersion and high or moderate density of particles, the result is a material with characteristics more or less hard rock, with a degree of compactness and/or variable cementation. They present in surface RmR_b values between 60 to 75 and they are considered as T2 terrain.

2.5.2 Subunit Vb

Pyroclastic materials loose or weakly cemented: Non-compact and easily collapsible. They form when magma fragments fall and settle near the eruptive center. There are two types if geochemical composition is considered: basaltic and salic. The smaller basaltics are called lapillis (2 to 15–20 mm), the largest are called scoria. Salic deposits (trachytes or phonolite composition) form pumice, light and porous. Both are considered as T3b terrain.

2.6 Unit VI: Volcanic breccias

This unit is associated with violent eruptive episodes of high explosivity. The final result is a chaotic and brecciated mass formed by blocks of different nature, generally very sharp. Grain size is variable. Matrix is fine and more or less cemented and occasionally very hard. These materials are very thick (up to hundreds of meters).

They exhibit characteristics of hard rock and sometimes of medium hardness. They present RmRb values of between 65 to 75 and they are considered land type T2 terrain.

2.7 Unit VII: Alluvial-colluvial deposits

These deposits are made of sand and very heterometric fragments ranging in size from centimeter to over a meter and rounded or subrounded forms. The matrix, of detrital nature, may be abundant or absent. They are soft or loose terrain type T3c.

2.8 Unit VIII: Coastal sands

This unit consists of beach deposits of loose basalt dark sand or silica clear sand or calcareous nature, extensions or sediment which have been transported by marine or aeolian process (dune formations). They appear along the coastline and ravine mouths. They have low to very low bearing capacity and are soft or loose terrain type T3c.

2.9 Unit IX: Clay and silt soils

These deposits consist of residual soils and lake sediments which are essentially clay and/or silty.

They are formed in the fund or semi-closed lake basins by sedimentation of fine or very fine clay-size detritus. They also could have formed by alteration by intense alteration of surface of rocky material. In both cases, the resulting material is usually silty or clayey nature. They are generally soft soil type T3j.

2.10 Unit X: Anthropic filled

They are defined as soft terrains type T3f, unfit to build buildings unless improvements or reinforcements carried out to increase their properties.

3 GEOTECHNICAL SURVEYS

Once the type of building and type of land are defined, guide criteria must apply in order to determine the minimum intensity and extent of the field exploration activities.

Recognition or surveys points relate to boreholes in this guide. Here, in this guide, boreholes are exploration drilling carried out by the rotation system with continuous extraction of intact cores.

3.1 Number of recognition points

At least three points of recognition is established and a maximum distance (dmax) between survey points. For the purpose of this guide, the maximum distances between points of recognition are showed in Table 4.

For T-3 terrains or when the survey is derived from another which has proved inadequate, recognition points are interspersed in the problem areas until they are sufficient to characterize the ground properly. Absence of singularities under the foundation level and excavation fronts must be checked in order to ensure the security of the building, construction process and neighboring buildings.

In geotechnical units III and IVb, the surveys will be carried out as Table 4 indicated, by the rotation system with continuous extraction of intact cores. Additionally, exploration will also be required under each load transfer element of the

Table 4. Maximum distance (dmax) between survey points.

Building type	T-1	T-2	T-3(*)
C-0, C-1	35	30	60
C-2	30	25	50
C-3	25	20	40
C-4	20	17	34

(*) Apply only for geotechnical units III and IVb.

Table 5. Minimum survey points (boreholes) and replacement percentage by continuous penetration tests, according to CTE.

Building type	Minimum survey points (boreholes)		Replacement (%)	
	T-1	T-2	T-1	T-2
C-0	–	1	–	66
C-1	1	2	70	50
C-2	2	3	70	50
C-3	3	3	50	40
C-4	3	3	40	30

Table 6. Minimum depth of prospecting.

Geotechnical units/ Building type	Minimum depth of prospecting (m)										
	I (T-3)	II (T-1)	III (T-3)	IVa (T-1)	IVb (T-3)	Va (T-2)	Vb (T-3)	VI (T-2)	VII (T-3)	VIII (T-3)	IX (T-3)
C-0	5	4	5	4	5	5	5	5	5	5	5
C-1	8	6	8	6	8	7	8	7	8	8	8
C-2	12	8	12	8	12	10	12	10	12	12	12
C-3	16	10	16	10	16	12	16	12	16	16	16
C-4	20	12	20	12	20	14	20	14	20	20	20

structure on the ground. In this case, the prospecting that exceeds the number of the application of Table 4, may be made by rotary-percussion drilling system, under the technical direction and supervision of an expert in the geotechnical survey.

In the event that the maximum distance (d_{max}) exceeds the dimensions of the study area, distances should be decreased until they reach the required minimum number of points. The maximum distance can be considered as the radius circle of the influence areas of the recognition points. The influence circle areas of the research points obtained must exceed 90% of the contact area with the ground.

In the case of buildings with floor area exceeding 10,000 m², the density of points on the surface excess can be reduced. This reduction should not exceed 50% of survey points obtained from applying the previous rule.

The initial number of survey points may be replaced by continuous penetration tests, in those geotechnical units that support this technique, in the percentage indicated in Table 5.

3.2 Depth of recognition points

The depth achieved will be one in which there is no significant settlements under the loads transmitted by the building. This depth can be that where net increase of tension in the ground, by the building's weight, is equal or less than 10% of the vertical effective stress, in that level point before building

construction. These considerations will be valid unless it has previously reached a geotechnical unit resistant (bedrock) such that the pressure applied on it by the foundation of the building produce significant deformations.

However, the minimum depth surveys has been established taking into account the peculiarities and problems associated with each of the geotechnical units and type of building, as shown in Table 6.

The depths indicated in Table 6 are referred to the final level of excavation. To these are added, if necessary, the thickness of artificial fillers or final depth of excavation to achieve the planned foundation level.

In the case of rotary-percussion drillings that can be made in Geotechnical Units III and IVb the minimum depths to reach are those given in Table 6 as well.

4 CONCLUSIONS

The geotechnical investigation guide for building in volcanic environments is a most useful document for architects, engineers and geotechnical experts. It is a reference document for the geotechnical investigation in volcanic environments, as it takes into account the peculiarities of these materials and applies the experience of many decades building on them.

The guide defines the different geotechnical units present in the Canary Islands and gives them geomechanical parameters. These units are similar to those found in other volcanic regions of the planet. Because in the Canaries is possible to find almost the entire spectrum of possible volcanic events, this guide is applicable to any volcanic environment.

The guide provides an operating procedure for each geotechnical unit and defines the minimum number of exploration and depth, depending on terrain types studied.

ACKNOWLEDGEMENTS

To Regional Ministry of Works and Transportation of the Government of the Canary Islands.

REFERENCES

Consejería de Obras Públicas, Vivienda y Transporte de la Región de Murcia, 2007. Guía de Planificación de Estudios Geotécnicos para la Edificación en la Región de Murcia Adaptada al Código Técnico en la Edificación.

Ministerio de Vivienda, 2006. Código Técnico de la Edificación (CTE).

Hernández-Gutiérrez, L.E., Rodríguez-Losada, J.A. 2006. Estimative rock mass parameters applied to the Canarian volcanic rocks based on the Hoek-Brown failure criterion and equivalent Mohr-Coulomb limits as a contribution in natural hazards. 300th Anniversary Volcano International Conference (GARAVOLCAN), Session 1. Garachico, Tenerife.

Rodríguez-Losada, J.A., Hernández-Gutiérrez, L.E. 2006. New geomechanical data of the Canarian volcanic rocks as a contribution for geophysics applied to the research in volcanic risk. 300th Anniversary Volcano International Conference (GARAVOLCAN), Session 2. Garachico, Tenerife. 22–26 de mayo de 2006.

Rodriguez-Losada, J.A., Hernandez-Gutierrez, L.E., Olalla, C., Perucho, A., Serrano, A., Rodrigo del Potro, 2007. The volcanic rocks of the Canary Islands. Geotechnical properties. Proceedings of the International Workshop on Volcanic Rocks W2. 11 ISRM Congress. Ponta Delgada (San Miguel, Azores). Session 1, Characterization of volcanic formations, 53–57.

Volcanic Rock Mechanics – Olalla et al. (eds)
© 2010 Taylor & Francis Group, London, ISBN 978-0-415-58478-4

Tunnel inventory of Grand Canary Island (Spain), geology and associated geotechnical problems

J.R. Jiménez & A. Lomoschitz
Department of Civil Engineering, University of Las Palmas de Gran Canaria, Canary Islands, Spain

J. Molo
Área de Carreteras, Consejería de Obras Públicas y transportes, Gobierno de Canarias, Spain

ABSTRACT: In the last two decades more than 30 road tunnels has been built in Grand Canary Island, using a variety of construction methods and reaching a total extension of 14.66 km. They are 156 to 1200 m long, with an average length of 458.20 m per tunnel. The inventory of tunnels includes basic design data (i.e., geometry, widths and number of lanes, velocity), geological formations and materials and the most relevant geotechnical problems. The article shows three tunnel examples which were excavated in different volcanic rocks: (1) Basanite pyroclast and lava flows in the Tafira ring road (GC-4); (2) Phonolite lava flow and agglomerates at La Laja (GC-1); and (3) Phonolitic, trachytic and rhyolitic ignimbrites and lava flows at Arguineguín-Puerto Rico stretch (GC-1). It is intended that this tunnel inventory could be useful for future projects and works in the Canary Islands.

1 INTRODUCTION

1.1 *Situation and interest of the topic*

The Island of Gran Canary is placed in the center of the Canary archipelago (Fig. 1). Grand Canary spreads over 243 km of coast, with long beaches of sand, placed predominantly in the south, and with areas of dunes and sandbanks now in the south of the island, and until a few years ago in the north of the island between La Isleta and Grand Canary.

Coves and beaches, big cliffs, extensive mountains, well-preserved craters and impressive ravines with subtropical vegetation and nature reserves are some of his many characteristics, which mark the big present changeability in Gran Canary.

1.2 *Situation and interest of the topic*

Until 1970 only existed in Grand Canary two tunnels, Tenoya's tunnel, 200 meters long, a section that shares a pedestrian sidewalk and has only one lane of traffic of alternative sense, and the tunnel of La Punta de La Laja, less than 100 meters length and two-way traffic, demolished in 1972 when he switched to four-lane highway GC-1.

In the early seventies the three tunnels were built from the C-821 between Arguineguín and Puerto Rico.

In 1981, entered service a tunnel 80 meters long, two lanes, and over 400 linear meters of artificial

Figure 1. Canary Islands. Localization of Grand Canary Island.

three-lane tunnel. Thus began the era of artificial tunnels to solve slope stability in big works.

But it was in 1990 when start in Gran Canary the tunnels Julio Luengo at the northern entrance to Las Palmas de G.C., and the two pairs of tunnels on the motorway GC-1 in Pasito Blanco and Arguineguín.

In 1994 the tunnel of the water maker was built, to three lanes of the roadway south of the GC-1 solution to support the future connection of the bypass to Las Palmas.

It is from 1998 when building the new tunnel Adolfo Cañas, for three lanes of the roadway itself cited above, a length of 1250 meters, including artificial tunnels in both access, thus saving the La Laja beach and rearrange the access to the south of the capital of the island. Linares, H. (1989).

2 GEOMETRY AND TECHNICAL CHARACTERISTICS

2.1 List of tunnels

The interest that reaches this high number of tunnels have been built on the island, is strengthened by the diversity of geological materials, volcanic and sedimentary, they have crossed.

This information is invaluable for future tunnel projects that will executed.

The aim of this paper is to show an inventory of tunnels built till this day, the design features used and the geotechnical properties of the materials traversed.

As an example we have taken three real examples of tunnels implemented on different lithologies of volcanic origin.

As a summary of tunnels implemented so far today in Grand Canary, it shows a tables of those in which shows the basic design data and descriptive projects.

3 GEOLOGICAL AND GEOTECHNICAL SETTINGS

3.1 General geology

Geological variability traversed by tunnels implemented is very diverse, from pyroclastic or Ignimbrites with basanite to phonolitic composition, trachytic and rhyolitic ignimbrite lava flows. Some sedimentary materials have been presents.

4 GEOMETRY AND TECHNICAL CHARACTERISTICS

4.1 Geology examples

Below are three examples of projects put in practice according to the geology in the project area.

4.2 Basanite pyroclasts and lava flows

Located in the Tafira ring road (GC-4), the double false tunnel Los Siete Lagares is around pyroclastic deposits and basanite lava flows of very dark shades.

The constructive method was executed by execution of the trench and subsequent execution of the tunnel structure through where stirrup rests triarticulated arch formed by prefabricated plates. Its length is 225 meters in two entrances, featuring 3 lanes in total. (Fig. 2). His execution was in 1999.

4.3 Phonolite lava flow and agglomerates

Situated just off the La Laja beach (GC-1), the tunnel named Adolfo Cañas crosses a powerful

Figure 2.   Los siete Lajares tunnels.

deposit of phonolite lava flows with volcanic agglomerates in the base.

The constructive method used was new austrian tunneling method, initiated by a particular stage of development and of smashes another basically doing a shotcrete support and bolts.

Were used from the outset two jumbo robots that allowed to drill a whole section.

Among the major geotechnical problems detected mainly highlights from the column stratification phonolite producing many flat roofs, (Fig. 3).

The correction used was rock reinforcement through the use of reinforced shotcrete in structural interaction with rock bolts of steel-fibre-reinforced shotcrete and Bernold type sheet system.

4.4 Phonolitic, trachytic, rhyolitic ignimbrites and lava flows

The stretch Arguineguín-Puerto Rico is a site of four twin tunnels bi-tube type unidirectional and one bidirectional, that in just two miles away is a point of reference for the entire island. The geology traversed is diverse, with a variety of materials Phonolitic composition and sedimentary deposits as conglomerates and slope deposits.

The twin tunnels Ingeniero Heriberto Linares, with up to 1188 meters in mine on the land side.

Explosives are used to dig through entire section. The reinforcement was performed with shotcrete and rock bolts, except for some short stretches where use trusses. For sizing was used the Active Structural Design, which uses the classification of Bieniawsky (1979) to estimate the properties of rock mass by laboratory tests and thus holds by calculating the stress-strain.

The tunnels entrances were performed with slopes. The inclination was 1(H):2(V).

Table 1. List of basic characteristics of implemented tunnels in Grand Canary Island. East zone.

Denomination	Longitude m	Constructive procedure	Geological formation	Geological materials	Geotechnicals problems
San José	570	No explosives. Trench.	Conglomerate/ pumice	Clasts, sands, clays	Detachments, water
Pico Viento	180	No explosives, Trench	Cemented conglomerate	Clasts, sands, clays	No problems
Santo Domigo*	275	Advance-destroy, Bernold	Cemented conglomerate	Clasts, sands, clays	No problems
Santo Domigo**	279	Advance-destroy, Bernold	Cemented conglomerate	Clasts, sands, clays	No problems
Pedro Hidalgo	464	Complete section	Cemented conglomerate	Clasts, sands, clays	No problems
Salto del Negro*	275	Advance-destroy, Bernold	Cemented conglomerate	Clasts, sands, clays	No problems
Salto del Negro**	283	Advance-destroy, Bernold	Cemented conglomerate	Clasts, sands, clays	No problems
Sabinal	156	Advance-destroy, Bernold	Cemented conglomerate	Clasts, sands, clays	No problems
Acceso a Tafira	253	Complete section	Cemented conglomerate	Clasta, sands, clays	No ploblems
Marzagán*	391	Advance-destroy, Bernold	Cemented conglomerate	Clasts, sands, clays	No problems
Marzagán**	371	Advance-destroy, Bernold	Cemented conglomerate	Clasts, sands, clays	No problems

* Sea side. ** Land side.

Table 2. List of basic characteristics of implemented tunnels in Grand Canary Island. South zone.

Denomination	Longitude m	Constructive procedure	Geological formation	Geological materials	Geotechnicals problems
Salvaje*	294	Advance-destroy, Bernold	Conglomerate	Phonolitic conglom.	Rocks fall
Salvaje**	289	Advance-destroy, Bernold	Conglomerate	Phonolitic conglom.	Rocks fall
Galeón*	405	Advance-destroy, Bernold	Conglomerate	No welded Ignimbrite	Rocks fall
Galeón**	405	Advance-destroy, Bernold	Trachytic/Riolitic	No welded Ignimbrite	Rocks fall
Pino Seco*	532	New Austrian T. Method	Phonolitic/Riolitic	Ignimbr./pyrocl. flows	Wedge-shaped rocks
Pino Seco**	407	New Austrian T. Method	Phonolitic/Riolitic	Ignimbr./pyrocl. flows	Wedge-shaped rocks
Balito*	349	New Austrian T. Method	Phonolitic/Riolitic	Ignimbr./pyrocl. flows	Wedge-shaped rocks
Balito**	349	New Austrian T. Method	Phonolitic/Riolitic	Ignimbr./pyrocl. flows	Wedge-shaped rocks
Ing. H. Linares*	1.162	New Austrian T. Method	Phonolitic/Riolitic	Ignimbr./pyrocl. flows	Wedge-shaped rocks
Ing. H. Linares**	1.184	New Austrian T. Method	Phonolitic/Riolitic	Ignimbr./pyrocl. Flows	Wedge-shaped rocks
El lechugal*	228	New Austrian T. Method	Phonolitic/Riolitic	Ignimbr./pyrocl. flows	Wedge-shaped rocks
El lechugal**	225	New Austrian T. Method	Phonolitic/Riolitic	Ignimbr./pyrocl. flows	Wedge-shaped rocks
Motor Grande	284	New Austrian T. Method	Phonolitic/Riolitic	Ignimbr./pyrocl. flows	Wedge-shaped rocks

* Sea side. ** Land side.

Table 3. List of basic characteristics of implemented tunnels in Grand Canary Island. North zone.

Denomination	Longitude m	Constructive procedure	Geological formation	Geological materials	Geotechnicals problems
Tenoya	294	Traditional method	Conglom./basaltic flow	Clasts and lavas	Rocks fall
Ing. J. Luengo*	593	Advance-destroy, Bernold	Conglomerate	Phonolitic conglom.	Rocks fall
Ing. J. Luengo**	613	Advance-destroy, Bernold	Conglomerate	Phonolitic conglom.	Rocks fall

* Sea side. ** Land side.

Figure 3.   Adolfo Cañas tunnel. Construction phase.

SECCIÓN TIPO TÚNEL BIDIRECCIONAL

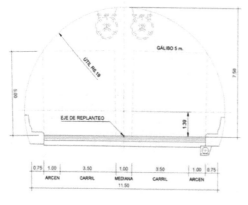

Figure 4.   Ingeniero H. Linares section tunnel view.

5 CONCLUSIONS

5.1 *Generals considerations*

The geometry and geomorphology of Grand Canary Island, has created a need to carry out work in the form of tunnels underground to allow for road improvements in communications between cities and touristic zones. The construction of such tunnels has enabled the study of the geomechanical behavior of many volcanic materials with different chemical compositions, the type of excavation and the construction method used in the preparation of each tunnel.

5.2 *Constructive methods and support elements*

The most used method has been the new austrian tunneling method, with advance-destroy and advance with complete section. The geotechnical problems more common have been the blocks fall in wedge and flat roofs due to the morphology of the gatecrashers and crossed deposits. The most used elements have been the bolts, the projected concrete and the revetment.

Materials have been considered favorable for excavation and execution phases, although some specific problems arise locally due to the solid state that may have different local properties. The variety of geological materials are a constructive reference for future tunnels.

REFERENCES

Peiró, R. 1997. Caracterización geotécnica de los materials volcánicos del archipiélago canario. *Tierra y Tecnología* 17(1): 45–49.

Linares, H. 1998. Túneles en Las Palmas de Gran Canaria. *III jornadas de carreteras. Geotécnia vial y túneles, Las Palmas de Gran Canaria, 16–17 April 2008.*

González de Vallejo, L. & Hijazo, T. 2008. Engineering geological properties of the volcanics rocks and soils of the Canary islands. *Soils and Rocks, Sao Paulo, January–April 2008.* 31(1): 3–13.

Tamames, B. & Fernández, F. 2002. Construcción de los túneles del tramo Arguineguín-Puerto Rico. *Ingeopres*, 104(1): 58–68.

Volcanic Rock Mechanics – Olalla et al. (eds)
© 2010 Taylor & Francis Group, London, ISBN 978-0-415-58478-4

Geotechnical characterization of El Verodal Tunnel in El Hierro, Spain

M.C. López-Felipe
Estudios del Terreno S.L., Spain

J.T. Fernández-Soldevilla
Obras Civiles de Tenerife S.L., Spain

ABSTRACT: El Hierro island, 0,2 million years old, is the youngest of the Canary Islands. Since it was recognized UNESCO Biosphere Reserve in 2000, the road infrastructure to be planned has to cause a minimum impact on the landscape of the island. This applies to El Verodal Tunnel to be built in the NW side of the island. Due not only to orography but also to administrative constraints, site studies are scarce. Moreover, the implementation of the existing rock mass classifications for volcanic rocks makes the design of the tunnel harder due to the great variability shown by these materials.

1 INTRODUCTION

The island of El Hierro with its 268 km² and 1500 m altitude is the westernmost, smallest and geologically youngest of the Canary Islands. Due to its geographical position at western edge of the known world, it was referenced as meridian 0 for several centuries. It is the emerged part of an oceanic volcano, mainly basaltic composition, whose submarine base stands on the Atlantic abyssal plain, about 3700–4000 m depth.

2 STRATIGRAPHY DESCRIPTION OF GEOLOGICAL UNITS

Field criteria, as well as geochronological and paleomagnetic criteria have been used for the general set of stratigraphy. Given these considerations, the following volcanic edifices can be considered:

2.1 Tiñor Edifice
2.2 El Golfo-Las Playas Edifice
2.3 Volcanism of the ridges or the structural axes
   Volcanism of the ridges ss
   Emissions that fill El Golfo valley
   Recent-subrecent emissions

2.1 *Tiñor Edifice (1,2-0,88 Ma) million years*

It is the oldest emerged area of the island. It is located in the NE area and inside El Golfo-Las Playas Edifice headwall of the landslide scarp. This edifice was mainly developed from NE rift.

There are 3 units to be considered: lower, middle and upper.

The lower section is characterized by a stack of thin lava interspersed with pyroclastic levels crossed by a strong network of dikes. The middle section, known as Tabular or "Plateau de San Andrés" consists of nearly horizontal lava flows. The upper section is characterized by the emission of pyroclastic rocks with subordinate lavas, known as Group Ventejis Picos-Moles Volcanoes.

Due to the rapid growth of the island, slope instabilities occurred. They gave place to the landslide of the west side of Tiñor Edifice, known as "The Tiñor landslide". Though it is currently hidden by the subsequent emission of the El Golfo-Las Playas Edifice and volcanism of the ridges, this landslide has been inferred from surface geological studies and information extracted from water galleries.

2.2 *El Golfo-Las Playas Edifice (545-176 Ka)*

The magmatic activity in the area restarts WSW Tiñor Edifice after a rest period of 350 ka. The resulting edifice is called "El Golfo-Las Playas Edifice", as it outcrops in this area.

This is the largest shield volcano in the evolution of the island with 20 km basal diameter and 2000 m high. However, the visible portion is about 600 m high. This is so because it was mostly plunged into the ocean, ("Landsliding of El Golfo") and because it is covered by recent eruptions.

Lava flows are nearly horizontal in its central area but reach strong dips towards the outer area.

Concerning stratigraphy, there are two units to be considered: the lower and higher: The lower unit is characterized by a predominance of pyroclastic and hydromagmatic deposits with subordinated basaltic lava flows and a strong intrusion of dykes. The higher unit has an accumulation of trachybasaltic and trachytic lava flows with subordinated lapilli intercalations.

This edifice began its process of gravitational destruction at the same time it was being built, firstly with the landsliding of Las Playas I, then El Julan, then again Las Playas II and finally the great collapse towards the N about 21,000 years ago: the landsliding of El Golfo. Avalanche deposits of these landslides have been identified on the seabed.

2.3 Volcanism of the ridges or the structural axes (158 Ka-2,5 Ka)

This third edifice develops from rifts that converge in the center of the island with angles of 120°. For

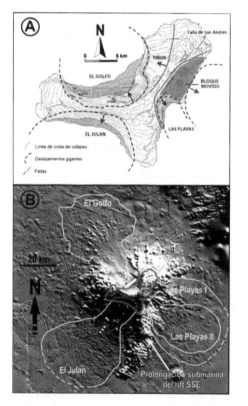

Figure 1. Landslides occurred in El Hierro. A) Map of the scars of the lanslide, indicating the resulting paleoshorelines after the collapses (modified from Carracedo et al., 1997; B) Image of the seabed around El Hierro obtained from multibeam bathymetry data (modified from Masson et al., 2002).

cartographic purposes there are 3 units: the dorsal emissions ss, emissions that fill El Golfo valley and recent and subrecent emissions.

The emission points in this edifice are those located along the structural axes.

Most of the emission points which filled the valley of El Golfo are located at the headwall of the landslide scarp. Lava emissions were very large and partly filled the depression, reaching depths of 200–300 m (data obtained from water galleries). They form the marine platform.

Recent and subrecent emisssions appear on the ends of the structural axes. They form sharp and rough scoriaceous lava flows (locally known as "malpaís") and well preserved volcanic cones.

3 GEOTECHNICS FOR THE CONSTRUCTION OF A TUNNEL

The "Project for the improvement of the HI-50 road, Section: Frontera-Sabinosa (Verodal connection)" involves the construction of a tunnel, known as "The Verodal Tunnel", 589.60 m long. Both ends will be artificial tunnels.

Only two rock drillings have been made, due to rough orography, one at the beginning and another at the end.

With this information along with geological mapping, it is inferred that the tunnel will affect two types of geotechnical units: basaltic pyroclasts from El Golfo-Las Playas Edifice (PBG) and volcanism of the ridges (PBD) as well as basaltic lava flows from El Golfo-Las Playas Edifice (CBG).

Two units can be distinguished within basaltic lava flows: central massive area and upper and lower scoriaceous areas.

3.1 Massive areas of the lava flows

The geotechnical characteristics of these types of materials from the viewpoint of geomechanics classification are shown below (Table 1).

Table 1. Bieniawski rock mass classification for massive areas of basaltic lava.

Bieniawski classifications (1989)		Score
Strength of intact rock material (MPa)	90	7
RQD	40%	6
Spacing of discontinuities	0.3–1	10
Condition of discontinuities	Rough undisturbed	19
Groundwater	Dry	15
Rating adjustment for discuntinuity orientation		10%
RMR		47

Table 2. Excavation and support proposed based on Bieniawski 1989.

RMR class	Excavation	Reinforcement		
		Bolts	Shotcrete	Steel sets
III.	Top heading and bench 1.5–3 m advance Complete support 20 m from face	Systematic bolts 4 m long spaced 1.5–2 m in crown and walls. Wire mesh in crown	5–10 cm in crown and 3 cm in walls	No

Therefore, it is considered a Class III rock, medium quality. The time of stability for an unsupported length of 10 m is 1 week. For this type of excavation (road tunnel) the kind of support to be used is shown below (Table 2).

Barton Rock Mass Classification is based on the RQD index and on a number of parameters such as: index of jointing, roughness, degree of alteration, coefficient for the presence of water and SRF (Stress Reduction Factor).

It gives a quality index "Q" which is not lineal (as RMR) but exponential, ranging from 0.001 to 1000, from very poor rock to exceptionally good. In this case, to determine the Q of Barton has been correlated with RMR Bieniawski, by the following expression: $E = e \exp RMR-44/9$.

The Q value obtained is 1.3, so that it would be classified as a rock of poor quality. From this value and with the ESR (Excavation Support Ratio) for this type of excavation (road tunnel) the kind of support to be used is 4: systematic bolting, shotcrete 40–100 mm: B + S.

3.2 *Toplayer and downlayer scoriae*

From a geotechnical point of view, scoriae are considered a granular soil with angular and subangular gravel with prickly rough surfaces involved in sands. This determines that there is good interlocking between them, giving greater compactness to the whole.

Geotechnical characteristics depend on its degree of compaction, associated in most cases with its geological age. Thus, scoriae from lava flows from the El Golfo-Las Playas Edifice show a high degree of welding, which confers certain compactness. On the other hand, scoriae from the volcanism of the ridges, have a low degree of welding so that they appear almost loose.

According to the classifications of Roman (1980) they are considered a weak or soft rock or a bad soil. Excavation in these loose materials carries a high risk of detachment. That is why inmediate support will be needed. Special treatments of reinforcement will have to be used.

3.3 *Basaltic pyroclastics*

In general, the geotechnical characteristics of pyroclastic deposits depend on the density of the welding between the particles, their imbrication's overlap and the degree of alteration of the deposit. The greater the overlap and welding the higher particle density and compactness of the deposit and, therefore, its bearing capacity.

The overlap is the number of contacts of a particle with surrounding particles. In turn, the weld is related to the percentage of areas of welded contacts respects to the total area of the particle.

Regarding outcrops, basaltic pyroclasts from El Golfo-Las Playas Edifice have a higher degree of welding that those from the volcanism of the ridges ss since they are so young that they have not undergone alteration processes to increase its compactness.

Given the appearance presented by these materials in the outcrops, they have been considered a priori as weak or soft rock or a bad soil according to Roman (1980). For these reasons, they will require support immediately after excavation as with the scoriae.

These treatments consist of installing a micropile umbrella grouting around the tunnel. If the area of loose pyroclasts continues after completion of the first umbrella, it would require the placement of successive umbrellas, with a minimum overlap between them of 2–3 m.

In areas where basaltic pyroclastics have a higher degree of welding (more compact) it is recommended to work with short advance in top heading (1 to 1.5 m) reinforcing with shotcrete (12–15 cm) and steel sets to prevent the collapse of the section.

4 USUAL PROJECTION OF THE SUPPORT OF A TUNNEL

Once the geometry of the tunnel for the use it is required is defined, its execution and supporting methods have to be defined, trying to ensure its safety.

The present basis for the design and definition of these activities go through a classification of the ground to be excavated, so that we can assume its behavior.

All rock mass classifications regarding the excavation of a tunnel are based on the assumption that the ground is homogeneous around the section concerned. This is so from the oldest by Lauffer, Terzaghi, Protodiakonov, etc. to the current by Bieniawski, Grimstad and Barton etc. The latter are usually used along with failure criterion (Hoek-Brown type or similar).

A number of parameters linked to the quality of the material to be excavated have to be defined as a unit: cohesion, friction angle, resistance to compression, fractures, families and guidance of diaclases, filled joints, shaped edges, presence of humidity.

Added to that are the characteristics of the work to be performed: dimensions, types and possible consequences of their defects (water works, roads, caves, nuclear plants, etc.).

On these premises, supporting is defined theoretically or empirically, backed by practice and the experience of well known authors.

As soon as an alteration of key assumptions such as homogeneity and flatness occur (eg proximity of the mouth, opening galleries across, crossing a field with a certain "homogeneity" to another with a different one, crossing faults or fractures, etc), the methodical rules become evanescent and a series of recommendations such as reducing advances, measuring and controlling convergence appear. According to them, the density or type of bolts, thickness and type of support, etc, will have to be redefined.

In short, everything will depend upon the judgement of the technicians, their experience and other conditioning such as financial circumstances, deadlines, etc.

In a volcanic environment, there is a succession of basaltic lava flows, from 1, 3 or 5 m thick, very variable in dips and directions along the drive. Along with them, there are layers of scoriaceous material of different thickness coexisting in a single advancing front.

Also, too often unfortunately, these discontinuous layers of basalt-scoria are interrupted by large bags of scoriae or pyroclasts. In some cases, these layers show some welding which gives them some "cohesion", but it can easily disappear with the vibrations of the excavation.

As it is easily understood, rock mass classifications provide very little information for these areas. This does not apply if the thickness of basaltic lava flow layers are very large related to the size of the work and loosely cover the entire section of the excavation and its surroundings. Moreover, scoriae

and pyroclasts are rather similar to grounds with high friction and little or no cohesion (welding).

Now we will approach the methodology used to advance in these areas. It has been applied in other projects such as Parador Tunnel in El Hierro in 1990, La Cumbre Tunnel in La Palma in 1998, Los Roquillos Tunnel in El Hierro in 2004, a tunnel in Timijiraque variant in El Hierro, which so far are having a satisfactory performance.

In the case we are dealing with we will only face two situations:

4.1 Progress in succession of basaltic lava flows (hard and soft elements)

Barton and Bieniawski Geomechanics classifications were used to massive areas of basaltic lava, from data obtained from rock drillings. We assume that the thickness of the lava flow can be taken into account for resistance. Considering the width of this excavation (~13 m), it will be considered a 10% of that width. With these classifications, we strengthen and improve the bearing capacity of these strata using empirical methods of supporting on their recommendations:

- Systematic bolting will sew the possible discontinuities, providing a strong "uniformity" and even some unstable element may be fixed.
- Fibre reinforced or unreinforced shotcrete, of various thicknesses depending on the characterization of materials.

This system, known as the New Austrian Tunneling Method NATM (and gives a bit of blush keep using the term "new"), takes into account the self-bearing capacity of the ground, plus the supports, to stabilize the stress and load that it transmits to the surrounding areas, and reaching stability, which is verified by measuring the convergences. They show if the thickness of shotcrete are sufficient or should be increased.

However, though with this method the excavation allows a smooth advance through this kind of grounds, bolting and shotcrete guarantee that those uplayer lava flows support the loads over them: in this case the weight of the scoriae between a lava flow and the next, loose elements among them, and so on.

Shotcrete, besides being a resistant element (whose main mission is to avoid an initial and uncontrolled decompression), provides stability to the walls against the active stress (usually very small in these cases) caused by the decompression and protects areas resulting from excavation, roof and walls, from the alteration that can lead to degradation of the bearing capacity.

Of course, although not very often, it may be that the lava flows are so fragile and weak that we

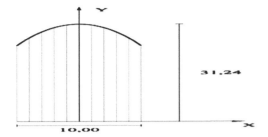

Figure 2. Parable of loads on the tunnel.

have to use stronger supporting methods: trusses, mesh, greater thicknesses of shotcrete and even rigid protective umbrella in advance.

4.2 Bag of pyroclasts more or less welded and other non-competent materials

In our opinion, the problem of supporting and bearing capacities according to soil mechanics classical hypothesis (Terzaghi, Protodiakonov and others) should be approached here. Unfortunately, geomechanics classifications are not applicable nowadays.

If objective data is available, and if the whole tunnel is dug through this homogeneous section, not only the application of computer calculations (finite elements, etc) but also the formulae of authors such as Protodiakonov may be valid to establish the bearing hypothesis. This applies not only to calculate the support (usually rigid) of the entire section of the tunnel, but also to calculate the stability of a slab on grade formed by the last basaltic lava flow over the crown, from where this homogeneous ground starts.

Here we consider both cases: that of the last lava flow, which we will find in the transition from lava flows to pyroclasts and the support to build inside pyroclasts.

In both cases the load to bear will be obtained from Protodiakonov method, which we consider very appropriate for this case.

The parable of charge on the tunnel, having an influence on the walls, with a pyroclastic density of 1.60 kg/cm^3 and a friction angle of 30° is:

$$y = -0.2765\ x^2 + 31.24$$

4.2.1 The case of the final lava flow as a carrier for these loads

From the data obtained, strength index and jointing, we get the calculation of the final lava flow, (derived from rock drilling #2, depth 102.80 to 104.60, from transition of lava flows to pyroclasts). Based on these and the recommendations of Bieniawski, Grimstad and Barton, bolts and shotcrete are used to approach homogeneity.

We assume then a beam with a simply supported beam with span of 10 m (very conservative hypothesis as its free span is really shorter due to the shape of the dome and also it might be considered at least some elastic embeddings at the ends).

The area of the parable of charge within 10 m of the alleged beam span is:

$$2 \cdot \int_0^5 y\ dx = 28358\ kg$$

representing a uniform overload of less than 28.940 kg/m. This load is increased by 60% (Protodiakonov recommended 50%).The maximum bending moment is found to be:

$$1.6 \cdot 28.94 \cdot 10^2/8 = 5,788,000\ mN$$

The average resistance to compression of the basalt is, according to tests made, 1 MPa. It is very likely to present a series of faults and joints which will confer an anisotropy, although this aspect of understanding the negative effects of these discontinuities, if it is not true fractures or cracks of considerable size, are not as fearsome as if we spoke of flexotraction or pure traction. Moreover, systematic bolting, which is designed following the indications of Bieniawski and Barton, tends to homogenize the rock mass. Given all this, we apply a load decrease coefficient of the resistance of 0.50, quite conservative, while also estimating the edge of the lava flow to prevent from appearing thrusts:

$$c = [(12 \cdot 578.80)/5000]^{\frac{1}{2}} = 1.18\ m$$

The borehole shows that this lava flow is approximately 1.60 m thick, though this number is not so reliable until checked.

The parameters that do offer actual margins of safety are:

Free span (10 m) of the lintel

- Double-supported beam hypothesis
- Density 1.60 to pyroclasts
- Friction angle 30°
- 1.60 load increase coefficient
- 2.0 load decrease coefficient (50%) of the uniaxial compressive strength.

We also have the added thickness of shotcrete which we define according to geomechanics classifications and which is is capable of absorbing substantial effort.

In this case of alternate lava flows the horizontal stress on the walls is much lower because the basaltic lava flows prevent vertical loads directly influence the horizontal thrusts.

4.2.2 *Advance in homogeneous bags or non competent material*

If one has to design a rigid support, a fairly common system is to use the formula of the thin tubes assuming an uniform stress on the perimeter of the tunnel.

We will continue to calculate the loads using Protodiakonov method and project with them the support.

The parable of vertical loads is the same as above, but here it is necessary to calculate the lateral stress, as we are immersed in a homogeneous field without basaltic lava flows interspersed.

The total pressure on the lineal metre of the tunnel at the walls is as follows:

$$E = \gamma m \cdot tg^2(45\text{-}\varphi/2)(2/(3tg\varphi))[(b+m\tan(45\text{-}\varphi/2)] + (m/2)$$

(m = height of tunnel 8.50 m, b = width 12.50 m.)

In this case, it is 110,390 kg/m with a maximum thrust point of 0.26 MPa. Protodiakonov recommends a safety factor of 1.2, i.e. 0.312 MPa, which we also adopted as uniform across the walls.

Volcanic Rock Mechanics – Olalla et al. (eds)
© *2010 Taylor & Francis Group, London, ISBN 978-0-415-58478-4*

Big Telescopes foundations in volcanic environments

W. Llamosas
Department of Construction Engineering, University of La Laguna, Tenerife, Canary Islands, Spain

ABSTRACT: The development of the astrophysics science requires new technical and more sophisticated tools to complete the observation tasks successfully. During the last years, some teams are developing bigger telescopes with the latest technologies, allowing to obtain better results in the astronomical observations. In this way, the Spanish Government and some partners developed and built a telescope with the biggest primary mirror of the world called "GRAN TELESCOPIO CANARIAS" (GRANTECAN) in La Palma, Canary Islands, Spain, being considered the biggest telescope of the world. As the engineer of the Enclosure Group, responsible of the Civil Work and Auxiliary Installations of the GRANTECAN project, I will show in this paper a summary of the geotechnical studies, a brief description of some of the mandatory requirements for the telescope pier foundations, and the final design to accomplish the foundations and structure project requirements.

1 INTRODUCTION

To develop the project and design of this telescope the enterprise Gran Telescopio Canarias, S.A. (GRANTECAN S.A.) was created. Its Project Office is responsible to manage all tasks needed to build this telescope, its future operation, management and new developments. These tasks include the development of technical specifications, execution of preliminary studies, definitive studies, design of some of the different telescope systems, contracting companies responsible for the implementation of other telescope systems and its operation. The Project Office was divided into the following groups: Enclosure, Telescope, Optics, Control, Instrumentation and Administration. The activities of the first two groups are directly related with the subject of this document. This paper presents a summary of the requirements starting, location chosen, preliminary and definitive geotechnical studies, and provides a brief description of the pier and enclosure building final solution.

2 POSSIBLE LOCATION. PRELIMINARY STUDIES SITE SELECTION

Of all the possible locations for the telescope, two were decided to be studied in more detail. These two possible locations are known as Site 1 and Site 2. (Fig. 1). Based on scientific parameters related to the quality of astrophysical observation, meteorological data and technical parameters

Figure 1. View of Site 1 and Site 2 in Roque de los Muchachos Observatory.

related to technical and economic feasibility of its implementation, the final location was chosen.

2.1 *Geological and geotechnical environments (8)*

Site 1 is located above the "Cono de los Muchachos" formed by the accumulation of pyroclastics and Site 2 is located on an area with a little inclination among the Telescopio Nazionale Galileo and the Roque de los Muchachos observatory residence, formed by the "lavas del Galileo". (8).

2.1.1 *"Cono de los Muchachos"*
Presents the classical structure and composition of the pyroclastic outfall fan cone. Although a part of the cone has been destroyed by erosive agents,

including ice in particular, it is still possible to detect the original position of the crater.

The component materials are fragmented layers and fairly continuous, with uniform thickness. The pyroclastics that form them, have grain size distribution and welding grade variables. This was confirmed by the geotechnical exploration program.

2.1.2 *"Lavas del Galileo"*

This lava partially covers and surrounds the "Cono de los Muchachos", forming an inclined plane where Site 2 is located.

The wall of the "Caldera de Taburiente" is a natural geological section of the subsoil structure, providing an overview and showing that the "Lavas del Galileo" are a succession of lave streams originating at some point in the "Caldera de Taburinte" and descending in northwest direction.

These lave streams have random laterally variations in thickness, variations in the relative proportions of slag roof and base and the central lava layer. Near the depression center line, a trend toward thickness increase is observed. They have suffered different disintegration grades and transformation in clays.

2.2 *Preliminary studies (4)*

The preliminary drilling program shows the realization of six mechanical boreholes with continuous coring sample retrieve. Were made five, two at Site 1 (one being a reserve, used depending on the results) and three at Site 2.

With the information obtained by the geological natural sections of the subsoil and the mechanical boreholes done, the following kind of materials are distinguished in this area:

Lava S, Pyroclastic mantle P, Unrecognizable stratum, Lava M, Pyroclastic mantle Q.

Some features are: Lava S partially weathered, pyroclastic mantle Q: typical red, Unrecognizable stratum: material that has suffered a high grade of transformation in clays so that is not possible to assign it to a particular lithological type. Its name results from the fact that its characteristics do not allow recognizing the parent. It is possible to distinguish that it is lava and not pyroclastic, Lava M: rock with interconnected vacuoles stretched parallel to the direction of flow, does not have visible crystals with a magnifying glass, Pyroclastic mantle Q: yellowish.

Finally, the studies were focused on Site 2 and three levels (7) were defined:

Upper level: It is composed by lava S, mantle P, Unrecognizable Stratum and some of the lava M.

It's an inhomogeneous mass soil, characterized by disintegration and transformation in clay, complete or in part, of the original rock. Exhibits a medium compactness and clays have medium plasticity that may experience significant volume changes.

Intermediate level: Composed by compact basaltic lava M with a maximum thickness of 6 m, 1 m thickness near the contact with the "Cono de los Muchachos", and increasing thickness when approaching the axis of the valley filled with "Lava del Galileo".

There are some cracks but given the little alteration of the material it is concluded that the cracks are closed.

Lower level: Lava M is supported on a field formed by clay probably very continuous because it was originally a pyroclastic mantle Q. The quality of the rock is improving with depth.

3 GEOTECHNICAL STUDY (1)

Once the results of the all preliminary studies (scientific parameters, geotechnical parameters, etc.) were compared it was decided to locate the GTC at Site 2.

Then, the geotechnical exploration program was focused on Site 2.

Being determined the probable location of the pier, the telescope building is located. In the exploration program, 6 boreholes are planned: one in the center of the pier, four around the pier that are located under the probable position of the telescope building and one in the center of the probable location of the auxiliary building.

These borehole locations help geophysical research surveys to determine the dynamic properties of the subsoil and geological continuity of the soil layers crossed by the boreholes, using Down hole and Cross hole testing.

The boreholes show that the explored area is composed of a layer of silty clay soil with a thickness of between 0.0 m and 0.40 m, a lava S layer with granular texture with a variable grade of weathering but generally thicker when deeper, with a thickness of between 6.0 m and 7.0 m, a layer of a pyroclastic mantle P with a maximum thickness of 1.0 m, an "unrecognizable" layer formed by lava highly weathered transformed into soils, that have fresher lava fragments inside (mostly of weathering grade III) and a stratum which is not continuous with a thickness of 7.5 m, a continuous and fresh stratum of vacuolar lava (Lava M), with an average thickness of between 3.0 and 3.5 m and a layer of pyroclastics (Mantle Q) of sand grain size in some areas with some grade of cementation .

To determine the geotechnical properties of the materials found in the field boring dynamic penetration tests were planned, samples (undisturbed when possible and disturbed) were taken to estimate the soil properties in laboratory and in the seismic geophysics exploration program.

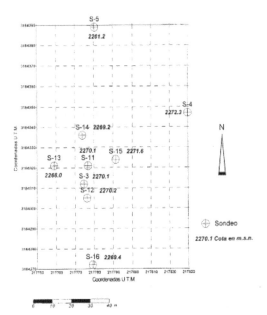

Figure 2. Boreholes location.

During the surveys granular nature of the soil is verified, with varying fines content, little or no plastics, soil or weathered rock found.

Then, it was nearly impossible to obtain a truly undisturbed sample of granular soil. To complete the geotechnical characterization of soils the results of SPT (Standard Penetration Test) and geophysical tests were used.

The description of the geotechnical characteristics of each stratum is as follows:

Lava S:
3 samples were taken and determinations of plasticity, soluble sulfate content, organic matter content and grain size distribution tests were performed.

None of them showed plasticity of the fine fraction.

The soluble sulfate content did not exceed 0.02%.

Organic matter was less than 1%.

The specific humidity is not representative of actual conditions in the ground due to water drilling.

As the samples were disturbed it was not possible to determine their density. The results of SPT and lava source of fine particles, associates a value of 20 to 22 kN/m³.

Pyroclastic mantle P:
Has not been sampled due to its thinness and discontinuity.

Unrecognizable Stratum:
4 samples were taken and determinations of plasticity, soluble sulfate content, organic matter content and particle size were made.

None of them showed plasticity of the fine fraction, soluble sulfate content, organic matter in significant quantities.

The specific humidity is not representative of actual conditions in the ground due to water drilling.

As the samples were disturbed, it was not possible to determine their density. Their comparison with other materials in situ associates a value of 20 to 22 kN/m³.

Lava M:
5 samples were waxed. Ultrasound measurements and compressive strength were performed on rock.

The compressive strengths had an average of 69.894 N/mm². (Min = 24.71 and Max = 153.06).

The average density was 26.4 kN/m³ (Min = 25.6 Max = 26.8).

The average speed of propagation of ultrasound is 3129 m/s, which implies a high density and low microfracturing.

Pyroclastic mantle Q:
The rock has a Q Barton index value of 13 which corresponds to a good rock.

The rock provides an index of Bieniawski RMR 1979 value of 63 which corresponds to a rock class II (good).

It is composed of rock fragments with gravel-sand grain size, with a little cementation grade that allows to find it in some zones as a soft rock.

2 samples were waxed. Ultrasound measurements and compressive strength were performed on rock. The compressive resistance gave values of 1.18 N/mm² and 1.36 N/mm².

The average density was 12.8 kN/m³.

The average speed of propagation of ultrasound was 675 m/s.

4 PROJECT REQUERIMENTS (3)

The pier is basically a concrete cylinder of 17.0 m outside diameter, 1.0 m thick and 7.20 m high, with two concrete slabs, one located at the base and the other at a height of 4.35 m with reference to the base.

The center of the telescope pier is located in UTM coordinates, X = 217,760, Y = 184,335, Z = 2265.

The pier stiffness and the pier first resonance frequency requirements are shown in Tables 1 and 2.

The pier foundation requirements were the following:

For any combination of loads, the differential seat between any two points of the pier must be less than 2 mm / year.

The maximum seats for the life of the facility should be smaller than 10 mm.

The foundation of the pier must be independent from other foundations and transmission of vibrations in adjacent foundations will be minimal or avoided.

Table 1. Pier stiffness requirements.

Stiffness	k1 N/rad	k2 Nm/rad	k3 Nm/rad
Stiffness	5×10^{10}	2.5×10^{12}	8×10^{11}

Table 2. Pier first resonance frequency.

Frequency	f1 Hz	f2 Hz	f3 Hz
Frequency	5	10	10

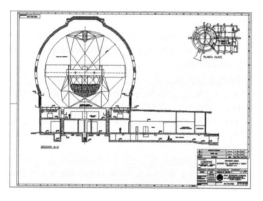

Figure 3. Enclosure and annex building section.

The foundation requirements of the telescope building were the following:

a) The foundation must be isolated from the pier foundation to minimize or prevent the transmission of vibration to it.

b) The differential settlement will allow the correct operation for opening and closing of the dome.

The following tolerances have been handled:
For the smooth running of the motor a maximum deviation of one per thousand was taken, this implies that the maximum differential seat between diametrically opposite points may be 50 mm.

Annex and Auxiliary buildings.

Has been adopted as follows:

Maximum seat during the whole lifetime of the installation: 20 mm.

Maximum differential between pillars adjacent seat: 10 mm.

5 FINAL DESIGN

Given the results of the geotechnical exploration program, it was decided to project a deep foundation for the telescope pier consisting of 40 cased cast in place piles resting on lava M. There were 32 piles on the external perimeter and 8 on the internal perimeter, internal diameter 0.65 m on reinforced concrete fck = 25 N/mm².

For the telescope building foundation a shallow foundation was designed. The inadequate soil was replaced by cyclopean concrete until finding natural soil with geotechnical properties needed to assure the achievement of technical requirements.

For the annex building a shallow foundation on natural ground surface was designed.

The foundations of these buildings are isolated.

6 STRUCTURE MODELING (5)

Different simplified 2D and 3D FEM calculation models were carried out to analyze soil-structure interaction of building foundations and the telescope pier foundation.

Different boundary conditions, different types of solids elements for the finite elements model, different types and dimensions of the isolation material were analyzed.

Finally, a numerical final model (56,226 nodes) was built using FEM, including the modeling of soil surrounding the foundation of the telescope pier, to achieve the static and dynamic stiffness requirements by the telescope pier. The software used was ANSYS Version 5.4 (Release). (Fig. 3).

In this model we analyzed the 75 modes of vibration and the influence of the isolation material placed to avoid transmission of vibrations.

In the pier finite elements model all the holes needed for installation passes and the entrance to pier were included. (Fig. 4).

The results obtained from numerical modeling achieve the specifications required in the

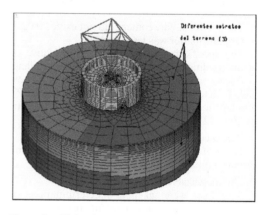

Figure 4. Finite elements model.

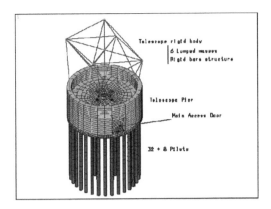

Figure 5.　Pier finite elements model with out soil.

"Civil Work Specification" and "Interface Telescope Pier" documents.

7 CONCLUSION

This study represents a reference for design and implementation of foundations of large telescopes on volcanic soils.

All the geotechnical studies, including site tests and laboratory tests, were done with the maximum precision possible that allows us to obtain reliable values of geotechnical parameters.

During the implementation phase of the pier foundation and of the telescope building correlation between the hypothesis assumed for geotechnical foundation design and the proposed project was found. The discrepancies found were not representative otherwise it would have been necessary to redesign the foundations.

It verifies the high variability of geotechnical parameters including static and dynamic ones for this area.

The results obtained from the numerical modeling achieve the specifications required in the "Civil Work Specification" document.

Today, the telescope is built and working, without problems due to foundations.

ACKNOWLEDGMENTS

The author is grateful to GRANTECAN SA enterprise by giving and authorizing the use of the referenced documentation and publication of this paper.

REFERENCES

Intecsa-Teno Ingenieros Consultores. 1999. Estudio Geotécnico de Detalle en el sitio del Emplazamiento del GTC. Vol. I y II. Grantecan Documents: 10–16, Apendix A Figure 1 (1).
LV Salamanca Ingenieros S.A., 2002. Civil Work and Instalation Project. Grantecan Documents: Plan DR-CW-05-A.dwg (2).
Llamosas W. 2002. Civil Work Specifications. Grantecan Documents. (3).
Marrero Nicolás. 1995. Informe geotécnico previo para la elección de la parcela de ubicación del Gran Telescopio Canarias. Grantecan Documents: 4, 6–10. (4).
Media Consultores de Ingeniería. S.L. 1999. Análisis por Elementos Finitos del Comportamiento Estructural del Pilar del Telescopio Grantecan. Grantecan Documents: 21, 23 Figure 4, 25 Figure 5. (5).
Media Consultores de Ingeniería. S.L.1999 Calculo de Respuesta Modal en la Interface Telescopio-Pilar. Grantecan Documents (6).
Muñoz C. et al. 1997. Estudio del sitio para el Gran Telescopio. Informe Final. Grantecan Documents (7).
Navarro, J. 1998. Descripción geológica de la zona del emplazamiento del GTC. Grantecan Documents: 1, 6–7, 11–13 (8).
Rodríguez Espinosa J. et al. 1998. Selección de la ubicación del GTC. Grantecan Documents: Figure 1(9).

Volcanic Rock Mechanics – Olalla et al. (eds)
© 2010 Taylor & Francis Group, London, ISBN 978-0-415-58478-4

Socorridos pumping station and water storage tunnel at Madeira Island

J.M. Brito, S.P.P. Rosa & J. Santos
Cenorgeo, Engenharia Geotécnica, Lda., Lisboa, Portugal

J.A. Sousa & A. Pedro
FCTUC, Coimbra, Portugal

ABSTRACT: The Socorridos pumping station and water storage tunnel are located at Madeira Island, Portugal, and are part of a system conceived to reutilize the water flow from the Socorridos hydroelectric plant. This facility includes an underground cavern with 26 m high, 12 m wide and 44 m length, and a tunnel 1250 m long with a storage capacity of 40,000 m³. The rock mass at the site is of volcanic origin with most of the excavation performed in a mass of volcanic breccia and basalts but with consolidated alluvia recent deposits located in the roof arch of the cavern. A description of the main characteristics of the project is presented and the results predicted by the design are compared with the values given by the instrumentation.

1 INTRODUCTION

The multi-purpose hydraulic system of Socorridos, located at Madeira Island, is composed of a vast network of reservoirs, tunnels and channels with the purpose of producing electric energy and supplying water to the region.

The Socorridos pumping station and water storage tunnel belongs to this system, which comprises (Fig. 1): i) a water storage tunnel, with 40,000 m³ storage capacity, which receives, during the day, the turbined water of Socorridos's hydroelectric power plant; ii) an underground pumping station including four vertical powerful pumps that, during low-consumption periods, mainly during the night, feed, in a period of six hours, the water back to a reservoir located 460 m above, in order to be reutilized by the turbines of the existent hydroelectric plant; iii) an auxiliary tunnel (with direct access to the water storage tunnel and to the lower half section of the pumping station; iv) a water gallery connecting the hydroelectric plant to the pumping station; and a sand discharge gallery associated with the sand extraction system conceived.

Due to its high complexity, particular care was assigned to the design and construction of the cavern for the pumping station (Fig. 2).

The main concerns were the geological and geotechnical characterization of the rock mass, the complete numerical modelling of the construction process and the observation scheme to be adopted during the execution of the relevant works (Cenorgeo, 2005).

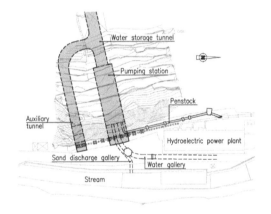

Figure 1. General plant of the hydroelectric system.

a)   b)

Figure 2. a) General view of the superior half section of Socorridos pumping station and auxiliary tunnel; b) interior view of the pumping station.

In this article, after a description of the project, special attention is given to the geotechnical characterisation and to the numerical modelling used for the design and to the confrontation of the predicted results with the observed ones during the construction works.

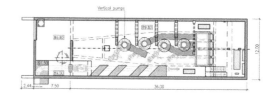

Figure 3. Ground level plant of the pumping station.

2 PROJECT CHARACTERISTICS OF THE PUMPING STATION

2.1 Location and geometry

The dimensions of the cavern (Figs. 3–5), defined to allow the placement of the four pumps (each with 16 m height), as well as all the needed accessories, are 26 m (height), 12 m (width) and 44 m (length). The pumping station is divided in a lower section, which includes the water reservoir and the sand extraction systems, and an upper section, for the equipments' operational area, namely the pump's engines and control room. The ground floor is levelled with the exterior embankment platform, allowing the access of heavy vehicles. The location of the pumping station inside the volcanic rock mass was conditioned by the level of the ground floor at the base of the slope, as seen in Figure 2a. Consequently the cavern has a maximum cover of 22 m.

In order to retain the sands and the fine particles that come from the hydroelectric plant flows, avoiding the pumps malfunction, two side channels were designed for sand extraction along the lower section of the pumping station (Fig. 4). The sand and the fine particles are then vacuumed to higher levels and returned to the river through the sand discharges gallery.

2.2 Geological-geotechnical characterization

Apart from the initial surface geological survey, the characterization of the rock mass included the execution of six rotary boreholes (ϕ76 mm). These allowed the recovering of rock samples for laboratory testing, the definition of the stratigraphy and the execution of six Lugeon tests and 20 dilatometer tests, in order to characterize the permeability and the deformability of the rock mass, respectively.

With the recovered samples, 11 uniaxial compression tests were performed to evaluate the mechanical characteristics of the rock material. Tests were also conducted to determine its unit weight and porosity.

The rock mass zoning was established according to the geological survey, the results obtained during the drilling and the subsequent testing results and the experience of the designer in projects in similar materials.

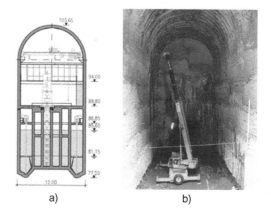

a)   b)

Figure 4. a) Transversal section of the pumping station; b) interior view of the pumping station in the end of the excavation.

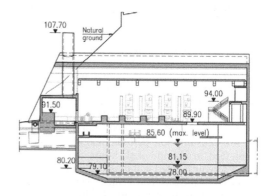

Figure 5. Longitudinal section of the pumping station.

Two distinct complexes with an almost horizontal interface were defined in the zone of the cavern excavation (Fig. 6). The upper complex is located in the roof arch of the cavern and is constituted by consolidated and heterogeneous alluvia recent deposits. The second, where most of the excavation is going to be performed, is composed of a mass of volcanic breccia and basalt.

The geotechnical parameters defined as being the most representative ones for the two complexes are presented in Table 1 (Cenorgeo, 2005).

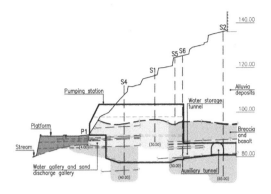

Figure 6.   Geological profile.

Table 1.   Geotechnical parameters.

Complex	γ (kN/m³)	ϕ' (°)	c' (kPa)	E (GPa)	ν	K_0
Alluvia	21	32	60	1	0,38	0,65
Breccia	22	36	80	1,5	0,38	0,65

2.3   Construction method and support system

Due to the geological and geotechnical characteristics of the hillside where the portal of the pumping station is located, the construction works started with its consolidation.

In order to execute this works a provisional rockfill with 8 m height and 1H/1V slopes, was constructed over the existing platform, with the rock material resulting from the excavation of the auxiliary tunnel.

When this phase ended, the excavation of the station was started from the top by levels. Limits to the progression of the excavation were prescribed according to the geotechnical characteristics of the materials found during the excavation.

Because the roof arch has a small cover and is totally located in alluvia, the excavation of the cross section in the upper part of the cavern had to be done in more than one stage, as seen in Figure 7. Firstly, a pilot tunnel was built with excavation steps of 1 m, which allowed to verify and validate the assumed geological and geotechnical conditions. Then, the complete top heading and the first bench were excavated, leaving a 12 m distance between each section.

The primary support used in the top heading consisted in 20 cm thick of shotcrete with fibre reinforcement and steel ribs (HEB200) spaced 1 m, supported at the base by 9 m long rockbolts (self-boring MAI R38N). For the pilot tunnel, it was used a support consisting of shotcrete with fibres

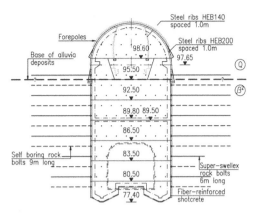

Figure 7.   Cross section of primary support and execution phases.

(15 cm thick) with HEB140 steel ribs spaced 1 m. All the upper zone of the cavern, excavated in recent deposits, was executed under the protection of a group of three sets of $\phi73$ mm SCH40 forepoles, spaced 30 cm, each with a total length of 12 m.

The lower part of the station was built in levels, 3 m high, allowing excavation steps of 3 m (12 m apart between different levels in the longitudinal direction). The primary support for this part of the structure consisted of shotcrete with fibres (20 cm thick) and a rockbolt mesh of 1,5 m × 2,0 m, to prevent the rock mass from decompressing, causing, eventually, blocks from falling. For the upper levels, self-boring, 9 m long MAI R38N rockbolts, were adopted, while for the lower levels, 6 m long Super Swellex bolts were chosen (Fig. 7).

For the top wall, it was used shotcrete with fibres (20 cm thick). Also rockbolts were applied in the same pattern and type as described for the transverse direction.

Considering the hydrogeological characteristics of the rock mass, a drainage and a waterproofing system was designed to the external contour of the upper section of the pumping station, comprising a PVC geomembrane (e = 2 mm) over a geotextile (500 g/m²) with draining functions. The collected water is directed by gravity to geocomposite drains of Stabidrain type (0,30 m × 0,04 m), placed throughout all the pumping station perimeter, being finally delivered into the station, in the storage area, by cross drains in PVC pipes with Ø90, spaced 3 m.

2.4   Final lining

The choice of the concrete class to use on the final lining and on the pumping station internal structures (C30/37) was determined by the water ph values (5,5 to 6,5), according to water physicochemical analysis. The pumping station final lining comprises reinforced concrete walls (e = 0,50 m), topped by a semi-circular vault (e = 0,50 m) with an inner section of 5,50 m radius. The total height is about 26 m, having a strut in the middle, materialized by the ground floor slab.

2.5 Instrumentation and observation

Considering the specific aspects of the project, the geotechnical and geological conditions of the site, and the spatial configuration of the cavern, an instrumentation plan was defined in order to allow: i) evaluation of safety during construction works; ii) comparison between the design assumptions and the observed behaviour; iii) extrapolation of the behaviour from the early stages of excavation to the later ones, in order to modify and adapt, if necessary, the construction methodology and the structural solutions according to the observed displacements of the rock mass.

In order to achieve this purpose, six instrumentation profiles were defined, with a distance of 7 m between them, which allowed measuring the following quantities: i) superficial settlements—24 settlement gauges; ii) sub-surface vertical movements—six extensometer rods with two reading points each; iii) lateral movements—four inclinometers; iv) and convergences—through the use of several survey points. The location of these instruments is shown in Figures 8 and 9.

3 PROJECT CHARACTERISTICS OF WATER STORAGE AND AUXILIARY TUNNELS

Water storage tunnel has 1250 m length, and a transversal section of 7 m × 5 m, in order to store 40,000 m³ of water. The longitudinal inclination is 0,01%.

The tunnel alignment was defined during the execution phase, searching the formations with the best geological and geotechnical characteristics in terms of stability and water-tightness, reducing also the associated costs; therefore several branches were executed, easing the vehicles circulation inside and allowing the creation of several work fronts (Fig. 10).

The auxiliary tunnel function is to give access to the water storage tunnel and to the pumping station, in order to execute maintenance and inspection works, being also used as an access during the construction phase. This tunnel has 73 m long and a cross-section of 5,5 m × 5,5 m, being horizontal in the initial 10 m, having then a 13,5% inclination up to the water storage tunnel.

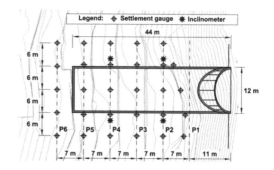

Figure 8. Plan of monitoring profiles.

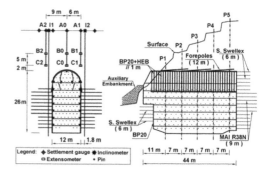

Figure 9. Geometry and instrumentation profiles: a) cross section; b) longitudinal profile.

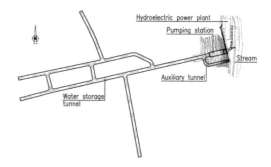

Figure 10. Water storage tunnel alignment.

The primary support was defined in the entire perimeter, working also as a final lining—in Table 2 the adopted solutions are defined, presenting in Figure 11 the ZG2 geotechnical zone corresponding sections. Concerning the execution phasing, the ZG1 zone was excavated in full face (in the areas with worst characteristics advances were made from 10 to 10 m in the higher half-section), while in the ZG2 and ZG3 zones the excavation was executed first in the higher half-section and only afterwards in the lower half-section—in these

274

Table 2. Tunnels primary support.

ZG1	Shotcrete with fiber reinforcement (0,15 m) and Swellex bolts (4 m), in the vault where necessary
ZG2	Shotcrete with fiber reinforcement (0,15 m) and systematic Swellex bolts (4 m), in the vault (2 m × 2 m mesh) and in the walls where necessary
ZG3	Shotcrete with fiber reinforcement (0,20 m in walls and in the vault of the water storage tunnel, 0,15 m in the other cases) and HEB160 metal ribs 1 m spaced

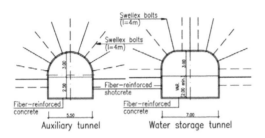

Figure 11. Tunnels lining-type for the ZG2 zone.

zones 2 m advances were made to ZG2 and 1 m in the ZG3 zone, keeping a distance of 20 m ahead of the excavation.

4 NUMERICAL MODELLING OF THE PUMPING STATION

4.1 General aspects

Due to the complex geometry and stratigraphy, 3D and 2D analyses were conducted in order to verify both the construction method and the chosen primary support. The 3D analyses were performed by the Finite Element Method software package PLAXIS 3D TUNNEL v1.2, which enabled the modelling of the soil-structure interaction and the complete excavation sequence. In Figure 12, the finite element mesh used for stage 127 is shown. The entire calculation of the structure included a total of 142 phases, taking almost 16 hours to run (29,000 elements mesh).

The behaviour of the rock mass was modelled by the Mohr-Coulomb model with the parameters presented in Table 1.

Particular care was given to the 3D modelling of the support where shell elements were used to model the vertical walls and the roof arch; solid elements were used to model the top walls and the forepoles and geotextile elements for rockbolts (Fig. 13).

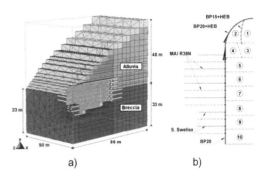

a)   b)

Figure 12. a) Finite element mesh (phase 127); b) adopted construction sequence.

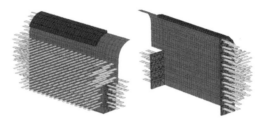

Figure 13. Primary support model of the pumping station.

Table 3. Rockbolts properties.

Designation	D_{ext} (mm)	E (mm)	A (mm²)	EA (×10³)	Q_{fail} (ton)
MAI R38N	37	9,5	717	150,6	50
S. Swellex	54	3,0	480	100,9	17

Table 4. Structural elements properties.

Designation	γ (kN/m³)	EA (×10⁶)	EI (×10³)	υ
BP20	24	5,80	19,3	0,2
BP15 + HEB	25	5,25	11,3	0,2
BP20 + HEB	25	7,44	31,3	0,2
Forepoles	–	1,60	85,3	0,2

Tables 3 and 4 present the properties of all materials used to model the primary support.

The performed calculations characterized the expected displacements, used to define the warning levels for the structure, the stress state during the construction sequence and the forces in the structural elements of the support and, in particular, in the rockbolts.

4.2 Initial stress state

Because of the irregular terrain profile, it is difficult to simulate in PLAXIS 3D TUNNEL the initial stress state. To overcome this difficulty, two solutions were studied. In the first one, the initial stress state was determined by the unit weight of the terrain (and K0) for the whole mesh. Then, the alluvia were excavated in sequence until the desired terrain profile was obtained. In the second solution, the initial stress state was only generated for the breccia, followed by the staged construction of an embankment until the desired profile was reached.

In Figure 14 the relative shear stresses for the two models are presented and it can be seen that the excavation procedure leads to higher stress levels and even to some plastification. The embankment model has, in general, lower stress levels, except near the portal zone where some of the stress points have reached the failure criterion.

4.3 *Comparison between the two models*

Even though the two models depart from different initial stress states, the results obtained after the excavation of the cavern were quite similar, especially for deformations.

The calculated vertical displacements for the last phase of the calculation are shown in Figure 15, where no significant differences between the two models can be pointed out. In both models, the maximum vertical displacement occurs at the top of the roof arch, in a section 15 m from the end of the cavern. Another important conclusion is that the excavation of the cavern doesn't seem to influence the stability of the hillside because no internal relevant displacements were obtained in the calculations.

From Figures 15 and 16 it is possible to conclude that the vertical displacements are similar for the two models, not only for the last phase but also during the complete excavation process. Another conclusion is that the largest percentage of displacements happens when the pilot tunnel is widened to the top heading. This effect tends to be reduced when the points are closer to the surface (point A0). After the conclusion of the top heading, it can be observed that the evolution of the displacements is almost constant in depth.

For horizontal displacements (Fig. 17), a difference, although very small, resulted from the two calculations. The larger deformations were obtained when the terrain profile was generated by excavation.

From the results of the 3D calculations, it can be concluded that rockbolts are indeed very important to maintain the stability of the vertical walls of the cavern. Due to this fact, several parametric studies were conducted in order to determine the influence of some parameters (both geotechnical and geometrical) in the rockbolt's mobilized axial force.

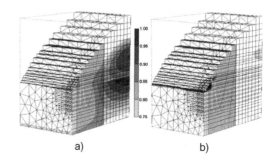

a)   b)

Figure 14. Relative shear stresses (initial stress state): a) excavation model; b) embankment model.

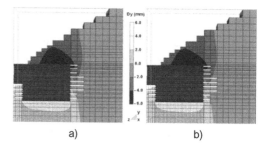

a)   b)

Figure 15. Vertical displacements—a) excavation model; b) embankment model.

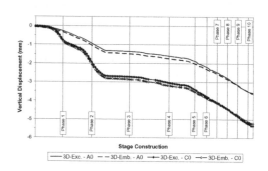

Figure 16. Evolution of vertical displacements—profile P4.

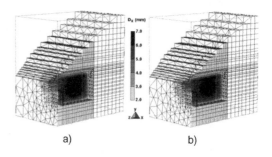

a)   b)

Figure 17. Transverse horizontal displacements: a) excavation model; b) embankment model.

276

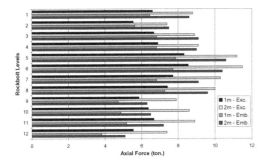

Figure 18. Axial force diagram for the rockbolts.

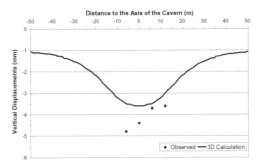

Figure 19. Surface settlement profile—Profile P2.

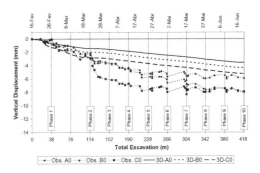

Figure 20. Evolution of vertical displacements (extensometer 0—profile P4) with time.

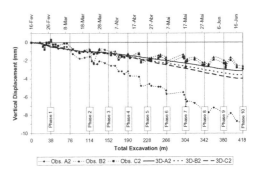

Figure 21. Evolution of the vertical displacements (extensometer 2—profile P4) with time.

In Figure 18 the result of one of these studies is shown for the two models. It can be concluded that if the rockbolt's spacing is changed from 1 to 2 m, the axial force becomes larger, but still far from the failure loads (Table 3). It can also be seen that the mobilized forces are equivalent in both models, mainly in the upper levels. In the lower levels, the excavation model originates larger values.

From the economical and structural point of view, this parametric study allowed to achieve an optimal solution for the bolt mesh.

5 COMPARISON BETWEEN THE PREDICTED AND OBSERVED RESULTS

As shown in the previous sections, no relevant differences between the two models were found and so the comparisons will only regard the observation data and the results given by the embankment model.

Figures 19 to 21 show both the vertical displacements yielded by the 3D calculation and the observation data from two profiles (P2 and P4).

The first figure has the results for profile P2. It can be concluded that the final displacements estimated by the numerical analysis agree well with

those that have been observed (the latter ones are only a bit larger).

Figures 20 and 21 shows the evolution of the vertical displacements for two distinct points of profile P4. At the axis (extensometer 0), the vertical displacements predicted by the calculation are smaller than the observed values. In extensometer 2, points B and C, the opposite occurs and the calculated displacements are larger. The biggest difference is found in point A, located at the surface, where the measured displacement was very large, especially when compared with the points at greater depths.

Finally, in Figure 22, the horizontal displacements recorded by the inclinometers located at profile P4 are compared with the results of the 3D numerical modelling.

In Figure 22, the transverse horizontal displacement (x direction) are shown and good agreement between the calculations and the measured values can be seen, especially in the excavation zone where a maximum displacement of 6,0 mm was obtained, against the 5,5 mm predicted by the numerical modelling.

The horizontal displacements in the longitudinal direction returned by the calculations were very small and in Figure 22 it is shown that they

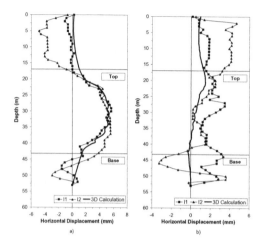

Figure 22. Inclinometer I1-2 a) x direction; b) z direction.

are in fact smaller than those obtained by the instrumentation.

6 CONCLUSIONS

The analysed project, concerning the Socorridos pumping station, presents a unique set of topographic, geotechnical, geological and geometrical characteristics. Due to this fact, special care was given to the geological and geotechnical characterization, to the numerical modelling of the construction sequence and to the observation system to monitor the construction works. After defining adequately the mechanical characteristics of the different formations, the 3D finite element analyses performed during design were able to predict with good accuracy the observed behaviour.

ACKNOWLEDGEMENTS

The authors would like to thank Empresa de Electricidade da Madeira for authorizing the publication of this article.

REFERENCES

Ambrósio, A., Mateus de Brito, J., Rosa, S., Santos, J., Almeida e Sousa, J. and Pedro, A. (2006). The Socorridos pumping station and water storage tunnel. *10th National Congress on Geotechnics*, Lisbon, Portugal (in portuguese).

Pedro, A., Almeida e Sousa, J., Mateus de Brito, J. and Ambrósio, A. (2006). Socorridos pumping station— Numerical modelling. *6th European Conference on Numerical Methods in Geotechnical Engineering*. Graz, Austria.

Mateus de Brito, J., Ambrósio, A., Rosa, S., Pedro, A. and Almeida e Sousa, J. (2006). Socorridos pumping station—Behaviour analysis. *III Luso-Brasilian Congress on Geotechnics*. Curitiba, Brasil (in portuguese).

Cenorgeo. (2005). Multi-propose hidraulic system of Socorridos. The Socorridos pumping station and water storage tunnel. *Final Project* (in portuguese).

Volcanic Rock Mechanics – Olalla et al. (eds)
© *2010 Taylor & Francis Group, London, ISBN 978-0-415-58478-4*

Road tunnel design and construction at Madeira Island

J.M. Brito, C.J.O. Baião & S.P.P. Rosa
Cenorgeo, Engenharia Geotécnica, Lda., Lisboa, Portugal

ABSTRACT: In the last two decades a new fundamental road network was undertaken at Madeira Island. Due to the vigorous relief of the Island, the great heterogeneity, the structural and lithological complexity of volcanic rock formations and to the land occupation, tunnels became dominant and a large spectrum of innovative structural tunnel solutions were developed. The objective of this paper is to give a global vision of the importance of this group of infrastructures, focusing in the new fundamental road network evolution and construction of road tunnels at the island and in the conception and design aspects, along with the description of the geological-geotechnical conditions, primary and secondary lining, phased construction and monitoring.

1 NEW FUNDAMENTAL ROAD NETWORK

In the last two decades, in order to improve the road accesses from Funchal city to the main villages, due to the increasing development, a new fundamental road network was undertaken at Madeira Island.

It was the most important investment in Madeira history, over 2000×10^6 €.

This accessibility plan was implemented by the construction of the priority highway (via rápida VR1) Ribeira Brava-Funchal-Caniçal in the south of the island (Fig. 1). This important highway, with 2+2 lanes, allows the liaison to the airport of Santa Cruz and to the port of Caniçal. The construction began in 1989 and was completed in 2004, with a cost of about 800×10^6 €.

Actually the fundamental road network of Madeira is also constituted by express ways (vias expresso—VE) with one lane in each direction plus one slow lane where necessary (Fig. 1).

With the ancient road network, also represented in Figure 1, to go from the capital Funchal to Porto Moniz in the Northwest, 3 hours were necessary. Today only 45 min are needed.

Actually the main regional road system length is about 200 km, with about 85 km in tunnels.

Figure 1. Fundamental road network of Madeira Island.

allow an acceptable level of comfortable accessibilities to the main villages of the island.

In order to achieve more adequate and economical solutions to overcome the topographic and geotechnical conditions, a large spectrum of innovative structural tunnel solutions were developed.

The great number of tunnels is a result of the geomorphology which imposes important restraints. While the European Directive (European Parliament, 2004) defines a maximum longitudinal pendant of 5% (except where geographically impossible), in Madeira exist tunnels with pendants from 8,5% (Pontinha tunnel) to a maximum of 13% (Pestana Júnior tunnel).

2 THE IMPORTANCE OF THE TUNNELS. CONCEPTION AND DESIGN ASPECTS

Due to vigorous relief of the island, the great heterogeneity, the structural and lythological complexity of volcanic rock formations and to the land occupation, tunnels became dominant, in order to

3 CONSTRUCTION EVOLUTION OF ROAD TUNNELS

At the first phase of the new roads, 20 tunnels were constructed to the West side of Funchal in direction to Ribeira Brava, with two important double tunnels, Quinta Grande and Ribeira Brava tunnels, and eight tunnels in the direction of the Airport. The landmark was the river João Gomes crossing (north of Funchal) with double tunnel João Gomes, cut and cover Jardim Botânico tunnel and the three lane Pestana Júnior tunnel. Figures 2 and 3 show these three tunnels in the interconnection to Funchal.

In 2000 there was a spectacular evolution on construction, with a total of 60 tunnels concluded, with two important tunnels, the Encumeada tunnel and the Norte tunnel, to the North of the island and the Santa Cruz tunnels with four lanes (Fig. 4). Airport prolongation works were also concluded in 2000. The roadway travel time from Funchal to Airport was reduced from 40 min to 15 min.

In 2004 the construction of more 48 tunnels was concluded, including the longest road tunnel of Portugal (Faial-Cortado) and the longest double tunnel of Portugal (Caniçal tunnel). In Figure 5 is shown the East portal with three lanes. The harbour of Caniçal was also constructed in 2004.

Currently 27 tunnels are under construction.

Figure 4.   Santa Cruz tunnels.

Figure 5.   Double Caniçal tunnel East portal with three lanes.

Table 1 shows the importance of tunnel's length when compared with the total length of the roads. We can observe that about 36% of the total length of the highway Ribeira Brava-Caniçal is crossed in tunnel. In a reduced length of this highway, between Machico and Caniçal, with 7,8 km extension, 66% of the road is underground. For the express ways this relation is about 52%.

The evolution of the number of road tunnels constructed in the last 16 years (between 1994 and 2010) is shown in Figure 6. On average about 10 tunnels are constructed a year.

In Figure 7, where is presented the number of tunnels versus the length of the tunnels, is interesting to verify that a number of 69 tunnels have a length greater 500 m (35% of total number of the tunnels), 11 tunnels have a length greater than 2000 m and two tunnels have a length greater than 3000 m.

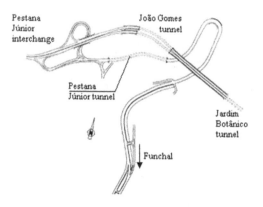

Figure 2.   Plan of João Gomes river crossing.

Figure 3. Aerial view of João Gomes river crossing.

4   GEOLOGICAL AND GEOTECHNICAL ASPECTS

The lenticular and layered structure of volcanic formations, including more or less disaggregated fractions, as well as frequently abrupt lythological

Table 1. Tunnel lengths and road lengths.

	Highway 2 × 2 lanes			Expressway 1 × 2 lanes			
	Ribᵃ Brava Funchal	Funchal Machico	Machico Caniçal	Machico Santana S. Vicente	Ribᵃ Brava S. Vicente Porto Moniz	Ribᵃ Brava Calheta Ponta do Pargo	Total
Total length of the road (m)	13995	20665	7770	34532	31805	33214	141911
Total number of tunnels (including double and secondary tunnels)	19	27	14	23	19	17	119
Total length of underground road (m)	5981	4378	5096	23002	16305	13078	67840
Total length of underground road/Total length of the road (%)	43	21	66	67	51	39	48

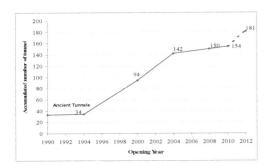

Figure 6. Evolution of the number of tunnels.

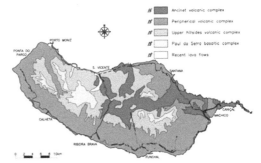

Figure 8. Plan of the volcanic complexes.

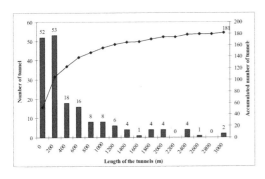

Figure 7. Number of tunnels versus length of the tunnels.

variations, provides a rock mass with a fairly heterogeneous character.

Madeira island comprises five volcanic complexes as shown in Figure 8.

The main complexes are β^1 and β^2.

The base volcanic complex β^1, the older one, is represented by different types of pyroclastic materials, with intercalations of basaltic lava, always weathered to very weathered. Pyroclastic formations are coarse, with big angular blocks, volcanic bombs, lapilli and ash. Sometimes the fine pyroclastic materials (tuffs with bombs) are dominant.

The post-miocenic volcanic complex β^2 is very heterogeneous and mainly consists of alternated and irregular layers of compact basalts and compact and disaggregated autoclastic breccias units with variable thickness, interbedded, as well as less significant compact tuffs. In general, the thick basalts correspond to the most representative geological formation and concern basaltic rocks, of high resistance, with gradual variation to vacuolar basalts, from slightly to very vacuolar and more or less breccious. They occur generally slightly weathered or fresh, with moderate to wide fractures. Fractured or disaggregated basalts are generally vacuolar or breccious formations, always more or less fractured or disaggregated, with weathering signals along the discontinuities or vacuols. Compact breccias occurred generally from moderately to very weathered, with moderate to close fractures. Disaggregated breccias occur from completely to highly weathered, with very close fractures. Tuffs are less representative and occur from moderately to highly weathered. Sometimes the compact and disaggregated breccias and the tuffs present significant thickness.

5 CROSS-SECTIONS

The cross-sections of the current tunnels have a minimum effective width between walls of 9,0 m for unidirectional tunnels (highway) and 9,6 m for bidirectional tunnels (expressways), with two lanes of 3,5 m each and variable heights, depending on the super elevation of the cross-section but with a minimum gabarit of 5 m.

The contour is formed by a semicircular arch roof vertically prolonged by the walls (horse-shoe cross section as shown in Figure 9). Only in the two old tunnels of the highway near Funchal (João Gomes and Jardim Botânico tunnels) the cross-section was formed by an horizontally elongated and inferiorly truncated ellipsis roof. More recently in a few tunnels it has been adopted a curved section (Pontinha tunnel).

Special cross-sections (Fig. 10) have been adopted in tunnels with three lanes and effective width of 12 m (Pestana Júnior tunnel) and four lanes and effective width of 18.5 m (Santa Cruz West and East tunnels).

Frequently it is necessary to include near the portals deceleration and acceleration lanes, which leads to wide cross-sections with three lanes in João Abel de Freitas tunnel and four lanes in Faial-Cortado Tunnel (Fig. 11). The largest wid-

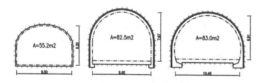

Figure 9.   Current cross-sections of tunnels.

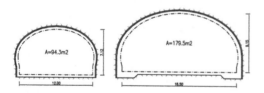

Figure 10.   Special cross-sections of tunnels.

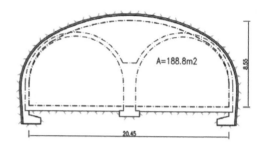

Figure 11.   Wide cross-sections of tunnels.

ening of this tunnel was necessary for reasons of execution of new roads on two phases near the Faial connection.

6   PRIMARY AND SECONDARY LINING

In general, based on lythological, structural and mechanical characteristics of the formations and of the corresponding longitudinal and cross-sectional geotechnical profiles, a geotechnical zoning on the rock mass is performed in the design stages which comprise three or four classes, being the behaviour of the material increasingly worse, corresponding to a good, medium, poor or very poor behaviour. For this geotechnical zoning a geomechanical classification is prepared, mainly based on Bieniawski (1989) and AFTES (2003), in which the quality of the roof, side walls and invert is predicted. This allows a classification of each zone in one of the five possible classes, to which is associated a certain rock mass quality.

This zoning cannot be considered accurate, due to significant lateral variation of the type and characteristics of the volcanic formations, along the tunnel length and the cross-section. The classification of a unique section could have variations, from good to very poor, whether we consider the roof, the walls or the invert. Nevertheless, it permits to establish at the design stage, with sufficient accuracy, the application zones for each type of support.

For the construction stage the Sequential Excavation Method has been used. In this method, commonly known by NATM, the definition of the type and the way of application of primary lining is a function of the knowledge of the rock mass behaviour from the observation and experience gained during execution. In the support's design it is considered that it is flexible enough to tolerate the possible deformations of the rock mass without absorbing great efforts.

The shotcrete has become the most recommendable support system by its initial high resistance, its easy and immediate application, its inter-action with the rock, establishing a sealing surface, and the possibility of being placed with a variable and reduced thickness, which allows a certain deformability necessary to the development of the discharge arch. The shotcrete is reinforced with metallic fiber or wire net, rockbolts or metallic ribs, with the advantage that these elements allow the increase of the mass arch thickness, thus increasing the resistance.

Due to the great heterogeneity, versatile construction methods have been adopted for a better adjust to the different geotechnical conditions. When the formations are ripable the dismount proceeds mechanically with hydraulic demolition hammers.

The compact basalts and compact breccias are generally in the limiar of ripability, making the use of explosives the most economical method to proceed with the excavation.

The geotechnical class ZG1 is mainly formed by compact basalt, by vacuolar compact basalt and by compact breccias. These are highly resistant, homogeneous formations, whose fracturation resulted from columnar disjunction produced by slow cooling of the magma and therefore do not significantly affects the stability. Consequently, class ZG1 is a good quality zone. In the current cross-sections, primary support includes only local swellex rockbolts associated with 0,05 m of shotcrete. Permanent support includes reinforced concrete 0,25 m thick (Fig. 12).

On the whole, class ZG2 is a moderate quality zone. It is constituted by resistant rocks, fractured basalts, fractured vacuolar basalts, consolidated breccias, with lenticular disaggregated breccias and tuffs and compact pyroclasts (mainly tuffs). Primary support includes swellex rockbolts with 4 m length spaced about 1,5 m by 1,5 m in the roof, associated with the immediate installation of a shotcrete shell 0,05 to 0,10 m thick. In sections with larger spans, like entry and exit slip roads,

bolts with 6 m long in a 1,5 m by 1,2 to 2 m mesh have been adopted, associated to shotcrete with 0,15 to 0,30 m thick (Fig. 13).

The permanent support is a reinforced concrete lining 0,25 m thick.

Class ZG3 is constituted by poor quality formations. It includes low resistance rocks, weathered and very weathered basalts, pyroclastic coarse formations with blocks, tuffs with bombs and ash and compact to medium consolidated soils. Primary support includes steel arches one meter spaced and steel fiber-reinforced shotcrete, 0,2 to 0,3 m thick.

Class ZG4 includes very poor quality formations as very weathered fine pyroclastic formations, slope deposits and sometimes ancient fills, with the presence of water. In the current cross-sections, primary support includes jet grout umbrella vault associated with a face reinforcement similar to primary support of class ZG3, basic underpinning of the half cross-section and an invert arch support. Permanent support is a reinforced concrete lining 0,4 m to 0,5 m thick. In the most unfavourable cases a definitive reinforced concrete invert arch has been adopted (Fig. 14).

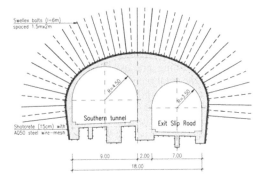

Figure 13.   Primary support for sections with large spans.

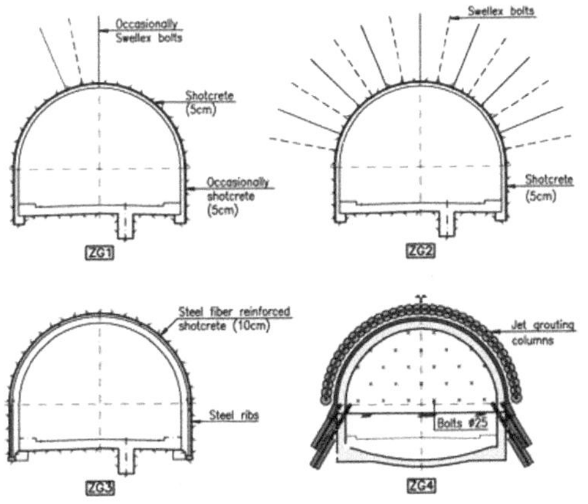

Figure 14.   Permanent support with invert arch.

Figure 12.   Primary support for different geotechnical classes.

The great lythological and structural heterogeneity of the rock mass has led to the conclusion that the design stage zoning should be confirmed during the work stage by a qualified technician to allow, in effective time, the adjustment of the different geotechnical zones ranges and to carry out rational and safe adaptations of the designed support to the effective conditions of the rock mass. This is based on the data collected during the excavations from the continuous geotechnical survey of the rock mass interested by the execution, from complementary boreholes and from the constant supervision and interpretation of the observation results.

Actually every tunnel has internal drainage and impermeabilisation systems. This is normally constituted by PVC geomembrane with 2 mm minimum thick and longitudinal drains in the base of each wall (Fig. 14).

7 CONSTRUCTION PHASES

Construction methods and adopted work planning in the current tunnels are mainly based on the acquired experience in other road tunnels excavated on similar rock masses. Particular aspects are taken into account such as tunnel dimensions, covering and whether it is an urban tunnel.

For the current cross-sections, the excavation and primary lining are usually executed in one phase with the total section in class ZG1 and two half sections, upper and lower, in classes ZG2 and ZG3. More partialized sections have been adopted in class ZG4 and for the widest tunnels.

As an example in East and West Santa Cruz tunnels, with 18,5 m width, considering that the longitudinal zoning is mainly conditioned by the type of rock mass existing at the level of the roof, it was considered that would be important to define that the tunnels excavation should begin by the execution of a pilot tunnel located at the central zone of the superior half section (Fig. 15).

In standard cross sections, excavation and primary support productivity is about five meter per day in class ZG2 and of two meter per day in class ZG3. Concreting with metallic formwork is about ten meter per day.

8 SUPPORT DESIGN

For the primary support design and for the secondary lining either in current sections as well as in the portal areas, generally the finite element method is used, making possible to model all work planning. The design rules associated with the Bieniawski geomechanical classification of rock masses for tunnels as well as AFTES recommendations about the specificity of the volcanic formations and the specific dimensions of the tunnels to be constructed are always taken into consideration. For the geotechnical zones, in which the possibility of a limited zone of the rock mass would detach from the tunnels roof into the applied support, some calculation verifications are also made by the hyperstatic reactions method (structural calculation program) in order to assess the maximum rock mass thickness that could be supported in that situation. For the rock mass (fractured basalts and compact breccias) vertical stress evaluation is based on Terzaghi, Bieniawski and Manuel Rocha methods. For soft and friable rocks and for soil mass (disaggregate breccias, pyroclasts and tuffs) vertical stress is determined by Terzaghi method.

3D structural calculation programs allow to modulate complex situations of the secondary lining as shown in Figure 16.

The geotechnical parameters are in general estimated based on the uniaxial compression resistance tests results of rock samples, as well as on the acquired experience on the same volcanic complex formations.

The conditioning parameter is the deformability modulus of the rock mass. In most cases, due to the impossibility of carrying out any representative "in situ" tests, the deformability is estimated very carefully. However on the most important tunnels, dilatometer tests are performed for the characterisation of the deformability of the rock mass.

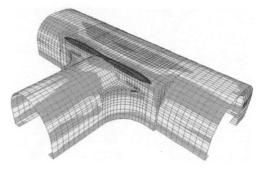

Figure 16. 3D FEM model for permanent support calculation.

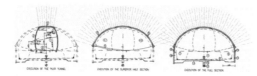

Figure 15. Santa Cruz tunnels. Construction phases.

In general, phreatic line is not considered in the design due to the permeable water of volcanic formations with good natural drainage conditions.

9 OBSERVATION, ALARM CRITERIA

In view of the fairly heterogeneous rock mass and taking in account the size of the tunnels, a thorough auscultation is made by means of the measurement of deformations of the surrounding rock mass, apart from the visual, continuous and systematic observation of the contour of the excavations and of nearby constructions, in order to detect possible anomalous behaviour. The main purposes are to assess the good planning application of the construction methods regarding the effective conditions of the rock mass, as well as the effectiveness of the designed structures, so as to compare the design models with reality and also to support possible design adaptations, improving the work safety balance.

The auscultation consists on the measurement of displacements, by methods such as the ones used in Santa Cruz tunnels and described next (Fig. 17): convergence readings with control of the absolute level between points in the contour of the excavation; readings of surface settlements through topographical marks control and readings of vertical displacements at the zone overlying the roof of the tunnels through nine rod strain meters (three strain meters per each profile located at the axis of the tunnel and on each wall).

The reading's periodicity of all the observation apparatus is adapted to the evolution of the observed displacements. The alarm criteria to be taken into account during the tunnels excavation is established based on calculation results performed

by the finite element method for the dimensioning of the primary support and of the final lining.

10 PORTALS

The shape of the portals depend on topographic, geological and geotechnical conditions and landscape. The portals must be planned to be harmoniously integrated in the surrounding environment.

Portals in compact basaltic rocks have been consolidated with rock bolts and mortar injection.

Portals in breccias and pyroclastic formations have been consolidated with rockbolts and shotcrete or reinforced concrete walls. Nowadays more frequently we use to cover the concrete or the shotcrete with basaltic stones like in the West portal of Faial-Cortado tunnel (Fig. 18).

For portals in difficult soil conditions, such as the case of the West portal of Caniçal double tunnel, when due to slope deposits and disaggregated tuffs with volcanic bombs, it was necessary to build in the upper zone a reinforced concrete wall with two level of anchorages over the tunnels and in the lower zone a reinforced shotcrete wall with a level of anchorages and various levels of bolts (Fig. 19).

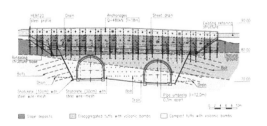

Figure 18. West portal of Faial-Cortado tunnel.

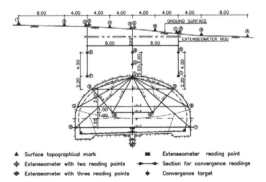

Figure 17. Current cross-section of location of the monitoring apparatus.

Figure 19. Caniçal double tunnel West portal retaining structure front view.

Table 2. Landmarks of tunnel construction.

Year	Tunnel	Main features
1992	Widening of ancient Caniçal tunnel (700 m)	Beginning of a new generation of Madeira island tunnels
1997	Quinta Grande and Ribeira Brava double tunnels	First double tunnels in Madeira from Funchal to Ribeira Brava (West side of the island)
1997	Pestana Júnior Tunnel (227 m)	First tunnel with three lances in Madeira. Excavation width of 13 m
2001	Encumeada (3080 m) and Norte (2097 m) tunnels	Important tunnels in the direction of the North of the Island
2001	West and East S. Cruz tunnels (135 m each)	Widest tunnels of Portugal (excavation width of 20,2 m). First two traffic direction tunnels with four lanes
2003	Faial-Cortado tunnel (3100 m)	Largest tunnel of Portugal. Largest widening (excavation width of 21 m, two plus two lanes)
2004	Caniçal tunnel (2100 m)	Largest double tunnel of Portugal

11 CONCLUSIONS

A reference is made for the important new fundamental road network implemented at Madeira Island in the last 20 years.

The percentage of tunnel length is unique, mainly due to vigorous relief of the island. More than about 150 road tunnels have been constructed, with a total length of 85 km. The principal landmarks of road tunnels construction are presented in Table 2.

Concerning the design, although considering the significant limitations associated with the nature of the volcanic rock mass of Madeira island, it is possible to improve the geological recognition in order to obtain the necessary guidelines to define the geotechnical zoning. It is also possible to plan the temporary support applying empirical methods and to define the characteristics to adopt in the calculation models, either by the finite element method, by taking into account the elasto-plastic behaviour of the rock mass, or by the hyperstatic reactions method. This is complemented with a thorough observation by a continuous geologic survey of the rock mass, adapted to the development of the excavations and with complementary boreholes, as well as with measurements of displacements inside the tunnels, at the surface or inside the surrounding rock mass. Therefore, it is possible to adapt, in effective time, the support designed in a conservative way, to the effective conditions of the rock mass, as well as to reduce significantly the development of the bracing of the full section. The observation design methodology and the supervision of the work execution have proved to be rather adequate and has allowed to adapt rationally and safely the predicted support, as well as to obtain a significant economy during the construction stage.

ACKNOWLEDGMENTS

The authors are greatly indebted to Estradas da Madeira of the Região Autónoma da Madeira for authorizing the publication of this chapter.

REFERENCES

AFTES, 2003. Caractérisation des massifs rocheux utile à l'étude et à la réalisation des ouvrages.

Baião, C., Brito, J.M., Rosa, S.P., Sousa, J.A., 2002. Tunnels of the new expressways of Madeira Island. The case of the Santa Cruz tunnels. *Eurock 2002*: 657–664.

Bieniawski, Z.T. 1989. *Engineering rock mass classifications*. New York: John Wiley & Sons.

European Parliament, 2004. Directive 2004/54/EC— Minimum safety requirements for tunnels in the Trans-European road network. *Official Journal of the European Union*.

Rodrigues, P., Brito, J.M., Baptista, J.P., Rosa, S., Bandeira, C., 2002. Geotechnical and layout aspects of the Pestana Júnior Tunnel (Funchal). *Eurock 2002*: 359–366.

Volcanic Rock Mechanics – Olalla et al. (eds)
© *2010 Taylor & Francis Group, London, ISBN 978-0-415-58478-4*

Geological and geotechnical conditions of human interventions in natural volcanic caverns: The outfitting of "Los Jameos del agua" Auditorium, Lanzarote, Canary Islands, Spain

C. Olalla
Escuela de Ingenieros de Caminos, Universidad Politécnica de Madrid, Madrid, Spain

A. Cárdenas
ENAC Ingenieros, Lanzarote Canary Islands, Spain

A. Serrano
Escuela de Ingenieros de Caminos, Universidad Politécnica de Madrid, Madrid, Spain

E. Pradera & D. Fernández de Castro
ICYFSA, Tres Cantos (Madrid), Spain

ABSTRACT: The possibility of using volcanic caverns associated with lava tubes is analysed according to their geological and geomorphological configuration and their dimensions. Each particular risk is defined and studied by considering its threat. The different natural risks are grouped into four categories: Group I: Structural instability zones due to hanged strata or extremely low width levels at the top of the cave. Group II: Medium and large block instabilities. Group III: Small block instabilities. Group IV: Surface weathering and instabilities in rock particles (sand or gravel).

In this chapter this methodology is applied to the well-known auditorium of "Los Jameos del Agua", located in Lanzarote. This volcanic island forms part of the Canary archipelago. This volcanic tube is visited by more than 800,000 people per year.

After an analysis of the different alternatives, the proposed technical solutions for each level of risk are described and the results of some calculations are shown.

1 INTRODUCTION

The auditorium located within the naturally-formed arch which is part of the "Jameos del Agua" complex to the north of the island of Lanzarote was closed for several years due to the instability observed in the solid rock mass of which the covering arch is comprised.

This arch forms part of a recent volcanic tube, originating in the La Corona mountain the origin

of which is located at approximately 10 km from the location of the object being studied. "Jameos del Agua" is located in the northern part of the island of Lanzarote at approximately 25 km from the city of Arrecife very close to the coast. The nearest town is Punta Mujeres approximately one kilometer away.

This is a unique landscape of exceptional natural beauty with a capacity for approximately 500 people. It is suitable for holding a wide variety of pubic events in the inside and boasts excellent acoustic conditions.

Given these circumstances, the Town Council of Lanzarote decided to commission a group of experts to study the problem and provide a report accordingly. The objective, from the point of view of the feasibility of the facilities for public use of the auditorium, was:

– to analyse the situation and possible evolution of Los Jameos del Agua,
– to identify the possible technical solution which would enable a reopening of these facilities in safe conditions for such use.

The conclusion reached was that the auditorium had various problems which were affecting its stability to a different extent. All of those, however, required undertaking specific action to considerably reduce the natural process contributing to its destabilisation and to allow opening it to the public whilst ensuring an adequate level of safety. The considerable patrimonial value of the complex required these actions to be defined and implemented whilst observing restoration and renovation requirements in such a way as to preserve the natural environment of which it is comprised.

2 GEOLOGICAL FRAME

In a previous paper the most important geological local features are described in detail. (Signorelli et al. 2007).

Geologically, the "Jameos del Agua" cave is formed by two different basaltic lava units:

– the Malpaís Lava Unit (MLU), and
– the Lava Tube Unit (LTU).

The MLU is present in the entire section of the cave and is formed by a succession of different sub horizontal (N10° E strike/1–5° ESE dip) lava members: Lower, Intermediate, Upper and Top lava subunits.

From the lower to the top lava members visually there is an increase in vesicularity and a decrease in the thickness of lavas, grading from aa lavas at the base to pahoehoe lavas at the top. The Lower malpaís member consists of massive grey lava flows surrounded by a thick and irregular layer of blocky and scoriaceous reddish lava rubble. The thickness of each massive lava flow is 3 m whereas the scoria zones range in size from 0.2 to 2 m thick. The intermediate, upper and top members are formed by a succession of pahoehoe lava flows. The lava flow has less vesicular nuclei at the base (Intermediate malpaís member) which grade into highly vesicular upper and top members.

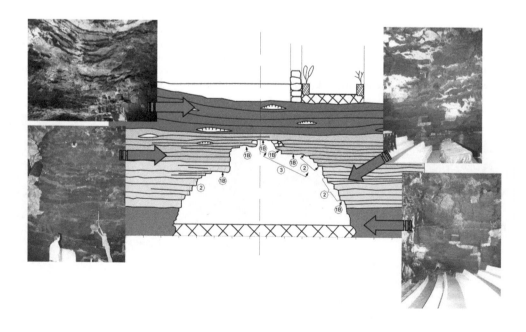

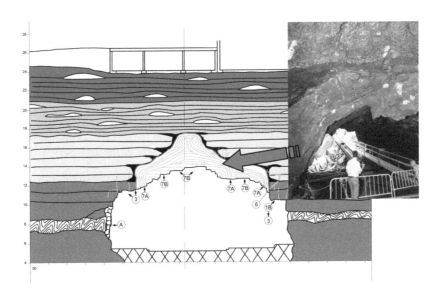

The thickness of the single lava flows decreases from the intermediate (1–1.5 m thick) to the upper and top lava members (up to 0.5 m thick). In the intermediate and upper lava members some inter-bedded scoriaceous layers are present. The upper and top lava members are characterized by several cavities made of a thin crust. When an artificial filling, covering the external roof of the cave was removed, a volcanic landscape made of highly vesicular, cavernous-type lava flows appeared, referred to as shelly pahoehoe lavas. Shelly pro-duces cavernous flows with fragile crusts. Other surface features (hornitos, pressure ridges, tumuli and others) that usually occur on pahoehoe lavas are not present here and may have been destroyed during the construction of the Tourist Centre.

Discontinuities among and inside the different basaltic units are a common feature. As the dif-ferent basaltic flows cool, shrinkage occurs and cracks form in the rock. Sometimes these disconti-nuities are responsible for unstable blocks of rock up to 1m across on the sides and the roof of the cave. Domed sections of the roof testify that col-lapses occurred in the past at the joint intersection of fractured rock. In any case it appears that the sides and roof of the cave have evolved in order to achieve more stable rock profiles.

The LTU is spatially limited to the backstage/upstage, the stage and the first 8 rows of seating.

Only in the zone of the backstage does the lava tube structure appear to be in its original state showing thin layers of lava with a concentric shell structure (like an onion) and levees (terraces mark-ing different stages of filling flows). The shell was formed where the lava chilled against the colder rock. Most of the roof appears to be a circular

stable arch. The tube profile here is in the shape of an "eight" (two superimposed circles). This profile is the result of the cooling of lava flow surfacing to produce a crust, beginning at the levees and grow-ing inward and downstream.

An active lava flow is actually a river of lava, a central stream of molten rock with levees of solid lava along its sides. As the lava continues to flow, the two sides start to form a roof across the flowing lava. When the two sides have completely covered the flowing lava, a lava tube is formed. Sometimes the flow sides form levees as the sides harden, and the top remain liquid. At the stage and in particular in the auditorium (first 8 rows of seat-ing) the LTU is spatially limited to the central part of the roof. Here the LTU presents features like adaptation folds, a fissure where the levees came together and glaze structures.

Most of the roof associated with the simultane-ous presence of LTU and MLU appears to be a broad and stable arch.

3 PROBLEMS DETECTED IN THE AUDITORIUM

3.1 Approach adopted

The problems detected in the area of study had a variety of pathologies of very different levels of magnitude which had to coexist:

– from the possible structural instability of com-plete packets of rock,
– to the possible "chineo", (detaching of parti-cles the size of sand or less), in areas which have deteriorated.

289

The criterion followed to decide on the treatments to be used was to group both problems and solutions depending on their order of magnitude or the size of the phenomenon to be studied.

Although the classification of the pathologies was carried out according to their magnitude, the extent of the need to treat them was the same when the objective was to guarantee the general stability of the Jameos Auditorium in such a way as to ensure that it could accommodate the public over relatively long periods.

Naturally, a severe structural fault would have catastrophic consequences, however identifying such a fault is relatively easy and the probability of its occurrence very low. Nevertheless, other faults which have an affect to a lesser degree are more difficult to detect and would have a higher probability of occurrence and their consequences would be far more serious.

In some ways the risk which defines the need and scope of the treatment of a specific type of problem would has to be reassessed in accordance with the probability of occurrence and the resulting damage. In this case the use of the complex as an auditorium, i.e. its use as a venue open to the public, requires the capacity for a large number of people, in a motionless situation, over long periods.

As such, it is absolutely essential to assess the risk of any type of incident occurring based on similar values and of lesser magnitude than those accepted for other types of constructions intended for human use.

3.2 First-level problems: Structural instability

3.2.1 General features

This is the most important type of problem. It is related to the possible collapse of all or part of the arch in the event of exceeding the structure's resistant capacity formed by the actual layers of rock in their natural state. This situation is particularly obvious in the entrance area to the auditorium where there is the additional problem of the rock thickness in the area closest to the entrance Jameos.

This is a laminar-type structure with horizontally structured layers of rock, superimposed which at the same time are not very well bound. Their resistant capacity depends to a large extent not only on the thickness and quality of the rock, depending on the span or the opening of the cavity, but on the degree and type of the existing fractures.

The stability conditions can turn into situations which are dangerous as the result of various causes amongst which it is worth highlighting the following:

– The increase in external loads which can be easily avoided by preventing access to the surface.
– Changes in the interior geometry of the arch which modify the stress when increasing the opening or the support conditions. This situation is, to a large extent, related to minor detachments or loss of blocks along the free edges of the layers.

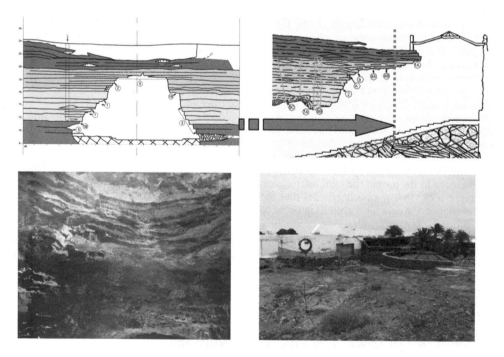

– The loss of resistance in the matrix rock of the mass or in the material in the discontinuities due to the fracturing or the progressive alteration of the material.

Throughout this process it must be taken into account, apart from the overall structural stability of the group, the problem presented by the existence of the open joints in the rock mass where resistance is fundamentally mobilised in the contacts between the irregularities of both sides. These circumstances give rise to the possibility that two bending (traction) areas exist in some parts of the layers and very high compression stress in the contacts of the rugosities. Another factor is alteration due to the attack capacity presented by the existence of the network of cracks and by the large specific surface area of the mobilised ledges.

This situation, which can be assumed to be general for the entire subterranean complex, is particularly critical in the entrance areas of the Jameos, where on a joint basis there are four unfavourable characteristics which act complementarily and are the following:

– The larger dimension of the cross sections of the cavern. This means a greater "equivalent" span between supports than in the rest of the cavern.
– Reduced thickness of the solid rock mass in the keystone of the corresponding section.
– An almost flat arrangement of the section roof due to the loss of the "shoulders" area. It is indeed this shape which is the clearest symptom of how critical the current situation is. It must be remembered that it has developed until becoming a situation in strict equilibrium.
– A greater ease of alteration due to the proximity of the free surface area.

3.2.2 Simplified calculations carried out to assess the condition of the Jameos

The first attempt to be made was that corresponding to a minimum cantilevering and opening which would be self-supporting and for which a simplification was used treating them as a built-in-free and bi-supported beam (maximum tractions in the centre of the opening) subjected to its own weight. These hypotheses do not take into account the three-dimensional nature of the problem, nor the fact that the numerous sub vertical fractures can make blocks become detached. Whilst aware of the degree of conservatism and simplification of the problem that has been assumed the objective was however essentially to have a reference framework of how the group functions.

For the purposes of the geomechanical characterisation of the upper unit and roof it can be considered as being the one only, allocating a

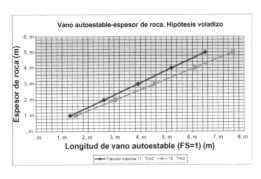

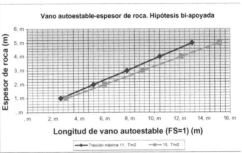

resistance to simple compression to these materials ranging between 11 and 15 MPa. This hypothesis is conservative if taking into account the results of all the tests; however this conservatism has to be used when there is a relatively limited amount of available information. With regard to the specific weight a value of 21.7 kN/m^3 is proposed, an average value for all the tests.

There is no information on resistance to simple traction. It is necessary to assess this based on normally accepted reasonable criteria and which usually allow assessing it as a tenth part of the resistance to compression i.e. between 1.1 and 1.5 MPa.

The cantilevers and openings observed were, in general, close to those calculated. This would imply a strict equilibrium of the structure of the solid mass which would require a reinforcement to allow separating the reinforced situation from the current one with regard to the safety factor. This reinforcement can be focused from the point of view of modifying the working hypotheses of the solid mass using anchoring modifying the thickness of competent rock, thus increasing, if possible, its average resistance to traction (a possible approach would be to incorporate an artificial stratum over the surface) or a combination of both.

3.3 Second-level problems isolated instability by average-sized blocks

These correspond to instability produced due to faults of lesser importance than previously mentioned. However, notwithstanding the physical risk that these pose for people, they present the danger that they could make the general structural weakness of the rocky medium previously studied spread or deteriorate over time by creating larger openings.

This represents a second level of problems. It was detected throughout practically the entire auditorium and would affect blocks with a size of approximately 50–60 cm.

In some cases the blocks have clearly open joints which could indicate the strict equilibrium of these. Stability would be produced by possible bridges or rock joint contacts with other adjacent blocks.

There is also the existence of another type of blocks in which the joints are totally or partially closed, however in which it is not possible to predict the real condition in interior areas where air bubbles or fractured or weakened areas could appear due essentially to the chaotic nature of the formation of these kinds of volcanic masses.

Finally, there are blocks formed by the strata of the different flows which are practically detached from the solid mass by way of "platforms". This lack of unity could be the result of material loss in the joints which surround them or of the presence of smaller caverns thus producing an empty area. Within this group there would also be platforms the formation of which could originate from the lava tube unit (LTU) which has produced a cover similar to "onion skin" the thickness and quality of which cannot be determined in depth.

In all of these cases, the hypothesis must be that of fastening it using systems which allow securing

the blocks under consideration with regard to their own weight. Although there is always a certain degree of friction between the planes, due to being unable to determine the extent of this, as well as the condition, it is recommendable to exert caution and not consider its effect as beneficial.

As such, the hypothesis would be that of a block with a volume of approximately 0.5 to 1 m^3, to own weight which would require a resistant force of approximately 20 to 40 tonnes. A safety factor of 1.6 was applied and a specific weight of 25 kN/m^3 was assumed, maximum value obtained in all the cases tested in the MTU intermediate units and the roof.

One of the premises, which will be discussed in more detail further on for this treatment, given its extent throughout the cave, is to avoid as far as possible, that these are visible once the action has been finished.

3.4 Third-level problems isolated instability by small-sized blocks

This type of problem, which could produce small detachments, could arise practically anywhere in the auditorium.

These blocks would be smaller in size, approximately 20 cm, i.e. 1 to 1.6 kN in weight employing an increase of 1.6 as a pessimistic hypothesis. Given the reduced size it would be difficult to apply isolated stabilisation solutions using borings since the area is too small to make a boring without causing the block to become detached.

Based on these premises, the system studied and ultimately used was directed at improving the joint between various blocks thus improving each of these individually. This could give the case of a specific area suffering from a second-level problem once all the smaller blocks have been joined.

3.5 Fourth-level problems superficial instability

These problems present the possibility that there are erratic and random detachments, isolated

elements, individual, with a variable size amongst which there are types such as sand, grit, gravel, coarse gravel or similar (which is generally referred to in civil works as "chineo").

As a general rule it can be affirmed that these types of detachments bear no relation to the structural stress of the cavern. They can occur at any point of its surface as the result of the individualisation of a fragment of rock or due to the loss of cohesion or the reduced resistance of some of the discontinuities or in the actual matrix of which it is comprised as a result of the alteration.

In this sense superficial-type treatments studied focused on consolidating the first centimeters of the material by increasing its cohesion and reducing the susceptibility of the material to erosion.

At the same time the treatment enabled the rock to continue being permeable, it "transpires" thus, more importantly and as a compulsory measure, its visual effect must be as invisible as possible or, in the event of being detected, be more concealed, blending into the area to which it is applied.

3.6 Anthropic actions

During the period that the auditorium was in open to the public various anthropic actions were undertaken directed at minimising minor instability, or to cover small gaps which had been appearing throughout the entire area of the complex.

These actions essentially consisted in fitting strips of stone, at times using materials from the area, (dark in colour) and at other times using light-coloured stone which affected the aesthetics of the surroundings. These last actions were more visually obvious given that they were emphasised due to their comparison with the rest of the walls as a result of the different colouring.

At times behind some of these strips mortar had been used for filling the gaps behind them. Neither the composition of these mortar fillings nor the depth they had reached could be accurately assessed.

It must be emphasised at times that these strips did not correspond as much with the attempt to fill and conceal gaps, but rather with the attempt to underpin areas with insufficient support and with the objective of concealing existing installations. Examples of this are cases of numerous cables and recesses for lights in the auditorium as well as for sound equipment.

Within this context it is worth highlighting the existence of a triple torsion protective mesh which was located in the area closest to the stage. This mesh extended from the keystone up to the sidewalls and the fixtures were clearly visible due to their galvanising.

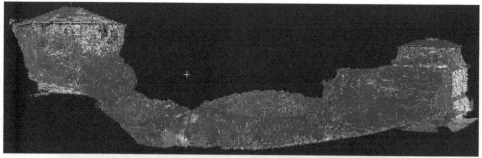

4 ANALYSIS OF SOLUTIONS

4.1 *General approach*

Once the various types of problems had been studied, the solutions to be adopted were analysed for the renovation of the auditorium. In order to do this it was very useful to classify the problems described in the previous sections of this paper, even although there were actions which aimed to solve various types of risk at the same time.

All the solutions considered shared the common denominator of the strict maintenance of the natural aesthetics of the volcanic materials affected, whilst changing their state as little as possible. In this sense tests had to be performed on site of all the treatments being considered for any level with the objective of verifying their suitability from an aesthetic point of view.

As already commented, the starting point for the focus of solving the problem was treating the complex as a patrimonial asset which had to be preserved, whilst constantly respecting the aesthetics of its natural surroundings.

This approach had to be observed without compromising the engineering actions which had to be undertaken to provide a solution to the various geotechnical problems detected in the complex and

which have been described in the previous sections of this paper. The harmonisation of both tasks had been considered in the project in such a way that the complex could be preserved and, if possible, would even see its aesthetics improved. Safety had to be guaranteed to the extent that it would be feasible to open it to the public whilst making the task of maintenance indispensable given the presence of a natural process the ultimate result of which is destabilisation.

At the same time, the work undertaken was undertaken without the presence of similar previous experience based on which approaches could be compared and solutions clarified. It was also ascertained that the different orders of magnitude of the problem needed to be graphically depicted in greater detail than those used up to that point. This was the reason for an additional detailed topographical survey using a laser-scanner which would enable obtaining a highly accurate topographical three-dimensional model of the entire Jameos.

4.2 *Objectives of the intervention*

With this group of determining factors an action and implementation protocol was developed with the following complementary and concurrent objectives:

– Restoration: This refers to the process of recovering the formal and aesthetic appearance whilst respecting the original structures and materials.
– Renovation: The objective of this is to recover the functionality of the complex without the respect for its originality being a premise.
– Conservation: This is understood as an intervention on the auditorium with a dual purpose of active conservation with the emphasis on its surroundings, or passive conservation with the objective of preventing the natural environment deteriorating any further.

This triple concept of the intervention process must be considered as a theoretical framework of illustrative reference and in keeping with the solutions to be adopted.

4.3 Catalogue of alterations and defects

Apart from the structural geotechnical problems of greater or lesser magnitude evident throughout the complex, which are not included within the scope of this section and which represent a classical mechanical problem inherent in geotechnical engineering, the catalogue of the most serious alterations which usually arise in the rock materials at the heart of the renovation, and the treatment of which must subsequently be considered, are the following:

– Jointing.
– Fracturing.
– Fragmentation.
– Displacements.
– Powdering.
– Alveolar erosion.
– Scaling.
– Formation of crusts or concretions on the surface.

All of the above were addressed taking into account the importance, scope and typology of each of them in the solutions adopted.

5 SOLUTIONS CONSIDERED IN LOS JAMEOS

The following sections discuss the solutions adopted for each problem in view of the approaches described above.

5.1 Structural solutions: Solutions type 1

These correspond to phenomena related to the horizontal arrangement of the numerous volcanic flows.

It is typical in this type of structural arrangement of volcanic flows to see the formation of stepped profiles in the areas of shoulders which results in the terrible situation of the work in cantilevering in the layers. In reality, apart from the influence of the alteration processes, the stepped areas comprise the part which is most sensitive to the geometrical modification of the profile due to the detaching of the layer edges which increases the effective opening subsequently increasing the risk of collapse of the roof.

These situations are always difficult to assess in terms of safety since, to a large extent, they depend on the quality of the rock and especially on the location and condition of the discontinuities many of which are not visible.

As such, in natural processes it is fundamental to accept that the situation produced at a specific time can be that of strict equilibrium at least in a certain area of the cavity which requires two types of actions.

– Adoption of actions which prevent further deterioration in the rock matrix and in the edges of the joints.
– Adoption of structural reinforcements which improve the safety margins until reaching acceptable levels suitable for the use for which the complex is intended.

From the perspective of this type of problem various areas were distinguished within the longitudinal development of the Jameos Auditorium:

– The entrance area.
– Areas with reduced thickness of rock of up to 2–3 meters.
– Areas which there is a sudden increase in the rock thickness up to 8 meters; however without the cover which constitutes the tube unit (MLU).
– Areas where the formation of the tube (MLU) appears covering the intermediate unit (LTU) like "onion skin".

It was decided to adopt reinforcement solutions which do not modify the natural appearance of the complex, whilst avoiding other visible structural solutions which would require great attention to be paid to adaptation to the environment, always subjectively.

5.1.1 Solutions type 1-A: Resin injections and bolts of 16 mm

A fundamental characteristic of this solution is that it originates from aesthetic reasons. This means that the treatment must be practically invisible with no possibility of using distributions panels. The surface of the rock must remain unchanged which is why the injection process must be undertaken with great care.

The aim of using this treatment is to achieve the compaction of the first meter of material thus creating a crown of treated material which represents a distribution surface, or ring, of the deepest braces (something similar to the results of sprayed concrete in a conventional tunnel). This also avoids the possibility of many of the medium-sized blocks falling and there is considerable improvement in the alteration action by sealing a large part of the joints found in the solid mass.

This treatment also enables treating, at least partially, second-level instability even although it is not considered as a complete guarantee of treatment for all the second-level blocks, especially those closest to the surface due to the limits of the effective access of the resin to the surface of the cavity.

The solution consisted in injecting high-viscosity resin manually using rotational drilling in a mesh ranging from 0.7 to 1 meter. These drills were selected with the smallest possible diameter to enable positioning the plug and cleaning the areas in which the injected resin was released to the outside whether via the joints of the actual rock or via the adjacent borings.

Once the injection is finished, a fiberglass rod with a diameter of 16 mm and a length of 1 meter is positioned inside each hole by way of top-up injection. As a last detailed action, the mouth of the drill is concealed using mortar made using the same injected resin mixed with stone from the Jameos which is ground beforehand.

The type of bolt planned is made from fiberglass because of its increased durability over time. This bolt is fully injected with epoxy-like resins thus guaranteeing attaching it to the rock.

5.1.2 *Solution type 1-B: Resin injections and bolts of 25 mm*

The treatment using long bolts measuring 4 m was also planned to be implemented systematically reaching as far as the sidewalls of the cavern, generally occupied by the lower unit of the massive basalts with a scoriaceous-type band.

In order to perform the injections, during the work phase it was recommended to position the plugs fully into the mouth of each hole without distribution panels being required which would be more difficult to conceal.

Given these characteristics, the diameter of the solid bar of 25 mm provides a working load of 170 kN and a weight which enables easily handling the bolt.

With regard to the geometry of the bolting it was recommended to have a maximum length of 4 meters in the areas where the rock was thick enough. In this sense, the areas with weaker rock require a special treatment directed at providing the bolting with an artificial stratum to enable subsequent anchoring to it.

The mesh planned will be 2×2 meters which means approximately between 8 and 10 bolts located in every 2 meters of Jameos.

With the two treatments considered in previous sections of this paper, complete treatment is given to the first-level problems and partially to second-level and third-level problems.

5.1.3 *Improved stratum for anchoring the bolts in the entrance area*

The aim of this solution is to create an artificial stratum of improved soil in the areas in which the rock thickness in the keystone was considerably reduced. It was considered that this treatment is required up to a profile where the thickness of the rock is practically 8 meters. The reinforcement is intended to prevent, in the long term, the combination of various vertical and horizontal groups of fracturing which could cause a large block to fall. In the event of a block with such characteristics becoming detached, the bolts would ensure that it remained attached.

The objective of this treatment is to create an artificial stratum to anchor the bolts and not, strictly speaking, to create a flagstone-like structural

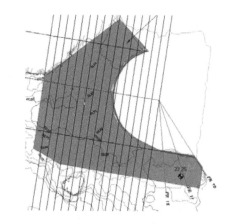

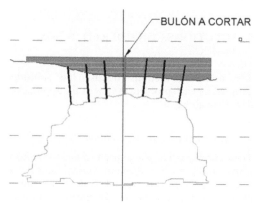

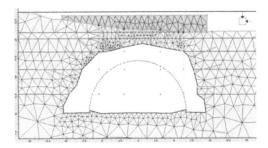 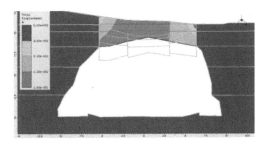

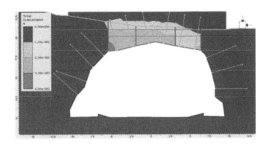

element. Nevertheless, to improve its behaviour with regard to possible tractions or fissures which could arise, a reinforced grating was included with three levels of bars measuring 25 mm in diameter.

A fundamental aspect analysed in this artificial stratum is its weight since it is located over an area which is particularly sensitive to possible overloads. With the objective of not applying overloads any greater than those which already existed, it was recommended to use concrete made of light aggregate which enables achieving very low densities and concrete with resistance of up to 25 MPa.

It was required that the stratum be constructed in one go to avoid any type of discontinuities which is why it was intended for the three levels of reinforcing to be positioned in one go.

As a precautionary measure the area located within the cave under the vertical of the improved stratum had to be dismantled while work was being performed on the surface and while waiting for the concrete to achieve its resistance over 28 days.

Likewise, to make a more simplified analysis of the effect of this stratum, two finite element models were made which included joining elements to reproduce the generation of one joint throughout the rock solid mass.

In the first calculation model the formation of the auditorium up to its present condition is reproduced in successive phases. In the final phase of the calculation a long-term hypothetical situation is simulated in which the resistant and deformational characteristics of the vertical joints are reduced until achieving the instability of the calculation model. The second model is exactly the same as the first. The only difference is that in the last phase the improved concrete stratum is added and the bolts to verify that this reinforcement manages to stabilise the arch of the auditorium.

5.2 *Isolated solutions: Solutions type 2*

These are solutions which aim to secure blocks from 10 centimeters up to 1 m³ which could become detached from the rock solid mass being in a strict equilibrium situation. This strict equilibrium can

be clearly seen in blocks which can be seen to have become detached from the solid mass. Nevertheless, those which show signs of unfavourable fractures and which have welding in their planes on the surface or not must also be considered as sensitive.

5.2.1 *Solutions type 2-A: Micro-seaming of blocks*

For the largest blocks of approximately 1 cubic meter a weight was estimated of approximately 40 kN. They had to be considered as entirely supported by the treatment to be used given that initially the behaviour and orientation of the fracture planes is uncertain, especially at depth with there could even be the possibility of gaps appearing (bubbles).

The treatment of these blocks was carried out partially with the injection treatment discussed in the previous solutions. This treatment, however, is complemented by using the micro-seaming of blocks and also by using fiberglass rods injected suing epoxy resins.

For diameters which are commercially available, the working load of the rods was calculated by using a safety coefficient of 1.6 (maximum value which is usually applied to conventional anchoring).

With the range of loads considered, the treatment was planned with a diameter of 8 mm, with the objective of making the hole as small as possible (approx. 2 mm in addition to the rod diameter). In the case of the largest blocks it would therefore be necessary to use two or three rods positioned in such a way as to take full advantage of the capacity of the rod (i.e. as parallel as possible to the

PUNTOS DE SOLDADURA

direction of movement of the unstable block which in general will be considerably vertical).

The maximum length commercially available for these types of rods is 2 meters which covers the range of lengths planned for use for this treatment.

5.2.2 Solution type 2-B: Chemical welding of smaller blocks

These are the cases where it was not possible to make holes in the blocks due to their small size (third-level problem), i.e. blocks of approx. 20 · 20 cm maximum. This would be equivalent to weights between 1 and 1.6 kN, with an increase factor of 1.6. It was considered joining the blocks by using so-called "chemical welding", which is basically the creation of bridges between different blocks using low viscosity epoxy resin weld beads applied manually.

This creates one larger block with which it could, at least theoretically, be possible to treat using micro-seaming. As occurred with treatment 2-A there is already a first previous action level since the generalised injections type 1a have already been performed. This action is therefore complementary in those cases in which there is no safety in the finish of the first action level.

5.3 Superficial solutions: Solutions type 3

This is the treatment which solves more minor problems i.e. this includes gravel-type sizes (between 2 and 50 mm). Logically it was considered that this could affect any point of the Jameos Auditorium due to the very small dimensions of the particles to be supported.

To find a solution to this problem it was planned to apply a compaction treatment to the surface of the rock where this phenomenon was expected to appear. A rock behaviour model was studied which assumed a material formed by a sandy and silty matrix containing inclusions of larger edges embedded in it (scoria). As from the gradual deterioration of this matrix "micro-cutting" could be produced to the edges embedded in the ground mass thus causing it to fall when finally a sufficient volume of the matrix has disappeared.

A rocky material exposed to the atmosphere deteriorates more on the surface than on the inside. This deterioration could be accompanied by the formation of hard surface crusts with low porosity.

Basically a compaction treatment must improve the mechanical resistance of the material, especially to traction, whilst improving its structure making water ingress more difficult and the circulation of saline or acidic solutions.

These products adhere to the walls thus reducing the empty space and creating connection points between these walls. In short, their objective is to act as a mineral cement, non-existent originally or lost due to alteration. The improvement in the mechanical properties of the rock depends, in this case, on the abundance of links or anchoring points which are established between the precipitated compound and the components of the stone material.

As a consequence of the compaction treatment there should be an increase in the resistance to alteration processes which conduct volume changes in the porous network of the material and which imply mechanical stress which affects the structure of the material (crystallisation of salts, etc.).

The compaction treatment considered had, as such, the objective of acting on the rock surface by increasing the cohesion of its matrix and increasing its resistance to weathering.

With the objective of complying with all the requirements a treatment is considered based on ethyl silicate as the active ingredient diluted in solvent in varying concentrations depending on the penetration desired.

With these kinds of products the thickness is between 5 and 10 centimeters which means that the ethyl silicate must be applied in variable concentrations in successive phases using a smaller or greater concentration. This enables restoring the material matrix by making it more resistant and less prone to weathering.

The selection of the product application method was carried out on site in such a way as to obtain as much penetration as possible in the material. This is a fundamental characteristic which conditions the effectiveness of the product. This penetration capacity is influenced, not only by the characteristics of the compaction agent, but by the product application mode, the type of solvent, the contact time, relative humidity and the pressure and temperature at which it is carried out.

A limit in the use of this type of treatment is the existence of salts on the rock surface which occurs in the case of the auditorium in some parts close to

Treatment name		Description	Type of check to be performed
Type 1	1-A	Resin injections from the inside	Test area of 2 m^2
	1-B	Fiberglass bolting	3 bolts including their injection and load test
Type 2	2-A	Micro-seaming of blocks	5 bolts spaced differently at various depths of 1 to 1.5
	2-B	Chemical welding of smaller blocks	Test area of 1 m^2
Type 3		Superficial solutions using binders and polyurethane membranes	Test area of 2 m^2
Type 4		Solutions for existing anthropic actions	In two existing actions

the stage. In these cases the project planned surface and isolated treatments based on paper pulp poultices. As such, it was recommended to plan a treatment that would cut the origin of the salts reaching the Jameos, although this treatment was not by any means the purpose of the project.

Another limit of the treatment refers to the edges which are already beginning to detach with open micro-fissures greater than 50–100 μm which cannot be treated with ethyl silicate due to very low viscosity hindering its penetration. For these edges the recommendable treatment is their elimination prior to treatment by thoroughly cleaning using compressed air. Afterwards, and as an additional measure to guarantee the effectiveness of the treatment, a second cleaning is proposed once the compaction treatment has been carried out.

As a complementary treatment directed at creating a film which could stabilise some edges superficially the spraying of a liquid polyurethane membrane was planned. In some ways this treatment would aim to perform a function similar to a very fine mesh since the largest sizes of stone will be stable due to the serious of actions at different levels already considered. This membrane has a resistance to traction of approx. 3.5 MPa which allows fixing edges which have already started to detach or those areas in which it is necessary to create "bridges" for micro-fissures of greater importance of 50–100 μm.

Some of these last surface treatments were modified on site for aesthetic reasons in some cases (shine in the polyurethane membrane) or due to the use of replacement techniques such as hydro demolition.

6 RECOMMENDED CONTROL AND SUPERVISION OF THE WORK IN THE PROJECT

The auscultation and supervision programme of the work was proposed as part of a series of test sections of each treatment along with the systematic supervision of the work. The following areas were proposed for the test sections.

Supervision and control actions were also recommended which had to be documented and systemised. The supervision team (geotechnical engineer and restorer) will be responsible for confirming on site the treatment to be applied in each case and for checking the final outcome of each one.

The working methodology planned was as follows:

– Division of the auditorium into "treatment rings" of 2 meters of treatment with a geotechnical survey and a photographic report of each one using cards or a dossier. This survey was to be detailed enough so as to enable a comparison to be made of the current condition and the final one.
– On site decision regarding the treatments to be used in each ring. In those considered systematic there will be an indication of the ideal areas for carrying out each one (holes, anchoring, etc).
– Supervision of the execution of each ring by carrying out a survey upon completion of work in order to verify the effect of the treatment applied, as well as the starting point for systematic maintenance during the operational phase.
– Check of the final treatment of each ring by stating whether new actions are required.

As a conclusion to this process it was planned to create a final report on the condition of the action in order to apply it to subsequent phases and to each of the sections.

As auscultation measures during the work a detailed topographical control was planned using keystone levelling and recording convergences in keystone and shoulders using prisms in each treatment ring. Likewise, in the area of the upper slabstratum a network of levelling reference marks was planned with a measurement frequency every 12 hours after removing the provisional propping of the arch.

All of the instruments, suitably concealed, were intended for continued use to enable supervision during the operational phase.

During the operation of the auditorium, once the actions described in the previous sections were completed and given its status as a "Cultural Asset", routines were devised for the control and supervision of the work carried out as well as maintenance programmes to guarantee its perfect conservation by using periodic controls aimed at checking effectiveness and ageing.

The maintenance routines will continue for the time deemed necessary in accordance with the conservation of the Asset.

The "Letter of 1987" defines maintenance as *"the set of actions programmed on a recurring basis, aimed at keeping the objects of cultural interest in optimum conditions and functionality, especially after they have undergone exceptional interventions for conservation or restoration purposes. The schedule and implementation of regular maintenance and control cycles to ascertain the conservation condition of an architectonical monument is the only guarantee to ensure that prevention is opportune and suited to the site with regard to the nature of the interventions and their frequency"*.

In short, during the operational phase there must be continuous supervision of the auditorium similar to that performed prior to this intervention, as a complement to the systematic care of it by specialised teams from the Town Council of Lanzarote which is considered as absolutely essential.

REFERENCE

Signorelli, S., Jover F.L., Pacheco, M.L., Zafrilla, S. and Cárdenas, A.. 2007. "The Jameos del Agua cave (Lanzarote, Canary Islands): Some morphological and geological features of a spectacular lava tube adapted to auditorium". *Volcanic Rocks—Malheiro & Nunes (Eds) Taylor & Francis Group, London, ISBN 978-0-415-45140-6*.

Volcanic Rock Mechanics – Olalla et al. (eds)
© 2010 Taylor & Francis Group, London, ISBN 978-0-415-58478-4

Study of lunar soil from terrestrial models (Canary Islands, Spain)

J.A. Rodríguez-Losada & S. Hernández-Fernández
Department of Soil Science and Geology, University of La Laguna, Tenerife, Spain

J. Martínez-Frías
Centro de Astrobiología (CAB), Spanish National Research Council (CSIC/INTA), Madrid, Spain

L.E. Hernández
Regional Ministry of Works, Government of the Canary Islands, Spain

R. Lunar Hernández
Departamento de Cristalografía y Mineralogía, Facultad de Geología, Universidad Complutense, Madrid, Spain

ABSTRACT: The Moon has a surface constituted mainly of basaltic materials. They are mostly vacuolar-like basalts that are also abundant in soils of volcanic origin on Earth. The geotechnical features of these rocks from the Canary Islands are supported by the basic characteristics deduced by NASA for this type of basalts. This paper deals with the geotechnical parameters of the lunar basalts taking into account the knowledge we have of the basalts from the Canary Islands used as terrestrial models as well as their suitability as building materials in future lunar bases. It is concluded that the lunar basalt, because of their abundance, ease of management and structural strength, is the best material for the construction of roads, tracks and even blocks with regolith for shielding of dwellings and facilities required on a lunar base.

1 INTRODUCTION

The exploration of a larger surface of the moon corresponding to the immediate environment to the lunar base will require that the people of the same have the means to facilitate their mobility. To achieve that displacement of one area to another of the lunar surface are the most appropriate not sufficient to have the best off-road vehicles adapted to lunar soil, but that means building roads or tracks for the movement of these vehicles can properly and higher speeds that would allow the dusty lunar soil. Furthermore, these tracks remedied, in part, the serious problem posed by lunar dust for transport vehicles and equipment and would create major maintenance costs.

Moreover, space transportation vehicles have a number of reactors that will cause the dusty lunar soil is lifted from the surface in large quantities, with the danger that entails, both for inhabitants of the lunar base as for equipment and facilities of the same. To limit these effects, we must also build tracks that allow the landing and takeoff of these space vehicles (must be far enough from the lunar base to avoid the effects of dust rising from the ground).

It is therefore necessary to build landing tracks and roads or access roads to the lunar base. And for their construction so that no other use of materials from natural resources of the Moon, as the transportation from Earth of the necessary materials would significantly increase the cost of installing the lunar base.

One of the key objectives for the development of a lunar base is the knowledge of the mechanical properties of lunar materials (Blacic 1985, Desai et al., 1992) and methods on the treatment of the lunar basaltic materials (Pletka 1993). The Moon has a very rich surface basaltic materials, especially in the lunar maria because they come from volcanic extrusions crack occurred at the surface and leave the magma through the cracks. This type of basalt vacuolar appearance due to strong gas release that occurred during their training, are abundant on the Moon and also on Earth, in soils of volcanic origin. The material used for the construction of these roads or highways must meet certain basic requirements: must be abundant, easy to handle and more suitable for work in terms of structural strength (to withstand the forces properly to be submitted lifetime). As physical and chemical experiments on a lunar basalt source are unable to carry out, this project will address the study of a vacuolar basalt from earth (particularly from the Canary Islands) with characteristics similar to

Table 1. Chemical compositions of selected samples (weight %).

Sample	EH-10	FV-19	FV-25	GC-52	LZ-22	TF-32	TF-40
SiO_2	41.69	50.53	49.34	43.69	50.76	42.63	46.31
Al_2O_3	11.38	16.34	13.37	13.81	13.53	11.83	16.33
CaO	11.35	7.09	9.36	9.34	9.45	11.07	9.40
Fe_2O_3	16.39	9.8	12.35	12.5	11.63	14.25	11.89
K_2O	1.14	2.5	0.68	1.48	0.66	1.35	1.90
MgO	10.11	1.99	9.76	9.41	9.23	12.02	4.99
MnO	0.19	0.20	0.16	0.18	0.15	0.19	0.21
Na_2O	2.53	4.8	3.19	3.85	2.89	2.80	4.17
P_2O_5	0.73	1.17	0.4	0.71	0.3	0.81	1.07
TiO_2	5.23	2.73	2.43	3.774	2.38	3.70	3.48
LOI	−0.09	2.89	−0.77	0.37	−0.65	−0.38	0.28
Total	100.65	100.04	100.27	99.11	100.33	100.27	100.03

those of lunar basalts. Specifically, it will attach some information about studies and physical evidence of such basalts that have been carried out by the Civil Engineering Laboratory of the Regional Ministry of Works (Government of the Canary Islands. Spain) in collaboration with the Centre for Civil Engineering Studies and Research, CEDEX, (Ministry of Civil Works, Spanish Government). This project that was called "Geotechnical characterization of the volcanic rocks of the Canary Islands" provided some of the most significant data that have been included here.

While this review, are also annexed a number of basic characteristics of this type of basalt, produced by NASA, which will complement the study of rocks from the Canary Islands. Also a paver robot is being designed to be capable of paving slabs with lunar basalt that has at its disposal.

The geotechnical project has characterized over three hundred samples, of which two thirds belong to Gran Canaria and Tenerife and the rest distributed among the remaining five islands (Rodriguez-Losada and Hernandez-Gutierrez 2006; Hernandez-Gutierrez and Rodriguez-Losada 2006; Rodriguez-Losada et al., 2007). Of these samples, only we focus on the vacuolar aphanitic basalts because they have similar features to those of lunar basalts known from samples taken on space missions.

We present data on the mineralogy, hardness and compressive strength indirectly deduced from tests of simple and quick implementation.

2 GEOCHEMISTRY AND MINERAL CHARACTERISATION OF SAMPLES

Of the total sample seven representative specimens of different lithotypes were selected.

Table 1 shows the analytical results, in which major elements appear expressed as oxides in weight percent.

To estimate the mineralogical composition a CIPW normative mineralogy calculation is used, based on the typical minerals that may be precipitated from a silicate molten, knowing the chemical composition. This CIPW Norm is especially useful for the study of volcanic rocks whose microcrystalline or glassy textures do not allow a modal mineral determination through a petrographic study.

Table 2 shows a summary of the CIPW Norm of all the analyzed rocks. It should be noted that because the iron content is in the form of ferric oxide, for calculation purposes it has been assumed a theoretical ratio of FeO to 60% in the basaltic rocks.

3 GEOTECHNICAL TESTS ON SELECTED SAMPLES

Several tests for geotechnical characterization of basaltic samples were carried out. Although there are no more significant data of the studied rocks, it is possible an indirect estimation of the most relevant data such as the compressive strength by means of other data easier to enforce. In addition, results of triaxial test on one selected sample are provided.

3.1 Sonic wave propagation

The speed of propagation of these waves depends on the elastic properties and the specific weight of the rock. This speed is affected by discontinuities in the rock so that, when analyzed the changes that have occurred in the wave, it can get much information

Table 2. CIPW Norm (weight %).

Sample	EH-10	FV-19	FV-25	GC-52	LZ-22	TF-32	TF-40
Orthoclase	6.74	14.78	4.02	8.75	3.90	7.98	11.23
Quartz	1.37	–					
Albite	12.76	40.53	26.99	16.95	24.45	9.74	25.39
Anorthite	16.32	15.37	20.15	16.04	22.01	15.73	20.22
Nepheline	4.69			8.46		7.56	5.36
Diopside	27.70	10.05	18.76	20.18	18.14	26.72	15.41
Hypersthene		0.86	12.23		18.44		
Olivine	10.42	1.02	5.45	11.55		14.91	5.44
Magnetite	9.50	5.67	7.16	7.25	6.74	8.26	6.89
Ilmenite	9.93	5.18	4.61	7.17	4.52	7.03	6.61
Apatite	1.73	2.75	0.95	1.68	0.71	1.92	2.53
Total	99.79	96.21	100.32	98.03	100.28	99.85	99.08

Table 3. Velocity of sonic waves (Vu).

Sample	TF-4	TF-32	TF-39	TF-40
Vu (m/s)	3818.72	5235.52	3835.8	5551.78
Sample	TF-41	TF-42	TF-48	LZ-7
Vu (m/s)	5768.15	5459.78	5338.55	2320.12
Sample	LZ-9	LZ-19	LZ-21	LZ-22
Vu (m/s)	4245.4	2420.7	2874.63	2254.69
Sample	LZ-9	LZ-19	LZ-21	LZ-22
Vu (m/s)	4245.4	2420.7	2874.63	2254.69

Table 4. Rebound index and equivalent uniaxial compressive strength.

Island (samples)	Rebound index (Average)	Equivalent compressive strength (MPa)
G. Canaria (2)	25.5	16.8
Tenerife (4)	43.3	48.9
Lanzarote (4)	52.5	65.5
El Hierro (2)	42.0	47.5
Fuertevent. (2)	49.5	59.5

Table 5. Point load test and equivalent uniaxial compressive strength.

Island (samples)	PLT (MPa) (Average)	Equivalent Compressive Strength (MPa)
Lanzarote (5)	2.54	50.8
El Hierro (2)	0.68	13.6
Fuerteventura (2)	5.00	100.0

on their characteristics (eg, the longitudinal velocity is greatly reduced by the existence of cracks).

The average values of sonic velocities obtained for basalt samples are shown in Table 3.

3.2 Rebound index by the Schmidt hammer test

Rebound index has been determined for samples of vacuolar basalts. Also, the equivalent compressive strength was deduced from the Schmidt hammer test. These values must be considered estimates, since this method of obtaining the compressive strength is not the most suitable due to the high margin of error involved.

The Table 4 reflects the values obtained by the hammer for some of the basaltic samples and its estimate equivalent in terms of compressive strength.

3.3 Resistance deduced from the point load test

In general, resistance is correlated with compressive strength, by:

$$\sigma_c = K \cdot I_s$$

where σ_c = Uniaxial Compressive Strength (in MPa); I_s = Point Load Test index; K = varies between 13 and 15 (Rodriguez-Losada et al., 2007).

The Table 5 shows the values of PLT obtained for some of the samples and the associated compressive strength.

Comparing the average values obtained here with those obtained in the hammer test appreciable differences are observed. As already indicated, this test is more reliable, thus the obtained values are closer to the true ones.

3.4 Resistance determination from the triaxial test

To get a more approximate estimate of the compressive strength that can have this type of basaltic

Table 6. Triaxial test at different confining pressures on a selected basaltic sample from Lanzarote.

Dens. g/cm³	2.2	2.2	2.1	2.1	2.3
σ_3 (MPa)	2.0	3.0	5.0	7.0	10.0
σ_1 (MPa)	59.7	79.0	68.5	72.5	96.1
Strain (%)	1.0	1.2	1.1	1.5	1.5

Table 7. Chemical composition of the simulated regolith, JSC-1 (average of three analysis) and of a real lunar basalt (weight %).

Oxide	JSC-1	S. d.	Lunar basalt 14163 "Apollo 14"
SiO$_2$	47.71	0.10	47.30
TiO$_2$	1.59	0.01	1.60
Al$_2$O$_3$	15.02	0.04	17.80
Fe$_2$O$_3$	3.44	0.03	–
FeO	7.35	0.05	10.50
MgO	9.01	0.09	9.60
CaO	10.42	0.03	11.40
Na$_2$O	2.70	0.03	0.70
K$_2$O	0.82	0.02	0.60
MnO	0.18	–	0.10
P$_2$O$_5$	0.66	0.01	–
LOI	0.71	0.05	–
Total	99.65		99.80

S.d.: Standard deviation.

rocks, there has been a triaxial test on a selected sample from the island of Lanzarote.

This test is the most versatile and common in the study of stress and deformation properties of rocks. It consists of subjecting a cylindrical rock sample to increasing axial stress to break, under different confining pressure. From the data obtained (Table 6) we estimate the compressive strength by using the Hoek-Brown criterion (Hoek et al., 2002).

For compressive strength from the data obtained by the triaxial test is to apply the failure criterion of Hoek-Brown. The equation defining this criterion is:

$$(\sigma_1 - \sigma_3)^2 = m_i \cdot \sigma_c \cdot \sigma_3 + \sigma_c^2$$

where σ_c = Uniaxial Compressive Strength (in MPa); σ_1 = major principal stress; σ_3 = confining pressure; m_i = constant material.

This equation can be likened to a regression line of type $Y^2 = A \cdot \sigma_3 + B$, where:

$$Y = (\sigma_1 - \sigma_3)^2$$
$$A = m_i \cdot \sigma_c$$
$$B = \sigma_c^2$$

Thus, known values of σ_1 and σ_3, we must determine the regression line (ie, the values of Y, A and B), from it, get the value of the RCS, σ_c. For this, first calculate the values of Yi for the five specimens, obtaining:

$$Y_1 = (59{,}74 - 2)^2 = 3333{,}91 \text{ MPa}^2$$
$$Y_2 = (79{,}02 - 3)^2 = 5779{,}04 \text{ MPa}^2$$
$$Y_3 = (68{,}48 - 5)^2 = 4029{,}71 \text{ MPa}^2$$
$$Y_4 = (72{,}47 - 7)^2 = 4286{,}32 \text{ MPa}^2$$
$$Y_5 = (96{,}05 - 10)^2 = 7404{,}6 \text{ MPa}^2$$

The regression line obtained for these five points is given by:

$$Y = 342{,}29 \cdot \sigma_3 + 3118{,}3$$

where

$$B = 3118{,}3 = \sigma_c^2$$
$$\sigma_c = 55{,}84 \text{ MPa}$$

This estimated value for the RCS is very similar to that obtained from testing at peak load, which was 54 MPa. Therefore, the estimated value of the compressive strength by that type of basalt is sufficient for the use of such rocks as building material for track or road pavement, especially considering that in the lunar surface there is one-sixth the gravity of Earth.

4 NASA THEORETICAL LUNAR REGOLITE

NASA has developed a material similar to the one existing in the lunar maria from a mixture of compounds derived from basaltic volcanic ash on earth (Willman et al., 1995). This simulated lunar regolith, called JSC-1 was subjected to various experiments and tests to check whether their characteristics are sufficiently similar to those of actual samples of lunar regolith gained in the past. After checking that the material is similar to the one from the moon, it will be useful to prove with it, different extracting processes to obtain important chemical elements such as oxygen and metals or to check its viability as raw material for the "in situ" construction of some structural elements in a lunar base.

4.1 Chemical composition of simulated regolith

This material closely resembles a regolith from the lunar maria with a low content of titanium and a certain percentage of crystals. The chemical composition of the simulated regolith was obtained by X-ray fluorescence. The samples were air-dried for two months prior to the analysis, separating the remaining material through a sieve of 177 μm. The chemical analysis results are reflected in Table 7.

4.2 Comparison of simulated regolith and real lunar soil

The simulated regolith JSC-1 is composed of basaltic ash with a typical composition of most terrestrial basalts (as described previously for vacuolar basalts from the Canary Islands). The lunar basalts and soils from lunar maria, are quite similar to the simulated regolith in terms of its main compounds. However, the lunar samples have no water and low contents of some volatile compounds such as Na_2O. In addition, lunar rocks were formed in highly reducing environments and contain iron as Fe_2+ and FeO (the Fe_2O_3 was only found as traces of the mineral hematite in some lunar meteorites, but not in samples from the space missions).

The minerals found in the majority of simulated regolith (plagioclase, pyroxene, olivine, ilmenite and chromite) are also characteristic of soils from v maria. Lunar soil also contains a significant proportion of crystals produced by the impacts of micrometeorites (formed under highly reducing conditions) with a certain amount of microscopic particles of native iron. By contrast, crystals of simulated regolith JSC-1 have a higher content of plagioclase and metal oxides than the lunar soil, but do not contain native iron. It should be noted that the micrometric textures of the lunar agglutinates are complex and may not accurately match with any terrestrial analogue, so the comparison is made under macroscopic point of view.

Other interesting geotechnical features of simulated lunar regolith compared to the real lunar soil are:

- Density: for JSC-1 has a value of 2.9 g/cm³, being in the range of the lunar soil, which gave values between 2.9 g/cm³ and 3.5 g/cm³.
- Internal friction angle: between 25° and 50° for the real lunar soil, while the value obtained for the simulated regolith is 45°.
- Internal cohesion: varies from 0.26 kPa and 1.8 kPa for lunar soil, while the value obtained for the simulated regolith is 1.0 kPa.

Accordingly, it can be concluded that the simulated regolith JSC-1 is a good chemical and mechanical analog of the real lunar soil. In fact, it is a material that has already been produced in large quantities to meet the needs required by scientists and engineers in their investigations.

4.3 Using simulated regolith JSC-1 for brickmaking

There have been several experiments with this material to check the possibility of sintering in small blocks or plates for the "in situ" construction on the Moon (Allen et al., 1992).

Table 8. Physical properties of molten lunar basal.

Molten density at 1473°K (g/cm³)	2.6–2.7
Solid density (g/cm³)	2.9–3.0
Water absorption (%)	0.1
Tensile Strength (MPa)	35
Compressive Strength (MPa)	54
Equivalente Mösh hardness	8.5
Specific heat (J/kg · K)	840
Melting temperature °K	1400–1600
Heat of fusion (J/kg)	$4,2 \cdot 10^5$ (±30%)
Thermal conductivity, solid (W/m · K)	0.8
Thermal conductivity, molten at 1500°K (W/m · K)	0.4–1.3
Surface resistivity (Ω · m)	10^{10}
Internal resistivity (Ω · m)	10^9
Sonic velocity, molten at 1500°K (m/s)	2300
Sonic velocity, solid at 1000 K (m/s)	5700

Basically, the experiment consisted on heating samples to 1100°C, after pretreatment to try to get the desired results. Initially, the regolith samples were compacted to reach a density of 2.45 g/cm³. Subsequently, the samples were sintered increasing their density to a maximum of 2.68 g/cm³. The results were a few blocks of 13.1 cm long, 5.6 cm wide and 4.6 cm in height, uniform and cracks-free (likely, the abundance of crystals on JSC-1 allowed an optimal adhesion between grains of the rock). Consequently, it has been demonstrated that with Earth analogues is achievable the use of lunar basalt as building material.

5 CONCLUDING REMARKS: APPLICATIONS OF THE LUNAR BASALT FOR "IN SITU" BUILDING

The lunar basalt, because of their abundance, ease of management and structural strength, is the best material for the construction of roads, tracks and even blocks for armour of dwellings and facilities for the lunar base that require it.

The raw lunar soil could be melted at a temperatures of 1550°K, then to be cooled and subsequently solidify in a hard and resistant material. Depending on how the cooling is done, the resistance will be different (NASA, 1980):

- Rapid cooling (within minutes): the basalt would become a liquid crystal polymer substance. It would be a hard material but slightly brittle, so cracks may propagate easily. If this option is selected, it would be necessary to divide the material into small plates to isolate potential fractures, allowing easier maintenance.
- Slow cooling (several hours): the basalt would became from a fully liquid state at 1570°K

into a solid, rather hard state at a temperature slightly less than 1370°K, in crystalline form. This method requires more time and energy, but will result in a material much harder and more durable, and should not be separated into such small plates, but could be prepared as a continuous surface (of more interest for lunar tracks).

In addition, the lunar soil has the adecuate characteristics for the manufacture of these basaltic blocks or plates because, unlike terrestrial basalts, are not affected by the weather or contamination by external agents. The Table 8 exhibits the major characteristics of these molten lunar basalts.

The pavement used in roads and highways on earth has between 15 cm–25 cm thick, and 30 cm for airport runways. Given that the gravity on the Moon is one sixth of the Earth and therefore the weights applied on the pavement will be six times smaller, the required thickness of pavement on lunar roads would be between 2.6 cm–4, 3 cm, while for landing runways might be sufficient with a thickness of 5 cm. To build these tracks, NASA has made several designs robots pavers, which would be responsible for smoothing the lunar soil to place the plates of basalt.

REFERENCES

Allen, C.C., Hines, J.A., McKay, D.S., and Morris, R.V. 1992. Sintering of lunar glass and basalt. In Sadeh W.Z., Sture, S., Miller, R.J. ed.: proceedings of the 3rd International conf on engineering, construction and operations in space (Space 92). Vol. 1 and 2: 1209–1218. Denver, Co, may 31–jun 04.

Blacic, J.D. 1985. Mechanical Properties of Lunar Materials Under Anhydrous, Hard Vacuum Conditions: Applications of Lunar Glass Structural Components. In: Lunar Bases and Space Activities of the 21st Century. Houston, TX, Lunar and Planetary Institute, edited by W.W. Mendell, 1985, p. 487.

Bletka, B.J. 1993. Processing of lunar basalt materials. In Lewis J.S., Matthews M.S., and Guerrieri M.L. eds 1993: Resources of Near-Earth Space, University of Arizona Press p. 325–350.

Desai, C.S., Saadatmanesh, H., and Allen, T. 1992. Mechanical properties of compacted lunar simulant using new vacuum triaxial equipment. In: Engineering, construction, and operations in space III: Space, 92; Proceedings of the 3rd International Conference, Denver, Co, May 31–June 4, 1992. Vol. 2 (A93-41976 17–12), p. 1240–1249.

Hernández-Gutiérrez, L.E., and Rodríguez-Losada, J.A. 2006. Estimative rock mass parameters applied to the Canarian volcanic rocks based on the Hoek-Brown failure criterion and equivalent Mohr-Coulomb limits as a contribution in natural hazards. 300th Anniversary Volcano International Conference (GARAVOLCAN), Session 1. Garachico, Tenerife. 22–26 de mayo de 2006.

Hoek, E., Carranza-Torres, C.T., and Corkum, B., 2002. Hoek-Brown failure criterion—2002 edition. In Proceedings of the Fifth North American Rock Mechanics Symposium, Toronto, Canada 1, 267–273.

NASA CP-2255. 1980. Advanced Automation for Space Missions: Proc. 1980 NASA/ASEE Summer Study held at the Univ. Santa Clara, CA, ed. R.A. Freitas, Jr. and W.P. Gilbreath.

Rodríguez-Losada, J.A., and Hernández-Gutiérrez, L.E. 2006. New geomechanical data of the Canarian volcanic rocks as a contribution for geophysics applied to the research in volcanic risk. 300th Anniversary Volcano International Conference (GARAVOLCAN), Session 2. Garachico, Tenerife. 22–26 de mayo de 2006.

Rodriguez-Losada, J.A., Hernandez-Gutierrez, L.E., Olalla, C., Perucho, A., Serrano, A., and Rodrigo del Potro, 2007. The volcanic rocks of the Canary Islands. Geotechnical properties. Proceedings of the International Workshop on Volcanic Rocks W2. 11 ISRM Congress. Ponta Delgada (San Miguel, Azores). Session 1, Characterization of volcanic formations, 53–57.

Willman, B.M., Boles, W.W., Mckay, D.S., and Allen, C.C., 1995. Properties of lunar soil simulant JSC-1. J. Aerospace Eng. 8(2): 77–87.

Volcanic Rock Mechanics – Olalla et al. (eds)
© 2010 Taylor & Francis Group, London, ISBN 978-0-415-58478-4

Geological and geotechnical conditions of the Machico-Caniçal highway

V.C. Rodrigues, S.P.P. Rosa, J.A.M. Brito & C.J.O. Baião
Cenorgeo, Engenharia Geotécnica, Lda., Lisboa, Portugal

ABSTRACT: The Machico-Caniçal highway, with 8 km long, is part of Madeira Island development program, connecting Funchal city to Caniçal village trough a continuous highway along the islands south coast. This highway crosses an extremely mountainous region, conditioned by the existence of two very heterogeneous volcanic complexes, generally covered by unstable slope deposits or by deep alluvial deposits. Those geological conditions lead to the execution of a wide range of civil engineering works. This paper presents the most relevant geological and geotechnical aspects identified along this highway, in particular in what special engineering structures (tunnels, bridges and viaducts) and retaining walls were concerned.

1 GENERAL LAYOUT CHARACTERISTICS AND MAJOR CIVIL ENGINEERING WORKS

The Machico-Caniçal highway, constructed between 2003 and 2004, is located on the eastern side of the Madeira Island, along the southern coastline, and it constitutes the connecting stretch between Santa Catarina Airport and Caniçal village (Fig. 1).

The stretch comprehended the construction of eight tunnels, five of them double, and four double bridges and viaducts (Cenor & Grid, 2002–2003). The tunnels had a total length of 9 750 m.

The bridges and viaducts comprised a total length of 1144 m.

An important set of retaining structures were also designed, with a total length of 2750 m, 1900 m corresponding to retaining structures equal to or higher than 8 m (Pereira *et al.*, 2004). In Tables 1A, 1B and 1C are presented the major characteristics of the special engineering structures (tunnels, bridges and viaducts) and retaining structures with height equal to or higher than 8 m.

2 FIELD INVESTIGATION

In the light of the unfavourable topographic conditions, of the occurring geological conditions (which were characterised by substantial heterogeneity and the presence, throughout practically the entire stretch, of thick and unstable slope deposits), of the acquired experience in these types of formations and also of the significance and number of special engineering structures (bridges and tunnels) and retaining walls involved, detailed geotechnical studies were carried out, based essentially on detailed geological surface surveys and deep survey consisting on rotational boring.

Up to 146 borings were made to investigate the 8 km length of the highway, comprising a drilling total of 3 071 m, which corresponds to 384 m of drilling/km.

Of all the drilling carried out, around 72% was made for the surveying of special engineering structures (bridges and tunnels), whilst the remaining 28% concerned the highway itself and the retaining walls. Of the survey works carried out for special engineering structures, an average of 203 m of drilling/bridge (21 m/support) and around 233 m/ tunnel (144 m/km) were made.

3 GEOLOGICAL AND GEOTECHNICAL CONDITIONS

The highway crosses formations of the Post-Miocene volcanic complex (β^2) and Mio-Pliocene volcanic complex (β^1) (Fig. 1).

The β^2 complex, of a substantially heterogeneous constitution, is formed by alternate layers of basaltic lava flows and pyroclastic materials, usually interstratified, also intercalated with generally less important layers of tuffs.

The β^1 complex, the oldest volcanic complex occurring on the island, as also a very heterogeneous constitution and is formed by a chaotic pile of coarse materials resulting from the projection of angular blocks, volcanic bombs and flows encased,

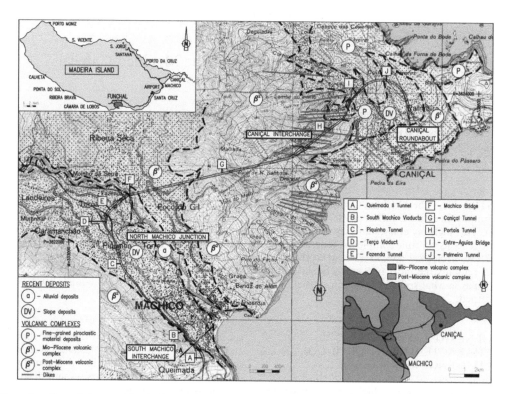

Figure 1.   General plant and Geological plant. Major civil engineering works constructed along the highway.

Table 1.   Major special engineering structures and support structures characteristics.

Tunnel	Width (m)	Cross section (m²)	Total length (m)	Max. covering (m)	Geological conditions
Queimada II	9	75	745	95	β^2
Access Road A	7	57	202	70	β^2
Piquinho	9	75	2×449	35	β^1 and β^2
Fazenda	9	75	2×163	30	β^2
A.Road E	7	57	185	40	β^2
Caniçal	9	75	2×2100	>300	β^1 and β^2
Portais	9	75	2×682	100	β^2
Palmeira	9.6	83	2×915	60	β^1 and β^2

A – Tunnels.

Table 1a.   Major special engineering structures and support structures characteristics.

Name	Total length (m)	Geological conditions	Type of foundation
South Machico Interchange Viaducts	60/60	β^2	Footings
Terça Viaduct	75/72	β^2	Footings and piles
Machico Bridge	215,9/207.1	β^2 and β^1	Micro piles
Entre-Aguas Bridge	234,4/219.4	β^2	Footings

B – Bridges and viaducts.

Table 1b. Major special engineering structures and support structures characteristics.

Wall	Type of structure	Length (m)	Max. height above found level (m)	Geological conditions
M01	Tieback wall	120	8	β^2
M02	Tieback wall	33	8	β^2
M03	Gravity wall	48	9	β^2
M04	Tieback wall	166	11	β^2
North abutment	Gravity wall	31	10	β^2
M05	Gravity wall	133	12	β^2
M12	Gravity wall	145	9	β^2 and β^1
M22	Gravity wall	69	9	β^2
TL2	Tieback covering	760	20	β^2
M1	Gravity wall	83	12	β^1 and β^2
M3	Gravity wall	65	12	β^2 and β^1
M8	Semi-gravity wall	165	8	β^1
M5	Gravity wall	84	12	β^1

C – Retaining structures with height equal to or higher than 8 m.

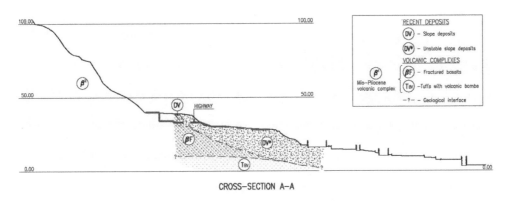

CROSS-SECTION A-A

Figure 2. Geological cross-section of the slope deposits at South Machico interchange.

to a more or less significant degree, in fine pyroclastic material. This complex is also occasionally intercalated with generally highly modified basaltic lavas (Rosa *et al.*, 1995, 1997). Throughout the Machico River valley, both complexes are practically covered either by alluvial deposits or by slope deposits.

The alluvial deposits are generally very coarse and heterogeneous, accumulated along the Machico River. These are constituted by rounded blocks or rolled fragments and pebbles of basalt, covered in a sandy-silt or clay-silt disaggregated matrix. This formation resulted from the deposit of material transported under torrential conditions either by the Machico River itself or its tributaries. These have been surveyed at depths range from around 6 to 18 m.

The slope deposits result from landslides and rockfalls from the declivous rocky slopes of the Machico River valley, and its accumulation along and at the base of those slopes.

These are very heterogeneous deposits comprehending fragments and blocks, predominantly of basalt, of various sizes, from blocks which range from 2 m in diameter to small angular or sub-rolled fragments covered in a clay-silt-sandy matrix of a brown-red dark colour. They are characterised by the fact that they show higher concentration of rocky blocks at the base. The matrix of these deposits consists essentially of highly plastic clays.

The water table, in most places, is situated between this formation and the volcanic substrate. These deposits are generally in limited equilibrium conditions due to their depth and geotechnical characteristics, to the substrate inclination on which they lay on and, above all, to water table variations. They easily become unstable, provoking landslide areas of more or less significance. They cover most of the slopes where the highway was implemented and gradually increased in depth to the base of the slope. These deposits have been surveyed and depths range from around 0.3 to 21 m (Fig. 2).

a. General view of the tieback slopes.

b. Details of a retaining wall

Figure 3.   South Machico Interchange.

a. Construction of the anchored wall.

b. General view of the portal.

Figure 4.   Western portal of the Caniçal double tunnel.

Given that one of the major obstacles to the implementation of the construction works was the thickness of slope deposits which covered substantial part of the highway, a geotechnical characterisation of these deposits was carried out in order to better understand their behaviour (Rosa *et al.*, 2004).

In the fine fraction, which controls the behaviour of these deposits, the lower percentage limit which passed through the #200 mesh was around 66%, the lower limits below 2 mm and 2 μ were 22% and 38%, respectively.

It was also obtained liquidity limit (LL) values between 56% and 122% and plasticity index (PI) values between 21 and 80%. The water content (w) and volume weight (γ) obtained were 28% and 18.4 kN/m³, respectively. For the particles density (G) values of between 2.53 and 2.90 were obtained. Sand equivalent (SE) ranged between 14 and 16%. Analysis carried out on the matrix using the x-ray diffraction method showed that the clay fraction is essentially comprised of montmorillonite (around 95%).

The undrained residual resistance was estimated based on the parameters in terms of actual tensions based on the Mohr-Coulomb rupture criteria.

It was obtained $cu_{res} = 14$ kPa for $c'_{res} = 0$ kPa and $\phi'_{res} = 14°$. However, to take into account the favourable contribution of gravel and blocks and the undulation of the rocky substrate, in terms of resistance to sheering, it was considered that favourable parameters could be adopted. As a result of this analysis it was adopted $c'_{res} = 0$ kPa and $\phi'_{res} = 16°$ and $cu_{res} = 20$ kPa.

4   EXAMPLES OF CIVIL WORKS AFFECTED BY THE SLOPE DEPOSITS

At the South Machico interchange, slope deposits vary in depth, reaching more than 20 m in localized areas.

In order to construct this interchange, a large amount of retaining structures were built, resorting to a wide range of solutions, in order to enable the construction of the highway and the respective connecting access roads to the existing regional road (Fig. 3).

The total length of this set of structures totals around 2750 m in length, with foundations based both directly and indirectly on the β^1 and β^2 volcanic complexes.

Of the civil works constructed the following could be emphasise, four anchored walls and three retaining walls set on jet-grouting columns.

At the Terça Viaduct the depth of slope deposits ranged from about 3 m on the northern side to 14 m on the southern side. This accentuated difference implicated that this viaducts foundations had to be mixed, with the northern abutment and central pillar sated directly on footings while the southern abutment was sated on pilings of about 15 m length.

In the western portal of the Caniçal double Tunnel, the thickness of the slope deposits lead to the construction of an anchored wall, with 65 m in length and 7 m in height, at the top of the front slope (Fig. 4).

5 CONCLUSIONS

The extremely rough geomorphology of the zone where this highway stretch was constructed, together with the occurrence of deep slope deposits with stability conditions close to the limit equilibrium, introduced substantial constraints to the surveying and mechanical characterisation, as well as to the establishment of basic criteria for the civil works design.

These problems required an important development of the geological and geotechnical studies, since the surveying phase, until the conclusion of the civil works, as well as the involvement of a multi-disciplinary group of technicians.

The technical support provided by geotechnical experts revealed to be essential, enabling not only the optimisation of the civil works design, but also the improvement off the safety procedures, in the light of the better understanding of the encountered conditions.

It also allowed the resolution of situations impossible to predict at initial stages and to carry out the necessary design adjustments, in useful time, as well as enabling an optimisation of human and material resources and an increased reliability of the constructed works.

ACKNOWLEDGEMENTS

The authors are grateful to the Regional Directorate of Infrastructures and Equipment of the Regional Government of Madeira and to the Tamega & Zagope & Tecnorocha & Somague construction consortium for the authorisation and help provided.

REFERENCES

Cenor & Grid. 2002–2003. *Machico-Caniçal highway*. Final project (in Portuguese).

Pereira, A.F., Baião, C.J., Freitas, A.R., Sousa, F.M., & Brito, J.M., 2004. Support structures and slope consolidation at South Machico interchange of Machico-Caniçal highway in Madeira Island. *9th National Congress on Geothecnics; Proceedings. Coimbra University. Coimbra. Portugal (in Portuguese)*.

Rosa, S.P., Brito, J.M., & Baptista, J.P., 1995. Geological and geotechnical conditions of Boa Nova-Cancela stretch of Funchal-Airport highway. *5th National Congress on Geothecnics; Proceedings. Coimbra University. Coimbra. Portugal* (in Portuguese).

Rosa, S.P., Brito, J.M., & Baptista, J.P., 1997. Geological and geotechnical conditions of civil works on new highways of Madeira Island eastern region. *6th National Congress on Geothecnics; Proceedings. LNEC. Lisboa. Portugal* (in Portuguese).

Rosa, S.P., Brito, J.M., Baião, C.J., & Freitas, Rodrigues, V.C., 2004. Geological and geotechnical conditions of the Machico-Caniçal highway. *9th National Congress on Geothecnics; Proceedings. Coimbra University. Coimbra. Portugal (in Portuguese)*.

Volcanic Rock Mechanics – Olalla et al. (eds)
© *2010 Taylor & Francis Group, London, ISBN 978-0-415-58478-4*

Geomechanical appraisal of the deformation potential of a deep tunnel in a volcanic rock mass

D. Simic
Ferrovial Agroman
Polytechnical University of Madrid, Madrid, Spain

J. López
Ferrovial Agroman

ABSTRACT: Ophiolitic complexes can be defined as an association of ultra basic volcanic rocks (ultramafic) and basic (mafic) constituents of the oceanic crust as a result of a phenomenon of abduction clash between continental plates. An intricate structure in which volcanic rocks are intruded in the direction of the schistosity of the metamorphic rocks is further complicated by large scale over thrusts which create tectonic melanges at the base of such mega structures, affecting it by secondary tectonic contacts and inverse faults. The whole entity is found in considerable tectonic disorder where packages of peridotites or pillow lavas of various sizes "float" inside a sheared shale-like mass. It is understandable that tunnelling in such a formation requires a good appraisal of the rock mass deformation potential as it will have a direct impact in the support behaviour, particularly in the deeper sections with more than 250 m overburden. This paper deals with the geomechanical characterisation of the heterogeneous rocks mass and the different models employed to simulate its behaviour during the tunnel excavation and support.

1 ORTHRYS MOUNTAIN OPHIOLITIC COMPLEX

The tunnel T-2 belongs to the E-65 Highway scheme, crossing the western slopes of "Paliovigla-Paliokazarma-Kazarma" ridge, belonging to the Othris massif in Central Greece.

The macro-geological unit in which the tunnel is located is called Orthrys Ophiolitic Complex, which has been widely studied from a genetic and geological point of view as well as from the geo-mechanical point of view by various authors (Marinos, Verroios, Foucault, etc.). This complex has been encountered in numerous significant projects during which the different geological units that comprise this formation unit have been studied.

Ophiolitic complexes can be defined as an association of ultra basic volcanic rocks (ultramafic) and basic (mafic) constituents of the oceanic crust that may appear arranged on the continental crust as a result of a phenomenon or obduction clash between continental plates. The Orthrys ophiolite complex shows an intricate structure in which the pillow lavas cut a serpentinized peridotite and gabbro volumes are rooted in the peridotites. Shear zones are common.

The general and synthetic outline of an Alpine ophiolitic complex is as follows:

- A top layer composed of sediments (mud and chert) and sedimentary rocks, in our case the layers of shale and chert.
- Below, a layer of compact pillow lavas with the presence of nodules of cherts and dykes intruded in the direction of the schistosity planes of metamorphic rocks. They form the basic or mafic section of the sequence
- Below, separated from the previous layer by a tectonic contact, a layers consisting of massive gabbros, layered gabbros and peridotites and tectonized peridotites is encountered. This section constitutes the ultramafic or ultrabasic part of the complex; which has the peculiarity of having textures similar to those of the sedimentary rocks. In basal layer, the rocks with texture and/or tectonic structure are at the bottom and separated from the massive formations by a transition area, usually by tectonic contact.

As ophiolites are associated with large-scale overthrusts, tectonic mélanges can be formed in the base and at the front of such megastructures. These ophiolitic mélanges contain ophiolitic rocks and other rocks of various paleogeographic origins; the whole entity being in considerable tectonic disorder with chaotic masses where blocks and packages of various sizes of any kind or rock (sedimentary or volcanic) "float" inside a sheared soil like mass.

Figure 1 shows the theoretical column type of an ophiolitic comp.

2 GEOLOGICAL UNITS: LITHOLOGY

The geological units encountered along the tunnel have been studied in the outcrops in the vicinity of the tunnel and also in the samples recovered from the boreholes executed along the tunnel. These geological units are shown in Figure 2, in a longitudinal geological section of the tunnel.

2.1 Basalts, dolerites and pillow lavas (Do, pla)

They consist mainly of green, grayish-green and grayish-red dolerites and basalts. The dolerites usually have veins of calcite and locally appear with serpentinized discontinuity surfaces, while the shales-cherts appear within the formation, mostly in the locations of the tectonic zones or as interlayers within the igneous formations. They appear slightly to moderately fractured (approximately two to three discontinuities per meter) and slightly to very weathered. In undisturbed or normal conditions they are very strong and form steep morphologies.

2.2 Gabbros and peridotites (p-g)

They consist of gabbros and peridotites moderately to strongly serpentinized, with gree to grayish-green color. They are distinguished from other ophiolitic formations by their holocrystalline texture. In general they are tectonized, resulting in moderate to significant fracturation at places, while they appear slightly weathered. The fresh formations are generally impermeable but the intensively fractured zones exhibit increased permeability.

2.3 Shale and chert (Sh)

They consist of red argillaceous shales with local interlayers of cherts, sandstone and thin-bedded limestone. At places interlayers and lenses of ophiolitic formations locally serpentinized, are evident, while continuous alternations of shales-cherts and ophiolitic thin-bedded formations are found also resulting in a mélange character.

The shales are usually thin bedded with thickness ranging from 1 cm to 10 cm approximately, strongly tectonized, folded and sheared with smooth to slickensided at places discontinuity surfaces and slightly to moderately weathered resulting in significant loosening of the formation near the ground forming mild morphological relief in the areas where they prevail.

This formation exhibits rather low permeability due to the dominance of the shales. Due to the intense tectonism, the mechanical properties of this rockmass are the lowest of all the formations that will affect the tunnels.

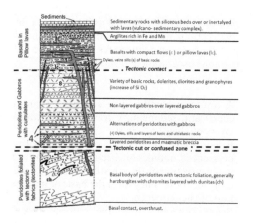

Figure 1. Synthetic column and a theoretical ophiolitic comple. Foucault and Raoult (1995).

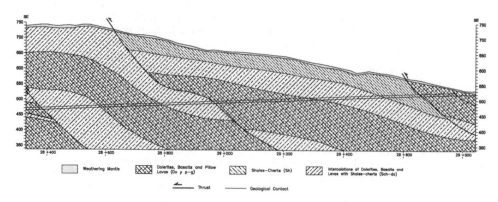

Figure 2. Geological profile along the tunnel.

2.4 Continuous shale alternations—cherts with dolerites and lavas (Sh-do)

The formation consists of red argillaceous shales with local interlayers of cherts, sandstones and thin-bedded limestone, in continuous alternations with dolerites and lavas. The formation appears highly tectonized, folded and sheared and often appears in the form of mélange.

The overall permeability is low although more permeable layers (dolerites and lavas) are present in continuous alternation with impermeable layers (shales-cherts). The mechanical properties depend on the layer with more presence.

Figure 3 shows the histogram with the distribution of different geological units as they appear in boreholes BS2O-1 to S2O-8.

3 GEOMECHANICAL CLASSIFICATION OF THE ROCK MASS

3.1 Criteria and methodology employed

The estimation of the Geological Strength Index (GSI) of the different geological units that will be encountered during the excavation of the tunnels has been based on the analysis of the samples recovered from B2O-1 to B2O-8, the results from the Uniaxial Compressive Strength Tests (UCS) and on those of the point load tests (PLT) as well as on the information given in the GIR report prepared by OK.

This analysis has been based among others on the recommendations given by E.Hoek and P.Marinos in their publication "*Variability of the engineering properties of rock masses quantified by the geological strength index: the case of ophiolites with special emphasis on tunneling (July 2006)*",

and also on the article "*Determination of residual strength parameters of jointed rock masses using the GSI system*" by M. Cai, P.K. Kaiser, Y. Tasaka and M. Minami and published in the International Journal of Rock Mechanics and Mining Sciences in September 2006.

The assumptions made in this study are the following:

– The result of the unconfined compressive test is equivalent, with a reasonable degree of accuracy, that obtained by multiplying the Point Load Test (Is) by 24, where Is = P/D^2 (where P is the load and D is the distance between them).
– It has been considered as more representative the results of those Point Load Tests in which the load was applied normal to the direction of the stratification and/or schistosity.
– the Geological Strength Index (GSI) has been estimated using the following expression:

$$GSI = RMR'_{89} - 5$$

where RMR'_{89} is the Bieniawski Index according to his 1989 proposal for solid for dry rock masses.
– When evaluating the condition of the discontinuities, it has been assumed that their condition has been affected by the drilling operations which tend to disturb their natural condition as well as to increase their aperture.
– Representative samples of the geological unit Sch-Do have been considered when evaluating the GSI from samples recovered from boreholes. Sub-division of the formation in homogenous section from a lithilogic point of view has been avoided.
– Due to the complexity of the Ophiolitic formation, it has been decided to reduce the number of geological units. This has been done in order to avoid a classification with many different formations which may lead to confusion and mistaken interpretation.
– The RQD and the spacing between the discontinuities have been evaluated in accordance with the latest recommendations and methodologies proposed by Bieniawski.
– The guidelines given by Hoek and Marinos in 2001 have been used when evaluating the GSI from boreholes.

3.2 Geotechnical characterization

3.2.1 Geotechnical Unit Do and p-g

The Do unit consists of dolerites and basalt. Dolerites often have serpentinized sections and frequent presence of shale and chert within this unit as intercalations, especially in the vicinity of the tectonic contacts.

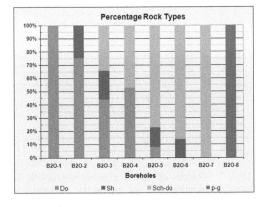

Figure 3. Distribution of different lithotypes in the boreholes B2O-1 to B2O-8.

The p-g unit only affects the North tunnel portal. This geological unit consists of gabbros and peridotities slightly to highly serpentinized. They can be are distinguished from the Do unit by its holocrystalline texture. These formations are treated as one unit due to their similar geo-mechanical characteristics.

In the analyses of boreholes that encountered this unit, the intercalations of shale and chert represent no more than 15% of the total length of the samples. The thickness of these intercalations is between 0.50 m and 3.00 m.

The thickness of the layer that consist mainly of dolerites (serpentinized or not), basalts, gabbros and/or peridotities is between 2.90 and 30.40 m. This formation normally presents in the order of two to three discontinuities per meter and are moderately to highly fractured in areas. The RQD rages from 12% and 77%, with an average value of 43%. Figure 4 shows the statistical analysis of RQD this formation.

The value of uniaxial compressive strength measured in specimens of dolerite, basalt, peridotite and gabbros ranges from 0.30 MPa to 134.10 MPa, with an average value of 25.0 MPa. The histogram in Figure 5 shows the distribution of the results obtained with this test.

The joints of discontinuity planes are smooth or slightly rough, sometimes lubricated, closed or slightly open (<1 mm) with hard filling material. They are generally slightly weathered or normal.

The GSI, as estimated from the borehole logs, varies between 39 and 66, with an average GSI of 50. Figure 6 shows the distribution of values obtained.

In figure 7 it is shown, as a conclusion, a comparison of the estimated GSI values with Marinos and Hoek (2001) recommendations. The main difference lies in the condition of the joints, significantly higher when obtained from boreholes and not outcrops, where the effect of weathering is much lower.

3.2.2 Sch-do geotechnical unit

These are layers of shale and/or argillites with intercalations of cherts, sandstone and thin-bedded siltstone in continuous alternations with dolerites and lavas. Its mechanical properties vary depending on the formation with greater presence, being layers of volcanic rocks (dolerites and lavas)

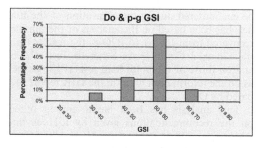

Figure 6. Frequency histogram of estimated GSI values from boreholes drilled in the Do & p-g unit.

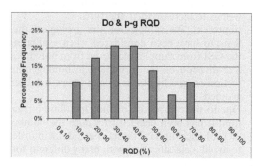

Figure 4. Frequency histogram of the UCS and PLT compressive strength test results in the Sch-do unit.

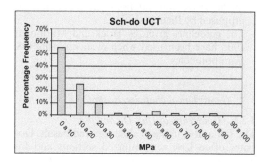

Figure 5. Frequency histogram of the UCS and PLT compressive strength test results in the Do & p-g unit.

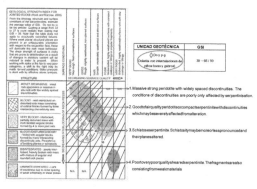

Figure 7. Ranges of variation of estimated GSI for the do geotechnical unit and p-g on the extracted core B2O-1 to B2O-8 and superimposed to those proposed by Marinos and Hoek rocks of varying qualities peridotitic, dolerite and/or serpentinites ofiliticos complexes.

of better geomechanical quality than those of the sedimentary rocks or metamorphic rocks (shales-chert and siltstone).

In order to estimate the GSI from the cores recovered from the boreholes, the samples have been divided into homogeneous sections according to lithological and geochemical criteria.

The range of RQD value assigned to the representative samples of the unit ranges from 10%–84%, with an average of 44%. There are two levels of values (see frequency histogram in Figure 8) depending on which lithologic formation prevails in the sample. In samples with higher percentage of shales, the RQD usually varies between 0% and 40%, whilst in samples with higher percentage of lava-basalts slightly fractured basalts, the RQD ranges from 70% to 100%.

The uniaxial compressive strength of intact rock is generally low, with an average value of 14.97 MPa and a range of 0.10 MPa to 87.36 MPa. Only 11% of the tested samples exceeded 30.0 MPa (see frequency histogram in Figure 9).

The joints of discontinuity planes are smooth or slightly rough, and slightly open to open (0.1 mm to 5.0 mm) with sandy or clayey infilling material.

The GSI estimated for this formation varies between 30 and 52, with an average GSI value of 47.

In boreholes B2O-3, B2O-4 and B2O-5, the layers of shale, sandstone and chert are more frequent than the volcanic lavas and basaltic. In addition, the rock is more fractured which leads to an average value of GSI 45 in this boreholes, with values ranging from 30 to 52.

On the other hand, boreholes B2O-6 and B2O-7 were dominated by basaltic volcanic rocks in which the degree of fracturing is smaller and the condition of the joints better than those in previous group. In this case, the average value of GSI is 50, with values ranging from 40 to 68.

Finally, in figure 10 it is shown a comparison of these GSI values with the ones proposed by Hoek and Marinos (2001).

3.2.3 Sh geotechnical unit

Unit Sh consists of sediments and sedimentary rocks (shales and cherts), associated with overthrusts of large scale. This formation is very tectonized and sheared. It is usual to find lenses of ophiolitic formations, serpentinized or not, due to their *mélange* character.

This formation is usually arranged in layers and/or well-defined bedding planes of 1 to 10 cm thickness, strongly tectonized, folded and sheared, with smooth to slickensided at places discontinuity surfaces and slightly to highly weathered. Due to the intense tectonism, the mechanical properties of this rockmass are the lowest of all the formations that will affect the tunnels.

The uniaxial compressive strength of intact rock ranges between 0.07 MPa and 16.72 MPa, with an average value 16.72 MPa. It is worth noting than 80% of the results are below 20 MPa, as shown in the frequency histogram of Figure 11.

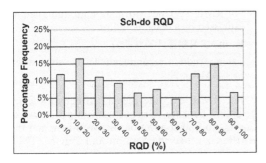

Figure 8. Frequency histogram of the RQD index in borehole logs corresponding to the Sch-do geologic unit.

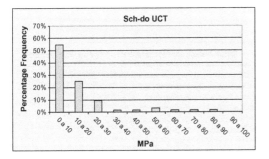

Figure 9. Frequency histogram of the UCS and PLT compressive strength test results in the Sch-do unit.

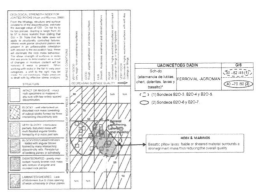

Figure 10. Ranges of GSI estimates for the Sch-do unit in the B2O-1 to B2O-8 boreholes superimposed to those proposed by Marinos and Hoek for pillow lavas with exfoliated and sheared zones (shales-chert) in ophiolitic complexes.

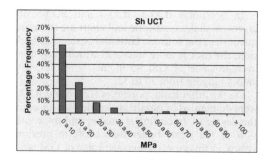

Figure 11. Frequency histogram of the UCS and PLT compressive strength test results in the Sh unit.

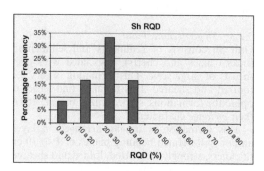

Figure 12. Frequency histogram of the RQD index in borehole columns corresponding to the Sh geological unit.

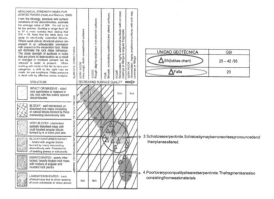

Figure 13. Display of the GSI assessments for the B2O-1 to B2O-8 boreholes corresponding to the Sh and Fault geotechnical units compared with the proposed range for serpentinites rocks with a marked schistosity and very low to low geomechanics quality ophiolite complex of Orthys by Hoek and Marinos.

The RQD is generally low, ranging between 0% and 36% with a average of 24% (see Figure 10).

Finally, in figure 13 it is shown a comparison of these values of GSI with the ones proposed by Hoek and Marinos (2001).

The same comparison has been carried out with the *Sch-do* unit. The GSI values estimated as estimated from the borehole logs have been superimposed on those proposed by Hoek and Marinos for the same type of rocks. The results are shown in Figure 12, where similar conclusions can be derived of the good agreement in general of both sets of data.

4 ROCK MASS DEFORMABILITY MODULI

The following table summarizes the values obtained from the laboratory tests of intact rock and the estimation and assessment of GSI, RMR, mi and other geomechanic parameters of the geological and geotechnical units identified along the tunnel.

Bieniawski proposed a simple correlation based upon RMR, after exhaustive in situ deformability tests planes in three projects, the Orange River Water Project (a dam founded in volcanic rocks and a tunnel excavated in shales), the Drakensberg Scheme (an hydroelectric central in a horizontal strata of layers of shales and sandstones) and the Elandsberg Scheme (another hydroelectric central):

– For RMR > 50 the deformation modulus can be expressed by a linear equation:

$$E\,rm = 2\,RMR{-}100\;(GPa)$$

– Later studies recommended its use when $55 < RMR < 90$.

For RMR < 55 the Bieniawski equation can't be used. There are subsequent investigation works that try to find connections between Erm and the RMR < 50, like the Serafim and Pereira (1983) equation, deduced from various in situ deformation tests in several dams in Asia:

$$E\,rm = 10^{(RMR-10/40)}\;(GPa)$$

TÚNEL T-2: litology, laboratory result tests and estimated geomechanic parameters					
Litología	DO	SCH - DO (1)	SCH - DO (2)	SH	F
GSI	50	45	50	35	25
mi	18	16	16	15	10
RMR	55	50	55	40	30
Q = e^(RMR-44)/9	3.3947	1.9477	3.3947	0.6412	0.2111
Ei (Mpa)	7 000	5 500	5 500	5 000	4 200
σ ci (Mpa)	25	16	22	15	10
σ cm (Mpa)	5.78	3.17	4.80	1.66	0.81

1. Boreholes B2O-3, B2O-4 and B2O-5.
2. Boreholes B2O-6 and B2O-7.

318

There are more empiric correlations between RMR and Erm, one of the most accurate maybe the Kayabashi, Golceoglu and Sonmez (2003) equation:

$$E\,rm = 0.0736\,e^{0.0715\,RMR}\ (GPa)$$

Another geomechanic index widely used is Q, proposed by Barton in 1974, its correspondence with RMR index is done by the following expression.

$$RMR = 9 * l\,n\,Q + 44$$

In a paper in 1996, Barton established the following correlation between the Q index and the rock mass deformability modulus.

$$E\,rm = 10 * Qc^{1/3}$$

where $Qc = Q * (\sigma_{ci}/100)$ and σ_{ci} is the unconfined compression strength of intact rock.

Hoek found that Serafim & Pereira equation is accurate in a good quality rock mass, but not in worse rock conditions. Based on this Hoek, Carranza and Corkum suggest use the σ_{ci} value to correct the deformation modulus.

$$E\,rm = (1\text{-}D/2)*(\sigma_{ci}/100)^{0.5} * 10^{(GSI\text{-}10)/40}\ (Gpa),$$
for $\sigma_{ci} < 100$ Mpa

where D is a disturb factor $(0 \leq D \leq 1)$.

Based in several data and tunnel auscultation measurements in Taiwan and China, in 2006 Hoek and Diederichs introduced an expression that assumes the following:

- It's based on the best data collection of measurements in tunnels.
- The maximum limit is given by deep tunnels (unconfined and unweathered rocks).
- That maximum limit is altered by D factor (Hoek & Carranza).
- Maximum Erm value is given by GSI between 90 and 100.
- It's for isotropic rock mass.
- To avoid Erm grows to infinite when GSI grows too it's introduced sigmoides functions (S shape).

The result is:

$$E_{rm} = E_i \left(0.02 + \frac{1-\dfrac{D}{2}}{1+e^{\frac{(60+15D-GSI)}{11}}} \right) GPa$$

where Ei is the intact rock deformation modulus.

TÚNEL T-2: Rock mass deformability modulus from empiric correlations

	DO	SCH-DO	SCH-OO	SH	F
Litology					
GSI	50	45	50	35	25
mi	18	16	16	15	10
RMR	55	50	55	40	30
$Q = e^{(RMR-44)/9}$	3.3947	1.9477	3.3947	0.6412	0.2111
Ei (Mpa)	7 000	5 500	5 500	5 000	4 200
$\sigma\,ci$ (Mpa)	25	16	22	15	10
$\sigma\,cm$ (Mpa)	5.78	3.17	4.80	1.66	0.81
Correlations between RMR and rock mass deformability					
RMR > 50 — Bieniawski (after 1977) $E_{rm} = 2\,RMR - 100\ *10^3\,MPa$	10 000	0	10 000	-20 000	-40 000
30 < RMR < 55 — Serafim & Pereira (1983) $E_{rm} = 10^{\frac{RMR-10}{40}}\ *10^3\,MPa$	13 335	10 000	13 335	5 623	3 162
Other correlations — Kayabashi, Golceoglu y Sonmez (2003) $E_{rm} = 0.0736\,e^{0.0715\,RMR}\ *10^3\,MPa$	3 756	2 627	3 756	1 285	629
Correlations between Q and rock mass deformability					
Barton (1996 y 1996) $E_{rm} = 10\,Q^{1/3}\ *10^3\,MPa\quad Qc = Q*(\sigma_{ci}/100)$	9 468	6 780	9 073	4 582	2 764
Correlations between GSI index and rock mass deformability					
Hoek, Carranza & Corkum (2002) for σ_{ci} < 100 Mpa $E_{rm} = (1-\frac{D}{2})\sqrt{\frac{\sigma_{ci}}{100}}\,10^{\frac{GSI-10}{40}}\ *10^3\,MPa$	5 000	3 000	4 690	1 633	750
0 < D < 1	0.0	0.0	0.0	0.0	0.0
Hoek & Diederichs (2006) $E_{rm} = E_i\left(0.02 + \frac{1-\frac{D}{2}}{1+e^{\frac{(60+15D-GSI)}{11}}}\right)$	2 150	1 230	1 690	567	251
0 < D < 1	0.0	0.0	0.0	0.0	0.0

The following table summarizes the values of the rock mass deformation modulus (Erm) deduced from the criteria explained above.

5 DEFORMATION POTENTIAL OF THE GEOTECHNICAL UNITS

Once the geotechnical parameters have been established in paragraph 4 and the stratigraphy along the tunnel estimated in figure 3, the deformation potential of the geotechnical formations at the tunnel depth can be ascertained by means of the Hoek-Brown failure criterion. According to Hoek and Marinos there is a relationship between the ratio $\sigma_{CM}/\gamma H$ and the tunnel convergency ε, where:

σ_{CM}: unconfined compressive strength of the rock mass.
γ: average specific weight of the rock mass
H: depth of tunnel
ε: % ratio of the tunnel radial deformation on the tunnel radius

If we denote $P_0 = \gamma H$, then:

$$\varepsilon = 0{,}2 \times (\sigma'_{cm}/P_0)^{\wedge\text{-}2}$$

$$\sigma'_{CM} = \sigma_{CI} \cdot \frac{(m_b + 4s - a(m_b/4 + s)^{a-1})}{2(1+a)(2+a)}$$

The following deformation potential classes are proposed by Hoek (2002).

319

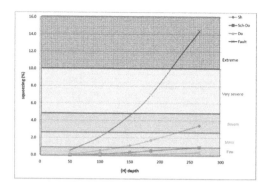

Figure 14. Deformation potential of the geotechnical units at different tunnel overburden.

Table 2. Range of potential problems, according to Hoek (2002).

Class	Deformation ε (%)	Potential problems
1	$\varepsilon \leq 1$	Few problems
2	$1 < \varepsilon \leq 2,5$	Minor squeezing
3	$2,5 < \varepsilon \leq 5$	Severe squeezing
4	$5 < \varepsilon \leq 10$	Very severe squeezing
5	$\varepsilon > 10$	Extreme squeezing

The figure 14 shows a sensitivity analysis of the Hoek criterion for the geotechnical parameters established in paragraph 4 for different tunnel depths.

6 CONCLUSIONS

The investigations carried out to appraise the deformation potential of a deep tunnel in this ophiolitic complex have shown that:

– Do and Sch-Do formations do not show squeezing problems even at the maximum tunnel depth.
– Sh formation shows only minor squeezing problems when encountered in the tunnel deeper than 150 m.

– The F formation, corresponding to the possible faults (which were not detected in any of the borings) shows squeezing behaviour when encountered deeper than 100 m. However, it must be noticed that according to the geological model described in paragraph 2 the inferred faults are hardly thicker than 10 m and, therefore, are not suitably modelled in the Hoek-Marinos scheme where it is considered that all rock mass is homogeneous with de parameters of the Fault.

As a general conclusion, given the stratigraphy along the tunnel established in Figure 3, the simplified analysis by Hoek-Marinos doesn't indicate foreseeable squeezing problems along the tunnel. However, the FLAC-3D finite difference analyses will be more accurate to model the rock mass behaviour to properly design adequate support means of the tunnel galleries.

REFERENCES

Bieniawski, Z.T. 1978. Determining rock mass deformability; experience from case histories. Int J. of Rock Mech & Min. Scio, Vol 15, pp 237–247.
Cai, M., P.K. Kaiser, Y. Tasaka and M. Minami. 2006. Determination of residual strength parameters of jointed rock masses using the GSI system.
Hoek, E., P. Marinos and M. Benissi. 1998. Applicability of the geological strength index (GSI) classification for very weak and sheared rock masses. The case of the Athens Schist Formation.
Kayabashi, A. and Gokceoglu, C. et al. 2003. Estimating the deformation modulus of rock masses; a comparative study. Int. J. of Rock Mech. & Min. Sci., Vol 40 pp 55–63.
Marinos, P., E. Hoek and V. Marinos. 2006. Variability of the engineering properties of rock masses quantified by the geological strength index: the case of ophiolites with special emphasis on tunneling.
Marinos P. and E. Hoek. 2001. Estimating the geotechnical properties of heterogeneous rock masses such as Flysch.
Steiakakis C. and O. Verroios. September 2005. Rock mass classification system (GSI) misuses in problematic rocks. Ricardo Z. Bieniawski. April 2003. New tendencies in rock mass characterization.

Volcanic Rock Mechanics – Olalla et al. (eds)
© 2010 Taylor & Francis Group, London, ISBN 978-0-415-58478-4

An access gallery to the underground Fuente Santa spring, La Palma, Spain

C. Soler
Dirección General de Aguas del Gobierno de Canarias, Spain

ABSTRACT: This paper presents the work done to recovey the historical Fuente Santa spring, well known at Middle Age due to its miraculous medical effects. The spring was buried by San Antonio volcano eruption in 1677, with debris, cinders and lava flows. A gallery was recently excavated under extreme and difficult natural conditions, a total lack of stability with granular materials involved. The direction of the axis of the gallery was initially unkown.

1 INTRODUCTION

The most complex excavation confronted by the Local Administration of the Canary Island refers to a hydraulic work was a 200-m long recovery gallery on the so-called Fuente Santa ("Holly Fountain") spring. This problem was not related to the creep of the excavated material but rather a total lack of stability in a gallery that was intended to be driven through a pile of debris representing a 150-m high overburden.

The Fuente Santa used to be a thermal spring emerging at sea level and at the foot of a coastal cliff with a height of 150 meters. (See Figure 1).

The water sprang forth among the boulders of a beach adjacent to the cliff and was impounded by two tidal ponds that because of their different degree of mixing with sea water evidenced various temperatures; the closest source to the point of blowing and therefore the hottest one, was named after San Lorenzo and the closest to the sea, to San Blas.

2 HISTORY

The cliff is topped by a plateau with a certain slope that reached elevation 500. It is known as *Cuesta de Cansado*, ("weary uphill"), an appropriate name to describe the "smoothness" of the surface. Such plain used to have fertile land where country houses, orchards, a town and a hermitage once were found. Crops of potatoes, mallow, vegetables and fruits provided enough resources to its inhabitants and they even supplied food for large number of pilgrims. They came from all Christian countries at that time from Europe and the American Continent arrived to enjoy miraculous cures. Prestige acquired by the spring was such that visitors raised

Figure 1. Profile of the ground with the Fuente Santa spring. It can be observed that thermal waters emerged at the foot of the cliff in a coastal plateau constituted by a boulder beach.

the *per capita* income of the islanders so as to be in the forefront of the archipelago. (See Figure 2).

Bonanza and prosperity lasted the whole 16th century and most part of the 17th century. During those times the Fuente Santa was visited by prestigious personalities such as Pedro de Mendoza y Luján, Alvar Núñez Cabeza de Vaca, Gaspar Frutuoso and fray Juan de Abreu Galindo.

The fame of the spring crossed the ocean and barrels of the precious water were sailed to America. It was believed that it healed skin diseases, rheumatism, arthrosis, elephantiasis, leprosy and syphilis, generating a health-related tourism in the island. But when richness flowed in abundance, when it was thought that health was not a concern to La Palma inhabitants and when holiness lied immersed in those pools at the southernmost tip of this island, suddenly ... everything came to an abrupt end. On Saturday November 13

Figure 2. An image can be observed as a perspective including, in addition to the coastal cliff and the tidal ponds, the plain known as Cuesta de Cansado.

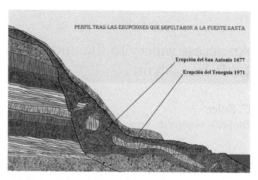

Figure 3. Soil profile showing how the Fuente Santa lies under 70 meters of debris, cinder and lava flows, and the coast line retracted 400 meters offshore.

of year 1677 the San Antonio volcano erupted, the same one leading the view of Cuesta de Cansado.

The eruption lasted three months, until the end of January of the following year, but the worst came in only one week after the start. In such a short time the 23 mouths that opened spewed lava flows that buried towns, houses, the hermitage, tilled fields and the worst that could have happened to La Palma: the burial of Fuente Santa spring under a 70-meter thick layer of debris, cinders and lava flows.

The process of vanishing of Fuente Santa is narrated by Nicolás de Sotomayor, special envoy of the Cabildo (Town Council) to Fuencaliente to write a report of the daily happenings during the eruption. Thanks to this detailed account we know that on November 23 one lava flow erupted at the foot of the volcano, one of the 23 mouths of hell that opened, carrying out a flow of stones and cinders resembling a drifting boulder. They reached to the edge of the cliff, at the foot of which the tidal ponds stood and a further thrust from the Averno made that all stone laden flow fall through the notches wearing away part of the cliff. Within a moment such boulder beach became flooded and both ponds disappeared under tons of debris. After such a ramp was formed, the lava flow, pushed by the thrust of the magmatic chamber, ran along the slope until the sea shore was reached thus enlarging the island under the continuous lava flow.

When the nightmare came to an end the landscape looked devastated: where the fields and houses of Cuesta de Cansado stood, only a dantesque panorama was left constituted by the volcanic malpais and no vestige remained that only three months before an unlimited richness and prosperity existed at that place.

But the worst was not only the loss of goods and lands; the unbearable was that the coast was

Figure 4. A perspective depicting the devastated landscape around Cuesta Cansado, where the town, hermit and agricultural fields disappeared.

not the same any more: the cliff walls had vanished among rock falls and were buried by the debris and lava flows. At the site formerly occupied by the Fuente Santa spring only a pile of rubble remained, attached to the former cliff and since the sea shore was now covered.

The coast line retracted 400 m offshore. The spring became buried under an avalanche of rock at a depth of 70 m of burden material. (See Figures 3 and 4).

The first attempt made to dig up Fuente Santa was made only ten years afterwards (1687) when remembrances were still fresh and the inhabitants of Fuencaliente could still recognize the original location, even after being buried, of the miraculous thermal spring. This try was followed by up to 16 attempts that already had to struggle against two serious problems. The first one was where to excavate, because with the exception of the first attempt, in the others nobody knew where the spring was buried; the second problem referred to the lack of stability of the material to be excavated.

During more than three centuries these two problems lead to 16 unsuccessful attempts to find it, equivalent to the same number of La Palma native generations that grew up since the spring disappeared in the 17th century.

Solving the first problem –to determine where it was buried- was compensated with the free interpretation of four stages that were conveyed from parents to children in the search of a cross that apparently marked the place of burial. Each attempt was believed to be the actual site where it lay buried and the mouth of a well was marked. At that moment the second problem was encountered: how to excavate through a soil constituted by stones not bigger than a boulder with any cohesion at all.

Right from the beginning the walls started collapsing toward the center of the excavation leaving in bare equilibrium the rest of the embankment that threatened to bury the workers as they dug deeper. All of the attempts had to be abandoned due to the unstable nature of the excavated walls and it was hardly possible to reach no more than one tenth of the depth required to encounter the sea level.

But this was no reason to continue trying, failure after failure, in the course of time but Fuente Santa remained buried. Illustrious La Palma inhabitants took part in those attempts, such as Juan Pinto de Guisla, Juan de Paz, Manuel Díaz Hernández, Juan Antonio Pérez Pino, Luciano Hernández, and Pedro Pérez Díaz, as well as continental geologists and engineers such as Lucas Fernández Navarro, Enrique Godet, Juan Gabala, and Juan A. Kindelán, and foreigners such as Antonio Joseph Palmerini and Leopold von Buch. Not all of them were material attempts but also the demands of islanders to recover the Fuente Santa arouse the interest of writers such as Viera y Clavijo, Alexander von Humboldt and more recently, the unequaled prose of Dulce María Loynaz.

3 THE PROJECT OF THE GALLERY

In the Christmas Season of 1995, the Mayor of Fuencaliente de la Palma requested the *Dirección General del Agua* of the *Consejería de Obras Públicas del Gobierno de Canarias* (Canary Islands Public Works Water Department) to carry out all necessary investigations to try to find out the Fuente Santa spring.

We started studying all historical documents where the morphology of the landscape was described when the spring water still erupted. We continued reading the attempts of those who preceded us and this is how, at the archives of the municipalities, of the cultural societies, of the museums and even at the National Archive of Madrid,

where we could learn about this history, a common one for mankind: the longing of people to recover a lost property. We surveyed the volcanic malpais, looking for something that we actually ignored, among other things the cross identifying the site. The land to be surveyed was a coastal stripe measuring three kilometers in length by four hundred meters in width fringe; apparently a too large area to find out a low flow rate spring even a thermal one. See Figure 5 where the San Antonio Volcano can be seen in the background and the Teneguía Volcano at closer range. In the foreground part of the old cliff can be observed that in the northern direction became buried under the lava flows of the San Antonio and the Teneguía volcanoes.

See Figure 6 where part of the old cliff can be also observed with its packages of lava flows dipping toward the sea with a small gradient, as well as the upper part covered by the volcanic malpais, and recent lava flows falling down the notches of the former cliff that was turned down and buried.

Figure 5. Aerial views of the southern zone where Fuente Santa spring could be found.

Figure 6. Photograph, taken from the west.

Once it was decided on the actions to be taken, investigations with rotary borings were done with core recovery so as to be able to take samples of the water at the coastal aquifer and to determine the morphology of the former coastline, the one that was buried, the one existing prior to the eruption of San Antonio volcano. A total number of five borings were done, three of them paid by the Local Government and two by private investors.

The water samples taken from the aquifer with sufficiently high temperature were considered as thermal, 29°, 36° y 42°C, respectively. They had an anomalous chlorine-carbonate ratio and an excessively high salinity content. In summary, the government-sponsored borings had found the spring water and the old cliff should be located at a distance of no more than 50 meters from the last boring. With only three borings it had been confirmed that the spring, although being buried, continued flowing under the same thermal degree. The possible area of flowing had been also delimited around the last boring evidencing the highest temperature.

With these results, a project was developed to excavate a gallery to find out and dig up the Fuente Santa spring. The alternative was a gallery rather than a well because the experience gained by our predecessors compelled to such a solution.

When asked on the direction of excavation, it wasdecided, as opposed to them that were not oriented to a specific point, to rely in temperature measurements and water analyses that should lead to its point of surging. To reach such an objective the invert elevation was planned just above the live equinoctial high tide elevation and upon advancing toward the cliff, by excavating the burden material spewed by San Antonio Volcano, we could make drain pots to collect water from the aquifer.

After reaching the old cliff, a transversal gallery branch should be excavated, parallel to the former sea shore until intersecting the flow of Fuente Sana. The expenditures amounted to 535.000 € that including everything necessary to excavate a 200-meters long gallery with a cross section of 2,5 × 2,5 meters, lining the walls with steel trusses spaced every one meter. Between each hoop and since the gallery was intended to be visited, it was decided not to install rows of plates but rather join the trusses with Ø 20-mm rods, packing the extrados with the same stones, the largest ones, recovered from the excavation of the face. (See Figure 7).

4   THE CONSTRUCTION OF THE GALLERY

That was the theoretical approach but the actual excavation taught that fulfillment of such objectives became more complicated than the predicted

Figure 7. Cross section of the recovery gallery of Fuente Santa spring showing the material through which the gallery was excavated.

goals. At the beginning, with such a lining system progress was made but it was slow and a length of 50 m could covered, after crossing cinders and piled stones produced by avalanches coming from different directions. That chaos was followed by the emerging of a lava flow that cover the full heading of the gallery, constituted by fine grain and very hard basalt that had to be excavated by means of explosives. Only ten meters further beyond, suddenly disappeared to encounter the loose stones once again. Cementation progressively decreased, and it became increasingly difficult to stabilize the heading so as to have an excavated length of one meter forward to install the following truss.

A shield-type of excavation was used, inserting 1,5-m long bars at the roof and shoulders of the cross section, slightly tilted to the front. The efficiency decreased but it could still cover a length of 100 meters, just under the boring that provided the reading of 42°C where a basalt flow appeared again, similar to the previous one and also with ephemeral nature because it vanished ten meters ahead and the unstable cinders were found again. However, prior to resuming the excavation, and talking advantage of the stability of both excavated basalt flows, two chambers were built (A and B, at a distance of 50 and 100 meters, respectively) and excavated through them under the soleplate until reaching the maximum equinoctial low tide elevation. (See Figure 8).

Two underground pools were therefore created whose water elevation varied with tides. They were filled twice a day with a mix of the sea water seeping from the sea shore and of the water from Fuente Santa spring that flowed from the subsoil; the latter with a larger proportion of thermal water than the former.

The first phase of the project came to an end. (See Figures 9 and 10).

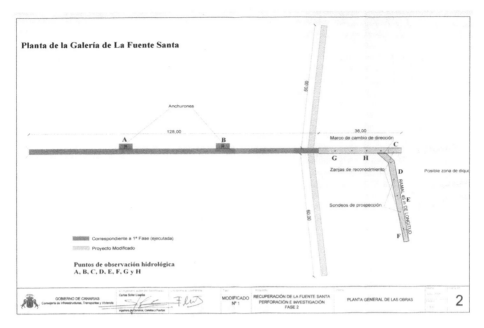

Planta de la Galería de La Fuente Santa

Anchurones

128,00

36,00

Marco de cambio de dirección

A

B

C

G   H

Zanjas de reconocimiento

D

Posible zona de diqu

RAMAL 45 m DE LONGITUD

Sondeos de prospección

E

F

60,00

60,00

Correspondiente a 1ª Fase (ejecutada)

Proyecto Modificado

Puntos de observación hidrológica
A, B, C, D, E, F, G y H

GOBIERNO DE CANARIAS
Consejería de Infraestructuras, Transportes y Vivienda

Carlos Soler Liogleta

Ingeniero de Caminos, Canales y Puertos

MODIFICADO
Nº 1

RECUPERACIÓN DE LA FUENTE SANTA
PERFORACIÓN E INVESTIGACIÓN
FASE 2

PLANTA GENERAL DE LAS OBRAS

2

Figure 8. The initial alignment of the project is shown, with two branches at each side after reaching the old cliff. The final alignment of the gallery is also depicted. The formed ponds are identified as A and B in the first phase, and C and F during the second stage.

Figure 9. Different views of the recovery gallery at Fuente Santa spring. To the left, the start of the operation showing the supporting structure constituted by steel trusses spaced every meter and connected by rods, packing the extrados with the same boulders (the largest ones) that were recovered from the heading excavation.

Figure 10. This photo shows the first pool and the chamber made to excavate it at a site with the boring where the 42°C temperature was recorded.

During the second stage a new excavation was started. Room temperature and the emission of carbon dioxide inside the gallery showed an increase and it became necessary to work by blowing fresh air in. The material encountered was still formed by loose stones with very variable grain size distribution and soon the lack of stability prevented

progress even with the use of a shield. A collapse of the heading finally convinced us that a change in the method was necessary.

See Figure 11 with the front of the gallery constituted almost exclusively by cinders with small-size particles and with certain cohesion, slightly higher than at the stones carried by the front of the lava flow.

To be able to continue with the excavation it became necessary to stabilize the roof, remove

Figure 11. An illustration of the material encountered at the heading of the excavation.

Figure 12. A collapse of the heading is shown that ceased when the natural slope of the materials involved reached 45 degrees.

the rock fall y fill the void left on top of the gallery prior to proceed further inside. This problem caused a delay of several months.due to collapse at the front (See Figure 12).

In Figure 13, can be observed the emerging of the grout as well as a slight stratification toward the right side wall.

We started injecting a water-cement grout in advance of the heading to be excavated. (See Figure 14).

The shield was left in place but the rods were replaced by galvanized steel pipes used to inject the water-cement mix. The stones were obtained at the job site and in this way, since the pipes remained embedded in the cement grout we could build a reinforced concrete dome overhead and above the trusses.

Grouting was quite sensitive in terms of pressure as well as dosing. If the mix was too thick and applied with low pressure, only a bulb surrounding the pipe could be obtained; whereas if the mix was very fluid and under a high pressure, a waterproof the flow of water from the spring and that could alter our route to the Fuente Santa could be obtained. Shortcrete was applied to the heading and the excavation could then proceed at a slow but uninterrupted rate.

Ten meters further from the beginning of this second phase, boulders began appearing at the foot of the heading. They were the largest particles found at the heading and therefore were used to pack the extrados of the gallery. These boulders were the pebbles of the former beach; the sea shore was and this was confirmed after finding barnacles seashells among the stones. Thickness of carried material progressively increased until almost reaching the height of the roof and subsequently decreasing in thickness. Only a dozen meters far inside the old cliff. was detected.

Figure 13. Change of the excavation procedure: cement grouting.

Figure 14. It can be observed the stabilization obtained using, in addition to grouting, the application of shortcrete, in this case at the roof and right side wall.

Thanks to the data being collected from the points where water was sampled, together with those obtained from drilling of the three inclined borings advanced from outside, at this stage of progress we were already aware that the Fuente Santa spring flowed to the right of the gallery and it was therefore possible to cancel the excavation of the left branch.

In Figure 15 it can be seen the yellowish cinder-type basalt flows forming the base of the cliff. To the left, the slightly stratified fill material spewed by San Antonio Volcano. The contact between both formations is a line starting at the lower left edge of the heading and continuing into the roof, at the right of the headstone.

To proceed with the right branch, it was neccessary to move away 20 meters from the heading and started excavating at the side wall along an alignment of 45 degrees with respect to the main gallery with the idea in mind of turning again an additional 45° subsequently and sited in a direction perpendicular to the original alignment, for the gallery to eventually reach a direction parallel to the coastline and just above the mean sea level.

The longing of moving farther away, because from the data of temperature and analysis of the water we knew that we were close to the spring emergence, made us forget precautions and this error was paid immediately: In January 2005 we had a collapse of the heading similar to that occurred the previous year. it can be observed that the material ceased to fall when it adopteds a maximum gradient roughly close to 45°. The grain size distribution of the stones forming the debris cone can be also appreciated. (See Figure 16).

But in this case, since it happened at the crossing section between the gallery and the branch, it became larger in volume and more dangerous, because the junction with the branch was one of the weakest points of the gallery since it had the largest clearance. It was necessary again to stabilize the roof by means of grouting, remove the rock fall and fill the void left above. Then the junction was reinforced by placing a frame constituted by IPN steel shapes; the trusses of the gallery and of the branch were supported against the upper cross beam. As a result of all these problems the excavation suffered a delay of about three months.

Once the cross section was stabilized work proceeded in the same 45° direction with respect to the main alignment; the material was found to be increasingly less stable due to the proximity to the old cliff. The ground excavated was the first one to be pushed by the momentum of the lava flow when falling down from the summit of the cliff. Because of this reason it was decided to add sand to the grout and increase further the pressure. The results demonstrated the sound judgment of the decision

Figure 15. Moment when the gallery reaches the old cliff that used to welcome countless sick persons who got in line to bathe at Fuente Santa, and was dug up after more than three centuries of being buried.

Figure 16. Collapse at the entrance to the branch.

and it was therefore possible to excavate the next ten meters before reaching again the old cliff.

The excavation rate was desperately slow; placement of each truss took more than a week provided no problem would develop. At that stage the cross section was again rotated 45° and the direction sought for was reached, parallel to the sea shore, to be followed until intersecting the flow of water of the Fuente Santa. Samplings were made at the soleplate as excavation progressed to measure the water temperature and to recover samples of water that were subsequently tested at the laboratory. This is how it could be realized that success had been achieved because the temperature progressively increased at the same rate as well as the content of bicarbonates and of CO_2.

Ten meters beyond the last break the heading of the gallery, with the soleplate at the boulder beach, encountered a thick dyke that covered a large proportion of the cross section. This was interpreted as a protruding dyke from its profile with respect to the old cliff.

In Figure 17 can be observed that it is not a single dyke. There is another one embedded therein having a more acid nature evidenced by a lighter color. The dyke is covered by the material fell down from the cliff some time before the arrival of the lava flows spewed by San Antonio Volcano. Loose stones remain stable because of grouting and shortcrete. Through the shortcrete layer the variable grain size distribution and the total lack of cohesion can be observed. At the contact section with the dyke, through which water from the injection

Figure 17. Heading of the gallery showing the volcanic dyke that conveys the water from the Fuente Santa to the coast.

Figure 18. Cross section of the recovery gallery of Fuente Santa showing the procedure used to excavate below the phreatic level.

flowed through, with no traces of cement, it can be appreciated that the nature of the boulders fails to correspond to cinders associated to volcanic lava flows. They are stones flows carried by the front of the lava flow that subsequently fell down from the top of the cliff.

The presence of the dyke gave confidence that the Fuente Santa spring was found; suddenly, all previous assumptions that provided keys to find out the spring were clarified upon contact with the dyke. From a hydrogeologic point of view such vertical volcanic structure carried water from the interior of the island, from the aquifer to the boulder beach . Such a dyke was the geologic structure that separated the thermal water at the side corresponding to the gallery from the water of the aquifer that hardly mixed with the thermal water, that should spring up at the opposite side of the dyke.

Up to that point, the temperature at the excavations made at the pebbles progressively increased, and it was reached the foot of the dyke with a water temperature of 45°. If the spring was there, after crossing the dyke, the temperature should decrease rapidly. We did so; after the gallery crossed the dyke and the soleplate was immediately excavated where a temperature of 30°C was detected; through the thickness of the dyke (four meters) a temperature gradient of 15°C was obtained.

The Fuente Santa spring was found: After more three hundred years it lay below our feet in front of the dyke.

Because the gallery should eventually display the spring, it was decided to excavate three more pools to allow the Fuente Santa waters flow through them.

The problem to be solved was that if the excavation through such cohesionless material was difficult to perform without it being saturated, it could be even more difficult when an attempt was made to excavate it under the phreatic water table. It was then decided to use injections of water, sand and cement. The problem was that an excessive increase of cohesion could prevent the flow of water into the pool because permeability could be lost upon increasing of the cohesion.

The problem was solved by drilling inclined borings at the side walls, in a downwards direction and separated. The objective was to build battered micropiles to hold the material together but at the same time allowing water to flow around them toward the pools. To execute this pile system a mix of water, cement and sand was injected, with the addition of setting additives. In this way, the grouting adopted under water the shape of bulbs surrounding the slotted injection pipe, therefore preventing cement from seeping into the aquifer.

In Figure 19 is depicted in the background, the old cliff covering the section at the right of the

Figure 19. The pool excavated below the phreatic level at the end of the main gallery.

Figure 20. Emergence of waters from Fuente Santa can be observed with its precipitation of aragonites due to the high temperature of the spring.

heading. Different beach pebbles can be also observed piled up at the extrados of the lining that can be identified as such because of their rounded shape.

Figure 21. Views of the beginning of the branch of the gallery where the frame constituted by IPN steel shapes lined with mortar to connect both alignments can be observed.

At the distance beyond, the dyke can be seen as well as the two additional pools where temperature decreases and induced the precipitation of mud that provided most of the curative virtues of Fuente Santa. (See Figure 20).

As opposed to the chambers excavated in sections that correspond with the outcroppings lava flows, in this section the materials lacked cohesion thus leading to stability problems, such as caving, that made it necessary to redesign the excavation procedure. At the end of both galleries the outcropping of the old cliff can be observed. (See previous Figures 19 and 20).

Figure 21 shows the applied method. The method. It consisted in building battered micropiles inclined toward the outside by a previous injection of cement grout through slotted steel pipes.

5 CONCLUSIONS

Finally, mention should be made that if with these works the Fuente Santa spring could be detected and unburied the reasons were not the actually the technological advances required to achieve this purpose; the difference was that, as opposed to our ancestors, we could resort to four disciplines.

The first one was history that with its accounts found at local and national archives we could find what we should look after and where to do it. The second one was Engineering that provided with the method of excavation, with a gallery rather than a well. Its election opened the way to be able to excavate through a pile of unstable and dangerous pieces of rocks, along a 200-m stretch and at the same time provided with a method capable of maneuvering and detecting the spring flow buried underground.

Then geology that showed how to tell apart the material spewed by the eruption from the original soils found at the site prior to burial of the spring. Last but not the least, Chemistry that through its cryptic language of cations and anions leads progressively to the source of Fuente Santa spring.

It was through the mixture of the four sciences, History, Engineering, Geology and Chemistry, that Fuente Santa could be found; Without any of them, or leaving out any of them, we would still be going around in circles below ground searching for something that used to belong to the island and that during three centuries was sought for with the same momentum applied by anyone in the search of his life.

ACKNOWLEDGEMENTS

I must recognize the work done by CORSAN-CORVIAM, supervised by Herminio Torres and Roberto Pareja, whose performance and professional knowledge made it possible to complete the 129-m long gallery and to build the modern pools of San Lorenzo and San Blas.

I must also thank the company SATOCAN that appointed Julián Mansilla as supervisor.

The grout supervisor was Manuel Jesús Rodríguez, who did a really good work.

The author wants to acknowledge the technician Manuel Fernández and the geologist Luis Hernández for their sound collaboration.

Volcanic Rock Mechanics – Olalla et al. (eds)
© *2010 Taylor & Francis Group, London, ISBN 978-0-415-58478-4*

Project and technical assistance to the retaining structures of Cabo Girão Tourist Resort

F.A. Sousa, C.J.O. Baião & J.M. Brito
Cenorgeo, Engenharia Geotécnica, Lda., Lisboa, Portugal

ABSTRACT: Due to the existing conditionings for the implantation of the different areas that comprise Cabo Girão Tourist Resort, at Madeira Island, namely the vigorous relief and the geological-geotechnical conditions, it was necessary to conceive an important group of retaining structures, using different solutions, varying from gravity and semi-gravity cyclopean concrete structures to reinforced concrete walls and soil nailed walls or nailed slopes. This paper presents the conception of the project in its different stages and the main activities developed under the scope of the special technical assistance provided during the execution phase.

1 CABO GIRÃO TOURIST RESORT GENERAL DESCRIPTION

The Cabo Girão Tourist Resort, with an approximated total area of 65000 m², is located 580 m above the sea level, at Madeira Island, comprising a multi-services area (N1 block), four residential areas (N2 to N5 set of blocks) and a leisure area (N6 block) (Fig. 1).

The total construction area of the tourist complex is of about 41100 m², placed between the areas defined by the regional road which crosses the complex.

The building of the N1 block, named the "main building" is a four storeys building, being two of them semi-underground storeys, and is located in the central area of the tourist complex. It is a leisure building, offering an exterior leisure area with two swimming pools and other equipments. The N2 to N5 blocks are placed around the main building, being the N2 block located west, comprising 5 buildings, and the N3 block located North, with twenty-one buildings, the N4 block located east, comprising a group of seven three-storeys buildings. The N5 block comprises twenty-one two-storey dwellings (six of them twin-houses) and is located in the southern area of the complex. The N6 block defines a leisure area with swimming pool and tennis court, being located at the west end of the N5 block (Fig. 1).

To create and make possible the implementation of these construction areas and respective infrastructures it was necessary to build several platforms on the existing terrain, which were held by significant excavation and landfill retaining structures.

A total of 40 retaining structures and slope consolidation structures were built, in a total length of about 3200 m (Figs. 1 and 2), which is twice higher than the one defined for the tourist complex streets' network, being therefore necessary to implement several different solutions, namely gravity and semi-gravity cyclopean concrete walls, reinforced concrete walls, soil nailed walls and soil nailed slopes. It is important to emphasize the influence of the retaining structures in the final cost of this type of work, of about 15%, revealing the importance of all the involved studies.

2 GEOMORPHOLOGY, GEOLOGICAL AND GEOTECHNICAL CONDITIONS

The resort is located on a long and angled slope oriented to the sea, with a north-south gradient, and abruptly cut off by the Cabo Girão cliff, with an unevenness of approximately 550 m. The average slope angle is between 20° and 30° and presented several old retaining walls, being crossed by a feeder road. The tourist complex was developed between levels 510 and 610 m, in an area with an unevenness of approximately 100 m.

The project paid special attention to the results obtained from the geotechnical investigation of the geological and geotechnical study (Cenorgeo 2003a), which included 19 rotation boreholes, executed throughout the resort area. According to this study one verified that some cover deposits occurred at the surface, resulting from the weathering process of the subjacent formations, which are mainly formed by volcanic bombs levels, covering mostly the resort northern area.

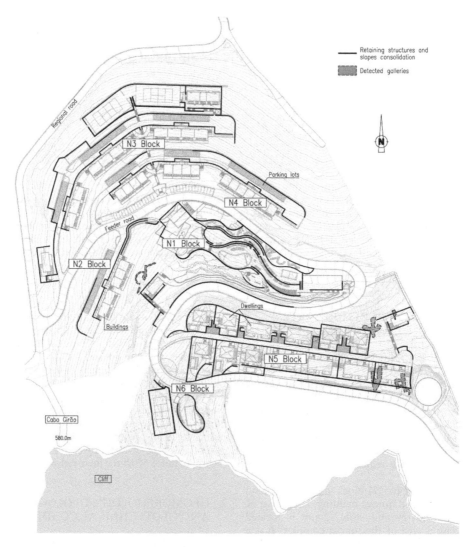

Figure 1. Cabo Girão tourist complex plant.

Figure 2. Cabo Girão aerial view (January 2006).

This formation presents a significant thickness, sometimes reaching 17 m and it's characterized, when exposed, to be easily erodible, provoking a quick change of its compaction and resistance characteristics, contributing to the existence of a weathered and decompressed upper horizon.

The outcrop appears only near the existent feeder road. In the southern zone, the outcrops are mainly of fractured basalts and disaggregated breccias, namely in the excavation slope adjoining the existent road. The geotechnical parameters adopted for these soils, formed mainly by tuffs, and derived from the tests results, were: density $\gamma = 19$ kN/m³, friction angle $\varnothing' = 30°$, cohesion $c' = 10$ kPa e base/soil friction angle tg $\delta'_f = 0,45$.

3 PROJECT

3.1 *Resort's layout*

The need to create platforms in a very angled slope in order to allow any construction—buildings, leisure areas and streets (Figs. 3 and 4), required that during the design phase particular attention was given to these structures' implantation, either in planimetry, either in altimetry, thus assuring an harmonious landscape integration. In fact, it was just by a straight interaction between the design team, particularly the one responsible for the geotechnical design, and the architecture office, responsible for the resort's layout, that it was possible to optimize the construction works volume associated to the masses handling and to the execution of the retaining structures, thus assuring the economical viability of Cabo Girão Tourist Resort's construction (Cenorgeo 2003b, Cenorgeo 2004).

Figure 3.   General view from the top of N3 block.

In fact, the philosophy that has driven the developed activity had for basis the necessity to obtain a balance between the excavation and land-fills solutions, hence assuring, besides the landscape integration, also the stability of the whole hill and the maintenance of all the foreseen structures, spaces and equipments.

As an example of the works developed during the project, it's presented some of the changes introduced in the architectural design resulting from the geotechnical conditionings:

– Elimination of the resort's internal circulation main street, taking advantage of the feeder road which crosses the resort, creating accesses for several blocks. As a matter of fact, the creation of a new internal street would imply, due to the area's topographical conditions, the execution of a significant number of retaining structures, as well as the execution of road and pedestrian overpasses on the existent road;

– New apartments' planimetry implantation, a change with an important incidence over N2 and N4 blocks, where initially there was a strong concentration of buildings, placed too near of an existent retaining structure and in an area with a significant slope angle, factors that would force major interventions. Therefore, all this structures were rearranged in N3 block, integrated in a smoother area at the hill's northern zone, in two main alignments, using mainly to landfill retaining structures, with a small height in the limit of the N3 block;

– Altimetry optimization of all blocks, structures and streets, to achieve an equilibrium between the parts which are above or below the ground level, and with the purpose to minimize the excessive height of some retaining structures, either exca-

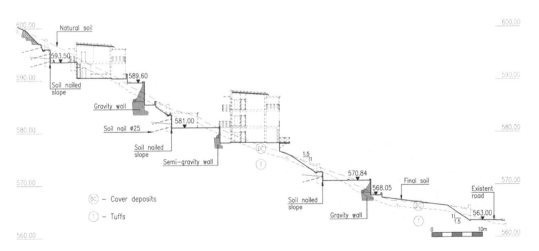

Figure 4.   N3 block cross-section.

vation or landfill type, and to guarantee then the necessary angles for the streets and for the access ramps, according to the legal regulations;
- Elimination of spaces between the buildings, thus achieving to have the same construction in a smaller area, reducing the extension of the retaining structures;
- Rotation of alignments of some streets and adjoining blocks, to a parallel direction of the one of the soil level curves, resulting in economical and well-integrated solutions, using to small height retaining structures, with particular incidence on N3 and N4 blocks (Fig. 5).

3.2 General descriptionof the retaining structures

When designing the solutions for the different retaining structures in all blocks, it has been considerated, besides all aspects already focused, all the conditionings, such as, geological and geotechnical, topographic, economical and execution issues (Cenorgeo 2004), which led to the need of using different solutions, namely gravity and semi-gravity cyclopean concrete walls, reinforced concrete walls, or nailed walls and soil nailing slopes.

Therefore, in the cases where the landfills to execute, in order to create the buildings or streets platforms, had, for a significant extension, an height higher than 4,5 m, it was chosen to execute retaining structures of gravity cyclopean concrete walls type or of semi-gravity walls type. In landfills which had an height under 5,5 m the choice was to execute reinforced "T" shaped concrete walls.

All these walls are founded made directly on the compact tuffs. In order to take advantage of the soil resistance, whenever possible, the semi-gravity type walls was adopted with a lower section in cyclopean concrete, with the concrete poured directly

Figure 5. View of the slope upper area, located in the limit of the N4 block.

onto the soil, and an upper section, corresponding to the landfill height, with a cantilever of reinforced concrete (Fig. 6).

Retaining structures with U-shaped section were also designed in order to include the cases where there was the need to guarantee the vertical interval between three platforms at different levels and reduced distance, making the execution of two parallels walls impossible. Therefore, the defined sections correspond, in general terms, to two parallel reinforced concrete cantilevers based on a common foundation element, in reinforced or cyclopean concrete (Fig. 7).

In cases of compact tuffs excavation where, due to geometric or architectural reasons, it was necessary to ensure a vertical excavation between two platforms and the referring massif was located under the level of the upper platform, it was decided to execute nailed walls with the higher section in "L"-shape, also in reinforced concrete, corresponding to the height where it was not possible to pour the concrete directly against the excavation face. The purpose of these walls was, in a way, to cover and abut the excavation to execute, and, in another way, to guarantee an identical finish to the one of the adjoining structures.

Considering the importance of the excavation slopes to execute, essentially the ones located in the limit of the different blocks, characterized by its height and gradient, either the future buildings construction on the platforms to support and the specific characteristics of those formations, the chosen option was a soil nailing slopes solution, with shotcrete and nails, which, besides allowing an adequate coating to abut and avoid the soil progressive disaggregation, is also a light retaining structure, flexible and economic, well integrated and that takes the maximum advantage of the resistance from the formations that will be excavated.

Since most of the soil nailing slopes to execute were very exposed, it was given a special concern minimizing their visual impact thus assuring the best landscape integration. Therefore, a color addictive was added in the sprayed concrete second layer, which final color was set in the execution phase, for the linings' adequate framing with the surrounding area.

4 TECHNICAL ASSISTANCE DURING EXECUTION PHASE

Giving continuity to the work concerning the preliminary and final design, and due to this work specificity, it was considered relevant to assure the follow-up and technical assistance to the execution of the numerous retaining structures, proceeding this way to the structures' adaptation and optimization.

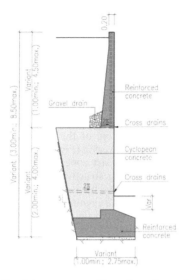

Figure 6. Retaining structure with semi-gravity section.

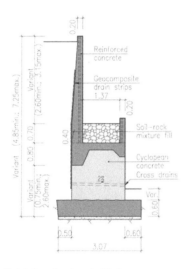

Figure 7. Retaining structure in U-shape.

This activity revealed itself very effective, allowing to proceed, in due time, to several design retaining structures' adaptation, taking advantage of the real geological and geotechnical conditions, that in some areas, revealed to be more favorable than the ones assumed during the design phase. On the other hand, it allowed to effectively solve some problems raised during the execution phase resulting from the local geological conditions.

Concerning to the importance of the technical assistance, in addition to other optimizations made, it could be stand out the reduction of the concrete volume in some walls, through the reduction of their foundations width and the cutback

on the soil nails quantity for the excavation slopes. This step was adopted based on the soil nails pull-out tests' results, defined under the final design scope, and visual inspection of the slopes to consolidate.

Several pull-out tests were executed evenly on the resort different blocks' areas, to measure the pulling-out force by length unit, fundamental parameter when designing soil nails (Freitas *et al.*, 2006). In most oh these tests was concluded that the pulling-out force by length unit was superior than the one considered in the design, information that allowed to adjust some slopes' soil nails meshes to 1,75V:1,5H and 2V:1,5H. This optimization,

335

as well as the reduction of the soil nails quantity defined in the natural slopes, translated into an economy for this type of works of about 1200 m, corresponding to 14% of the previewed total.

During the works execution six natural water storage galleries were found, from which only two had already been identified but without any survey being made due to the dense vegetation cover. A detailed study was made to each one of these galleries with the purpose to develop solutions that could guarantee their stabilization in order to integrate them in the resort, as an attraction point. In some cases there was the need to make some architectural adjustments, in others it was necessary to full fill with cyclopean concrete.

The water storage gallery detected in the resort southern zone, near the existent reservoir, located under a landfill retaining wall and under a dwelling yet to build, with an H-shape (Fig. 1) is emphasized due its importance and size.

To allow future visits to the gallery its reinforcement was analyzed, having as basis the need to keep the N5 block's definition unchanged.

Hence, a solution of columns in reinforced concrete was executed to support the landfills necessary for the dwellings platforms, as well as to support the retaining structure which foundations, in the entrance area of the detected gallery, had over 1.30 m of covering (Fig. 8).

In the gallery east zone, due to the proximity with the retaining structures foundations, 3 steel micropiles were executed. In the west zone, where the wall was on the alignment of one of the gallery cavities, a cyclopead concrete fill was made in this "corridor", along of the wall width.

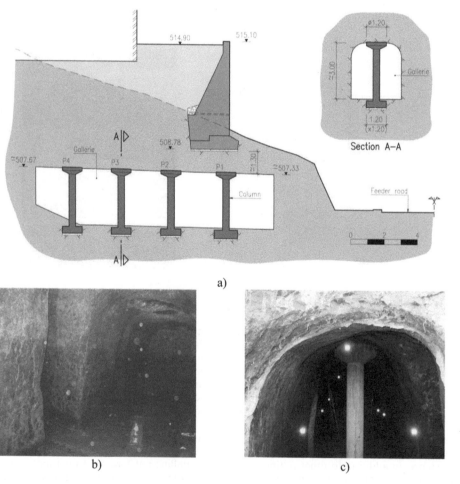

Figure 8.   a) Solution for the gallery found at N5 block; b) View of the gallery before the columns execution; c) View of the gallery entrance, after the columns execution.

336

Finally, we emphasize, under the technical assistance scope, the detailed analysis of the instrumentation reports with weekly readings. Due to the numerous retaining structures to execute it has been chosen to monitor only the most important, the ones with a higher total height and which eventual misbehavior could induce major damages, optimizing this way the monitor plan to implement, considering enough the installation of 70 topographical marks to observe the retaining structures and soil' horizontal and vertical displacements in the surface.

The results analysis revealed an adequate behavior of the structures, and no instability was detected.

In fact, the maximum horizontal displacement registered at the retaining walls was of about 3 to 4 mm for the higher structures, with approximately 8.0 m height, and in the excavation structures the measures never reached 0,1% of the excavation height, with an average height of about 6.0 m.

5 CONCLUSIONS

The Cabo Girão Tourist Resort retaining structures and all the resort's global design led to, in all project phases, a strong intervention and cooperation between the Architecture and Engineering teams. In fact, the optimization of all solutions in the design phase became a significant task, assuring the necessary integration taking into account the existing conditionings, all the more, since the retaining works corresponded to 15% of the work total value.

In such a large and multidisciplinary project, it was essential the compatibility, not only between the different retaining structures adopted, but also between the different civil engineering specialties, being the work organization crucial in order to simplify the execution process.

Associated with this significant coordination activity it is also important to point out the extreme importance of knowing the existent geological conditions already in the design phase, which allowed taking the most out of the formations mechanical characteristics and of the local conditions in order to define the different kind of retaining structures to use.

In the technical assistance phase the basis of the project were validated and the solutions adjusted to the real conditions that were found during the works. These adjustments sometimes resulted in significant work reductions and, subsequently, in some costs savings.

ACKNOWLEDGEMENTS

The authors would like to thank to Holiday Property Bond (Cabo Girão Resort Developer), to Mota-Engil (Contractor) and to all parties involved, for their important technical contribution for the success of this tourist development and for the permission to disclose the elements presented in this paper.

REFERENCES

Cenorgeo 2003a. *Cabo Girão Tourist Resort. Geological-Geotechnical Study.* Final Project (in portuguese).
Cenorgeo 2003b. *Cabo Girão Tourist Resort. Retaining Structures.* Prelimary Studies (in portuguese).
Cenorgeo 2004. *Cabo Girão Tourist Resort. Retaining Structures.* Final Project (in portuguese).
Freitas, A.R.J., Baião, C.J.O., Brito, J.A.M. 2006. Retaining structures of special slopes. The case of the new expressways in Madeira Island. *10th National Congress on Geotechnics. Lisboa. Portugal* (in portuguese).

Volcanic Rock Mechanics – Olalla et al. (eds)

Retaining structures in Machico-Caniçal expressway at Madeira Island

F.A. Sousa, A.R.J. Freitas, M.F.M. Conceição & C.J.O. Baião
Cenorgeo, Engenharia Geotécnica, Lda., Lisboa, Portugal

ABSTRACT: Given the conditions found along the entire expressway that connects the city of Machico and the village of Caniçal, on the Madeira Island, specifically the topography, the urban occupation and the highly unfavorable geological-geotechnical conditions, characterized by the existence of slope deposits of significant depth with very poor resistance characteristics, different solutions had to be used for the retaining structures, specifically gravity walls, reinforced earth retaining walls with jet-grouting foundation, anchored pile walls and soil nailed walls and slopes.

1 GENERAL DESCRIPTION

The Machico-Caniçal expressway starts in the Queimada zone at the entrance to Machico, runs around this city to the West and then continues, for another 7 km, ending at the Caniçal roundabout. The main retaining structures, which are described in this chapter, are located along a 700 meter stretch known as the Machico Sul Interchange, which includes the initial stretch of the expressway and some access branches to the city of Machico, and a stretch known as the Caniçal Interchange/Caniçal Rouondabout at the exit of the Portais Tunnel.

The Machico Sul Interchange includes a total of 18 retaining structures and the stabilization for an important excavated slope; in addition, 3 significant retaining walls had to be built to enable a road bed at the exit to the Portais Tunnel.

The retaining structures required developing an important set of different solutions, specifically gravity walls, reinforced earth retaining walls with jet-grouting foundation, anchored pile walls and soil nailed walls and nailed slopes, with some situations requiring a combination of two or more of these solutions. Table 1 shows the retaining structures designed, which extend for some 2350 meters. Figure 1 is a drawing of the expressway along the above mentioned areas and the location of the structures.

2 GEOLOGICAL AND GEOTECHNICAL CONDITIONS

The area along which the expressway runs is characterized by a rather hilly topography with sheer slopes at the base of which are found slope deposits reaching a depth of 20 meters (Cenor & Grid 2002b, Rosa *et al.*, 2006). These deposits are the result from the breakdown of adjacent rocky slopes and the accumulation of aggregates along the slopes and their bases, and for this reason they are very heterogeneous. They are made up of fragments of various types of rock, predominantly basalts, with varying size. Some are boulders as large as 2 meters in diameter, more concentrated at greater depths, and some are cobblestones and angular or semi-rounded pebbles involved in silt-clay-sand.

The matrix in which these deposits are found is made up essentially of high plasticity montmorillonite clay with a residual friction angle around $14°$. The geotechnical characteristics of these deposits are particularly unfavorable, sometimes showing evidence of recent slides, which required the use of special retaining solutions creating major problems for completing the work.

Underlying the slope deposits are very heterogeneous post-miocene β^2 and myo-pliocene β^1 complexes. The post-miocene β^2 consists of alternating deposits of basaltic lava with differing amounts of brechoid pyroclastic materials and volcanic tuffs, and the mio-pliocene β^1 is primarily made up of different levels of tuffs and volcanic bombs.

The expressway is intersected by small waterways that favor, not only superficial erosion, but deep water infiltration, especially where the slope deposits contact the rock mass.

The geotechnical parameters of the supported soil and foundation considered in the design projects are shown in Table 2. These have been selected based on the results of the tests conducted as part of the geological and geotechnical study (Cenor & Grid 2002b), and considering to the experience acquired during construction and monitoring of numerous similar structures under comparable situations.

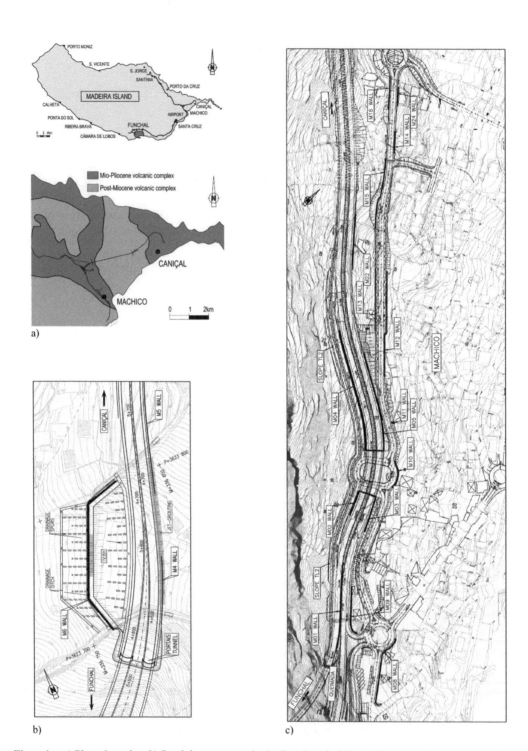

Figure 1.   a) Plants Location; b) Retaining structures in the East Portal of Portais Tunnel; c) Retaining structures in the Machico Sul Interchange.

Table 1. Overall characteristics of the retaining structures.

Stretch	Designation	Type of retaining structures	Approximate length (m)	Maximum height above foundation (m)
	M01	Soil nailed wall	120.0	8.2
	M02	Soil nailed wall	32.6	8.3
	South Abutment	Gravity wall in cyclopean concrete	30.5	7.2
	M03	Gravity wall in cyclopean concrete	48.0	8.5
	M04	Soil nailed wall	165.5	10.9
	North Abutment	Gravity wall in cyclopean concrete	30.5	9.9
	M05	Gravity wall in cyclopean concrete	133.4	11.8
	M08	Reinforced earth retaining wall with jet-grouting foundation	56.6	5.4
		Anchored pile wall	39.4	5.1
	M09	Semi-gravity wall	67.9	4.8
Machico Sul Interchange	M10	Reinforced concrete wall with jet-grouting foundation	39.0	3.9
	M11	Reinforced concrete wall	4.4	2.9
	M12	Gravity wall in cyclopean concrete	144.6	9.0
	M13	Anchored wall	32.0	4.2
		Gravity wall in cyclopean concrete	53.4	2.8
	M14	Reinforced earth retaining wall with jet-grouting foundation	30.1	3.5
	M15	Anchored pile wall	98.2	6.2
	M16	Anchored pile wall	44.0	7.4
		Gravity wall in cyclopean concrete	20.0	6.0
	M22	Gravity wall in cyclopean concrete	69.2	9.0
	M24	Reinforced concrete wall	25.0	3.3
	TL2	Soil nailed slope	760.0	20.0
	M4	Reinforced concrete wall with jet-grouting foundation	50.0	7.6
East Portal of Portais Tunnel	M5	Reinforced concrete wall with jet-grouting foundation + Gravity wall in cyclopean concrete	84.0	11.8
	M6	Anchored pile wall + Gravity wall in cyclopean concrete	125.9	7.0

Table 2. Geotechnical parameters of supported and foundation materials.

Parameter	Slope deposits	Tuffs	Weathered tuffs with volcanic bombs	Compacted tuffs with volcanic bombs	Disaggregated breccias	Compacted breccias	Fractured basalts
Density γ (kN/m³)	19	19	19	19	20	21	22
Friction angle $\varnothing'$ (°)	12 a 16	28	25	30	36	38	40
Effective cohesion c' (kPa)	5	15	15	20	10	25	50
Undrained shear strenght cu (kPa)	20	–	–	130	–	–	–

3 CONCEPTION AND OVERAL DESCRIPTION OF THE SOLUTIONS DEVELOPED

The solutions studied for the various retaining structures took into consideration the geological and geotechnical, topographic, land occupation, economic and executive aspects (Cenor & Grid 2002c). The most important constraints were geological and geotechnical in nature due to the presence of poor resistance slope deposits along the entire expressway, oftentimes unstable and of significant depth.

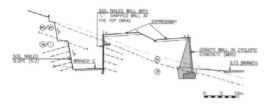

Figure 2. Typical cross section next to the base of the slope.

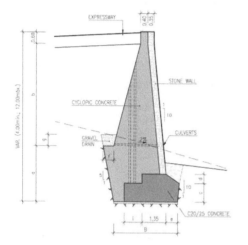

Figure 3. Gravity walls in cyclopean concrete.

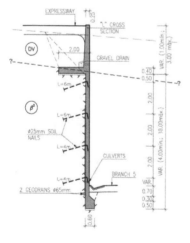

Figure 4. Soil nailed walls with a L-shaped section wall at the top.

3.1 Machico Sul Interchange

Where it was necessary to build retaining structures for road bed embankments located in areas with shallower slope deposits capable of being removal, the option was made for gravity walls in cyclopean concrete supported directly on the rock bed, thus keeping wall deformation and consequently embankment and road bed deformation at a minimum. Such structures offer the additional advantage in that they are quickly and easily built and their cost is lower than other structures of a similar height, such as reinforced concrete walls. This solution was generally used for expressway structures, since the route alignment runs along the base of rocky cliffs where slope deposits are shallower, as can be seen in the cross section shown in Figure 2. The height of this set of walls (M03, M05, M12 e M22) (type and cross section can be seen in Figure 3) varies between 4.0 and 12.0 meters.

Where it was necessary to make deep excavations into the rock face and because of the geometry of the route, required an inclination close to the vertical, soil nailed walls were placed in order to protect and stabilize the excavation. This is a fast, flexible and low cost solution that takes advantage of the resistance of the rock mass itself.

In locations where nailed walls were to be build and where the road bed was at an elevation above the top of the massif with a maximum difference in elevation of around 3.0 meters, a reinforced concrete wall (L-shaped cross-section) was placed at the top of the nailed walls to support the embankments required to retain the road bed. This solution was only required in excavations associated with the expressway (Fig. 2).

Soil nailed walls (M01, M02, M04) with a L-shaped section as shown in Figure 4 are made up of a 0.25 meter thick reinforced concrete wall poured directly onto the excavations face with heights varying between 4.0 and 10.0 meters and with 1 to 4 levels of nails using 4.0 to 6.0 meter long, 25 mm diameter steel rods.

In situations where excavations of more than 3.0 meters were required in slope deposits and where, on the one hand, the massif was located at about the same elevation as the road bed and, on the other, there were space limitations due to the proximity to buildings at the top of the excavation, an anchored wall solution was employed so as to control and limit deformations. The only anchored wall included (M13) has a maximum height of 4.2 meters and is made up of 2 rows of reinforced concrete panels and 1 level of 480 kN capacity anchors.

The various branch roads are placed at elevations lower than the expressway as slope deposits are quite a bit deeper making them impossible to be completely removed so as to develop suitable conditions for placing the foundations of the different retaining structures. Thus the structures built in these locations required deep foundations as illustrated in Figure 5.

342

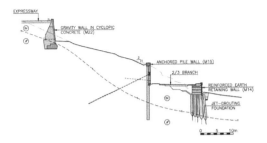

Figure 5. Typical cross section removed from the base of the hillside.

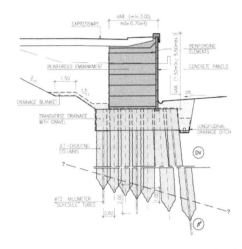

Figure 6. Reinforced earth retaining walls with jet-grouting foundation.

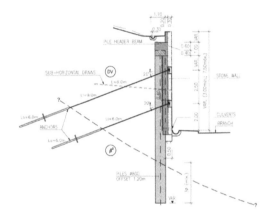

Figure 7. Anchored pile wall.

In situations where embankments required retaining structures higher than 3.0 meters the solution selected was reinforced soil walls (M08 and M14) or reinforced concrete walls (M10) with Ø1000 millimeter diameter jet-grouting piles normally running through to the bed rock. In particular, reinforced earth retaining walls are characterized as being highly resistant to static and seismic soil movements and capable of adapting to unfavorable foundations, thus being considered a suitable solution for the specific conditions expected.

This type of structure, which the cross section is shown in Figure 6, has the additional advantage of being quick and easy to build at a fairly competitive cost and blending well with the surroundings.

Where the situation required excavating slope deposits to a depth of more than 3.0 meters and where the rock mass was located at an elevation below the road bed the solution selected was an anchored pile wall with Ø800 millimeter piles set with 1 or 2 levels of permanent anchors with load capacities between 480 kN and 720 kN.

The primary objective of using this type of retaining structure (M08, M15 and M16), of which a cross section can be seen in Figure 7, is to control the deformation of soil behind the structure and consequently deformations in buildings located oftentimes very close to the top of the structure. This solution is also quick and easy to build.

The concrete piles were placed at a distance of 1.20 meters from each other, running from the natural surface of the terrain, in most cases into the rock mass to a depth of no less than 3 times the diameter of the pile. However, in locations where the slope deposits at the level of the pile penetration depth were deeper than about 6.0 meters, a decision was made not to drive all of the piles into the rock mass but instead to use a solution of alternating suspended and deep piles, thus optimizing costs. Altogether the walls projected include piles varying in length from 5.0 meters to 20.0 meters.

The solution selected for containing excavated slopes in the volcanic formations was soil nailing structures made of shotcrete and soil nails defined as required by the nature and mechanical characteristics of the formations to be excavated, complying with the behavior observed in natural slopes of similar characteristics.

This solution enables not only a sufficient sheath to confine and avoid the progressive erosion of the slopes, but also provides a light weight, flexible and low cost coating associated with the ground and that takes advantage of its resistance.

In general two types of walls were considered: Type "a" walls, denser and more resistant, to be used in the more weathered formations or those that are more easily broken down, made up of two layers of shotcrete not less than 5 centimeters thick, each one around an electro-welded mesh such as AQ50, and lighter type "b" walls, used for slopes in more compact formations, made up

Figure 8. Overview of the Machico Sul Interchange.

Figure 9. Overview of M6 wall in the start of construction.

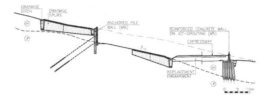

Figure 10. Typical cross section for the stretch following the Portais Tunnel.

of one layer of shotcrete not less than 5 centimeters thick sprayed onto an electro-welded mesh type CQ30.

Soil nails are made of A400NR steel rods 25 or 32 millimeters in diameter, sealed into 3" (0.076 millimeter) holes with cement slurry. Rods vary in length based on the geological-geotechnical characteristics and the geometry of each group of slopes to be stabilized.

It should be noted that only one excavated slope, in its more weathered and disaggregated areas, was treated. This slope is about 760 meters long and as high as 20 meters. Figure 8 shows a view of the construction and the entire length of this excavation slope (TL2).

3.2 Caniçal interchange/Caniçal roundabout

Immediately after the East exit of the Portais Tunnel, the expressway runs along an area of slope deposits of significant depth, at times up to 10.0 meters. These deposits show evidence that they are evolving as buildings located in the village of Machico, close to the area where construction occur, show signs of significant damage that can be attributed to movement in this mass of slope deposits.

For this reason, in addition to ensuring the stability of the retaining structures developed to build the highway, an attempt was made to substantially improve the overall stability of the mass of existing slope deposits, thus minimizing future damage to these buildings.

Therefore the solution found was to build the retaining structure (M6 wall) away from the expressway, thus reducing the slope deposits to be supported and the mass potentially instable by the partial removal of it. Because of the depth of the deposits in this area it was impossible to support the retaining structures by shallow foundations making necessary the use of deep foundations.

With an excavation deeper than 3.0 meters into slope deposits, an anchored pile wall was defined, with 1 or 2 levels of anchors made of Ø800 reinforced concrete pilings placed at 1.0 meter inter-

vals driven from the surface of the natural terrain (Fig. 9). Their free height varied between 3.5 meters and 6.5 meters. One or two levels of permanent anchors were included, with load capacities of 600 and 720 kN and 35° from horizontal, to be built whenever pilings, with 20.0 and 28.0 meters long, as it can be seen at Figure 10.

Along the wall, where the rock mass is located at elevations significantly lower than the bottom of the excavation, it was necessary to improve ground properties in order to ensure an adequate behavior of the structure, in particular its ability to mobilize the passive resistance. Thus, given that jet-grouting technology was available on site, slope deposits were treated using this technology along a stretch some 30 meters long and 4.0 meters wide, constructing jet-grouting pilings driven down into the rock mass. Jet-grouting treatment used Ø1000 millimeter diameter columns set in a 1.0 meter (longitudinal to the wall) × 7.0 meter (transverse to the wall) grid.

Two areas of the soil mass, each about 2.0 meters, were left untreated at the base of the treated area so as not to avoid natural percolation in the soil which could raise the groundwater level behind the wall.

Given the importance of water to the resistance of this formations, and because the retaining structure was designed taking into consideration soil drainage behind the structure, a suitable drainage system was designed to effectively lower the groundwater level in the area. Thus, in addition to designing drainage solutions for the retaining wall itself, so as to collect water efficiently, an overall surface drainage system was designed.

The main function of which is to ensure suitable drainage of waters that rise up behind the pile wall, as not only is it important to avoid hydrostatic pressure, to which the drainage sSen that the resistance characteristics of the terrain are highly dependent on the existence of water, and thereby avoiding significant earth pressure that are hard to determine and control.

In this way, a wall drainage system was designed made up of geocomposite strips drains placed between the piles. The water collected drains to the outside in culverts that discharge into half-pipe lined culverts built into the edge of the road at the same elevation as the base of the excavation and connected to the general drainage system.

A drainage culvert made up of a collector built in perforated concrete connected to a geotextile wrapped gravel drain was built behind the top of the anchored pile wall over which runs a half-pipe intercepting culvert. In addition, drainage spurs were designed, perpendicular to the wall and connected to the drainage ditch located behind the wall. Complementing this surface drainage system it was defined a drainage ditch parallel to the wall and incorporating a collector in porous concrete inserted in a geotextile enclosed gravel drain with a top culvert, at a relatively large depth, that discharge the water collected at the extremities (Figs. 10–11).

The ground in front of the pile wall was defined taking into account esthetics considerations, with the option made to include a 4.0 meter wide sidewalk in front of the wall, with the gap to the road elevation overcame by a slope with 1V:6H inclination.

For this slope a surface drainage system was designed similar to the one used behind the curtain, made up of drainage spurs connected to a drainage ditch that in turn runs into a porous concrete collector. A 2.5 meter wide ditch is connected to this drainage ditch. All of the drainage elements discharge the water collected into a Ø1200 PH.

Regarding the structures to be built along the right side of the expressway for retaining the road embankments, two solutions were studied foreseeing the site's condition. Thus where the slope deposits were so deep as to make it impossible to drive foundations for retaining structures into the massif, the recommendation was to build reinforced concrete walls with Ø1000 millimeter diameter jet-grouting piles normally driven into the bed rock so as to ensure adequate foundation (Fig. 10).

Where the slope deposits were not deep, enabling foundations for retaining structures to be

Figure 11. Overview of M6 wall in East portal of Portais Tunnel.

driven into the bedrock, the recommendation was to build gravity type walls of cyclopean concrete. Because of the height of the embankments and the quality of the foundation materials, these walls provide good guarantees against deformation and consequently limit any deformations in the embankments and in the pavement overlying.

4 CONCLUSIONS

The design of the retaining and stabilization structures for the Machico-Caniçal expressway was subject to constraints of diverse nature, primarily due to the existence of areas with thick slope deposits with very poor resistance characteristics and densely occupied areas all along the expressway. The combination of these two factors led to the use of a several solutions. To each case it was given individual attention in order to adapt the type of solution designed and the construction methods so as to obtain economically and technically adequate solutions.

It should be stressed that it is essential to fully understand the geological conditions during the conception/design phase, thus enabling an appropriate characterization of the mechanical characteristics of the volcanic formations in order to allow the optimization of the solutions designed.

ACKNOWLEDGEMENTS

The authors would like to thank the Social Equipment Department of the Regional Government of Madeira (the Client), the Zagope/Engil Consortium (the Contractor) and all other stakeholders involved in the enterprise under analysis, both for the technical contributions they have made to its success and for the authorization to disclose the elements that make up this paper.

REFERENCES

Cenor-Grid 2002a. *Machico-Caniçal Expressway. Machico Sul Insterchange. Road building Project.* Final Project (in portuguese).

Cenor-Grid 2002b. *Machico-Caniçal Expressway. Machico Sul Insterchange. Geological-Geotechnical Study.* Final Project. (in portuguese).

Cenor-Grid 2002c. *Machico-Caniçal Expressway. Machico Sul Insterchange. Retaining Structures.* Final Project. (in portuguese).

Rosa, S.P., Rodrigues, V.C., Brito, J.M., Baião, C.J. 2004. Geological and Geotechnical conditions of the Machico-Caniçal Expressway. *9th National Congress on Geotechnics. Aveiro. Portugal* (in portuguese).

Volcanic Rock Mechanics – Olalla et al. (eds)
© *2010 Taylor & Francis Group, London, ISBN 978-0-415-58478-4*

Tindaya Mountain Cavern: Art and underground engineering

J. Ramos Gómez
Iberinsa, Madrid, Spain

ABSTRACT: Chillida's visionary artwork to create a large Space in the Tindaya Mountain represents a big technical challenge. On the one hand, dimensions of the span and flat profile of the cavern roof are unique in underground engineering. On the other hand, the transmission of the artist's idea and the realization of an aesthetic concept combining art, technique and nature in a natural way will require an innovative support solution.

1 INTRODUCTION

1.1 Initial idea

The Tindaya Project is the posthumous work of the sculptor Eduardo Chillida, who died in 2002. In 1994, he proposed the creation of a large sculpture within Tindaya Mountain, on the island of Fuerteventura (Canary Islands, Spain).

Tindaya Mountain rises near the sea from a plain to a height of nearly 400 meters and is one of the main beds of pre-historic prints on the island (Figure 1). The idea of Chillida was to form a sculpture without materials creating a place inside a mountain that would offer men of all races and colours a great Space dedicated to tolerance.

The sculpture is based on the concepts of space, scale and light and consists of excavating a large cavern of approximate dimensions 65 m long, by 45 m wide and 40 m high. Two light shafts connect the main space with the surface and are oriented to capture the rotation of the sun and the stars providing natural lighting. Moreover, a horizontal tunnel leaving from the cavern will give visitors a view to the sea and horizon. The tunnel, of 15 m wide and 15 m high, is also the principal access to the Space.

Chillida wanted to mark the interior of the mountain and to create a Space that will frame our vision of the landscape by the entrance tunnel and the light shafts (Figure 2).

1.2 The project

The project of the Tindaya Cavern was directed by the architect Lorenzo Fernández Ordóñez (Estudio Guadiana) in collaboration with the architect Daniel Díaz Font, who led a multidisciplinary team consisting of the engineering firms OVE ARUP &

Figure 1. Tindaya mountain.

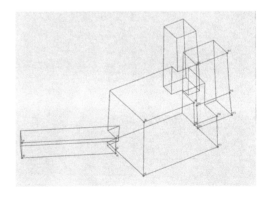

Figure 2. Geometry of the space.

PARTNERS and SCOTT WILSON PIESOLD and also external consultants consisting of Professor Dr. Evert Hoek and the Spanish Engineering Consulting IBERINSA.

In the field campaign of the project the following enterprises were involved: Técnicas Especiales de Perforación (TEP), Sorein, In Situ Testing and MeSy. The laboratory tests were executed at the Mining School of the Polytechnic University of Madrid and at the Geological Faculty of the University of Oviedo. The calculations were done by using the codes FLAC^{3D} and 3DEC and were realized by Itasca Spain.

The project was divided in 3 phases:

– Phase I (2003): Study of Alternatives based on initial fieldwork without intrusive investigations on the mountain.
– Phase II (2005): Geotechnical investigation with boreholes and laboratory and in situ tests in order to confirm the technical viability of the project.
– Phase III (2006): Detailed General Design.

Chillida wanted that Tindaya Cavern were a sculptural empty-space which respected the natural rock mass of the mountain (Figure 3). Thus, the artistic concept required that all the inner faces of the cavern, the shafts and the entrance tunnel would be of natural rock with no visible support elements and no shotcrete coating. The artistic requirement for the rock surface to be exposed involves special challenges for the engineering in addition to the underground design challenges. Furthermore the Tindaya Mountain presents many environmental and preservation constrains which also dictate the design and construction methods.

After initial studies, the Space was situated within the mountain in essentially sound rock

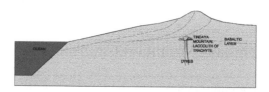

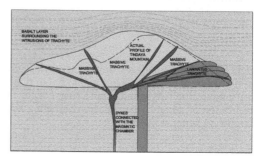

Figure 4. Laccolith of trachyte and mountain slides.

mass, avoiding the main geological weakness (fracture planes or dykes). In addition the sculpture was situated at the greatest possible distance from the area where ancient prints were found to guarantee their strict preservation.

2 GEOLOGY OF TINDAYA MOUNTAIN

The island of Fuerteventura was formed about 48 million years ago by big volcano shields, which accumulated gradual layers of basalt during 22 million years. At the place where actually the Tindaya Mountain is situated, about 12 million ago a laccolith of trachyte was built due to an initial intrusion of this material through dykes connected with the magmatic chamber (Figure 4).

Afterwards a second intrusion of trachyte produced an elevation of the basalt layer in the circumference. During the Miocene a series of mountain slides and mountain side erosions of the big volcanoes uncovered the laccolith of trachyte located in its interior and which now represents the actual Tindaya Mountain.

The rock types which occur in the Tindaya Mountain are trachyte syenite, which is dominant and called trachyte for conveniente, and dykes of basalt and gabbro. The majority of the dykes could contain weak and weathered rock with groundwater seepage.

3 GEOTECHNICAL PROPERTIES

3.1 *Field campaign*

The scale of the site investigation was unusually large in order to minimize the risk of encountering

Figure 3. Vision of the Tindaya cavern.

unknown ground conditions in the reduced space of the mountain where the cavern will be located. The field campaign comprised 15 rotary cored investigation boreholes (vertical and inclined) to a total of 1.650 m in length and also a complete laboratory and in situ program of testing with down-hole and cross-hole geophysical tests, dilatometer tests, hydro-fracture stress tests and permeability tests.

Thirteen of the fifteen boreholes were tested using geophysical optical and acoustic televiewer probes in order to analyse the discontinuities (3.000 joints were studied). In addition, 2 surface geophysical techniques were employed (electrical resistivity and seismic refraction) to obtain geological subsurface information and depth of weathering.

The field campaign has provided enough data to have a complete geological model of the Tindaya Mountain.

3.2 Geotechnical design assumptions

The cavern and associated structures are located in relativity massive trachyte body with late stage steep dipping basalt dykes intruding the trachyte. The distribution of the dykes and its thickness, strike and dip extent is somewhat variable, although a dominant NE structural trend can be observed.

The dominant trachyte is competent and strong (UCS = 40–120 MPa), with a high porosity (9–19%) and potentially abrasive (Cerchar Abrasivity Index of 0.6–1.3). Trachyte presents a banding of hydrothermal origin, although it was found that this has a little influence on the mechanical behaviour of the rock.

The distribution of joints appears relatively uniform within the massive trachyte body with four dominant joint sets. As seen in the borehole logs, less than 4% of the joints are infilled, such that the roughness of the joints will govern shear strength. Analysis of the joint shear strength indicates that representative values are: JRC = 6, JCS = 60 MPa, $\phi_b = 30°$, dn = 7 to 10° and $k_n = k_s = 200$ MPa/m.

The dykes appear to be highly fractured at the surface and generally also at depth. The width of basalt dykes and weathering adjacent trachyte is usually 3 m (Figures 5 and 6).

It is apparent that weathering of Tindaya stone may occur in a relatively short period of time, not geological time, and seems to be more susceptible in areas exposed to moisture.

In situ stresses in the rock mass are higher than expected with a horizontal to vertical stress ratio of approximately 2.

Based on the geotechnical tests as well as on rock mass characterization, Table 1 resumes the geotechnical assumptions which have been taken into account for the design.

Figure 5. Basalt dyke.

Figure 6. Model (Estudio Guadiana) showing the dykes crossing the Space. The majority will intersect the entrance tunnel.

Table 1. Geotechnical parameters.

	Trachyte	Basalt*
γ [kN/m³]	23	23–27
UCS [MPa]	62.6 (40–120)	66,7
σ_t [MPa]	4.5	–
m_i	18	–
E_{mass} [GPa]	10–13	5–13
RMR	60–90	–
Q	10–50	2–50
GSI	65–80	50–80

* Few samples suitable were obtained due to the poor quality of the rock.

3.3 Hydrogeology

Fuerteventura is a dry island, located within the Sahara Dry Belt area. In Tindaya Mountain, measurement of borehole water levels and the known springs on the northern and western flanks of the mountain indicate that there is a water table roughly following the topography. Springs are supplied by long term flow through the fracture network in the rock mass. The maximum water level is inferred to lie just above the roof level of the cavern.

Measured permeability of the rock mass is between 1×10^{-5} to 5×10^{-7} m/sec approximately, even in more fractured zones adjacent to dykes. In situ measurements of the intact rock material permeability however are much lower as expected, at about 1×10^{-10} m/sec.

4 SUPPORT DESIGN

4.1 Initial concepts

The rocks in Tindaya mountain are strong brittle materials and, with a cover of approximately 50 m, the stability of the cavern will be defined almost entirely by structurally controlled blocks and wedges created by the intersection of discontinuities in the rock mass.

The most important factor to stabilize the blocky rock mass is the great increase in strength that comes with the application of a modest amount of confinement. The installation of a systematic array of tensioned anchors is the method of applying the confinement necessary to retain the interlocking and, hence, the strength of the rock mass.

Cavern support will increase the natural shear resistance along the dykes, master joints and minor joint planes. The support will secure the rock in the cavern roof and, moreover, will transfer the load above the cavern roof to the adjacent rock mass without overloading the cavern walls.

The support idea is to excavate a series of small galleries above the cavern roof and from these galleries to install reinforcing elements before excavation of the cavern commences. These reinforcing elements will induce compressive stresses in the rock mass immediately above the cavern roof that will provide the confinement required to give this rock the strength to span the 50 m cavern roof.

Three initial support concepts were analyzed in the project:

1. Reinforced rock plate in the roof cavern with horizontal galleries and inclined anchors between these galleries and cavern roof (Figure 7).

2. Horizontal pre-stressed thin plate in the roof cavern with anchors from 2-way galleries. A mesh of horizontal ground anchors installed from galleries aligned in two directions parallel to the long and short axes of the cavern (Figure 8).

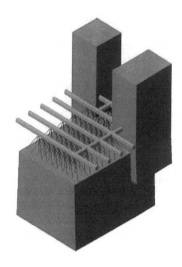

Figure 7. Sketch of the reinforced rock plate support.

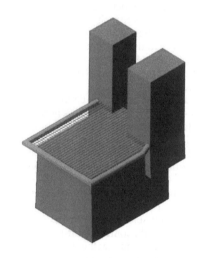

Figure 8. Sketch of the horizontal thin plate support.

3. Reinforced rock arch in the roof cavern. A series of reinforced rock arches formed from ground anchors installed along five curved galleries above the cavern roof. Three arches would be included in the pillar between the light shafts (Figure 9).

4.2 Calculation with 3DEC

While FLAC3D (continuum medium) was used to investigate the stress concentrations and to replicate the in situ ground stresses determined during hydrofracture tests, the final support design was carried out using the three dimensional program 3DEC (discontinuous medium) of Itasca Consulting

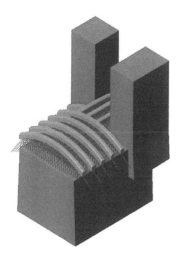

Figure 9. Sketch of the reinforced rock arch.

Group. This program treats the rock mass as a series of blocks bounded by joint surfaces.

The three initial support concepts were modelled by Itasca Spain. The model included the mountain slopes and the main underground works (cavern, light shafts and entrance tunnel) with a simplified representation of the mapped joints and dykes. While, in common with similar programs, 3DEC has a limit on the complexity of the model that can be run, 8 dykes and 34 master joints were selected to intersect the cavern and entrance tunnel in the model. The rock mass between the discontinuities was modelled as non-deformable blocks.

The strength of the master joints between the blocks is assumed to have Mohr-Coulomb strength parameters, with zero cohesion and tension. The shear strength along the discontinuities relies on friction, dilation and the normal force acting across the discontinuity.

The excavations and support were wished in place, apart from a sequential excavation of the cavern roof area. The excavation was assumed to be in dry on the basis that the rock mass will be drained by the initial excavations (pilot gallery in entrance tunnel and roof galleries) and the recharge of the groundwater table will be low in the arid climate of the island.

4.3 Hybrid concept

The modelling showed that the rock mass could be supported with the three initial solutions, however, the reinforced roof plate would required exceptionally high maximum axial loads in the support. The option with only horizontal anchors was inadequate because the orientation of the anchors with respect to the joints and dykes would result with

the anchors operating predominantly in shear and it was considered that this support system would carry a high risk of shear failure.

The arch concept was the preferred for the long term support of the cavern roof. However, it was necessary to introduce some changes. The number of anchors would need to be increased to reduce the individual anchor loads (there were anchors with loads >200 t). In addition, it was recognized that horizontal and inclined anchors would be required to control the potential relaxation of the subvertical joints towards the shaft walls and would be easy to install between the light shafts.

These conclusions required the development of a hybrid support concept combining the initial three concepts. The final idea was to create a reinforced rock arch within the ground above cavern roof with curved galleries excavated from the light shafts and anchors installed around the galleries. Longer ground anchors would also be installed from the galleries to support the rock mass between the arches and the cavern roof. In addition, rock bolts would be installed during excavation of the cavern roof to support the little blocks between the ground anchors.

4.4 3DEC model sensitivity and results

The hybrid solution for the support was modelled in 3DEC and the base case model was investigated by comparing the displacements and support loads for the following cases:

– Two sequences of excavation
– Water table: An analysis representing the long term recovery of the groundwater table close to the cavern roof level was carried out.
– Seismic loading
– Joint orientation: The position of key subvertical joints was varied by 10° in strike and/or dip
– Joint strength: The dilation angle was reduced from 10° to zero in single degree decrements
– Rock block deformability: As a final check, the initial assumption of rigid blocks was changed to deformable blocks

The base case model was run initially without support. This model showed that those areas of the cavern roof adjacent to the light shafts walls were particularly sensitive to the failure of columns of rock bounded by vertical joint planes and the shaft walls.

While the hybrid concept offers a solution to the support of the Tindaya Space, the stability achieved has shown to be sensitive to a variety of factors, in particular variation in critical master joint orientations. It is thus considered that the support will need to be validated or modified using the joint data collected during the excavation of the Space.

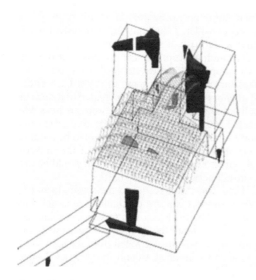

Figure 10.   3DEC result print.

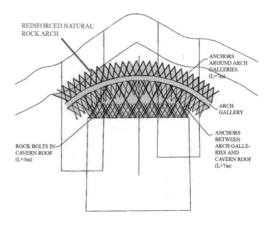

Figure 11.   Roof cavern support concept.

4.5   *Cavern permanent support*

The solution for the roof cavern permanent support consists of curved arch galleries above the cavern roof from which anchor radiate to obtain a reinforced rock arch within the ground (Figure 11). The complete cavern permanent support includes:

– 7 m long anchors forming the arches radiate around the galleries, such that the galleries are situated within an annular arch of strengthened rock.
– 14 m long anchors supporting the ground between the arches and cavern roof.
– Rock bolts of 3 m in length installed from the roof of the cavern during its excavation to support little blocks.
– Rock bolts of 7 m in length installed from the walls of the cavern.
– The rock pillar between the two light shafts will be supported above the cavern roof with sub-horizontal stranded ground anchors (L = 23 m). Additionally radiating bolts will be installed from a gallery extending into the pillar from the rock arches to support the pillar above these ground anchors.

The maximum working axial load required for anchors is 88 t in the area between the arch galleries and cavern roof. In the sub-horizontal anchors between the light shafts the maximum working axial load is 60 t. The load of rock bolts in cavern roof and anchors around the arch galleries was of 30 t.

4.6   *Light Shafts and entrance tunnel permanent support*

The joints are predominantly subvertical, such they form rock columns bounded by joint release

surfaces. If unrestrained, the release surfaces will relax into shafts, reducing the shear strength along these surfaces. This effect becomes worse by non-circular shaft profiles.

The support solution in light shafts consists of 10 m long anchors with working loads of 76 t and rock bolts of 7 m length and working load of 30 t.

The entrance tunnel will be supported with 7 m long rock bolts.

5   CONSTRUCTION METHOD

Basically, the construction sequence of the Space is the following:

– Construction of portal and pilot gallery of entrance tunnel.
– Pilot gallery will bifurcate to the base of each light shaft and pilot shafts will be excavated upwards ("raised") from the pilot gallery to ground surface using an Alimak system. Pilot gallery and pilot shafts will be supported with temporary fibreglass rock bolts and, if it is necessary, shotcrete. A third Alimak shaft could be excavated for safety reasons.
– A test gallery will be excavated off the pilot gallery and within the future space of the cavern to prove excavation, support and surface finishing techniques and to perform in situ and laboratory tests.
– Excavation of the complete section of light shafts from the ground surface.
– Two access galleries (2.5 m wide and 3 m high) will be excavated from the light shafts when the light shafts floors reach the galleries level.

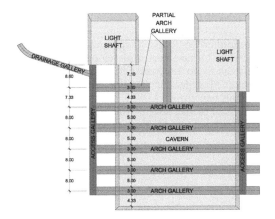

Figure 12. Plan view of galleries above roof cavern.

Subsequently, will be excavated the arch galleries (3 m wide and 3.5 m high) between the access galleries. A partial arch gallery will be also excavated between the light shafts. All the muck will be cleared through the pilot shafts to the pilot gallery. A drainage gallery (2 m wide and 3 m high) will be excavated from the mountain slope to the north access gallery (Figure 12).

– The support in the arch galleries will be installed and operative before excavation of the cavern roof starts.
– The cavern roof will be excavated as a series of parallel headings from the light shafts and rock bolts and anchors will be installed as the headings are advanced.
– The remaining cavern volume will be excavated as a series of benches and the constructor will choose to simply remove the rock as spoil or recover blocks for commercial use.
– The entrance tunnel will be excavated to full size from the cavern to the portal.

Explosives will not be permitted in the portals areas (entrance tunnel and light shafts) to avoid damage to the mountain slopes. The portal of the entrance tunnel will be the unique access to the works in the Space in order to reduce environmental problems in the mountain. Although mechanical excavation will be used in portals, the underground works will be excavated using drill and blasting techniques.

6 SPECIAL TREATMENTS AND SURFACE FINISHING

The artistic concept of Chillida´s sculpture will require that all the inner faces of the Space will be of natural rock with no visible support elements. Rock surfaces with half-barrels after blasting, anchors, bolts, steel mesh or shotcrete will not be acceptable finish. Thus, it will be necessary to apply some special treatments and surface finishing techniques. These methods are the following:

– The excavation in all the Space (cavern, light shafts and entrance tunnel) will leave a skin rock, at least 50 mm thick, between the excavation and finished roofs and walls. This will subsequently be cut off using a diamond wire to create a smooth finished rock surface.
– All the permanent support (bolts and anchors) will be countersunk with holes to a depth below the finished surface to prevent being cut by the diamond wire.
– The rock bolt and anchor holes will be reamed with a thin-walled core barrel and the rock cores will be retained.
– The threaded ends of the rock bolts will be used to hold pulleys around which the diamond wires will run.
– Discs of rock will be cut from the retained rock cores and cemented into the reamed sections of the rock bolt and anchor holes using a suitable rock adhesive.
– The cavern and entrance tunnel floors will be ground.
– Weathered dykes will be cut back, drained and supported with a shotcrete or concrete infill.
– Discontinuities containing seasonal and long term ground water flows will intersect the light shafts and cavern walls. Drainage pipes will not be permitted and the inflows and seepages will need to be intersected and fully sealed with grout.
– Finally, the cavern walls and roof will be cleaned with high pressure water jets and geotextile will be fixed to the finished surfaces and sprayed with a resin coat. The geotextile and resin will be used to secure the rock surfaces between the anchors and rock bolts (steal mesh or shotcrete are not aesthetically acceptable). If resin and geotextile are not sufficient to support the little blocks of rock, additional small (1 m long) rock bolts (stainless steels pins) will be installed and cement grouted. It is possible that resin and geotextile will not be required in the walls of the Space.

7 MONITORING

The galleries above the cavern roof will remain open to provide permanent access to the cavern support and instrumentation points. The instrumentation will monitor the rock mass and ground anchor support during cavern excavation and for the long term performance of the Space.

The quantity of instrumentation will be very significant and will require a remote "real-time" recording system and a long time monitoring. The pres-support and excavation of the cavern roof will be critical and the instrumentation will focus on the monitoring of the support and the movement of the rock mass above the rock arches and between the rock arches and cavern roof.

ACKNOWLEDGEMENTS

The author would like to acknowledge and thank everyone to have part in developing this project, particularly to Lorenzo Fernández Ordóñez, the Project Manager, for the kind permission to publish this paper.

Special thanks also to Wolfgang Kreiner for his collaboration to write this article.

Volcanic Rock Mechanics – Olalla et al. (eds)
© 2010 Taylor & Francis Group, London, ISBN 978-0-415-58478-4

Author index

T - #0209 - 071024 - C0 - 246/174/20 - PB - 9780415584784 - Gloss Lamination